Stimuli-Responsive Gels

Stimuli-Responsive Gels

Special Issue Editor

Dirk Kuckling

MDPI • Basel • Beijing • Wuhan • Barcelona • Belgrade

MDPI

Special Issue Editor
Dirk Kuckling
Department of Chemistry Paderborn University
Germany

Editorial Office
MDPI
St. Alban-Anlage 66
Basel, Switzerland

This edition is a reprint of the Special Issue published online in the open access journal *Gels* (ISSN 2310-2816) from 2015–2018 (available at: http://www.mdpi.com/journal/gels/special_issues/stimuli-gels).

For citation purposes, cite each article independently as indicated on the article page online and as indicated below:

Lastname, A.A.; Lastname, B.B; Lastname, C.C. Article title. *Journal Name* **Year**, *Article number*, Page Range.

ISBN 978-3-03897-210-5 (Pbk)
ISBN 978-3-03897-211-2 (PDF)

Cover image courtesy of Dirk Kuckling.

Contents

About the Special Issue Editor

Dirk Kuckling is Professor of "Organic and Macromolecular Chemistry" at the Department of Chemistry of the Paderborn University, Germany. He studied chemistry at Christian-Albrechts-University of Kiel, Germany and received his PhD degree in 1994 in the field of synthetic organic chemistry. In 1994, he joined the Institute for Macromolecular Chemistry of the Dresden University of Technology as a postdoc and started his own research group in 1995. In 2001/2002 he worked as visiting assistant professor at the Department of Chemical Engineering at Stanford University, USA. After his "Habilitation" in 2004, he was appointed in 2008 as full professor at the Paderborn University. The main focus of his work is the synthesis of novel functional polymers. For this purpose different polymerization techniques are used to prepare polymers with a designed structure and functionality. Such polymers were applied as actuators and sensor materials for microfluidic and biomedical applications.

gels

MDPI

Editorial

Stimuli-Responsive Gels

Dirk Kuckling

Organic and Macromolecular Chemistry, Department of Chemistry, Paderborn University, Warburger Str. 100, D-33098 Paderborn, Germany; dirk.kuckling@uni-paderborn.de

Received: 4 July 2018; Accepted: 6 July 2018; Published: 9 July 2018

Although the technological and scientific importance of functional polymers have been well established over the last few decades, the most recent focus that has attracted much attention concerns stimuli-responsive polymer gels. These materials are of particular interest due to their abilities to respond to internal and/or external chemo-physical stimuli; such responses are often large and macroscopic. Aside from the scientific challenges of designing stimuli-responsive polymer gels, the main technological interests concern numerous applications, ranging from catalysis in microsystem technology and chemo-mechanical actuators to sensors. Since the phase transition phenomenon of hydrogels is theoretically well understood, advanced materials based on predictions can be prepared. Since the volume phase transition of hydrogels is a diffusion-limited process, the size of the synthesized hydrogels is an important factor. Consistent downscaling of the gel size will result in fast smart gels with sufficient response times. To apply smart gels in microsystems and sensors, new preparation techniques for hydrogels have to be developed. For upcoming nanotechnology, nano-sized gels as actuating materials would be of great interest. Finally, new design concepts for tough polymer gels are of interest for overcoming the mechanical shortcomings of conventional gels.

This Special Issue includes 17 papers covering a wide range of subjects including thermo- and pH-responsive hydrogels, functionalized materials, supramolecular stimuli-responsive structures, composite hydrogels, sensors, and biomedical applications. Together, these contributions not only provide an excellent overview of the current state-of-the-art in the field, but also point out exciting challenges and opportunities for future work. In a number of reviews, the most recent findings on "Stimuli-Responsive Gels" are compiled in this book, also supplemented by original contributions.

Thermo-responsive gelling materials constructed from natural and synthetic polymers can be used to provide triggered action and therefore are employed in customized products such as drug delivery and regenerative medicine products, as well as in other industries. Some materials enable Arrhenius-type viscosity changes based on coil to globule transitions. Others produce more counterintuitive responses to temperature change because of agglomeration induced by enthalpic or entropic drivers. The contribution of M. Joan Taylor et al. reviews recent developments in these thermo-responsive gels (doi:10.3390/gels3010004). Poly(vinyl caprolactam) (PNVCL) is one of the most important thermo-responsive polymers because it is similar to poly(N-isopropyl acrylamide) (PNIPAM). The use of PNVCL instead of PNIPAM is considered advantageous because of the assumed lower toxicity of PNVCL. The review by Chang-Sik Ha et al. focuses on the recent studies on PNVCL-based stimuli responsive three-dimensional hydrogels (macro, micro, and nano) for biomedical applications (doi:10.3390/gels2010006). In order to study the impact of a novel polyphenolic-based multiacrylate on the properties of N-isopropylacrylamide (NIPAM) hydrogels, J. Zach Hilt et al. synthesized a series of novel temperature-responsive hydrogels with varying contents of chrysin multiacrylate (ChryMA) (doi:10.3390/gels3040040). It was shown that the incorporation of ChryMA decreased the swelling ratios of the hydrogels and shifted their lower critical solution temperature (LCST) to a lower temperature, which could be attributed to the increased hydrophobicity being introduced by the ChryMA. Kuckling et al. synthesized new functional monomers based on vanillin (doi:10.3390/gels2010003). The incorporation of a dialkyl amino group compensated for the hydrophobic effect of the vanillin monomer. PNIPAM hydrogel thin films with transition temperatures in the physiological range could be obtained.

Stimuli-responsive cationic polymers and hydrogels are an interesting class of "smart" materials that respond reversibly to changes in the external pH. This reversible swelling-shrinking property brought about by changes in external pH conditions makes these materials useful in a wide range of applications such as drug delivery systems and chemical sensors, which are summarized in a review by G. Roshan Deen et al. (doi:10.3390/gels4010013). A specific class of responsive polymer-based hydrogels, which are formed through the association of oppositely charged polyion segments, is highlighted in a review by Christine M. Papadakis and Constantinos Tsitsilianis (doi:10.3390/gels3010003). The underpinning temporary three-dimensional network is composed of hydrophilic chains (either ionic or neutral) physically crosslinked by ion pair formation arising from the intermolecular polyionic complexation of oppositely charged repeating units (polyacid/polybase ionic interactions). Due to the weak nature of the involved polyions, these hydrogels respond to pH and are sensitive to the presence of salts. Adding poly(ethylene glycol) (PEG) during the preparation of hydrogels, followed by removal after polymerization, is shown by Björn T. Stokke et al. to improve the swelling dynamics of DNA hybrid hydrogels upon specific ssDNA probe recognition, which is peculiar for these hydrogels (doi:10.3390/gels1020219). Supramolecular stimuli-responsive polymer gels were constructed by Charles-André Fustin and Jean-François Gohy et al. from heterotelechelic double hydrophilic block copolymers that incorporate thermo-responsive sequences (doi:10.3390/gels1020235). The dynamic mechanical properties of this particular class of materials are studied by rheological experiments, showing that structurally reinforcing the micellar network nodes leads to the precise tuning of the viscoelastic response and yield behavior of the material. Finally, the release behavior of glucagon-like peptide-1 (GLP-1) from a biodegradable injectable polymer (IP) hydrogel was reported by Yuichi Ohya et al. (doi:10.3390/gels3040038). This hydrogel shows temperature-responsive irreversible gelation due to the covalent bond formation through a thiol-ene reaction.

A further review by Paula I. Soares et al. presents recent advances in stimuli-responsive gels with an emphasis on functional hydrogels and microgels, highlighting the high impact of stimuli-responsive hydrogels in materials science (doi:10.3390/gels4020054). A final review by Michael J. Serpe et al. is dedicated to the development and application of hydrogels and hydrogel particles for sensing small molecules, macromolecules, and biomolecules (doi:10.3390/gels2010008). For usage in microsystems, the preparation, and hence the characteristics, of these hydrogels (e.g., degree of swelling, size, cooperative diffusion coefficient) are key features, and have to be as reproducible as possible. Richter et al. focuses on the preparation reproducibility of PNIPAAm hydrogels under different conditions, and investigates the influence of oxygen and the UV light source during the photo polymerization process (doi:10.3390/gels2010010). T. Randall Lee et al. describes the encapsulation of gold nanoshells (~160 nm in diameter) within a shell of temperature-responsive poly(N-isopropylacrylamide-*co*-acrylic acid) (P(NIPAM-*co*-AA)) using a surface-bound rationally-designed free radical initiator in water for the development of a photo thermally-induced drug delivery system (doi:10.3390/gels4020028). The diameter of the P(NIPAM-*co*-AA) encapsulated nanoshells decreased as the solution temperature was increased, indicating the collapse of the hydrogel layer with increasing temperatures. Further, the surface plasmon resonance peak of the hydrogel-coated nanoshells appeared at ~800 nm, which lies within the tissue-transparent range that is important for biomedical applications. Including a silica core into thermo-responsive PNIPAAm microgels enhances their mechanical robustness. Birgit Fischer et al. showed that by varying the concentration gradient of the crosslinker, the thermo-responsive behavior of the core-shell microgels can be tuned with three different temperature scenarios (doi:10.3390/gels3030034). The further synthesis of stimuli-responsive colloidal nanocomposite hydrogels is reported by Avinash J. Patil et al., achieved by exploiting non-covalent interactions between anionic cellulose nanocrystals and polycationic delaminated sheets of aminopropyl-functionalized magnesium phyllosilicate clays (doi:10.3390/gels3010011). The preparation of hydrogel nanocomposites containing silver nanoparticles with a size of 15–21 nm is presented by G. Roshan Deen et al., achieved by diffusion and in-situ chemical reduction in chemically crosslinked polymers based on N-acryloyl-N'-ethyl piperazine (AcrNEP) and NIPAM (doi:10.3390/gels1010117). The polymer chains of the hydrogel network enabled the control and stabilization of silver nanoparticles without the need for additional stabilizers. The silver nanoparticles present in the nanocomposite offered additional physical crosslinking, which influenced media diffusion and penetration velocity. Finally, the preparation of PNIPAM hydrogels containing carboxymethylcellulose (CMC) and CMC/Fe_3O_4 nanoparticles is reported by Andrea Atrei et al. (doi:10.3390/gels2040030). FESEM images show that the CMC in the PNIPAM hydrogels induces the formation of a honeycomb structure. This surface morphology was not observed for pure PNIPAM hydrogels prepared under similar conditions. Both PNIPAM/Fe_3O_4 and PNIPAM/CMC/Fe_3O_4 hydrogels exhibit a superparamagnetic behavior.

Review

Thermoresponsive Gels

M. Joan Taylor *, Paul Tomlins and Tarsem S. Sahota

INsmart group, School of Pharmacy Faculty of Health & Life Sciences, De Montfort University, Leicester, LE1 9BH, UK; tomlinspaul@gmail.com (P.T.); ssahota@dmu.ac.uk (T.S.S.)
* Correspondence: mjt@dmu.ac.uk; Tel.: +44-116 250 6317

Academic Editor: Dirk Kuckling
Received: 26 September 2016; Accepted: 16 December 2016; Published: 10 January 2017

Abstract: Thermoresponsive gelling materials constructed from natural and synthetic polymers can be used to provide triggered action and therefore customised products such as drug delivery and regenerative medicine types as well as for other industries. Some materials give Arrhenius-type viscosity changes based on coil to globule transitions. Others produce more counterintuitive responses to temperature change because of agglomeration induced by enthalpic or entropic drivers. Extensive covalent crosslinking superimposes complexity of response and the upper and lower critical solution temperatures can translate to critical volume temperatures for these swellable but insoluble gels. Their structure and volume response confer advantages for actuation though they lack robustness. Dynamic covalent bonding has created an intermediate category where shape moulding and self-healing variants are useful for several platforms. Developing synthesis methodology—for example, Reversible Addition Fragmentation chain Transfer (RAFT) and Atomic Transfer Radical Polymerisation (ATRP)—provides an almost infinite range of materials that can be used for many of these gelling systems. For those that self-assemble into micelle systems that can gel, the upper and lower critical solution temperatures (UCST and LCST) are analogous to those for simpler dispersible polymers. However, the tuned hydrophobic-hydrophilic balance plus the introduction of additional pH-sensitivity and, for instance, thermochromic response, open the potential for coupled mechanisms to create complex drug targeting effects at the cellular level.

Keywords: thermoresponsive; micelle; hydrogel; organogel; UCST; LCST; multi-stimulus; drug delivery

1. Introduction

During the past decade or so there has been interest in gelatinous materials (and their precursors such as xerogels) that respond to a change in their local environment e.g., specific ligands, pH and temperature among others, some of which are being considered as exploitable in, for example, advanced drug delivery systems. This represents a step change from inert polymer systems where the mechanism for delivery was passive diffusion through or dissolution from a matrix or reservoir. Of these materials, the focus here is on gels with thermoresponsive properties [1–3].

Gels are soft semi-solid materials that consist of components including those that act as the liquid dispersion medium (hereafter the "solvent") and the gelling agent (gelator) of which the former is generally numerically the greater [4]. The distinction between thickening and gelling is often taken to be rheologically defined as the case when the ratio of loss to elastic modulus values (tan delta) is less than unity for a gel [5] but colloquially, soft masses formed by aggregates, as discussed below, are often referred to as gelling phenomena. The solvent molecules penetrate a hydrocolloidal network formed by the gelator, which itself can be of several types, some imparting elasticity while others, including inorganic particulate gels, easily permit disruption. Some gels, as in hydrogels, can crosslink physically or be covalently cross-linked [5–8] in which case, as in this review, the crosslinking agent can be seen as a gelling agent. Other systems can have a mixture of transient and permanent bonding e.g., lightly covalently stabilised with receptor-mediated glucose responsive gels [9]. Bridging molecules and particles create two- and three-dimensional aggregates of various structured types such as micelles (as discussed below) that are very viscous so-called gel phases (irrespective of tan delta value). Consequently, they depend on characteristics like molecular size, shape, relative hydrophilicity and ionic nature that maintain solvation. The latter is critical and opposes processes of crystallisation, precipitation and collapse, some of which can be reversible and form the basis of stimulus response. The viscoelastic characteristics of such materials depend on influences on both the solvent and gelator, but the permeability is strongly influenced by the usual factors i.e., partition (if the solvent is not water, for example) and the diffusion coefficient [10]. The latter depends on the microviscosity (i.e., the solvent), the interactive binding forces from the gelator and the porous nature of its resulting network. For the latter, both the extent (porosity) and the tortuosity are important especially for diffusing molecules approaching the size of the pores. The latter can change if the factors controlling the inter- and intra-molecular gelator junctions are affected by ambient conditions and this can also be exploited for responsiveness.

Not surprisingly in terms of biocompatibility, many thermoresponsive pharmaceutical gels consist of hydrophilic polymers in aqueous systems, although there are examples of apolar systems [11–13]. In either case, self-assembly structures including clays and polymers, combine physical weakness with a dynamic equilibrium that allows change and repair and like the solvation status, this has relevance to the responsiveness of the material to stimuli. By comparison, hydrogels and thermosets have a reinforced inter-chain structure that is usefully stronger but will not reform in many cases, if breached. Some non-rigid covalently bonded structures, however, are well known to be capable of stimulus responses because of elasticity, the ability to exchange solvent with the external environment or because responsive components have been designed in. In addition, there is increasing ability to fine tune the responsiveness to specific temperature tolerances [14]. New materials include dynamic covalent bonding and vitrimers of aqueous and non-aqueous type, some of which are within the scope of this review being composites with interesting intermediate characteristics conferring an ability to respond to temperature change [15]. The physical form of gels encompasses bulk and particulate materials including organogels, fast-responding nanogels, aerogels, xerogels and cryogels [16], which are peripherally included in this review, but may not always be biocompatible at an in vitro stage of development.

Sensor and actuation properties occur in gelatinous materials as uniquely simultaneous events. Physical and chemical influences provoke interactions involving solvent and gelator molecules. This can achieve a variety of results including viscosity change, current generated by redox, colour conversion and allosteric transformation in receptors. Utility can be found in controlled delivery of effector molecules [17] but also, for example, in healing [18], regenerative medicine [19,20], tissue engineering [21,22] and smart encapsulation of cells [23]. Chemo-electrical and gravi/volumetric differences can be coupled to accomplish work [24] and there are examples in, among others, responsive muscle-like contracting materials [25,26].

It has already become possible to achieve spatial targeting to tailor treatment by highly specified drug and gene delivery [26–29] via active docking by molecular partners at cellular and subcellular levels. The epitome would be the continuous and detailed response of materials to body, tissue or cellular changes in an imposed homeostasis that revolutionises treatment for diseases that continually metamorphose. One could imagine not only the accurate closed loop delivery of medication in diabetes and other conditions by using ligand-specific materials, but also the meticulously apposite auto-adaptation to clonal changes in metastatic disease.

There are recent reviews that cover aspects of the biomedical use of variously smart materials [30–33]. The way forward, as captured in the examples given in this article, is to use new synthetic, analytical and biological techniques to increase the robustness, cyclic reliability, precision, appropriateness and speed of response. Sometimes responsiveness can be achieved by maximising the surface area using nanostructures and so these aspects will also be included here, with the focus being on thermosensitivity singly or in combination with other traits.

2. Thermosensitivity, Thermoresponsiveness and Phase Transition Rheology

Many polymers, including natural ones like xanthan gum, starch, gellan, konjac, carrageenans, collagen, fibrin, silk fibroin, hyaluronic acid and gelatin [34–39], form gels by several main mechanisms as discussed by Gasperini [40]. These are often usefully temperature sensitive in terms of viscosity change, but thermoresponsive design will increasingly be combined with synthetic features that relate to specific monomeric inclusion that may transduce to a thermal signal or act independently. However, innovative types are also often thermally responsive in terms of outputs other than viscosity or volume, examples being opacity, colour (thermochromic), hydrophobicity and electroconductivity [41–44]. While viscosity changes underpin this review, these other features will be mentioned where relevant. The speed of response often depends on the physical form, principally the surface area and thus much emphasis can be found on nanoforms and this will also be explored. Thermoresponsiveness as a term is used to imply a critical change over a small temperature range, as opposed to a progressive thermosensitivity [45]. Honed individuality is widely represented in the literature and, in general, the variety of polymers with thermoresponsiveness is now huge [1,31,46]. In common with smaller molecules, polymers can exhibit not only glass transitions, but phase separation behaviours, at low and/or high temperature, somewhat confusingly known as Upper and Lower Critical Solution Temperatures, respectively. These transition lines distinguish solution the phase from undissolved two-phase systems that can cause a useful subsequent action. Values for these parameters are variously tabulated for a range of substances including polymers [47,48], with Aseyev focusing very comprehensively on non-ionic structures [49].

2.1. The Upper Critical Solution Temperature (UCST)

One such transition is the UCST which can be designed in to occur within a zone around body temperature and the effect on, for example, viscosity can therefore be exploited for medicine, including personalised drug delivery [50] sensing and feedback systems [51] as in Figure 1 that depicts an injection solution that gels on cooling to below a UCST designed for close to body temperature.

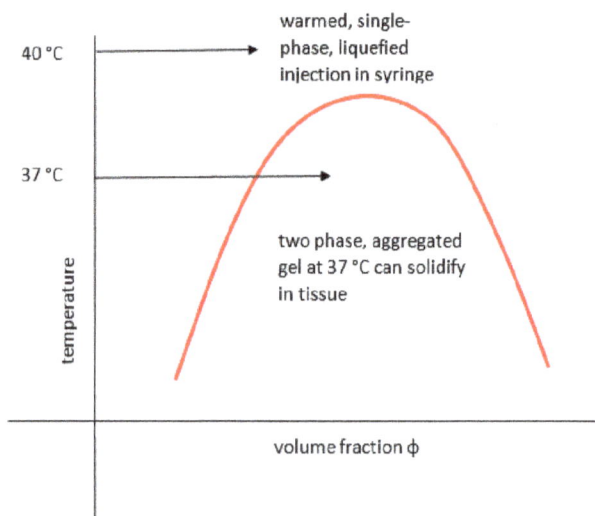

Figure 1. Gelling of injection below its upper critical solution temperature (UCST).

The UCST profile on a temperature-composition (volume fraction) plot for a polymer or blend is the less commonly documented transition type and is defined by the gelator coming out of solution *below* values on a critical separation line known as the spinodal curve, where in addition, a sensitive corridor exists between that and the so-called co-existence (or binodal) curve (Figure 2). These two curves are coincident at their apex, the critical point that represents the infinitely large molecular weight. This is often erroneously termed the cloud point, which is the temperature at which a sample becomes opaque at *any* concentration point on the critical curve. There are two mechanisms for phase separation, namely spinodal decomposition and nucleation plus growth. In practice, phase separation is measured by cloud point, but using turbidity as the criterion depends on the temperature change kinetics, sample geometry, analytical technique (e.g., light, neutron or x-ray scattering), the contrast between the two components, as well as viscosity and flocculation physics. Haas published his 1970 work on both poly(N-acryloyl glycinamide) and gelatin, where the Flory–Huggins and Flory–Rehner theories were applied to the study of the rubbery state of these UCST-type materials to gain insight into their crosslinking mechanisms. Poly(N-acryloyl glycinamide) was also used by Boustta for murine in vivo systems with controlled release over two to three weeks, of neutral and ionic drugs exemplified by model materials of a variety of molecular weights [52,53]. It should be clarified from the outset that there is a difference between in vitro demonstration of pharmaceutical potential in terms of function or biological engineering and the translation to a working medical product for which the safety, efficacy and advantage must be paramount, as discussed by Vert [54]. Safety and medical relevance have several facets including being compatible in pharmacological and inflammatory terms, the latter itself being divisible into molecular, morphological and topological aspects. For example, aggregated and fibrous polymeric forms such as carbon nanotube gels, popular for in vitro study [55], could be toxic in vivo because of biological processes [56,57]. Likewise, implanted materials will normally induce some degree of inflammatory response [58] related to a complex set of responses involving protein deposition, myeloid cell recruitment and fibroblastic walling off. The efficacy of drug delivery is related to the whole body pharmacokinetic constraints and also to any changes imposed on the pharmacodynamic properties of the drug carried, an example being the release of drug from covalent bonding to a carrier. These complexities are not always well mimicked by non-biological systems; however, the direct relevance difficulties they pose do not negate the value of simple systems to study mechanisms in vitro, on the basis that complexity may need to be built step by step for understanding. Consequently, the investigation of a wide range of polymer systems even outside biological tolerances is needed for progress.

In colloidal solution, in a so-called "good" solvent, the polymer molecule is in a loosely coiled state (a in Figure 3). This state exists between the LCST and UCST theta points, where this range can be demonstrated, as explained later in this article. For the UCST case, when cooled to the theta point (b in Figure 3), the solvent passes through the point of minimal polymer solvation as it tends towards "poor" (c in Figure 3). At this latter stage, the polymer molecules and the solvent are enthalpically driven to associate preferentially with themselves, as quantified by the Flory χ parameter. Solvation then fails, taking the system below the UCST curve.

The assumption is that the UCST is above the freezing point of the solvent, and aqueous systems displaying UCST have therefore not often been found, though this may change with advances in polymer design. Where this two-phase state can be demonstrated, i.e., a compacted polymer sphere that is by definition not interacting with solvent, the viscosity might be expected to reduce. However, hydrophobic aggregation may instead produce interesting gelling systems (see below).

The solvated polymer configuration is assumed to move randomly within the confines of a spherical locus but the model is unlike other random walk simulations because it cannot involve time-space superimposition. Therefore, in predictive models, the theoretical volume of the contracted but soluble "globule" model has had to be increased to allow for this.

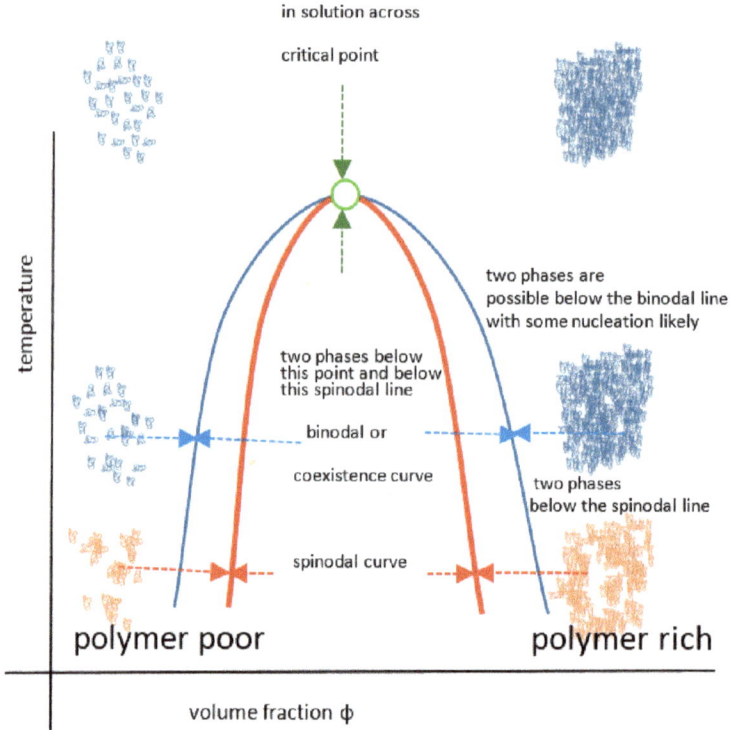

Figure 2. The UCST detail showing spinodal and binodal curves.

a) expanded coil found between UCST and LCST theta points where solvent defined as "good"

b) theta point (upper or lower), still one phase and polymer minimally perturbed by solvent

c) fully contracted globule, two phases and solvent now excluded and described as "poor"

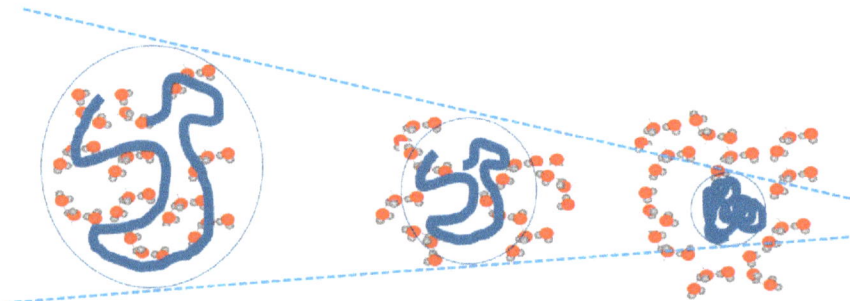

Figure 3. Relationship of the polymer form with temperature for a polymer showing UCST behaviour.

Cooling the coiled molecule towards the critical curve (but above Berghmans point (BP) on the glass transition (Tg) curve (see Figure 4)) first reaches the so-called "upper" theta point (θ_u), as defined by Flory for polymers, which lies above the spinodal curve.

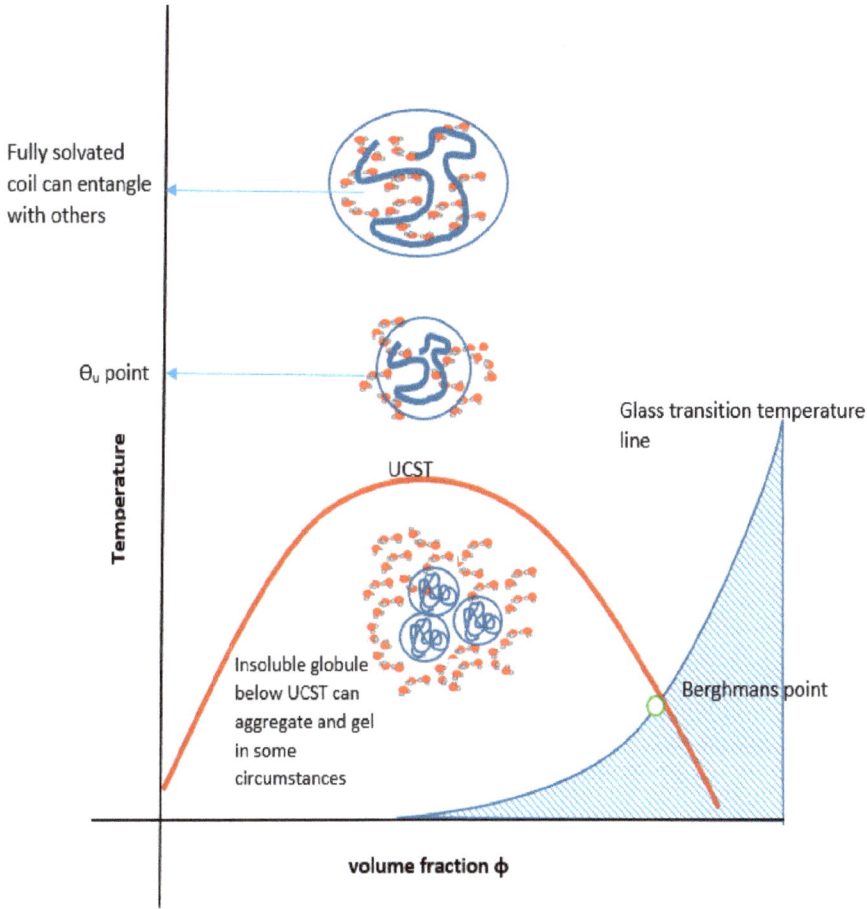

Figure 4. UCST, glass transition (Tg) and Berghmans point.

Using the Mark Houwink plot of log molecular mass vs. log intrinsic viscosity, η allows the determination of the exponent α which takes a value of 0.5 at the theta point θ_u [59]. At θ_u, the entropic solvent swelling effect on the polymer is equal and opposite to the enthalpic energy of polymer-polymer interaction which is a contracting influence. The detail of the thermodynamics have been reviewed elsewhere [60,61].

Above the θ_u temperature value, interactions between extended, solvated coils usually form larger structures especially if branched, relying on van der Waals, electrostatic, hydrogen and occasionally metal coordination bonding [62]. Inter-chain junctions comprise entanglements and knots [60]. At rest, entanglements are in dynamic equilibrium but maintain a three-dimensional assembly imposing a viscoelastic status detected by oscillatory rheometric techniques. They are easily disrupted by rotational stress, often showing a pseudoplastic drop in viscosity with increasing rotational frequency at a given temperature. Further heating of the extended coil pushes the system towards LCST behaviour (see below). Materials may exhibit a true UCST, above which the solvent and solute phases are miscible but, since other phase changes are recognised, the temperature (T) polymer volume fraction (φ) profile should be numerically described before applying the UCST and LCST terms [31].

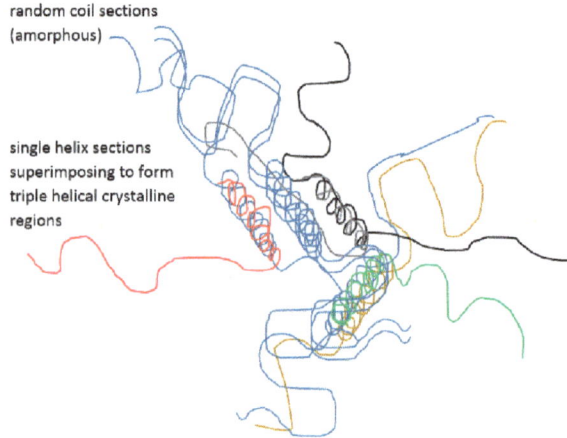

random coil sections
(amorphous)

single helix sections
superimposing to form
triple helical crystalline
regions

Figure 5. Gelatin coil and helix 'crystalline' regions.

Particular materials, such as gelatin and agarose, deserve attention in the discussion of UCST. Gelatin has long been used in a variety of pharmaceutical products, ranging from the simple suppository which is required to liquefy at body temperature to skin preparations which may be required to solidify after applying the melt. Regardless of how familiar these are, with gelatin featuring strongly in recent biomedical research, such as nanofibres and copolymers, and despite well-established observations and facts, including those of Haas [52,63,64], there remains confusion about the thermal behaviour mechanism. It is well known that cooling a hot concentrated gelatin melt allows single and then triple helices to be formed between coils (Figure 5), in the gradual desolvation process [65–67] as the solvent becomes effectively "poorer." This means that, as in the case of simpler coil to globule-forming polymers, the polymer-polymer chain interactions gradually outweigh those of the solvent-polymer, whereby solvent is progressively excluded and a helix results for gelatin. The helices become further intertwined to resemble the parent collagen and the resulting junctions predispose to some crystallinity within the larger network that may retain considerable amorphous regions [68]. For gelatin, at high concentrations (>15% w/w), an elastic gel is therefore formed around 35 °C, depending on several factors including concentration and thermal history. Viscosity hysteresis in a heating-cooling cycle indicates the time dependence of the solid structural arrangements so that fast-cooled gels will contain mainly amorphous material, whereas a slowly cooled system allows the helical rods to align to form distributed crystalline regions. Badii established some years ago that for slowly formed solid gelatin thermograms recorded through reheating indicate a mobilisation of amorphous material in addition to a melting endotherm for the present crystalline regions. Conversely, the latter are missing from a fast-cooled sample [68,69]. Haas reported that gelatin and poly(N-acryloyl glycinamide) (PNAGA) crosslinking and solidification differed in that crystallinity was a major feature only in gelatin [52], thereby adding weight to the idea that gelatin does not display a real UCST. However, gelatin cooling through the solidification point is often termed a UCST phenomenon perhaps analogous to heating poly(N-isopropylacrylamide) (PNIPAm) above the LCST [70,71] as is discussed later. Interestingly, although the helix formation is part of the progressive desolvation of the gelatin coil, in general the temperature journey to the θ value dense sphere, towards either LCST (temperature rise) or UCST (temperature fall), means that the coil form and attendant solvent interactions are minimised. This may not be the confirmed case for the helices at the gelatin solidification temperature across the gelling concentration range. In gelatin solutions that were chosen to be diluted to gel, i.e., 0.25 to 1% w/w, the cloud points (referred to as both Flory or θ_u temperatures, despite clearly not being at an apex concentration) were reported by Gupta to be at about 15–17 °C, i.e., much lower than gelling temperature. Gupta was unable to demonstrate an explicit UCST where insolubility occurred [72] and it therefore seems yet to be clarified if the gelation at higher concentrations is a glass transition or a UCST.

Seuring points out that although in general UCST materials have been identified less often for medical use than LCST (see below), there is room for design and trial of promising new polymeric candidates [73]. The UCST condition is enthalpically driven, as explained above, deriving from strongly exclusive bonding within the solvent and within the polymer. In general, polar materials displaying explicit UCST-related cloud points can be divided into several types, including HB-UCST that rely on H-bonding, those with supramolecular crosslinking additives and C-UCST that involve Coulombic forces. The latter include zwitterionic interactions and relate to ionic materials and milieu [73]. To be widely useful, the working range for a UCST should arguably be in the 0–60 °C range (narrower if for tissue use), operate at low ionic strength, be predictable, reversible, repeatable and sharp. It should be independent of concentration, counterion or pH, suggesting that HB-UCST materials may be more relevant. However, development has been slow and Seuring cites cases such as the previously mentioned H-bonding PNAGA, for which ionic sensitivity is so great that its UCST is often suppressed (and was long unrecognised) due to acidic impurities active at even low concentration [73]. Approaches to producing polymers useful for their USCT were to minimise or neutralise ionic groups, maximise the H-bonding capacity, design against hydrolysis and use copolymers with regular repetition. Thus, for example, a tri-stimuli responsive injectable algin-aminocaproic acid thixogel that formed a sol on heating was recently reported by Chejara [74]. Maji presents a study of L-serine-based zwitterionic polymers with a UCST that is relevant for biomedical use, although the low pH would appear to be a potential problem in tissues for a depot dose [75]. Ji has looked at bespoke polypeptide UCST (and LCST) polymers for cell toxicity in the targeted delivery of cytokines [76]. Some examples that are not explicitly UCST but appear to be broadly similar in that heating produces a gel to sol transition, include work by Gao [77] who describes a low molecular weight phenylboronic acid material that in aqueous polyethylene glycol solution forms a fibrous gel when cooled to 37 °C. It was used with doxorubicin such that after intratumoural injection of the warm, liquefied preparation, this drug became encapsulated in the gel structure that formed in the tissues and could sustain delivery, maintaining effectiveness but reducing general toxicity compared to plain carrier.

Yao describes a fibrous supramolecular, multi-stimuli metal-organic gel system [78]. This is stabilised by hydrophobic and π–π interactions in addition to a coordination component with lead nitrate in low molecular weight benzimidazol ligands. The aqueous, thermoresponsive metal-organic gel formed a sol on heating, as well as being pH and chemically sensitive, and has been used to scavenge dyes from waste water. Conceivably, the same technology could be used in systems for removing poisons from plasma.

Examples of UCST behaviours can be found in polar and apolar solvent systems in a range of disciplines. The classical case of polystyrene in cyclohexane is discussed later in this review in a wider context, but some viscosity adjusters in automotive mineral oils displaying UCST-related phenomena may be transferable to wide temperature range equipment such as pumps, or to design of heat-sensitive organogels that might find application for drug delivery. Thus, engine oil formulation choices must be temperature tuned to prevent the polymer additive to fall below the UCST line where it would become insoluble. Around the theta point (θ_u), one type of additive, poly(alkylmethacrylate), is a minimally oil-solvated globule that therefore does not significantly increase the viscosity of the oil. This is an advantage allowing the oil to assist in the cold starting of engines. At higher temperatures, the additive becomes better solvated and assumes a coil form. Each additive molecule is able to mesh with others and increase the viscosity of the oil so that it can lubricate and load-bear well at high temperatures. The oil itself will have a falling viscosity with rising temperature (normal Arrhenius relationship) but the fall will be reversibly mitigated by the expanding additive because of its globule to coil transition [79], thereby forming an impressive self-adjusting or closed loop system.

Organogels are gelling systems that relate to the above oil additive example, because they are often responsive with a UCST and thus fit within the current remit, having the possibility of biomedical use [80–82]. Organogels can be covalently crosslinked (see below with hydrogels), but again, as for the polar equivalent, they can be weaker, more transiently stabilised structures formed in dispersion sometimes from low molecular weight gelators [83,84] in apolar solvents that include hydrocarbons such as hexane, esters such as isopropyl myristate and vegetable oils, the latter being suitable for dosage forms. Low molecular weight may include the optically responsive (e.g., anthracenes) and respond to a heat source. These could be H-bond or comprise π-π stacking systems and can self-assemble in this way in vivo [85,86] and thus have imaging potential.

Some organogel formulations form a category that is intermediate with hydrogels, where there is a small proportion of water in the solvent mix. In these, nanotubular structures arise from reverse micellar networks filled with water to form polar channels [87]. Some organogels are not explicitly described as having UCST values but may reasonably be assumed to have similar phase transitions, such as in the following dually responsive cases. Zang describes [88] a novel multi-stimuli responsive organogel containing salicylidene Schiff base derivative (cholesterol 2-(3,5 di-tert-butyl-2-hydroxybenzylideneamino)acetate) that self-assembles into nanofibres that undergo a reversible gel to sol transition on heating, where additionally the gel but not the sol has strong fluorescence, associated with the fibre formation. The material is also zinc and possibly fluoride ion responsive [41]. Yang describes an azobenzene gelator in an apolar solvent that can be reversibly triggered by heating, UV excitation or by shear [88]. A dually active organogel system for injected thermoresponsive chemotherapy has also been proposed [89]. In this study, bridged pillarene dimers with a guest linker in the gel state can also emit fluorescence and thus act as a tissue-imaging agent when delivering temperature-targeted drugs as the triggered sol. These simple UCST-type organogels should be distinguished from LCST organogels, bigels and crosslinked organogels which are discussed later.

2.2. The Lower Critical Solution Temperature (LCST)

This is the more frequently encountered circumstance than UCST for polymers and their blends. When heating above the LCST, dissolved polymers like PNIPAm at their critical gel concentration (CGC) will exhibit aqueous insolubility, being above cloud points on the LCST curve where aggregation of desolvated polymers can occur [90]. There is an analogous theta point, the θ_l just below the LCST curve [91] as is reviewed below. The underlying process of contraction of the extended coil to a globule as the solvation decreases [92] on heating is documented for many polymers including, for example, dextran [59,93], and bears discussion in this context. The globular form depends to some extent on molecular weight, such that the typical hydrophilic polysaccharide dextran of molecular weight 200 kDa, is described as assuming a coil of classic Flory spherical shape in water. Above this molecular size for dextran, Masuelli notes that the extensive branching causes a conformational change to an ellipsoid with a detectably increased value of chain stiffness parameter (defined as log viscosity to temperature ratio, $d(\ln [\eta])/dT$). The normal Arrhenius relationship whereby the dextran gel viscosity falls on heating towards θ_l is explained by the size and shape contraction as the solvent becomes again poorer and polymer-polymer interactions predominate (as was the case for the upper theta point θ_u). Antoniou describes the contraction as entropic and dependent on the loss of solvent-related interaction with dextran OH and ether –O– hydrogen bond acceptor groups [59]. The Mark-Houwink α value tended towards 0.5 as the temperature was raised from 20 to 40 °C, again indicating the globule state. Güner had previously reported the θ_l for dextran to be 43 °C and stated that this was related to LCST behaviour [94], but it is also well known that polymers with an overwhelmingly hydrophilic character will not demonstrate an actual LCST. Other materials will not show water insolubility under normal pressure conditions e.g., if it is above boiling point [95]. Masuelli reported Mark-Houwink α values of *below* 0.5 for dextran heating which appear to suggest conditions between the θ_l and an LCST, but this anomalous value was evidently due to applying linear polymer modelling to a hyperbranching system as referred to above.

Aqueous polymers for which the LCST is below the boiling point of water can progress to insolubility on heating beyond their θ_l (analogously to the UCST case). This large group includes many well-known examples of pharmaceutically useful polymers such as a variety of substituted *N*-acrylamides typified by PNIPAm and some substituted celluloses and chitosans and the wider poloxamer family to name but some [96]. The most often reported of these, PNIPAm, is widely reviewed elsewhere and debated as to its biocompatibility. Two recent papers have reviewed and measured this in biological and ophthalmic systems and reported that toxicity, while measurable, may be cell-type and time dependent [31,97].

Li reports a cyclodextrin-poloxamer system with LCST suitable for medical application such as drug and gene delivery [98]. For non-crosslinked polymers with relatively hydrophobic groups, progressive exclusion of the water solvent on heating beyond the LCST creates an entropically driven hydrophobic self-association. In the case of PNIPAm this implies the isopropyl groups where some of the hydrophilic pendant amides project to interact with bulk water through the hydrophobic collapsed backbone and isopropyl core [99]. Thus the LCST, like the UCST depends, for linear and branched chains, on the dominance of hydrophobic groups [93]. The coil to globule transition including the dehydration of the coil and additional aggregation such as for PNIPAm has been variously proposed as both single and multistage models [30,100–102]. It should be noted that between the temperatures of 20 and 32 °C, PNIPAm shows the more typical Arrhenius viscosity reduction with transition from coil to globule before the sharp rise at the spinodal curve [103] where aggregation and, at high concentrations, gelling occurs as described. The literature does contain examples of LCST insolubility that occurs without gelling such as the case of threonine-based chiral homopolymers that precipitate reversibly but with profiles that are very pH dependent [104]. Normally, however, the process of the phase change is dependent on the conditions and the structural detail so that, for example, Costa looks at adjusting the LCST value of PNIPAm with salts and pH while Silva describes the similar behaviour of hydroxypropylmethylcellulose (HPMC) [101,105]. Shi discusses a PNIPAm biconjugate with azobenzene and rhodamine that upon irradiation changes conformation reversibly, thus modulating the coil-globule transition temperature as a result [106]. That all polymers with LCST insolubility and gelling behaviour undergo a slower rehydration when temperatures fall although the hysteresis is the subject of debate. The aggregation step as in PNIPAm was reported early as fully reversible, though reports to the contrary continue to emerge [107–109]. Jeong [110] long ago described the factors important for the kinetics of reversible systems of this sort. However, they are additionally dependent on the total surface area, so that subdivided systems such as micelles and nanogels re-equilibrate faster (see below for these). In one study with gold nanoparticles coated with PNIPAm, the rationale is explored for the conditions needed for the polymer to influence aggregation and the importance of excess PNIPAm [111]. Polymers that respond with heat-induced agglomeration-related viscosity rise are dubbed "negatively thermoresponsive." The most useful candidates in the flurry of design possibilities for drug delivery are those that gel predictably near body temperature, so that materials can enter the body cold (rather than hot as is possible in UCST systems) and will gel at or before 37 °C. Therefore, when Yang discusses alginate beads with an LCST of 55 °C, these show potential but need modification to make the upper temperature achievable safely for the drug delivery proposed [112]. In a study that indicates how unpredictable design can be, Yu constructed core shell nanogels with opposite charges, using modified but thermosensitive PNIPAm. The electrical neutrality was independent of temperature and this system was proposed as a novel in situ gelling system [113].

Lastly in this section and in analogy to the UCST cases, there are simple (i.e., non-crosslinked) organogels with LCST-type behaviour, such as the dual-stimuli homopolymers of poly(7-methacroyloxycoumarin) that exhibit a tunable LCST separation in chloroform among other solvents [114]. Some organogels are structured particulates, such as the ciprofloxacin-carrying shell core stearic acid-alginate type described by Sagiri [115], are at least temperature sensitive if not defined explicitly as LCST.

2.3. Micellar Systems with UCST and/or LCST

Developments involve the self-assembly of multiblock copolymers with amphiphilic properties to form micelles with UCST properties as well as those with the more common LCST profile. Many of these produce thermoresponsive gelling systems that have biomedical prospect.

2.3.1. Micellar UCST Systems

The synthesis techniques are now refined to an extent that numerically defined structures with thermoresponsive blocks and grafts can be produced to tune UCST polymers to required ranges (depending on proposed utility). These are extensively reviewed by others [49,116–120]. Fuijihara [121] describes micellar diblock copolymers with pendant ureido groups. These are in aqueous solution of the monomeric form above the UCST curve but assemble below it at 32 °C to form micelles with a core of poly(2-ureidoethyl methacrylate) or PUEM and shell of poly(2-methacroyloxyethylphosphorylcholine) or PMPC. In the general case, desolvated micelles may then further aggregate to form a non-covalent gelatinous network below the UCST transition, the gelling temperature depending on the hydrophilic-hydrophobic sections that form shell and core respectively. Ranjan [122] has returned to a well-studied surfactant Triton-X that consists of polydisperse preparation of isooctylphenoxy-polyethoxyethanols. Solutions of below 35% v/v are viscous solutions but at higher concentrations are homogeneous gels (35%–60% v/v) or heterogeneous melts (>60% v/v) with a previously unrecognised UCST separating the 35% v/v sol and gel domains at about 25 °C. The model here was a nucleation such that the nucleus radius was 5 nm in the gel phase below the UCST, but 4 nm in the solution and the supra-UCST sol region. Many other such micelle examples can be found, including the synthesis of a methoxy-poly(ethylene glycol)-*block*-poly(acrylamide-*co*-acrylonitrile) (mPEG-*b*-poly(AAm-*co*-AN)) amphiphilic copolymer tunable series with UCST, and tested in vitro for responsive delivery of doxorubicin on cultured tumour cells [123].

Yuan [124] has reported the synthesis and characterisation detail of two series of ethylcellulose graft (EC-g-) copolymers showing opposing thermoresponsive behaviours. Structurally, each ethylcellulose backbone has a graft comprising a poly(ε-caprolactone) (PCL) component in a block structure, the first with poly(2-dimethylaminoethyl methacrylate) to form EC-g-(PCL-*b*-PDMAEMA) and the second, a quaternised version, with poly[3-dimethyl(methacryloyloxyethyl) ammonium propanesulfonate] to form EC-g-(PCL-*b*-PDMAPS). Of these two, the first demonstrates an LCST, while the quaternised structure gives a UCST response (see Figure 6).

For the EC-g-(PCL-*b*-PDMAPS) system, a sharp increase in transmittance and a sharp decrease in hydrodynamic radius was observed above 32 °C. Thus, unlike the conventional coil to globule shrinkage in some UCST systems, interactions of many micelle types are revealed by size expansion at temperatures below the UCST and gels may form. In the EC-g-(PCL-*b*-PDMAPS) case, quaternised groups collapse as expected but form aggregates of this kind, whereas above the transition temperature, extended micellar coils form analogously to the conventional model. An opposite case with LCST for the non-quaternised version is also described in this paper.

Because of the restrictions in aqueous design of UCST systems in general, as referred to above, binary solvent mixtures have been proposed such as the ethanol-aqueous systems discussed by Zhang [96]. These create possibilities in areas not possible in totally aqueous systems and any additional toxicity of ethanol is justified by Zhang as remaining suitable for medical and personal care products in view of the advantage of how finely adjustable such systems can be. He reviews the underlying theory of cosolvency and non-cosolvency as well as several examples of micellisation in the ethanol-water system such as versions of poly(methyl acrylic)-*b*-polystyrene (PMA-*b*-PS) block copolymers with a PS core and UCST values above that of PMA itself. The importance in the physical manifestation of thermal transitions is demonstrated by the double hydrophobic polystyrene$_{88}$-*b*-poly(methyl methacrylate$_{80}$) (PS$_{88}$-*b*-PMMA$_{80}$) polymer that has been developed. It also self-assembles at low concentration in 80% ethanol into micelles. However, at higher concentration (1%), it gels as a UCST response due to proximity of the large radius of gyration for PMMA structures [125].

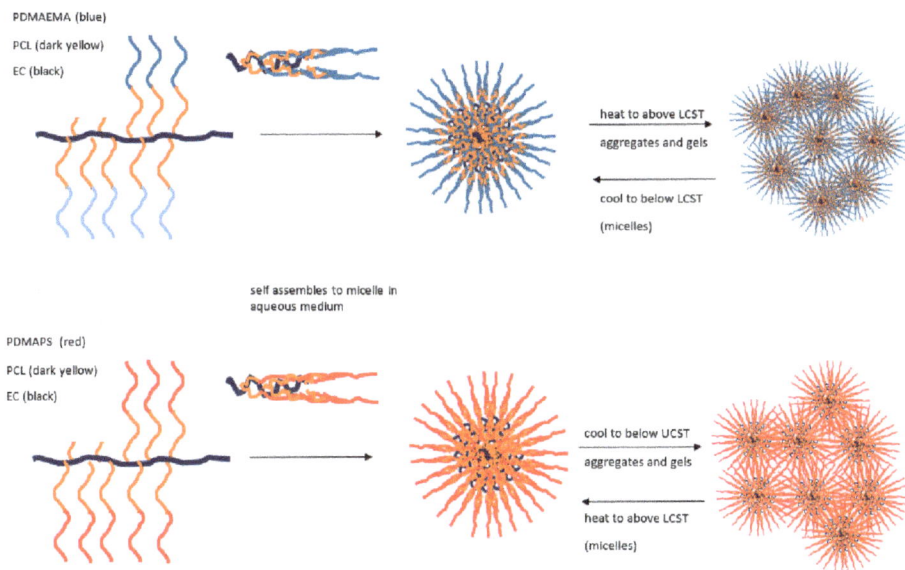

Figure 6. Related EC-g-copolymer aqueous micelle systems that gel when aggregated beyond LCST or UCST respectively. Adapted from [124].

2.3.2. Micellar LCST Systems

At the critical micelle concentration (CMC), raising the temperature again dehydrates the structures by destabilising the equilibrium hydrogen bonding that favours the micelle. This fosters increased hydrophobically stabilised entanglement in the micelle core which contracts, typically leaving hydrophilic groups in the corona to interact with the aqueous solvent. The relative hydrophobicity of the structure determines the CMC and the LCST (often referred to as the Critical Micelle Temperature or CMT) [126,127]. Above the latter, the polymeric solute comes out of solution but may simply maintain a collapsed micellar state. The more dominant the hydrophilic component, the higher the LCST, although some less hydrophilic components may be subjugated to allow particular moieties such as PNIPAm to have a dominant effect on the LCST [27] and subsequent gelling of an aggregated micellar system. Not all insolubles may aggregate and gel [128] but in general this LCST behaviour has the appearance (but not the thermodynamic drive) of being a mirror image of the micelles discussed above with UCST properties and has the general advantages of LCST systems.

A classic example of micellar LCST behaviour is the poloxamer group, patented in the 1970s. These triblock amphiphiles are based on polypropylene oxide and polyethylene oxide (PEG$_{2-130}$-POP$_{15-67}$-PEG$_{2-130}$). Poloxamers form micelles by internalising the relatively hydrophobic PPO segment and, though often described as spherical, may be cubic, cylindrical or elongated (worm-like) depending on concentration among other factors. Poloxamer F-127 has an LCST close to body temperature and has been applied to drug delivery and a variety of other medical uses such as nanoparticle coatings, cell targeting with cytotoxics and delivery of phototherapeutic agent like chlorin e6 [129]. Gelling makes them potentially useful for localised injection and for cavity instillation [127,130]; much as has been suggested for PNIPAm, so that solutions of around 20% *w/w* can be syringed when cold and can be used to fill wound cavities as conforming liquid bandages that solidify in situ. These have been proposed as delivery agents for local anaesthetics, cytotoxics, antibiotics and anti-inflammatories.

Despite the interesting properties of poloxamer, many alternative micellar systems have been proposed because, for example, one of the problems with poloxamer and similar structures such as tetronics is that the gel phase forms at relatively high concentration (20%–30%) [106]. Poloxamers are described as erodible but not biodegradable, and so concentration is a concern among other potential health issues that have been well recognised for many years [131]. Since that time, a variety of well-defined systems have arisen of which the following are examples. A polylactide-polyethylene glycol-polylactide (PL-PEG-PL) series was synthesised in a one-step procedure that avoided toxic coupling agents. It clearly has similarities to poloxamer [132] but unlike poloxamer, was biodegradable, though performed for several weeks before being cleared when trialled as an anti-adhesion cavity filling protective for bowel injury treatment. Micelles underwent phase transition to a gel within 2 min in vitro and in vivo and compared well with hyaluronic acid gels described as standard for the purpose. Hydrogel and organogel formulations with polylactide have been reviewed [133] and include the covalent and non-covalent use of D and L stereoisomers with glycolic acid in copolymers with PEG, polycaprolactone, polyurethane and others.

The commercial use of polylactide has been reviewed [120] and includes Oncogel® which is a PL-PEG-PL formulation as that considered above with co-formulated paclitaxel for injection as a sol that undergoes conversion to a gel depot.

Chitosan is a low toxicity, natural water soluble copolymer of the saccharides β-(1-4)-linked D-glucosamine and N-acetyl-D-glucosamine and is the partially acetylated form of insoluble chitin of crustacean shell origin. It is protonated in acidic milieu and can be quaternised in the trimethylamine form. It is not naturally thermoresponsive but in combination with other materials can become so and thus chitosan-β-glycerophosphate systems are widely reported [134]. In the present context, several examples can be found in recent literature of chitosan copolymers used to form micelles with an LCST. Thus, for example, grafting chitosan to PNIPAm produces nanomicelles used with a payload of curcumin, the LCST being tunable between 38 and 44 °C [135]. A triblock graft polyacrylic acid (PAA) copolymer series of phthaloylchitosan-PNIPAm-PAA has been prepared by RAFT polymerisation [136]. These polymers form micelles that have LCST transition temperatures that are adjustable between the crucial temperatures of 34–40 °C by chain length, branching, pH and concentration, with the sharp, reversible transitions at pH 4. Other chitosan-containing micellar systems have sensitivities to other stimuli, such as redox [137], and are mainly outside the scope of this review.

Cellulose-based materials, e.g., hydrophobically modified hydroxyethylcellulose (HMHEC), can form nanomicelles with a hydrophobic core loaded with the light-emitting reporter poly(9,9-dioctylfluorene) [138]. As explained earlier, it cannot be assumed that materials that have a micellar LCST and thus lose solubility above it, will necessarily gel [139]. In the HPC-g-PDMAEMA case, above LCST and at high pH, the micelles seem to destabilise rather than further aggregate and gel. Chen also reports a triblock (polycaprolactone–PNIPAm-β-substituted alanine) non-aggregating micelle system which was used to carry doxorubicin and the light-activated material meso-tetraphenylchlorin (m-TPC). This is thermoresponsive with an LCST that was tunable with the introduction of the pH-sensitive moiety (alanine) because of the protonation of the carboxylate groups. The proposal was that at body temperature and neutral pH, the micelles would be stable, but in the mildly acidic tumour environment, the structures are merely destabilised and this allows drug release [140].

Micellar systems in general are widely featured in the recent literature partly because of the increasing ability to synthesise their component complex polymers. Particular micelle synthesis designs for LCST are attributable to He [141] who describes a facile, one-pot RAFT polymerisation method to create block copolymers of pentafluorophenyl acrylate (PFPA) and methyl salicylate acrylate (MSA) esters that form dually responsive micelles. He reports utilising the ester selectivity towards a series of aliphatic amines to modify the backbone with amino groups while maintaining the block shape. This elegant investigation produced micelles that were pH and temperature responsive, indicating the increasing ability to design smart structures. However, in this particular case, with LCST values around 65 °C, they were not directly suitable for in vivo use because the aim should be to inject coolly and for body temperature to gel the formulations. These investigative studies pave the way for more directly relevant materials with respect to tissue use. Similarly, other problems centre on solubility issues such that Kim has designed a dually responsive micelle that is composed of a PNIPAm chain with random single inclusions of malachite green that confer photoresponsive qualities, potentially suitable for imaging and possibly phototherapy, in addition to the LCST. The micelle can aggregate with rising temperature but, due to the low solubility of the complex below the LCST, the concentration restraint may have somewhat inhibited gelling in this particular design, inviting developmental modification [142].

Landmarks in thermogelling micellar design include the characterisation of systems where, for example, Weiss showed micelles progressing to further aggregation via two LCST-related cloud points where contraction of the core occurred at the lower temperature LCST, whereas multimicellar aggregation took place above the second [143]. A decade before, Xu had shown a RAFT polymerised NIPAm-PDMA micelle system had a similar accretion stage associated with a second LCST, that could be opposed by dense branching [144]. However, in another complex RAFT product, temperature-induced core-corona and core shell corona stages have been documented as associated with the multiple critical temperature behaviour. A copolymer was thus formed of PEG and a vinylphenyl (V) component to give rise to a diblock section (mPEGV) comprising short PEG brush on an essentially polystyrene (PS), i.e., the pendant phenyl on linear backbone (see Figure 7). This was then further bonded to a terminal linear PNIPAm to give poly(mPEGV$_{466}$)$_{18}$-b-PNIPAm$_{60}$ [145].

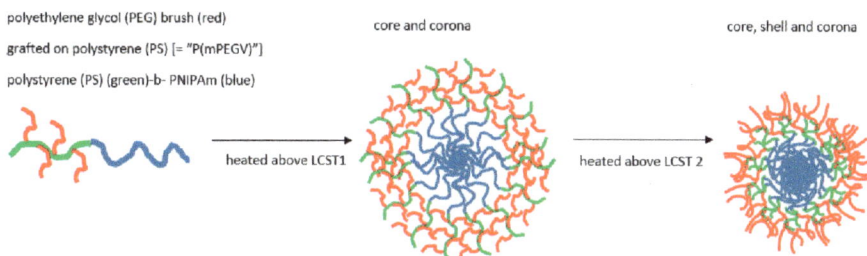

Figure 7. Dually responsive (i.e., two sequential stages LCST) micelle formation from poly(mPEGV$_{466}$)$_{18}$-b-PNIPAm$_{60}$ in water. Adapted from [145].

On raising the temperature through a first and second LCST value above which gelling may occur, the preliminary collapse was of the PNIPAm core producing a corona of PS-PEG, followed by the second of the PS moieties leaving a corona of PEG surrounding a PS shell enclosing the original core of PNIPAm.

Liposomal particles are superficially similar to micelles and are briefly included here. They have a bilayer and more permanent structure than the dynamic nature of micelles. In the simple spherical form, at low concentration, they can be constructed as unilamellar or as a set of concentric lipid bilayer walls assembled typically from a tightly packed lecithin derivative. The lipids are capable of undergoing gel–sol transitions which can be tuned to just above body temperature. However, the term "gel" is being used here in a different sense than for polymers and micelles, in that the region inside each bilayer at temperatures below a transition becomes less mobile. Thus, gelling occurs within the nanodimensioned bilayer and is therefore separate from most of the examples tackled in this review. However, it is also well known that for phospholipids and the like, the relationship between micellar, lamellar and non-lamellar lipid phases involves agglomerates, such as the hexagonal and cubic liposomal forms that overlap with the gelled micellar amphipathic polymers discussed above [146].

2.4. Materials Displaying Both a UCST and LCST

All partially miscible polymer liquid or polymer/polymer mixtures may have, at least in theory, both a UCST and an LCST, there being a theoretical continuum (Figure 8). In practice, they may not be identifiable experimentally.

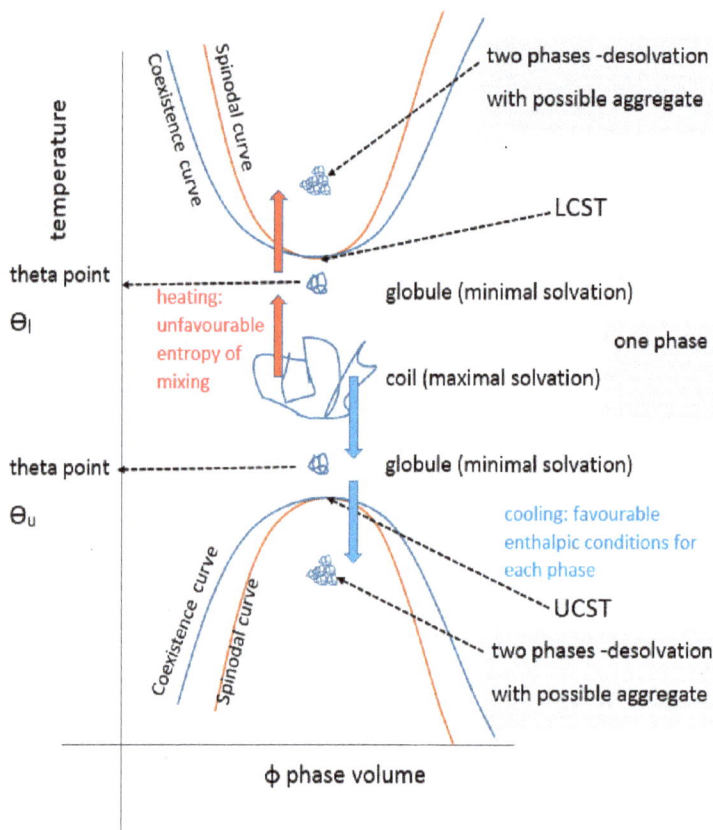

Figure 8. Unified LCST and USCT showing both theta points.

There are many examples of mixtures where the LCST has been calculated to be above the degradation temperature of one or more of the components and for many other systems the theoretical UCST is below the glass transition temperature of one or more of the components. There are also many examples where the kinetics of phase separation are too slow to monitor experimentally due to the high viscosity of the system. This might be resolved by adjusting the conditions (e.g., molecular weight, temperature, pressure, ionic moieties or environment, pH or valency of coordination ions). The general case for materials displaying both critical regions has been fully reviewed recently by Clark and Lipson who present several theoretically related phase diagrams, including the extreme where the upper and lower critical solution temperatures are superimposed or where, with poorer solvents, the curves are joined in an hourglass pattern of immiscibility, for critical pressure and molecular weight values [92] (see Figure 9). The systems presented by them are organic solvent-based including a star-variant of PS in cyclohexane. Both theta values (θ_u and θ_l) have been reported for a few materials [147]. As far back as 1971, the subject was addressed by Siow for polystyrene in acetone and polyisobutylene in benzene [148] and yet there is sometimes confusion between the Flory θ_u and the theta type equivalent θ_l for LCST. Aqueous systems that exhibit both are rarer than inorganic solvent-based ones, some like PEG actually display a closed solubility loop. The polystyrene in the cyclohexane system is the archetypal example with the theta points found at 33 and 210 °C [148], where the polymer has assumed the compressed sphere, there being an extended coil between these temperatures. More UCST cases are, however, emerging and Seuring points out that the aqueous solutions of poly(vinyl methyl ether) (PVME), poly(vinyl alcohol) (PVA) and poly(2-hydroxyethyl methacrylate) (PHEMA) display both UCST and LCST, although for these the UCST is below normal pressure water freezing point [73]. Zhu, however, reports the synthesis of a family of zwitterionic homo-, co- and terpolymers that have this property as the linear non-aggregate [149], although the systems may aggregate without truly gelling and are complex with some overlap as discussed elsewhere [92,150] (see also Figure 3). Balu reports a protein-polymer of the 16-resilin type (further discussed below), that displays both LCST and UCST and reversible gelling [151].

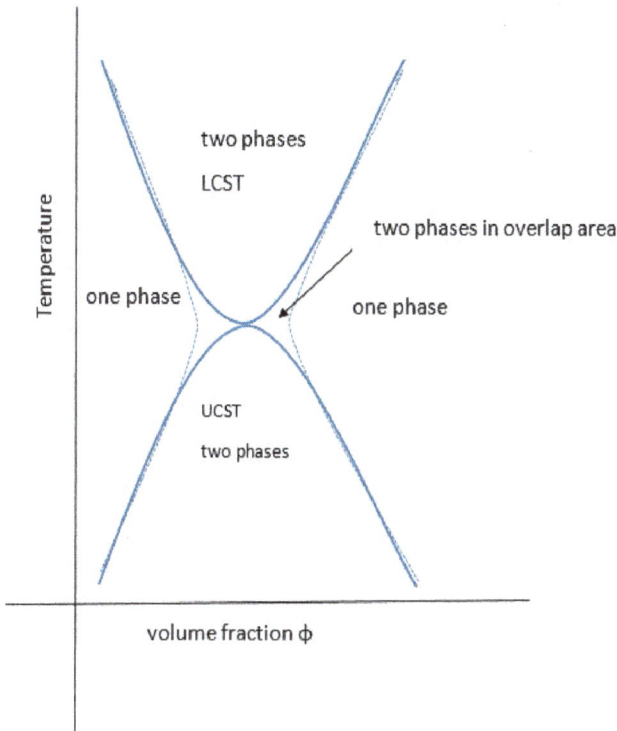

Figure 9. Hourglass pattern of some combined LCST and UCST behaviours.

Micellar systems can also be found that do this. For binary solvent micellar systems, Zhang discussed some block copolymers of 2-phenyloxazoline (PhOx, hydrophobic) and methyl (or ethyl) oxazoline (MeOx or EtOx hydrophilic) incorporated into the copolymer in gradient form [96]. One of these, PhOx$_{50}$ EtOx$_{50}$ in ethanol water 40:60 mix, produced a micellar system that is an example of a material exhibiting both UCST and LCST that was fully reversible. Fine-tuning of an aqueous alcohol binary solvation of poly(diethylene glycol ethyl ether acrylate)-poly(methyl acrylate) (PDEGEA-PMA) block polymers has been shown by Can to produce so-called schizophrenic systems that not only have both LCST and UCST behaviour but, in the process, also exhibit a micelle form reversal where the PMA component can be in either core or corona [152] (Figure 10). Zhang reports a similar but aqueous system [153], while Shih has produced hemocompatible micelles of poly(*N*-isopropylacrylamide)-*block*-poly(sulfobetaine methacrylate) (PNIPAAm-*b*-PSBMA) block copolymers using ATRP, such that they were non-fouling and anticoagulant between 4 and 40 °C and had potential in blood-contacting roles.

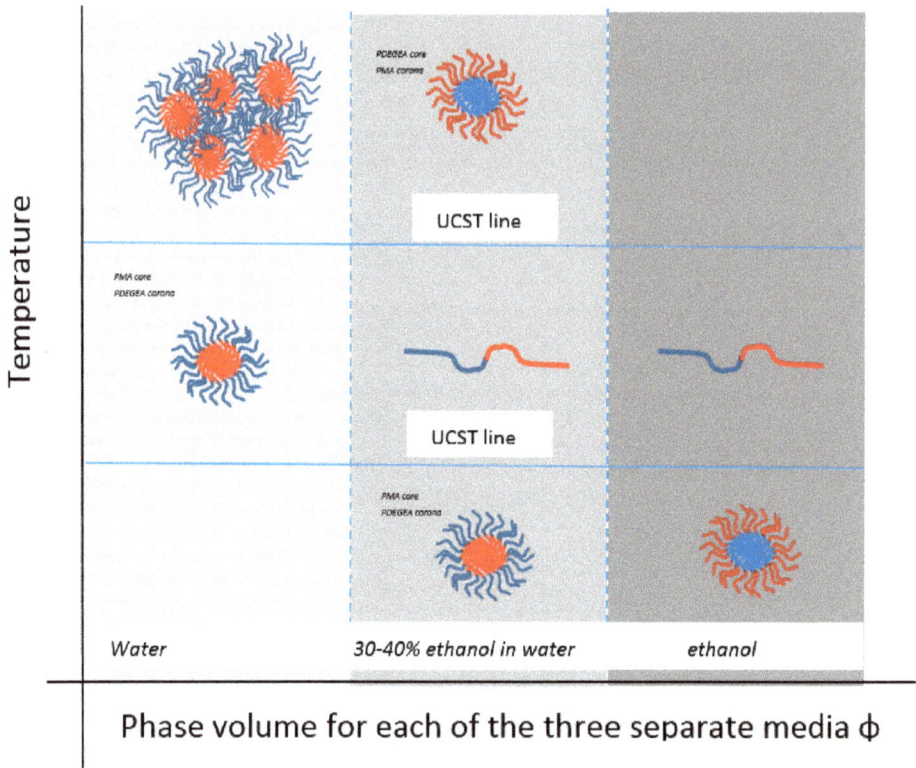

Figure 10. Schizophrenic (reverse) PDEGEA-PMA micelles in ethanolic solution. Adapted from [152].

3. Hydrogels and Crosslinked Organogels Including Particulate Forms

Hydrogels are three-dimensional, elastic lattices of hydrophilic polymer, absorbing rather than dissolving or dispersing in water, the swelling capacity depending on the detail of the crosslinks and components. Many natural and synthetic polymers can be used to form hydrogels which by definition are aqueous [154]. They can be non-ionic, ionic or zwitterionic [155]. Some authors have included non-covalently stabilised semi-solid materials in this class and thus it might embrace gelatin, alginate, pectin and agarose if not a wider range of less elastic mucilages [156]. However, others would restrict the definition to covalently stabilised structures on the basis that otherwise the solvent-induced swelling properties that are usually associated with the immersion of hydrogels would ultimately lead to molecular dispersion via an irreversible gel to sol process. The limitation of this restriction is in the production of covalently bonded micro- to nanogels that do disperse (but not dissolve—see below); however, this review assumes hydrogels have permanent bonding. Advantages of a covalently stabilised structure in pharmaceutical formulation is in the protection of drug molecules, the imposition of both simple and complex release kinetics and the possibility of building phase volume change as a stimulus response, rather than gel to sol change as in dynamic bonding. Hydrogels, like their linear counterparts, comprise component monomers with a range of properties allowing tailoring, but in addition have multipurposed crosslinking agents so that the end product can have bespoke responsiveness in a robust system. One range of continuing innovative copolymeric hydrogels comprises a degree of hydrophobicity associated with the monomeric complement, for example using particular acrylics and also derivatives of amino acids such as valine [157], lipids such as cholesterol [158] and semi-synthetic polysaccharides such as chitosan adjuncts, starch [159] and modified cellulosics [160]. These can impose a volume-phase transition temperature (VPTT) on the hydrogel analogous to the critical solution phase temperature of a monomeric system.

In a review by Mah, crosslinked hydrogels comprising monomers with UCST characteristics contract on cooling below the VPTT relating to the monomeric UCST [161]. They swell on heating above it and are known as positive thermophilic or positive temperature-sensitive hydrogels. Attempts have been made to model temperature-induced unconstrained and constrained swelling [162]. Negative thermophobic or negative temperature-sensitive hydrogels, like crosslinked PNIPAm, contract on heating above the VPTT relating to the LCST and swell below it. For hydrogels, contraction happens with the exclusion of water and the expansions require a source of water to occur. The volume changes are consistent with the globule to coil to globule theoretical continuum of the monomeric systems, with the analogous enthalpic and entropic driving forces (Figure 11). An aqueous hydrogel system rarely if ever demonstrates both types of VPTT although, as described above, there are examples of dual critical solution temperatures for linear and branched equivalent structures.

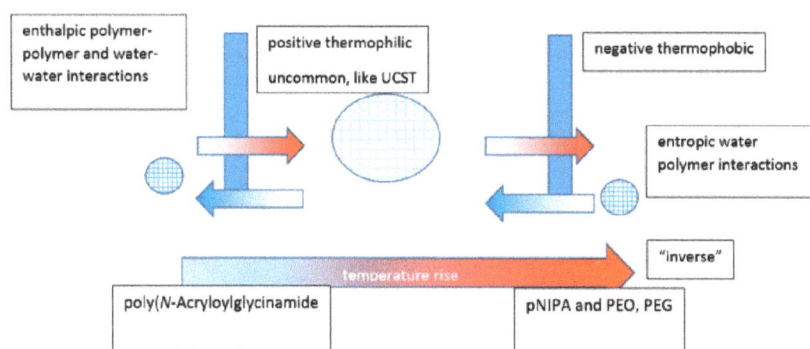

Figure 11. Swelling characteristics of hydrogels.

Transitions of this sort can be smooth and continuous or sharp. Much depends on transient bonds between polymer chains and polymer-solvent bonds, such as H-bonding, rather than the covalent bonds holding the hydrogel structure together [161]. Hydrogels can be categorised in various ways. They can be manufactured as block and graft types, each with a variety of cross-linker molecules. Graft copolymers perform most rapidly, particularly comb-like versions [163], and any one of the monomeric components can be thermoresponsive although thermophilic behaviour is quite difficult to engineer because of the effect of polymerisation on water solubility of the monomer.

Temperature-induced shrinking, as with covalently crosslinked PNIPAm hydrogels, is found for many materials with additions to the category including, for example, poly(N-acroyl-N-propyl piperazine) polymers [164]. Casolaro describes cobalt and ferrous magnetic nanoparticles incorporated into thermoresponsive vinyl hydrogels, based on the α-amino-acid residues phenylalanine and valine, to deliver doxorubicin into HeLa cells [165]. There are also other similar cases for crosslinked magnetic nanogels [166,167].

Hydrogels can be constructed of the one homo or copolymer network but it is also possible to build more than one network that coexist in the same volume (interpenetrating networks or IPNs). Alzari compared thermoresponsive terpolymers with interpenetrating networks, comprising N-isopropylacrylamide (NIPAm), hydroxyethyl methacrylate (HEMA), 2-acrylamido-2-methylpropane sulfonic acid (AMPSA) copolymerised with varying ratios of the latter two. The IPNS were synthesised by two methods, including frontal polymerisation. The thermoresponsive swelling was quite different for the two, with the high AMPSA concentration terpolymers swelling strongly above a UCST (which seems to be wrongly ascribed to thermophobic behaviour). Some IPNs were unresponsive irrespective of concentration, but others shrank presumably with LCST characteristics though only if the HEMA-AMPSA network was pre-synthesised and the PNIPAm polymerised in situ i.e., the reverse method did not perform this way [45]. A cyclically thermophobic (swells with cooling) semi-IPN of starch within a PNIPAm structure was reported by Dragan [168] as a system to release diclofenac sodium in temperature-dependent pulses that were additionally influenced by pH. It was postulated that these might be useful for pyrexic spikes because of the apposite temperature range of the system. The cyclodextrin (CD)-enhanced drug loading of ibuprofen and the 35 °C responsive zone was proposed as a tool for pulsed delivery, for a similar proposed purpose by Wang [169]. A superficially similar theme is to be found in work by Wei [170], who investigated injections of a CD polymer to provide sustained release in a rabbit arthritis model. This, however, is not a hydrogel but a micellar poly(ε-caprolactone)-poly(ethylene glycol)-poly(ε-caprolactone) copolymer (PCEC-β-CD), for which the CD was host to indometacin. A further difference between this and Wang's work is that this is an essentially LCST system where the micelles gel as previously described when raised to body temperature and is thus not like the UCST ibuprofen hydrogel for which additional release might be triggered [169]. Wang also reported a positively responsive (thermophilic i.e., shrinking with cooling and thus relating to UCST) hydrogel synthesised by grafting poly(acrylic acid) (PAAc) to maleic anhydride—cyclodextrin (MAH-β-CD) and forming an IPN with polyacrylamide (PAAm). Schmidt [171] summarises the incorporation of CD into many structure types including hydrogels and micelles, whereby they can temper the shape and content as well as crystallinity of the polymer being synthesised by providing protective in roles as mediators in polymerisation or post-polymerisation. For hydrogels, CDs can be used as crosslinkers if the accepting group fits. In fact, fit is possible for multiguest accommodation. In the last decade, many papers have built on this CD crosslinking concept including an injectable, bioadhesive complex with gelatin and a pH and temperature-responsive type [172,173]. A superbly graphically illustrated review by Arunachaleam [174] considers cyclodextrin and polyrotaxane structures in responsive gels. Polyrotaxanes comprise one or more circular host molecules, such as a cyclodextrin, threaded onto a dumbbell-shaped linear backbone, so that the host molecule cannot escape. Complex construction is possible and the variety is huge. Some representatives of the group are thermoresponsive gels, such as a hydrogel formed from dibenzylammonium-terminated linear polycaprolactones crosslinked in this way with a tetravalent DB24C8 crown-bearing structure.

In general, some of the applications for such hydrogels can involve tissue interaction and the disadvantages of covalent systems are in the potency of any unbound multifunctional monomers in terms of biotoxicity and reactivity with the matrix content. Hennink has reviewed hydrogel synthesis methodology with particular reference to this aspect and also to implants of proteins which are themselves vulnerable if incorporated during covalent bond formation or before the unreacted monomer is removed. However, the subject, as was the case for micelles, is fast developing [7,175,176]. Where delivery or other incorporation of protein molecules is involved, protein to pore size ratio is often important. Either gradual hydrogel degradation can be or protein diffusion can be required, and these can be stimulus responsive including to temperature [177–180]. This implies that parts of the network might need to be labile so that local conditions can foster hydrolytic processes to proceed either chemically or by enzyme action to slowly release the protein payload and preferably to produce harmless excretable products.

Similar apolar (crosslinked organogel) systems can be formed. For example, some are mesoporous such as one described by Helgeson formed from thermogelling of nanoemulsions. The gelling here can be a reversible process or, using crosslinkable gelators, can be photo-polymerised and in some cases converted into a strong but soft material with permanent nanoporous characteristics [181]. A crosslinked valine-containing pH-sensitive hydrogel has been fabricated from an organogel intermediate [182]. Sagiri discusses bigels defined here as a high temperature emulsification of a gelatin hydrogel and a stearic acid-based organogel of sesame oil or soybean oil. The bigels are said to have aggregates whereas superficially similar emugels were true, finely dispersed droplets in the continuum. However, although these showed endothermic peaks at 44 and 85 °C, no claim was made for thermoresponsiveness [80]. A similar system has recently been described by Singh [183,184].

4. Proteins

Proteins are also thermosensitive and capable of sol-gel change although the mechanisms involved are less clear cut with folded active sites formed and combinations of covalent and non-covalent bonds. The distinction between polymer gels and particulate gels (e.g., metallic hydroxides) finds a resonance with protein gels that have occurred via aggregation [185–187]. The protein gelatin is exceptional in many ways, acting more like some polysaccharides (e.g., carrageenan, agar, agarose) reversibly melting rather than irreversibly unfolding and coagulating with increased temperature like myofibrillar and ovalbumins. The sol formation for gelatin is a denaturation process and the helical formation is part of its remarkable refolding. In general, gelation of proteins more often involves irreversible disulphide bond rupture. Gelatin is again unusual in that it has none, probably contributing to the reversibility of its structure compared with others in which the thermoresponsive utility is limited. Gelatin itself forms a physically entangled gel, but recently, Yue built on a wealth of work where gelatin has been methacrylated to give thermoresponsive gels with a wide range of mechanical properties and fabricated products [188]. Heating proteins more often results in a different behaviour, with molecular changes on several levels, affecting tertiary and secondary structure. Unfolding at the endothermic denaturation point is entropy driven and as remaining crystalline regions melt, a mobile amorphous state may then, unlike the gelatin case, foster interactions that then cause irreversible gelling at a transition possibly identical to an LCST [99]. This follows the formation of apolar aggregates which may themselves be disulphide-stabilised from free thiols or lysine amine groups. Artificial, biotechnically formed elastin-type polypeptide structures with the repeating unit Valine-Proline-Glycine-Xaa-Glycine, where Xaa represents a non-proline "guest" amino acid, mimic natural analogues throughout the biological world. These materials exhibit phase transition behaviours that are specific to structure but are triggered by stimuli, including temperature and pH, and thus they share the characteristics previously discussed. When triggered, for example by temperature, insolubility and aggregation occur that involve a helical interaction of molecules with hydrophobic bonding. This property can be conferred on a conjugate used, for example, in ligand sensing, movement, biochemical processes, drug delivery, etc. In terms of the latter, it can be honed to a temperature differential said to distinguish some tumour cells, thus allowing cytoplasmic retention of drugs conjugated to the peptide polymeric structure [189,190]. Similar properties can be found in silk-elastin-like proteins (SELPs) for which Xia has proposed production methodology [191]. Preliminary work with these materials produced materials where the aggregation was non-reversible. In recent work, Glassman used valine as the principal guest amino acid and arrested the spinodal separation process previously leading to the heat-induced coacervation characteristic of the class. The process can be controlled and produces reversibly thermoresponsive materials that can be toughened for use in biomedical settings among others [192,193].

5. Thermoresponsive Vitrimers and Composites

In contrast to most hydrogels, covalent thermal curing can be engineered, such as for conventional thermosetting polymers exemplified by polyurethane-forming components bearing functional groups.

These have mechanical strengths that generally far outweigh those of even the toughest types of hydrogels [194]. However, they are also resistant to solvents so they represent the opposite end of a spectrum of polymer types that would find little application in diffusional drug delivery systems like implants and are themselves outside the scope of this review.

However, an intermediate type of material has been developed comparatively recently—vitrimers i.e., a polymer that has aspects of vitrification [195]. These have been described as mechanically strong but, when heated, they have gradually active associative exchange mechanisms [196] that occur over a wide quasi-glass transition temperature range, thus combining features of thermoplastic and thermosetting materials [197–199]. This means that they have the same characteristics and can be self-healing and also be reshaped, which has implications for sustainability and also for recycling [15,200,201]. Fortman describes poly(hydroxyurethanes) that can be activated by heat and stress and proposes their utility on the basis of low toxicity and catalyst-free reparability [202]. A medical example is a heat-activated aqueous solution that creates nanomolecular bridging of tissue via a gelling process similar to medical cyanoacrylates, but without the solidity or the sting. Popular press science articles refer to this being relevant to situations where suturing is impractical. Interestingly, an all-DNA vitrimer-like material has been described in which star-shaped sequences of designer single-stranded DNA bind via sticky ends to network upon cooling and can be used as encapsulation devices. These are more fragile than many synthetic vitrimer types, but still have the property of continual self-healing with parameters that can be tuned catalytically [203].

The remaining thermoresponsive polymer category in this review are composites with shape memory [204] imposed by forming above a transition temperature T_{trans} that is greater than the glass transition (Tg) or the melting temperatures (Tm). These elastic compounds are typically improved in terms of aggregation, strength and thermoresponsiveness by adding nanosized rods, films and fibres of metal oxides, graphine and other carbon materials, noble metals and natural polymers such as cellulose. Such materials are useful in many settings, as discussed by Pilate, including smart medical devices [205]. In many envisaged uses, external heating would be less convenient than internally generated heat using the fillers to convert other energy forms, such as light, electrical, magnetic, chemical or acoustic. Carbon black, for example, dispersed through the chosen polymer gels [206] converts electrical energy to heat and creates the shape change which is temporary provided the temperature does not exceed T_{trans} [205]. A synergistic activity between carbon black, carbon nanofibres and nickel strands, whereby electrical connectivity was enhanced, optimised the heat transfer that underlies the reversible contractile behaviour of proposed artificial muscles [207]. In a variant, where gold nanorods were incorporated into poly(sodium 4-styrenesulphonate), the temperature rise was induced by near-infrared light which can penetrate tissue and was able to trigger the release of doxorubicin conjugated to the gold while a gold nanocage-hyaluronic acid platform was similarly photothermally employed [208,209]. Wang has a somewhat similar system to deliver plasmid-derived genetic material to the cytosol from endosomes via gold nanorods exposed to photothermal excitation at 808 nm [210].

6. Concluding Remarks

Thermoresponsive materials have been the subject of many published studies and have been reviewed regularly, reflecting the explosion in interesting and complex molecular structures such as the synthetic methodology, specifically Reversible Addition Fragmentation chain Transfer (RAFT) and Atomic Transfer Radical Polymerisation (ATRP). The focus in this review has been the viscosity changes as these polymeric molecules respond to temperature change, with the accent on upper and lower critical solution temperatures (UCST and LCST). Where possible, these have been applied to biomedical utility, especially drug delivery, medical sensing and the combination of the two for smart systems wherein sensing and actuation are combined. Lessons can be learned from the use of such materials in other fields where they exhibit UCST, LCST and related behaviour, using low molecular weight gelators, single polymeric molecules and their aggregates. There is much published on the LCST behaviour and its possible exploitation for medicine. Unusually, we have explored UCST materials more fully because the area has been sparse and yet there may be a rich seam to be mined in organogels and self-assembled hydrophobic materials such as micelles.

Covalently stabilised gels give further opportunity for actuation both mechanically and by combination with other inputs (pH etc.) and outputs (thermochromic, conductive etc.) and, given that acceptable pharmacologically active molecules are scarce, the use of novel developments in thermoresponsive materials is poised to enable targeting and activation in fields such as cancer, diabetes and immune-provoked inflammatory disease. The dream of a continuously responding dose form for mutational change is a distant but achievable goal.

There are, however, concerns about safety mostly in the areas of toxic leaching of monomers, failure to degrade, inflammatory tissue response and infection. The concepts of precision of dose rate, reproducibility of short and long timescales and dose dumping are also important in this regard, especially for the delivery of potent materials. In this review, we have also briefly addressed the porous behaviour of permanent gels including the relevance of pore size to solute ratio, enzyme vulnerability design and tissue stresses on block implants. We have also contextualised the design and use of particulates and of self-healing, shape-adjusting materials that have dynamic covalent bonding intermediates between the permanence of covalent bonding and the transience of self-assembly.

Acknowledgments: The work contributing to this has been supported by the Edith Murphy Foundation and by NEAT grant K024 (NIHR, UK). There was no specific allocation for publishing in open access.

Author Contributions: M. Joan Taylor, Tarsem S. Sahota and Paul Tomlins wrote and edited the review paper and M. Joan Taylor collated and finalised it.

Conflicts of Interest: The authors declare no conflict of interest.

References

1. Klouda, L. Thermoresponsive hydrogels in biomedical applications: A seven-year update. *Eur. J. Pharm. Biopharm.* **2015**, *97*, 338–349. [CrossRef] [PubMed]
2. Aguilar, M.R.; San Román, J. 1—Introduction to smart polymers and their applications. In *Smart Polymers and Their Applications*; Aguilar, M.R., Román, J.S., Eds.; Woodhead Publishing: Cambridge, UK, 2014; pp. 1–11.
3. Matanović, M.R.; Kristl, J.; Grabnar, P.A. Thermoresponsive polymers: Insights into decisive hydrogel characteristics, mechanisms of gelation, and promising biomedical applications. *Int. J. Pharm.* **2014**, *472*, 262–275. [CrossRef] [PubMed]
4. Zhang, J.; Zeng, L.; Feng, J. Dynamic covalent gels assembled from small molecules: From discrete gelators to dynamic covalent polymers. *Chin. Chem. Lett.* **2016**, in press. [CrossRef]
5. Saha, D.; Bhattacharya, S. Hydrocolloids as thickening and gelling agents in food: A critical review. *J. Food Sci. Technol.* **2010**, *47*, 587–597. [CrossRef] [PubMed]
6. Nazir, A.; Asghar, A.; Aslam Maan, A. Chapter 13—Food Gels: Gelling Process and New Applications. In *Advances in Food Rheology and Its Applications*; Ahmed, J., Ptaszek, P., Basu, S., Eds.; Woodhead Publishing: Cambridge, UK, 2017; pp. 335–353.
7. Hennink, W.E.; van Nostrum, C.F. Novel crosslinking methods to design hydrogels (Republished 2012, first published 2002). *Adv. Drug Deliv. Rev.* **2012**, *64*, 223–236. [CrossRef]
8. Goponenko, A.V.; Dzenis, Y.A. Role of mechanical factors in applications of stimuli-responsive polymer gels—Status and prospects. *Polymer* **2016**, *101*, 415–449. [CrossRef]
9. Sahota, T.S.; Sawicka, K.; Taylor, M.J.; Tanna, S. Effect of varying molecular weight of dextran on acrylic derivatised dextran and concanavalin A glucose-responsive materials for closed-loop insulin delivery. *Drug Dev. Ind. Pharm.* **2011**, *37*, 351–358. [CrossRef] [PubMed]
10. Silva, A.S.G.; Pinheiro, M.N.C. Diffusion Coefficients of Timolol Maleate in Polymeric Membranes Based on Methacrylate Hydrogels. *J. Chem. Eng. Data* **2013**, *58*, 2280–2289. [CrossRef]
11. Esposito, E.; Ravani, L.; Mariani, P.; Puglia, C.; Mazzitelli, S.; Huang, N.; Cortesi, R.; Nastruzzi, C. Gelified reverse micellar dispersions as percutaneous formulations. *J. Drug Deliv. Sci. Technol.* **2016**, *32*, 270–282. [CrossRef]
12. Feng, G.; Wang, H.; Yang, Y. Diffusion and Release of the Guest Molecules in Supramolecular Organogel. *Acta Chim. Sin.* **2008**, *66*, 576.

13. Wang, D.; Zhao, J.; Liu, X.; Sun, F.; Zhou, Y.; Teng, L.; Li, Y. Parenteral thermo-sensitive organogel for schizophrenia therapy, in vitro and in vivo evaluation. *Eur. J. Pharm. Sci.* **2014**, *60*, 40–48. [CrossRef] [PubMed]

14. Constantinou, A.P.; Georgiou, T.K. Tuning the gelation of thermoresponsive gels. *Eur. Polym. J.* **2016**, *78*, 366–375. [CrossRef]

15. Chabert, E.; Vial, J.; Cauchois, J.P.; Mihaluta, M.; Tournilhac, F. Multiple welding of long fiber epoxy vitrimer composites. *Soft Matter* **2016**, *12*, 4838–4845. [CrossRef] [PubMed]

16. Ganesan, K.; Dennstedt, A.; Barowski, A.; Ratke, L. Design of aerogels, cryogels and xerogels of cellulose with hierarchical porous structures. *Mater. Des.* **2016**, *92*, 345–355. [CrossRef]

17. Teotia, A.K.; Sami, H.; Kumar, A. 1—Thermo-responsive polymers: Structure and design of smart materials. In *Switchable and Responsive Surfaces and Materials for Biomedical Applications*; Zhang, Z., Ed.; Woodhead Publishing: Oxford, UK, 2015; pp. 3–43.

18. Zhu, Y.; Hoshi, R.; Chen, S.; Yi, J.; Duan, C.; Galiano, R.D.; Zhang, H.F.; Ameer, G.A. Sustained release of stromal cell derived factor-1 from an antioxidant thermoresponsive hydrogel enhances dermal wound healing in diabetes. *J. Control. Release* **2016**, *238*, 114–122. [CrossRef] [PubMed]

19. Zhu, Y.; Jiang, H.; Ye, S.; Yoshizumi, T.; Wagner, W.R. Tailoring the degradation rates of thermally responsive hydrogels designed for soft tissue injection by varying the autocatalytic potential. *Biomaterials* **2015**, *53*, 484–493. [CrossRef] [PubMed]

20. Van Hove, A.H.; Burke, K.; Antonienko, E.; Brown, E., III; Benoit, D.S.W. Enzymatically-responsive pro-angiogenic peptide-releasing poly(ethylene glycol) hydrogels promote vascularization in vivo. *J. Control. Release* **2015**, *217*, 191–201. [CrossRef] [PubMed]

21. Akiyama, Y.; Okano, T. 9—Temperature-responsive polymers for cell culture and tissue engineering applications. In *Switchable and Responsive Surfaces and Materials for Biomedical Applications*; Zhang, Z., Ed.; Woodhead Publishing: Oxford, UK, 2015; pp. 203–233.

22. Lin, J.B.; Isenberg, B.C.; Shen, Y.; Schorsch, K.; Sazonova, O.V.; Wong, J.Y. Thermo-responsive poly(N-isopropylacrylamide) grafted onto microtextured poly(dimethylsiloxane) for aligned cell sheet engineering. *Colloids Surfaces B Biointerfaces* **2012**, *99*, 108–115. [CrossRef] [PubMed]

23. Smink, A.M.; de Haan, B.J.; Paredes-Juarez, G.A.; Wolters, A.H.; Kuipers, J.; Giepmans, B.N.; Schwab, L.; Engelse, M.A.; van Apeldoorn, A.A.; de Koning, E.; et al. Selection of polymers for application in scaffolds applicable for human pancreatic islet transplantation. *Biomed. Mater.* **2016**, *11*, 035006. [CrossRef] [PubMed]

24. Khan, Z.U.; Edberg, J.; Hamedi, M.M.; Gabrielsson, R.; Granberg, H.; Wågberg, L.; Engquist, I.; Berggren, M.; Crispin, X. Thermoelectric Polymers and Their Elastic Aerogels. *Adv. Mater.* **2016**, *28*, 4556–4562. [CrossRef] [PubMed]

25. Ismail, Y.A.; Martínez, J.G.; Al Harrasi, A.S.; Kim, S.J.; Otero, T.F. Sensing characteristics of a conducting polymer/hydrogel hybrid microfiber artificial muscle. *Sens. Actuators B Chem.* **2011**, *160*, 1180–1190. [CrossRef]

26. Ge, D.; Qi, R.; Mu, J.; Ru, X.; Hong, S.; Ji, S.; Linkov, V.; Shi, W. A self-powered and thermally-responsive drug delivery system based on conducting polymers. *Electrochem. Commun.* **2010**, *12*, 1087–1090. [CrossRef]

27. Ward, M.A.; Georgiou, T.K. Thermoresponsive Polymers for Biomedical Applications. *Polymers* **2011**, *3*, 1215–1242. [CrossRef]

28. Priya James, H.; John, R.; Alex, A.; Anoop, K.R. Smart polymers for the controlled delivery of drugs—A concise overview. *Acta Pharm. Sin. B* **2014**, *4*, 120–127. [CrossRef] [PubMed]

29. Shim, M.S.; Kwon, Y.J. Stimuli-responsive polymers and nanomaterials for gene delivery and imaging applications. *Adv. Drug Deliv. Rev.* **2012**, *64*, 1046–1059. [CrossRef] [PubMed]

30. Lemanowicz, M. Thermosensitive aggregation under conditions of repeated heating-cooling cycles. *Int. J. Miner. Process.* **2015**, *144*, 26–32. [CrossRef]

31. Gandhi, A.; Paul, A.; Sen, S.O.; Sen, K.K. Studies on thermoresponsive polymers: Phase behaviour, drug delivery and biomedical applications. *Asian J. Pharm. Sci.* **2015**, *10*, 99–107. [CrossRef]

32. Casado, N.; Hernández, G.; Sardon, H.; Mecerreyes, D. Current trends in redox polymers for energy and medicine. *Prog. Polym. Sci.* **2016**, *52*, 107–135. [CrossRef]

33. Li, J.; Stachowski, M.; Zhang, Z. 11—Application of responsive polymers in implantable medical devices and biosensors. In *Switchable and Responsive Surfaces and Materials for Biomedical Applications*; Zhang, Z., Ed.; Woodhead Publishing: Oxford, UK, 2015; pp. 259–298.

34. Bilanovic, D.; Starosvetsky, J.; Armon, R.H. Cross-linking xanthan and other compounds with glycerol. *Food Hydrocoll.* **2015**, *44*, 129–135. [CrossRef]

35. Bhatia, M.; Ahuja, M.; Mehta, H. Thiol derivatization of Xanthan gum and its evaluation as a mucoadhesive polymer. *Carbohydr. Polym.* **2015**, *131*, 119–124. [CrossRef] [PubMed]

36. Kim, J.; Hwang, J.; Kang, H.; Choi, J. Chlorhexidine-loaded xanthan gum-based biopolymers for targeted, sustained release of antiseptic agent. *J. Ind. Eng. Chem.* **2015**, *32*, 44–48. [CrossRef]

37. Bassas-Galia, M.; Follonier, S.; Pusnik, M.; Zinn, M. 2—Natural polymers: A source of inspiration. In *Bioresorbable Polymers for Biomedical Applications*; Perale, G., Hilborn, J., Eds.; Woodhead Publishing: Cambridge, UK, 2017; pp. 31–64.

38. Zhu, B.; Ma, D.; Wang, J.; Zhang, J.; Zhang, S. Multi-responsive hydrogel based on lotus root starch. *Int. J. Biol. Macromol.* **2016**, *89*, 599–604. [CrossRef] [PubMed]

39. Karolewicz, B. A review of polymers as multifunctional excipients in drug dosage form technology. *Saudi Pharm. J.* **2015**, *24*, 525–536. [CrossRef] [PubMed]

40. Gasperini, L.; Mano, J.F.; Reis, R.L. Natural polymers for the microencapsulation of cells. *J. R. Soc. Interface* **2014**, *11*. [CrossRef] [PubMed]

41. Zang, L.; Shang, H.; Wei, D.; Jiang, S. A multi-stimuli-responsive organogel based on salicylidene Schiff base. *Sens. Actuators B Chem.* **2013**, *185*, 389–397. [CrossRef]

42. Seeboth, A.; Lotzsch, D.; Ruhmann, R.; Muehling, O. Thermochromic polymers—Function by design. *Chem. Rev.* **2014**, *114*, 3037–3068. [CrossRef] [PubMed]

43. Mojtabavi, M.; Jodhani, G.; Rao, R.; Zhang, J.; Gouma, P. A PANI–Cellulose acetate composite as a selective and sensitive chemomechanical actuator for acetone detection. *Adv. Device Mater.* **2016**, *2*, 1–7. [CrossRef]

44. Li, X.; Gao, Y.; Serpe, J.M. Stimuli-Responsive Assemblies for Sensing Applications. *Gels* **2016**, *2*, 8. [CrossRef]

45. Alzari, V.; Ruiu, A.; Nuvoli, D.; Sanna, R.; Martinez, J.I.; Appelhans, D.; Voit, B.; Zschoche, S.; Mariani, A. Three component terpolymer and IPN hydrogels with response to stimuli. *Polymer* **2014**, *55*, 5305–5313. [CrossRef]

46. Ward, M.A.; Georgiou, T.K. Multicompartment thermoresponsive gels: Does the length of the hydrophobic side group matter? *Polym. Chem.* **2013**, *4*, 1893–1902. [CrossRef]

47. Flory, P. *Selected Works of Paul J Flory*; Mandelkern, L., Mark, J., Suter, U., Yoon, D.Y., Eds.; Stanford University Press: Redwood City, CA, USA, 1985; Volume 1.

48. Kawaguchi, H. *Biomedical Applications of Hydrogels Handbook*; Ottenbrite, R., Park, K., Okano, T., Eds.; Springer Science and Business Media: New York, NY, USA, 2010.

49. Aseyev, V.; Tenhu, H.; Winnik, F.M. Non-Ionic Thermoresponsive Polymers in Water. In *Self Organized Nanostructures of Amphiphilic Block Copolymers II*; Müller, H.E.A., Borisov, O., Eds.; Springer: Berlin/Heidelberg, Germany, 2011; pp. 29–89.

50. Coelho, J.F.; Ferreira, P.C.; Alves, P.; Cordeiro, R.; Fonseca, A.C.; Góis, J.R.; Gil, M.H. Drug delivery systems: Advanced technologies potentially applicable in personalized treatments. *EPMA J.* **2010**, *1*, 164–209. [CrossRef] [PubMed]

51. Generalova, A.N.; Oleinikov, V.A.; Sukhanova, A.; Artemyev, M.V.; Zubov, V.P.; Nabiev, I. Quantum dot-containing polymer particles with thermosensitive fluorescence. *Biosens. Bioelectron.* **2013**, *39*, 187–193. [CrossRef] [PubMed]

52. Haas, H.C.; Chiklis, C.K.; Moreau, R.D. Synthetic thermally reversible gel systems. III. *J. Polym. Sci. A-1 Polym. Chem.* **1970**, *8*, 1131–1145. [CrossRef]

53. Boustta, M.; Colombo, P.; Lenglet, S.; Poujol, S.; Vert, M. Versatile UCST-based thermoresponsive hydrogels for loco-regional sustained drug delivery. *J. Control. Release* **2014**, *174*, 1–6. [CrossRef] [PubMed]

54. Vert, M. Polymeric biomaterials: Strategies of the past vs. strategies of the future. *Prog. Polym. Sci.* **2007**, *32*, 755–761. [CrossRef]

55. Pourjavadi, A.; Doulabi, M. Multiwalled carbon nanotube-polyelectrolyte gels: Preparation and swelling behavior for organic solvents. *Solid State Ion.* **2014**, *257*, 32–37. [CrossRef]

56. Louro, H.; Pinhão, M.; Santos, J.; Tavares, A.; Vital, N.; Silva, M.J. Evaluation of the cytotoxic and genotoxic effects of benchmark multi-walled carbon nanotubes in relation to their physicochemical properties. *Toxicol. Lett.* **2016**, *262*, 123–134. [CrossRef] [PubMed]

57. Aillon, K.L.; Xie, Y.; El-Gendy, N.; Berkland, C.J.; Forrest, M.L. Effects of nanomaterial physicochemical properties on in vivo toxicity. *Adv. Drug Deliv. Rev.* **2009**, *61*, 457–466. [CrossRef] [PubMed]

58. Mooney, J.E.; Rolfe, B.E.; Osborne, G.W.; Sester, D.P.; van Rooijen, N.; Campbell, G.R.; Hume, D.A.; Campbell, J.H. Cellular Plasticity of Inflammatory Myeloid Cells in the Peritoneal Foreign Body Response. *Am. J. Pathol.* **2010**, *176*, 369–380. [CrossRef] [PubMed]

59. Antoniou, E.; Tsianou, M. Solution properties of dextran in water and in formamide. *J. Appl. Polym. Sci.* **2012**, *125*, 1681–1692. [CrossRef]

60. Maffi, C.; Baiesi, M.; Casetti, L.; Piazza, F.; de Los Rios, P. First-order coil-globule transition driven by vibrational entropy. *Nat. Commun.* **2012**, *3*, 1065. [CrossRef] [PubMed]

61. Oh, S.Y.; Chan Bae, Y. Role of intermolecular interactions for upper and Lower Critical Solution Temperature Behaviors in polymer solutions: Molecular simulations and thermodynamic modeling. *Polymer* **2012**, *53*, 3772. [CrossRef]

62. Deshmukh, P.K.; Ramani, K.P.; Singh, S.S.; Tekade, A.R.; Chatap, V.K.; Patil, G.B.; Bari, S.B. Stimuli-sensitive layer-by-layer (LbL) self-assembly systems: Targeting and biosensory applications. *J. Control. Release* **2013**, *166*, 294–306. [CrossRef] [PubMed]

63. Van Nieuwenhove, I.; Salamon, A.; Peters, K.; Graulus, G.; Martins, J.C.; Frankel, D.; Kersemans, K.; de Vos, F.; van Vlierberghe, S.; Dubruel, P. Gelatin- and starch-based hydrogels. Part A: Hydrogel development, characterization and coating. *Carbohydr. Polym.* **2016**, *152*, 129–139. [CrossRef] [PubMed]

64. Xu, J.; Zhao, Z.; Hao, Y.; Zhao, Y.; Qiu, Y.; Jiang, J.; Yu, T.; Ji, P.; Liu, Y.; Wu, C. Preparation of a Novel Form of Gelatin With a Three-Dimensional Ordered Macroporous Structure to Regulate the Release of Poorly Water-Soluble Drugs. *J. Pharm. Sci.* **2016**, *105*, 2940–2948. [CrossRef] [PubMed]

65. Sahoo, N.; Sahoo, R.K.; Biswas, N.; Guha, A.; Kuotsu, K. Recent advancement of gelatin nanoparticles in drug and vaccine delivery. *Int. J. Biol. Macromol.* **2015**, *81*, 317–331. [CrossRef] [PubMed]

66. Zorzi, G.; Seijo, B.; Sanchez, A. *Handbook of Polymers for Pharmaceutical Technologies, Structure and Chemistry*; Thakur, V.K., Thakur, M.K., Eds.; Wiley: Hoboken, NJ, USA, 2015; Volume 1.

67. Duconseille, A.; Andueza, D.; Picard, F.; Santé-Lhoutellier, V.; Astruc, T. Molecular changes in gelatin aging observed by NIR and fluorescence spectroscopy. *Food Hydrocoll.* **2016**, *61*, 496–503. [CrossRef]

68. Badii, F.; Martinet, C.; Mitchell, J.R.; Farhat, I.A. Enthalpy and mechanical relaxation of glassy gelatin films. *Food Hydrocoll.* **2006**, *20*, 879–884. [CrossRef]

69. Badii, F.; MacNaughtan, W.; Farhat, I.A. Enthalpy relaxation of gelatin in the glassy state. *Int. J. Biol. Macromol.* **2005**, *36*, 263–269. [CrossRef] [PubMed]

70. Navarro, G. Chapter 5—Temperature-Sensitive Pharmaceutical Nanocarriers in Smart Pharmaceutical Nanocarriers. Torchilin, V., Ed.; Imperial College Press: London, UK, 2016; pp. 143–177.

71. Shikanov, A.; Domb, A.J. Chapter 23—Polymer-based drug delivery systems in Focal Controlled drug delivery. Springer Science & Business Media: New York, NY, USA, 2014; pp. 511–556.

72. Gupta, A.; Mohanty, B.; Bohidar, H.B. Flory temperature and upper critical solution temperature of gelatin solutions. *Biomacromolecules* **2005**, *6*, 1623–1627. [CrossRef] [PubMed]

73. Seuring, J.; Agarwal, S. Polymers with upper critical solution temperature in aqueous solution. *Macromol. Rapid Commun.* **2012**, *33*, 1898–1920. [CrossRef] [PubMed]

74. Chejara, D.R.; Mabrouk, M.; Badhe, R.V.; Mulla, J.A.S.; Kumar, P.; Choonara, Y.E.; du Toit, L.C.; Pillay, V. A bio-injectable algin-aminocaproic acid thixogel with tri-stimuli responsiveness. *Carbohydr. Polym.* **2016**, *135*, 324–333. [CrossRef] [PubMed]

75. Maji, T.; Banerjee, S.; Biswas, Y.; Mandal, T.K. Dual-Stimuli-Responsive l-Serine-Based Zwitterionic UCST-Type Polymer with Tunable Thermosensitivity. *Macromolecules* **2015**, *48*, 4957–4966. [CrossRef]

76. Ji, Y.; Zhu, M.; Gong, Y.; Tang, H.; Li, J.; Cao, Y. Thermoresponsive Polymers with Lower Critical Solution Temperature- or Upper Critical Solution Temperature-Type Phase Behaviour Do Not Induce Toxicity to Human Endothelial Cells. *Basic Clin. Pharmacol. Toxicol.* **2016**. [CrossRef] [PubMed]

77. Gao, W.; Liang, Y.; Peng, X.; Hu, Y.; Zhang, L.; Wu, H.; He, B. In situ injection of phenylboronic acid based low molecular weight gels for efficient chemotherapy. *Biomaterials* **2016**, *105*, 1–11. [CrossRef] [PubMed]

78. Yao, H.; You, X.; Lin, Q.; Li, J.; Guo, Y.; Wei, T.; Zhang, Y. Multi-stimuli responsive metal-organic gel of benzimidazol-based ligands with lead nitrate and their use in removal of dyes from waste-water. *Chin. Chem. Lett.* **2013**, *24*, 703–706. [CrossRef]

79. Covitch, M.J.; Trickett, K.J. How Polymers Behave as Viscosity Index Improvers in Lubricating Oils. *Adv. Chem. Eng. Sci.* **2015**, *5*, 134–151. [CrossRef]

80. Sagiri, S.S.; Singh, V.K.; Kulanthaivel, S.; Banerjee, I.; Basak, P.; Battachrya, M.K.; Pal, K. Stearate organogel-gelatin hydrogel based bigels: Physicochemical, thermal, mechanical characterizations and in vitro drug delivery applications. *J. Mech. Behav. Biomed. Mater.* **2015**, *43*, 1–17. [CrossRef] [PubMed]

81. Ibrahim, M.M.; Hafez, S.A.; Mahdy, M.M. Organogels, hydrogels and bigels as transdermal delivery systems for diltiazem hydrochloride. *Asian J. Pharm. Sci.* **2013**, *8*, 48–57. [CrossRef]

82. Zhu, M.; Xu, Y.; Ge, C.; Ling, Y.; Tang, H. Synthesis and UCST-type phase behavior of OEGylated poly(γ-benzyl-L-glutamate) in organic media. *J. Polym. Sci. A Polym. Chem.* **2016**, *54*, 1348–1356. [CrossRef]

83. Hanabusa, K.; Suzuki, M. Development of low-molecular-weight gelators and polymer-based gelators. *Polym. J.* **2014**, *46*, 776–782. [CrossRef]

84. Kumar, P.; Kadam, M.M.; Gaikar, V.G. Low Molecular Weight Organogels and Their Application in the Synthesis of CdS Nanoparticles. *Ind. Eng. Chem. Res.* **2012**, *51*, 15374–15385. [CrossRef]

85. Guo, H.; Jiao, T.; Shen, X.; Zhang, Q.; Li, A.; Zhou, J.; Gao, F. Binary organogels based on glutamic acid derivatives and different acids: Solvent effect and molecular skeletons on self-assembly and nanostructures. *Colloids Surface Physicochem. Eng. Asp.* **2014**, *447*, 88–96. [CrossRef]

86. Mateescu, M.A. 1—The concept of self-assembling and the interactions involved. In *Controlled Drug Delivery*; Mateescu, M.A., Ispas-Szabo, P., Assaad, E., Eds.; Woodhead Publishing: Cambridge, UK, 2015; pp. 1–20.

87. Ilbasmis-Tamer, S.; Unsal, H.; Tugcu-Demiroz, F.; Kalaycioglu, G.D.; Degim, I.T.; Aydogan, N. Stimuli-responsive lipid nanotubes in gel formulations for the delivery of doxorubicin. *Colloids Surfaces B Biointerfaces* **2016**, *143*, 406–414. [CrossRef] [PubMed]

88. Yang, R.; Peng, S.; Hughes, T.C. Multistimuli responsive organogels based on a reactive azobenzene gelator. *Soft Matter* **2014**, *10*, 2188–2196. [CrossRef] [PubMed]

89. Song, N.; Yang, Y. Stimuli-responsive fluorescent supramolecular polymers based on pillarenes for controlled drug release. *J. Control. Release* **2015**, *213*, e137. [CrossRef] [PubMed]

90. Tanaka, F.; Katsumoto, Y.; Nakano, S.; Kita, R. LCST phase separation and thermoreversible gelation in aqueous solutions of stereo-controlled poly(*N*-isopropylacrylamide)s. *React. Funct. Polym.* **2013**, *73*, 894–897. [CrossRef]

91. Chu, B. *Scattering Techniques Applied to Supramolecular and Nonequilibrium Systems*; Chen, S.H., Chu, B., Nossal, R., Eds.; Springer: New York, NY, USA, 2012.

92. Clark, E.A.; Lipson, J.E.G. LCST and UCST behavior in polymer solutions and blends. *Polymer* **2012**, *53*, 536–545. [CrossRef]

93. Masuelli, M.A. Dextrans in Aqueous Solution. Experimental Review on Intrinsic Viscosity Measurements and Temperature Effect. *J. Polym. Biopolym. Phys. Chem.* **2013**, *1*, 13–21.

94. Güner, A.; Kibarer, G. The important role of thermodynamic interaction parameter in the determination of theta temperature, dextran/water system. *Eur. Polym. J.* **2001**, *37*, 619–622. [CrossRef]

95. Rao, M.K.; Rao, S.K.; Ha, C. Stimuli Responsive Poly(Vinyl Caprolactam) Gels for Biomedical Applications. *Gels* **2016**, *2*, 6. [CrossRef]

96. Zhang, Q.; Hoogenboom, R. Polymers with upper critical solution temperature behavior in alcohol/water solvent mixtures. *Prog. Polym. Sci.* **2015**, *48*, 122–142. [CrossRef]

97. Lima, L.H.; Morales, Y.; Cabral, T. Ocular Biocompatibility of Poly-*N*-Isopropylacrylamide (pNIPAM). *J. Ophthalmol.* **2016**, *2016*, 5356371. [CrossRef] [PubMed]

98. Li, J. Cyclodextrin-based self-assembled supramolecular hydrogels and cationic polyrotaxanes for drug and gene delivery applications. *J. Drug Deliv. Sci. Technol.* **2010**, *20*, 399–405. [CrossRef]

99. Graziano, G. On the temperature-induced coil to globule transition of poly-*N*-isopropylacrylamide in dilute aqueous solutions. *Int. J. Biol. Macromol.* **2000**, *27*, 89–97. [CrossRef]

100. Tang, Y.; Liu, X. Collapse kinetics for individual poly(*N*-isopropylmethacrylamide) chains. *Polymer* **2010**, *51*, 897–901. [CrossRef]

101. Costa, M.C.M.; Silva, S.M.C.; Antunes, F.E. Adjusting the low critical solution temperature of poly(*N*-isopropyl acrylamide) solutions by salts, ionic surfactants and solvents: A rheological study. *J. Mol. Liq.* **2015**, *210*, 113–118. [CrossRef]

102. Wang, M. A single polymer folding and thickening from different dilute solution. *Phys. Lett. A* **2015**, *379*, 2761–2765. [CrossRef]

103. Milewska, A.; Szydlowski, J.; Rebelo, L.P.N. Viscosity and ultrasonic studies of poly(*N*-isopropylacrylamide)—Water solutions. *J. Polym. Sci. B Polym. Phys.* **2003**, *41*, 1219–1233. [CrossRef]

104. Shoji, K.; Nakayama, M.; Koseki, T.; Nakabayashi, K.; Mori, H. Threonine-based chiral homopolymers with multi-stimuli-responsive property by RAFT polymerization. *Polymer* **2016**, *97*, 20–30. [CrossRef]

105. Silva, S.M.C.; Pinto, F.V.; Antunes, F.E.; Miguel, M.G.; Sousa, J.J.S.; Pais, A.A.C.C. Aggregation and gelation in hydroxypropylmethyl cellulose aqueous solutions. *J. Colloid Interface Sci.* **2008**, *327*, 333–340. [CrossRef] [PubMed]

106. Shi, Y.; Yang, J.; Zhao, J.; Akiyama, H.; Yoshida, M. Photo-controllable coil-to-globule transition of single polymer molecules. *Polymer* **2016**, *97*, 309–313. [CrossRef]

107. Boutris, C.; Chatzi, E.G.; Kiparissides, C. Characterization of the LCST behaviour of aqueous poly(N-isopropylacrylamide) solutions by thermal and cloud point techniques. *Polymer* **1997**, *38*, 2567–2570. [CrossRef]

108. Zhou, K.; Lu, Y.; Li, J.; Shen, L.; Zhang, G.; Xie, Z.; Wu, C. The Coil-to-Globule-to-Coil Transition of Linear Polymer Chains in Dilute Aqueous Solutions: Effect of Intrachain Hydrogen Bonding. *Macromolecules* **2008**, *41*, 8927–8931. [CrossRef]

109. Zhang, L.P.; Noda, I.; Wu, Y. Quantitative comparison of reversibility in thermal-induced hydration of poly(N-isopropylacrylamide) and poly(N-isopropylmethacrylamide) in aqueous solutions by "concatenated" 2D correlation analysis. *Vib. Spectrosc.* **2012**, *60*, 200–205. [CrossRef]

110. Jeong, B.; Gutowska, A. Lessons from nature: Stimuli-responsive polymers and their biomedical applications. *Trends Biotechnol.* **2002**, *20*, 305–311. [CrossRef]

111. Jones, S.T.; Walsh-Korb, Z.; Barrow, S.J.; Henderson, S.L.; del Barrio, J.; Scherman, O.A. The Importance of Excess Poly(N-isopropylacrylamide) for the Aggregation of Poly(N-isopropylacrylamide)-Coated Gold Nanoparticles. *ACS Nano* **2016**, *10*, 3158–3165. [CrossRef] [PubMed]

112. Yang, L.; Shi, J.; Zhou, X.; Cao, S. Hierarchically organization of biomineralized alginate beads for dual stimuli-responsive drug delivery. *Int. J. Biol. Macromol.* **2015**, *73*, 1–8. [CrossRef] [PubMed]

113. Yu, T.; Geng, S.; Li, H.; Wan, J.; Peng, X.; Liu, W.; Zhao, Y.; Yang, X.; Xu, H. The stimuli-responsive multiphase behavior of core-shell nanogels with opposite charges and their potential application in in situ gelling system. *Colloids Surfaces B Biointerfaces* **2015**, *136*, 99–104. [CrossRef] [PubMed]

114. Sato, E.; Masuda, Y.; Kadota, J.; Nishiyama, T.; Horibe, H. Dual stimuli-responsive homopolymers: Thermo- and photo-responsive properties of coumarin-containing polymers in organic solvents. *Eur. Polym. J.* **2015**, *69*, 605–615. [CrossRef]

115. Sagiri, S.S.; Singh, V.K.; Banerjee, I.; Pramanik, K.; Basak, P.; Pal, K. Core-shell-type organogel-alginate hybrid microparticles: A controlled delivery vehicle. *Chem. Eng. J.* **2015**, *264*, 134–145. [CrossRef]

116. Strandman, S.; Zhu, X.X. Thermo-responsive block copolymers with multiple phase transition temperatures in aqueous solutions. *Prog. Polym. Sci.* **2015**, *42*, 154–176. [CrossRef]

117. Wagner, H.J.; Sprenger, A.; Rebmann, B.; Weber, W. Upgrading biomaterials with synthetic biological modules for advanced medical applications. *Adv. Drug Deliv. Rev.* **2016**, *105*, 77–95. [CrossRef] [PubMed]

118. Cheng, W.; Gu, L.; Ren, W.; Liu, Y. Stimuli-responsive polymers for anti-cancer drug delivery. *Mater. Sci. Eng. C* **2014**, *45*, 600–608. [CrossRef] [PubMed]

119. Chang, X.; Ma, C.; Shan, G.; Bao, Y.; Pan, P. Poly(lactic acid)/poly(ethylene glycol) supramolecular diblock copolymers based on three-fold complementary hydrogen bonds: Synthesis, micellization, and stimuli responsivity. *Polymer* **2016**, *90*, 122–131. [CrossRef]

120. Jain, A.; Kunduru, K.R.; Basu, A.; Mizrahi, B.; Domb, A.J.; Khan, W. Injectable formulations of poly(lactic acid) and its copolymers in clinical use. *Adv. Drug Deliv. Rev.* **2016**, *107*, 213–227. [CrossRef] [PubMed]

121. Fujihara, A.; Shimada, N.; Maruyama, A.; Ishihara, K.; Nakai, K.; Yusa, S. Preparation of upper critical solution temperature (UCST) responsive diblock copolymers bearing pendant ureido groups and their micelle formation behavior in water. *Soft Matter* **2015**, *11*, 5204–5213. [CrossRef] [PubMed]

122. Ranjan, R.; Rawat, K.; Bohidar, H.B. Interface versus bulk gelation and UCST in hydrophobically assembled TX-100 molecular gels. *Colloids Surface Physicochem. Eng. Asp.* **2016**, *499*, 113–122. [CrossRef]

123. Huang, G.; Li, H.; Feng, S.; Li, X.; Tong, G.; Liu, J.; Quan, C.; Jiang, Q.; Zhang, C.; Li, Z. Self-assembled UCST-Type Micelles as Potential Drug Carriers for Cancer Therapeutics. *Macromol. Chem. Phys.* **2015**, *216*, 1014–1023. [CrossRef]

124. Yuan, H.; Chi, H.; Yuan, W. Ethyl cellulose amphiphilic graft copolymers with LCST–UCST transition: Opposite self-assembly behavior, hydrophilic-hydrophobic surface and tunable crystalline morphologies. *Carbohydr. Polym.* **2016**, *147*, 261–271. [CrossRef] [PubMed]

125. Hoogenboom, R.; Rogers, S.; Can, A.; Becer, C.R.; Guerrero-Sanchez, C.; Wouters, D.; Hoeppener, S.; Schubert, U.S. Self-assembly of double hydrophobic block copolymers in water-ethanol mixtures: From micelles to thermoresponsive micellar gels. *Chem. Commun. (Camb)* **2009**, 5582–5584. [CrossRef] [PubMed]

126. Liu, T.; Hu, S.; Liu, D.; Chen, S.; Chen, I. Biomedical nanoparticle carriers with combined thermal and magnetic responses. *Nano Today* **2009**, *4*, 52–65. [CrossRef]

127. Caramella, C.M.; Rossi, S.; Ferrari, F.; Bonferoni, M.C.; Sandri, G. Mucoadhesive and thermogelling systems for vaginal drug delivery. *Adv. Drug Deliv. Rev.* **2015**, *92*, 39–52. [CrossRef] [PubMed]

128. Gibson, M.I.; O'Reilly, R.K. To aggregate, or not to aggregate? considerations in the design and application of polymeric thermally-responsive nanoparticles. *Chem. Soc. Rev.* **2013**, *42*, 7204–7213. [CrossRef] [PubMed]

129. Biswas, S.; Kumari, P.; Lakhani, P.M.; Ghosh, B. Recent advances in polymeric micelles for anti-cancer drug delivery. *Eur. J. Pharm. Sci.* **2016**, *83*, 184–202. [CrossRef] [PubMed]

130. Karavasili, C.; Fatouros, D.G. Smart materials: In situ gel-forming systems for nasal delivery. *Drug Discov. Today* **2016**, *21*, 157–166. [CrossRef] [PubMed]

131. Ruel-Gariépy, E.; Chenite, A.; Chaput, C.; Guirguis, S.; Leroux, J. Characterization of thermosensitive chitosan gels for the sustained delivery of drugs. *Int. J. Pharm.* **2000**, *203*, 89–98. [CrossRef]

132. Shi, K.; Wang, Y.L.; Qu, Y.; Liao, J.F.; Chu, B.Y.; Zhang, H.P.; Luo, F.; Qian, Z.Y. Synthesis, characterization, and application of reversible PDLLA-PEG-PDLLA copolymer thermogels in vitro and in vivo. *Sci. Rep.* **2016**, *6*, 19077. [CrossRef] [PubMed]

133. Basu, A.; Kunduru, K.R.; Doppalapudi, S.; Domb, A.J.; Khan, W. Poly(lactic acid) based hydrogels. *Adv. Drug Deliv. Rev.* **2016**, *107*, 92–205. [CrossRef] [PubMed]

134. Zhou, H.Y.; Jiang, L.J.; Cao, P.P.; Li, J.B.; Chen, X.G. Glycerophosphate-based chitosan thermosensitive hydrogels and their biomedical applications. *Carbohydr. Polym.* **2015**, *117*, 524–536. [CrossRef] [PubMed]

135. Sanoj Rejinold, N.; Sreerekha, P.R.; Chennazhi, K.P.; Nair, S.V.; Jayakumar, R. Biocompatible, biodegradable and thermo-sensitive chitosan-g-poly(N-isopropylacrylamide) nanocarrier for curcumin drug delivery. *Int. J. Biol. Macromol.* **2011**, *49*, 161–172. [CrossRef] [PubMed]

136. Zhang, K.; Wang, Z.; Li, Y.; Jiang, Z.; Hu, Q.; Liu, M.; Zhao, Q. Dual stimuli-responsive N-phthaloylchitosan-graft-(poly(N-isopropylacrylamide)-block-poly(acrylic acid)) copolymer prepared via RAFT polymerization. *Carbohydr. Polym.* **2013**, *92*, 662–667. [CrossRef] [PubMed]

137. Guerry, A.; Cottaz, S.; Fleury, E.; Bernard, J.; Halila, S. Redox-stimuli responsive micelles from DOX-encapsulating polycaprolactone-g-chitosan oligosaccharide. *Carbohydr. Polym.* **2014**, *112*, 746–752. [CrossRef] [PubMed]

138. Qiu, X.; Hu, S. "Smart" Materials Based on Cellulose: A Review of the Preparations, Properties, and Applications. *Materials* **2013**, *6*, 738–781. [CrossRef]

139. Qiu, Y.; Park, K. Environment-sensitive hydrogels for drug delivery. *Adv. Drug Deliv. Rev.* **2012**, *64*, 49–60. [CrossRef]

140. Chen, C.; Wang, Y.; Hung, C. In vitro dual-modality chemo-photodynamic therapy via stimuli-triggered polymeric micelles. *React. Funct. Polym.* **2016**, *98*, 56–64. [CrossRef]

141. He, L.; Shang, J.; Theato, P. Preparation of dual stimuli-responsive block copolymers based on different activated esters with distinct reactivities. *Eur. Polym. J.* **2015**, *69*, 523–531. [CrossRef]

142. Kim, S.; Joseph, V.S.; Hong, J. Dual stimuli-responsive copolymers comprising poly(N-isopropylacrylamide) and poly(cyano malachite green). *Colloids Surface Physicochem. Eng. Asp.* **2015**, *476*, 8–16. [CrossRef]

143. Weiss, J.; Laschewsky, A. Temperature-induced self-assembly of triple-responsive triblock copolymers in aqueous solutions. *Langmuir* **2011**, *27*, 4465–4473. [CrossRef] [PubMed]

144. Xu, J.; Luo, S.; Shi, W.; Liu, S. Two-stage collapse of unimolecular micelles with double thermoresponsive coronas. *Langmuir* **2006**, *22*, 989–997. [CrossRef] [PubMed]

145. Su, Y.; Li, Q.; Li, S.; Dan, M.; Huo, F.; Zhang, W. Doubly thermo-responsive brush-linear diblock copolymers and formation of core-shell-corona micelles. *Polymer* **2014**, *55*, 1955–1963. [CrossRef]

146. Koynova, R.; Tenchov, B. Transitions between lamellar and non-lamellar phases in membrane lipids and their physiological roles. *Open Access Biochem.* **2013**, *1*, 1–9.

147. Sundarajan, P.R. *Physical Properties of Polymers Handbook*, 2nd ed.; Mark, J.E., Ed.; Springer: New York, NY, USA, 2007.

148. Siow, K.S.; Delmas, G.; Patterson, D. Cloud-Point Curves in Polymer Solutions with Adjacent Upper and Lower Critical Solution Temperatures. *Macromolecules* **1972**, *5*, 29–34. [CrossRef]

149. Zhu, Y.; Batchelor, R.; Lowe, A.B.; Roth, P.J. Design of Thermoresponsive Polymers with Aqueous LCST, UCST, or Both: Modification of a Reactive Poly(2-vinyl-4,4-dimethylazlactone) Scaffold. *Macromolecules* **2016**, *49*, 672–680. [CrossRef]

150. Caneba, G. *Free-Radical Retrograde-Precipitation Polymerization (FRRPP): Novel Concept, Processes, Materials, and Energy Aspects*; Springer: Berlin/Heidelberg, Germany, 2010.

151. Balu, R.; Dutta, N.K.; Choudhury, N.R.; Elvin, C.M.; Lyons, R.E.; Knott, R.; Hill, A.J. An16-resilin: An advanced multi-stimuli-responsive resilin-mimetic protein polymer. *Acta Biomater.* **2014**, *10*, 4768–4777. [CrossRef] [PubMed]

152. Can, A.; Zhang, Q.; Rudolph, T.; Schacher, F.H.; Gohy, J.; Schubert, U.S.; Hoogenboom, R. Schizophrenic thermoresponsive block copolymer micelles based on LCST and UCST behavior in ethanol–water mixtures. *Eur. Polym. J.* **2015**, *69*, 460–471. [CrossRef]

153. Zhang, Q.; Hong, J.; Hoogenboom, R. A triple thermoresponsive schizophrenic diblock copolymer. *Polym. Chem.* **2013**, *4*, 4322–4325. [CrossRef]

154. Borgogna, M.; Marsich, E.; Donati, I.; Paoletti, S.; Travan, A. *Polysaccharide Hydrogels: Characterization and Biomedical Applications*; Matricardi, P., Alhaique, F., Coviello, T., Eds.; Pan Stanford Publishing: Singapore, 2016.

155. Ahmed, E.M. Hydrogel: Preparation, characterization, and applications: A review. *J. Adv. Res.* **2015**, *6*, 105–121. [CrossRef] [PubMed]

156. Haseeb, M.T.; Hussain, M.A.; Yuk, S.H.; Bashir, S.; Nauman, M. Polysaccharides based superabsorbent hydrogel from Linseed: Dynamic swelling, stimuli responsive on–off switching and drug release. *Carbohydr. Polym.* **2016**, *136*, 750–756. [CrossRef] [PubMed]

157. Roy, S.G.; Kumar, A.; De, P. Amino acid containing cross-linked co-polymer gels: PH, thermo and salt responsiveness. *Polymer* **2016**, *85*, 1–9. [CrossRef]

158. Engberg, K.; Waters, D.J.; Kelmanovich, S.; Parke-Houben, R.; Hartmann, L.; Toney, M.F.; Frank, C.W. Self-assembly of cholesterol tethered within hydrogel networks. *Polymer* **2016**, *84*, 371–382. [CrossRef]

159. Vakili, M.R.; Rahneshin, N. Synthesis and characterization of novel stimuli-responsive hydrogels based on starch and L-aspartic acid. *Carbohydr. Polym.* **2013**, *98*, 1624–1630. [CrossRef] [PubMed]

160. Joshi, S.C. Sol-Gel Behavior of Hydroxypropyl Methylcellulose (HPMC) in Ionic Media Including Drug Release. *Materials* **2011**, *4*, 1861–1905. [CrossRef]

161. Mah, E.; Ghosh, R. Thermo-Responsive Hydrogels for Stimuli-Responsive Membranes. *Processes* **2013**, *1*, 238–262. [CrossRef]

162. Drozdov, A.D.; Sanporean, C.-G.; Christiansen, J.D. Modeling the effects of temperature and pH on swelling of stimuli-responsive gels. *Eur. Polym. J.* **2015**, *73*, 278–296. [CrossRef]

163. Cong, H.; Zheng, S. Poly(*N*-isopropylacrylamide)-block-poly(acrylic acid) hydrogels: Synthesis and rapid thermoresponsive properties. *Colloid Polym. Sci.* **2014**, *292*, 2633–2645. [CrossRef]

164. Deen, G.R.; Mah, C.H. Influence of external stimuli on the network properties of cationic poly(*N*-acryloyl-*N'*-propyl piperazine) hydrogels. *Polymer* **2016**, *89*, 55–68. [CrossRef]

165. Casolaro, M.; Casolaro, I.; Bottari, S.; Del Bello, B.; Maellaro, E.; Demadis, K.D. Long-term doxorubicin release from multiple stimuli-responsive hydrogels based on α-amino-acid residues. *Eur. J. Pharm. Biopharm.* **2014**, *88*, 424–433. [CrossRef] [PubMed]

166. Heidarinasab, A.; Ahmad Panahi, H.; Faramarzi, M.; Farjadian, F. Synthesis of thermosensitive magnetic nanocarrier for controlled sorafenib delivery. *Mater. Sci. Eng. C* **2016**, *67*, 42–50. [CrossRef] [PubMed]

167. Zhang, Z.; Song, S. Thermosensitive/superparamagnetic iron oxide nanoparticle-loaded nanocapsule hydrogels for multiple cancer hyperthermia. *Biomaterials* **2016**, *106*, 13–23. [CrossRef] [PubMed]

168. Dragan, E.S.; Apopei Loghin, D.F.; Cocarta, A.; Doroftei, M. Multi-stimuli-responsive semi-IPN cryogels with native and anionic potato starch entrapped in poly(*N,N*-dimethylaminoethyl methacrylate) matrix and their potential in drug delivery. *React. Funct. Polym.* **2016**, *105*, 66–77. [CrossRef]

169. Wang, Q.; Li, S.; Wang, Z.; Liu, H.; Li, C. Preparation and characterization of a positive thermoresponsive hydrogel for drug loading and release. *J. Appl. Polym. Sci.* **2009**, *111*, 1417–1425. [CrossRef]

170. Wei, X.; Lv, X.; Zhao, Q.; Qiu, L. Thermosensitive β-cyclodextrin modified poly(ε-caprolactone)-poly(ethylene glycol)-poly(ε-caprolactone) micelles prolong the anti-inflammatory effect of indomethacin following local injection. *Acta Biomater.* **2013**, *9*, 6953–6963. [CrossRef] [PubMed]

171. Schmidt, B.V.K.J.; Hetzer, M.; Ritter, H.; Barner-Kowollik, C. Complex macromolecular architecture design via cyclodextrin host/guest complexes. *Prog. Polym. Sci.* **2014**, *39*, 235–249. [CrossRef]

172. Feng, Q.; Wei, K.; Lin, S.; Xu, Z.; Sun, Y.; Shi, P.; Li, G.; Bian, L. Mechanically resilient, injectable, and bioadhesive supramolecular gelatin hydrogels crosslinked by weak host-guest interactions assist cell infiltration and in situ tissue regeneration. *Biomaterials* **2016**, *101*, 217–228. [CrossRef] [PubMed]

173. Yang, K.; Wan, S.; Chen, B.; Gao, W.; Chen, J.; Liu, M.; He, B.; Wu, H. Dual pH and temperature responsive hydrogels based on β-cyclodextrin derivatives for atorvastatin delivery. *Carbohydr. Polym.* **2016**, *136*, 300–306. [CrossRef] [PubMed]

174. Arunachalam, M.; Gibson, H.W. Recent developments in polypseudorotaxanes and polyrotaxanes. *Prog. Polym. Sci.* **2014**, *39*, 1043–1073. [CrossRef]

175. Gregory, A.; Stenzel, M.H. Complex polymer architectures via RAFT polymerization: From fundamental process to extending the scope using click chemistry and nature's building blocks. *Prog. Polym. Sci.* **2012**, *37*, 38–105. [CrossRef]

176. Ahmed, M.; Narain, R. Progress of RAFT based polymers in gene delivery. *Prog. Polym. Sci.* **2013**, *38*, 767–790. [CrossRef]

177. Lynam, D.; Peterson, C.; Maloney, R.; Shahriari, D.; Garrison, A.; Saleh, S.; Mehrotra, S.; Chan, C.; Sakamoto, J. Augmenting protein release from layer-by-layer functionalized agarose hydrogels. *Carbohydr. Polym.* **2014**, *103*, 377–384. [CrossRef] [PubMed]

178. Wu, J.; Xiao, Z.; He, C.; Zhu, J.; Ma, G.; Wang, G.; Zhang, H.; Xiao, J.; Chen, S. Protein diffusion characteristics in the hydrogels of poly(ethylene glycol) and zwitterionic poly(sulfobetaine methacrylate) (pSBMA). *Acta Biomater.* **2016**, *40*, 172–181. [CrossRef] [PubMed]

179. Yom-Tov, O.; Neufeld, L.; Seliktar, D.; Bianco-Peled, H. A novel design of injectable porous hydrogels with in situ pore formation. *Acta Biomater.* **2014**, *10*, 4236–4246. [CrossRef] [PubMed]

180. Dutta, S.; Samanta, P.; Dhara, D. Temperature, pH and redox responsive cellulose based hydrogels for protein delivery. *Int. J. Biol. Macromol.* **2016**, *87*, 92–100. [CrossRef] [PubMed]

181. Helgeson, M.E.; Moran, S.E.; An, H.Z.; Doyle, P.S. Mesoporous organohydrogels from thermogelling photocrosslinkable nanoemulsions. *Nat. Mater.* **2012**, *11*, 344–352. [CrossRef] [PubMed]

182. Roy, S.G.; De, P. Swelling properties of amino acid containing cross-linked polymeric organogels and their respective polyelectrolytic hydrogels with pH and salt responsive property. *Polymer* **2014**, *55*, 5425–5434. [CrossRef]

183. Singh, V.K.; Banerjee, I.; Agarwal, T.; Pramanik, K.; Bhattacharya, M.K.; Pal, K. Guar gum and sesame oil based novel bigels for controlled drug delivery. *Colloids Surfaces B Biointerfaces* **2014**, *123*, 582–592. [CrossRef] [PubMed]

184. Singh, V.K.; Pal, K.; Banerjee, I.; Pramanik, K.; Anis, A.; Al-Zahrani, S.M. Novel organogel based lyotropic liquid crystal physical gels for controlled delivery applications. *Eur. Polym. J.* **2015**, *68*, 326–337. [CrossRef]

185. Amin, S.; Barnett, G.V.; Pathak, J.A.; Roberts, C.J.; Sarangapani, P.S. Protein aggregation, particle formation, characterization & rheology. *Curr. Opin. Colloid Interface Sci.* **2014**, *19*, 438–449.

186. Liu, L.; Qi, W.; Schwartz, D.K.; Randolph, T.W.; Carpenter, J.F. The Effects of Excipients on Protein Aggregation During Agitation: An Interfacial Shear Rheology Study. *J. Pharm. Sci.* **2013**, *102*, 2460–2470. [CrossRef] [PubMed]

187. Li, K.; Zhong, Q. Aggregation and gelation properties of preheated whey protein and pectin mixtures at pH 1.0–4.0. *Food Hydrocoll.* **2016**, *60*, 11–20. [CrossRef]

188. Yue, K.; Trujillo-de Santiago, G.; Alvarez, M.M.; Tamayol, A.; Annabi, N.; Khademhosseini, A. Synthesis, properties, and biomedical applications of gelatin methacryloyl (GelMA) hydrogels. *Biomaterials* **2015**, *73*, 254–271. [CrossRef] [PubMed]

189. Kowalczyk, T.; Hnatuszko-Konka, K.; Gerszberg, A.; Kononowicz, A.K. Elastin-like polypeptides as a promising family of genetically-engineered protein based polymers. *World J. Microbiol. Biotechnol.* **2014**, *30*, 2141–2152. [CrossRef] [PubMed]

190. Hoffman, A.S. Stimuli-responsive polymers: Biomedical applications and challenges for clinical translation. *Adv. Drug Deliv. Rev.* **2013**, *65*, 10–16. [CrossRef] [PubMed]

191. Xia, X.; Xu, Q.; Hu, X.; Qin, G.; Kaplan, D.L. Tunable Self-Assembly of Genetically Engineered Silk–Elastin-like Protein Polymers. *Biomacromolecules* **2011**, *12*, 3844–3850. [CrossRef] [PubMed]

192. Glassman, M.J.; Olsen, B.D. Arrested Phase Separation of Elastin-like Polypeptide Solutions Yields Stiff, Thermoresponsive Gels. *Biomacromolecules* **2015**, *16*, 3762–3773. [CrossRef] [PubMed]

193. Glassman, M.J.; Avery, R.K.; Khademhosseini, A.; Olsen, B.D. Toughening of Thermoresponsive Arrested Networks of Elastin-Like Polypeptides to Engineer Cytocompatible Tissue Scaffolds. *Biomacromolecules* **2016**, *17*, 415–426. [CrossRef] [PubMed]

194. Costa, A.M.S.; Mano, J.F. Extremely strong and tough hydrogels as prospective candidates for tissue repair—A review. *Eur. Polym. J.* **2015**, *72*, 344–364. [CrossRef]

195. Pritchard, R.H.; Redmann, A.; Pei, Z.; Ji, Y.; Terentjev, E.M. Vitrification and plastic flow in transient elastomer networks. *Polymer* **2016**, *95*, 45–51. [CrossRef]

196. Denissen, W.; Winne, J.M.; Du Prez, F.E. Vitrimers: Permanent organic networks with glass-like fluidity. *Chem. Sci.* **2016**, *7*, 30–38. [CrossRef]

197. Snijkers, F.; Pasquino, R.; Maffezzoli, A. Curing and viscoelasticity of vitrimers. *Soft Matter* **2016**. [CrossRef] [PubMed]

198. Yang, Z.; Wang, Q.; Wang, T. Dual-Triggered and Thermally Reconfigurable Shape Memory Graphene-Vitrimer Composites. *ACS Appl. Mater. Interfaces* **2016**, *8*, 21691–21699. [CrossRef] [PubMed]

199. Smallenburg, F.; Leibler, L.; Sciortino, F. Patchy particle model for vitrimers. *Phys. Rev. Lett.* **2013**, *111*, 188002. [CrossRef] [PubMed]

200. Obadia, M.M.; Mudraboyina, B.P.; Serghei, A.; Montarnal, D.; Drockenmuller, E. Reprocessing and Recycling of Highly Cross-Linked Ion-Conducting Networks through Transalkylation Exchanges of C–N Bonds. *J. Am. Chem. Soc.* **2015**, *137*, 6078–6083. [CrossRef] [PubMed]

201. Pei, Z.; Yang, Y.; Chen, Q.; Wei, Y.; Ji, Y. Regional Shape Control of Strategically Assembled Multishape Memory Vitrimers. *Adv. Mater.* **2016**, *28*, 156–160. [CrossRef] [PubMed]

202. Fortman, D.J.; Brutman, J.P.; Cramer, C.J.; Hillmyer, M.A.; Dichtel, W.R. Mechanically activated, catalyst-free polyhydroxyurethane vitrimers. *J. Am. Chem. Soc.* **2015**, *137*, 14019–14022. [CrossRef] [PubMed]

203. Romano, F.; Sciortino, F. Switching bonds in a DNA gel: An all-DNA vitrimer. *Phys. Rev. Lett.* **2015**, *114*, 078104. [CrossRef] [PubMed]

204. Pei, Z.; Yang, Y.; Chen, Q.; Terentjev, E.M.; Wei, Y.; Ji, Y. Mouldable liquid-crystalline elastomer actuators with exchangeable covalent bonds. *Nat. Mater.* **2014**, *13*, 36–41. [CrossRef] [PubMed]

205. Pilate, F.; Toncheva, A.; Dubois, P.; Raquez, J. Shape-memory polymers for multiple applications in the materials world. *Eur. Polym. J.* **2016**, *80*, 268–294. [CrossRef]

206. Gan, S.; Wu, Z.L.; Xu, H.; Song, Y.; Zheng, Q. Viscoelastic Behaviors of Carbon Black Gel Extracted from Highly Filled Natural Rubber Compounds: Insights into the Payne Effect. *Macromolecules* **2016**, *49*, 1454–1463. [CrossRef]

207. Aabloo, A.; de Luca, V.; di Pasquale, G.; Graziani, S.; Gugliuzzo, C.; Johanson, U.; Marino, C.; Pollicino, A.; Puglisi, R. A new class of ionic electroactive polymers based on green synthesis. *Sens. Actuators A Phys.* **2016**, *249*, 32–44. [CrossRef]

208. Arunkumar, P.; Raju, B.; Vasantharaja, R.; Vijayaraghavan, S.; Preetham Kumar, B.; Jeganathan, K.; Premkumar, K. Near infra-red laser mediated photothermal and antitumor efficacy of doxorubicin conjugated gold nanorods with reduced cardiotoxicity in swiss albino mice. *Nanomedicine* **2015**, *11*, 1435–1444. [CrossRef] [PubMed]

209. Wang, Z.; Chen, Z.; Liu, Z.; Shi, P.; Dong, K.; Ju, E.; Ren, J.; Qu, X. A multi-stimuli responsive gold nanocage-hyaluronic platform for targeted photothermal and chemotherapy. *Biomaterials* **2014**, *35*, 9678–9688. [CrossRef] [PubMed]

210. Wang, F.; Shen, Y.; Zhang, W.; Li, M.; Wang, Y.; Zhou, D.; Guo, S. Efficient, dual-stimuli responsive cytosolic gene delivery using a RGD modified disulfide-linked polyethylenimine functionalized gold nanorod. *J. Control. Release* **2014**, *196*, 37–51. [CrossRef] [PubMed]

Review

Stimuli Responsive Poly(Vinyl Caprolactam) Gels for Biomedical Applications

Kummara Madhusudana Rao [1], Kummari Subba Venkata Krishna Rao [2,3] and Chang-Sik Ha [1,*

[1] Department of Polymer Science and Engineering, Pusan National University, Busan 609 735, Korea; msraochem@gmail.com (K.M.R.)

[2] Department of Chemistry, Yogi Vemana University, Kadapa 516 003, Andhra Pradesh, India; drksvkrishna@yahoo.com (K.S.V.K.R.)

[3] Department of Chemical Engineering and Material Science, Wayne State University, Detroit, MI48202, USA

* Correspondence: csha@pnu.edu; Tel.: +82-51-510-2407

Academic Editor: Dirk Kuckling

Received: 13 November 2015; Accepted: 13 January 2016; Published: 25 January 2016

Abstract: Poly(vinyl caprolactam) (PNVCL) is one of the most important thermoresponsive polymers because it is similar to poly(N-isopropyl acrylamide). PNVCL precipitates from aqueous solutions in a physiological temperature range (32–34 °C). The use of PNVCL instead of PNIPAM is considered advantageous because of the assumed lower toxicity of PNVCL. PNVCL copolymer gels are sensitive to external stimuli, such as temperature and pH; which gives them a wide range of biomedical applications and consequently attracts considerable scientific interest. This review focuses on the recent studies on PNVCL-based stimuli responsive three dimensional hydrogels (macro, micro, and nano) for biomedical applications. This review also covers the future outlooks of PNVCL-based gels for biomedical applications, particularly in the drug delivery field.

Keywords: stimuli responsive; gels; poly(vinyl caprolactam); biomedical applications

1. Introduction

The development and applications of novel biomaterials play an important role in improving the treatment of diseases and the quality of health care. The focus of recent research on polymeric biomaterials is on producing new materials, including those with improved biocompatibility, mechanical properties and responsiveness. Polymeric biomaterials have been applied to medicine which includes controlled drug delivery systems, coatings of tablets, artificial organs, tissue engineering, polymer-coated stents, dental implants, and sutures [1,2]. This is a major step towards the development of polymeric therapeutic devices from newly synthesized polymeric materials with desired properties.

1.1. Stimuli Responsive Gels

The ability to swell and shrink in the presence and absence of aqueous media, respectively, is the most characteristic property of hydrogels [3]. Two important factors affect the swelling behavior of hydrogels; the hydrophilicity of the polymer chains and the crosslink density [3]. Hydrogels with stimuli responsive properties can be prepared by incorporating some stimuli-responsive co-monomers, either into the backbone of the network structure or as pendant groups. These hydrogels have the ability to swell, shrink, bend, or even degrade as a response to an external signal [3]. These stimuli-responsive hydrogels are also called "intelligent hydrogels". They swell and shrink reversibly with small changes in the environmental conditions, such as pH, temperature, electric field, ionic strength, and type of salt [3].

Hydrogels have been used extensively in the development of smart delivery systems. In these systems, a polymeric matrix can protect a drug from hostile environments (such as low pH and enzymes), while controlling drug release when the gel structure is changing in response to environmental stimuli [4–7]. Bioactive molecules can be trapped easily by an equilibrium swelling method or by simply mixing before the formation of hydrogels. The drug can be released via a range of mechanisms triggered by various stimuli [3]. The recent interest in these systems has focused on hydrogels in the form of macro-, micro-, and nanogels because of their promising applications in drug delivery systems, cell encapsulation and enzyme immobilization [8,9]. Temperature-responsive hydrogels especially in micro or nano forms show a volume phase transition at a specific temperature that causes a sudden change in the solvent state. The incorporation of thermoresponsive polymers with other responsive moieties can exhibit dual responsive properties. The present review focuses mainly on the recent progress of the thermoresponsive behavior of PNVCL and their stimuli responsive gels for biomedical applications.

1.2. Importance of Thermoresponsive Poly(Vinyl Caprolactam)(PNVCL)

Most polymers are soluble in aqueous media when heated, but some water soluble polymers may precipitate from the solution upon heating [10]. This unique property is characteristic of polymers that dissolve when cooled and phase separate when heated above a certain temperature, known as a lower critical solution temperature (LCST) [10,11]. The LCST is mainly dependent on the hydrogen bonding between water molecules and the structure of functional monomer units of polymers. Phase or reversible conformational changes take place for thermosensitive polymers over small variations of temperature [11]. Various methods to determine sol-gel transition or phase change have been known, including cloud point measurement [12–14], differential scanning calorimetry (DSC) [15] and rheology [12]. The value of the transition temperature may vary slightly depending on the experimental method used, as different stages of the gelation process may be measured by each technique [14]. The most important thermoresponsive polymer is poly(N-isopropylacrylamide) (PNIPAM) [16–19], which undergoes a sharp coil-to-globule transition in water at 32 °C. Below this temperature, a hydrophilic state changes to a hydrophobic state. Water molecules associated with the side chain isopropyl moieties are released into the aqueous phase when temperature increases above the critical point. PNIPAM is easily accessible by radical polymerization with variable architectures such as block copolymers [20–22], gels [23–25] or grafted surfaces [26,27]. Therefore, PNIPAM-based materials are excellent candidates for applications in biosensing, controlled delivery and tissue engineering [28]. Figure 1 lists the important thermoresponsive polymers and their structures. Compared to PNIPAM, some other N-substituted poly(acrylamide)s exhibit similar behaviors in aqueous solutions. For example, poly(N,N'-diethylacrylamide) (PDEAAM), poly(N-cyclopropylacrylamide) and poly(N-ethylacrylamide) have been reported displaying LCSTs in the range from 30 °C to 80 °C [29]. Vinyl ether monomers can also be used to generate thermosensitive polymers. For example, poly(methyl vinyl ether) (PMVE) has a transition temperature at 35 °C, which renders it an interesting candidate for biomedical applications [30].

N-vinyl caprolactam (NVCL) has been known to possess noteworthy properties for biomedical applications, e.g., solubility in water and organic solvents, high absorption ability, and a transition temperature within the setting of these applications (33 °C) [31]. Even though the LCST of PNVCL solutions are close to PNIPAM, however, there are significant differences in the thermodynamic and molecular mechanisms underlying the phase transition. In contrast to PNIPAM, PNVCL possesses a classical Flory–Huggins thermoresponsive phase diagram. The phase transition behavior depends on the molecular weight of PVCL and the solution concentration [32,33]. This unique feature allows for controlling the temperature sensitivity of the polymer by varying its molecular weights. There are two stages in the phase transition of PNVCL in aqueous solution. The hydrogen bonding transformation is predominant at the first transition stage below the LCST and hydrophobic interaction is predominant at the second stage above the LCST [34]. A "sponge-like" structure may be formed for PVCL mesoglobules due to the absence of topological constrains as well as self-associated hydrogen bonds that could further continuously expel water molecules upon increasing temperature [32]. When the polymer chain length or polymer concentration is increased, the LCST of PNIPAM and PNVCL decreases [35,36]. In general, hydrophobic compounds decrease the LCST, whereas hydrophilic and charged compounds increase LCST of polymers due to the strong interactions between water and hydrophilic or charged groups [37]. LCST disappears when those polymers contain too many hydrophilic comonomers [36]. Salts are known to lower the LCST of PNIPAM and PNVCL [35,37–39]. Moreover LCST of PNVCL is known to decrease by addition of a small amount of alcohol [36]. Anionic surfactants usually prevent phase separation of PNVCL solutions when heating, while the hydrophobic interactions of anionic and cationic surfactants would lead them to bind to PNVCL. The behavior of PNVCL looks like a polyelectrolyte upon binding of the surfactant which means the polymer coil swells. As the surfactant concentration increases, the transition temperature increases [40]. Lau and Wu reported that the phase transition temperature of PNVCL depends on the molecular weight of the polymer [41]. Tager *et al.* showed for PNVCL with Mw = 5×10^5 g/mol the change of the phase transition temperature from 32 °C to 34 °C depending on the concentration of the polymer in solution [42]. Meeussen *et al.* estimated the phase diagrams for polymers with different molecular weights. Moreover, cloud point measurements and theoretical calculations of PNVCL were carried out [31]. On the other hand, the lack of popularity among researchers for PNVCL compared to PNIPAM in previous years has likely been due to the affinity of polymerization of NVCL in a controlled manner because the polymerization kinetics is very difficult to measure. Unlike PNIPAM, the hydrolysis of PNVCL would not produce small amide compounds that are unwanted in biomedical applications [35]. Despite this, PNVCL have attracted attention for use in the biomedical field by considering the excellent properties such as biocompatibility [43]. This unique feature together with its overall low toxicity, high complexing ability and film forming properties enables its use in many industrial and medical applications, in particular in the biomedical field [34,35]. So far, various types of PNVCL based temperature responsive carriers such as micelles and vesicles have been reported and used for drug delivery applications [44–47]. However, the uses of hydrogels, especially in the form of micro/nanogels, are more advantageous in the biomedical field than other carriers since covalently crosslinked micro-/nanogels exhibit improved stability as compared to non-crosslinked carriers during *in vivo* delivery, leading to minimal premature release of bioactive agents [44–47]. However, chemical crosslinking also raises increasing concerns on the potential toxicity of those covalent crosslinking residuals. However, the purification step is a major concern to avoid such toxicity of hydrogels. In addition, the hydrogels can be used not only for drug delivery but also immobilize the enzymes or cells and are useful for biomedical applications.

1.3. Biocompatibility of PNVCL

Biocompatibility is the most important property in biomedical applications. PNVCL is not yet an FDA approved polymer. Henna *et al.* studied cytotoxicity of thermosensitive polymers like poly(N-isopropylacrylamide), poly(N-vinylcaprolactam) and amphiphilically modified poly(N-vinylcaprolactam). They have found that PNVCL is non-toxic in cellular contact during short exposure times, and enhanced cellular attachment is achieved with the PNVCL coated particles (compared to PNIPAM) [43]. Results show that PNVCL and PNVCL grafted with poly(ethylene oxide)(PEO) were well tolerated at all polymer concentrations (0.1–10 mg/mL) after 3 h of incubation at room temperature and at 37 °C. However, when the temperature increased to 37 °C (above LCST of PNVCL), cytotoxicity effects (8%) were also improved. In another study PNVCL-b-PEG was also found not to be toxic to human endothelial cells (E.A. hy EA.hy926) in the concentration range from 0 to 400 µg/mL [48]. The cytotoxicity of PNVCL, PNVCL-b-PVP on two cell lines (HeLa, human cervical carcinoma cells; and HEK293, human embryonic kidney cells) were negligible for all the tested samples for both cell lines compared with the negative control for polymer concentrations between 0.1 and 1 mg/mL [49].

Poly(*N*-isopropylacrylamide) (PNIPAM)

Ploy(*N*-vinylcaprolactam) (PNVCL)

Poly(*N,N*-diethylacrylamide) (PDEAM)

Poly(*N*-ethylmethacrylamide) (PNEMAM)

Poly(methyl vinyl ether) (PMVE)

Poly(2-ethoxyethyl vinyl ether) (PEOVE)

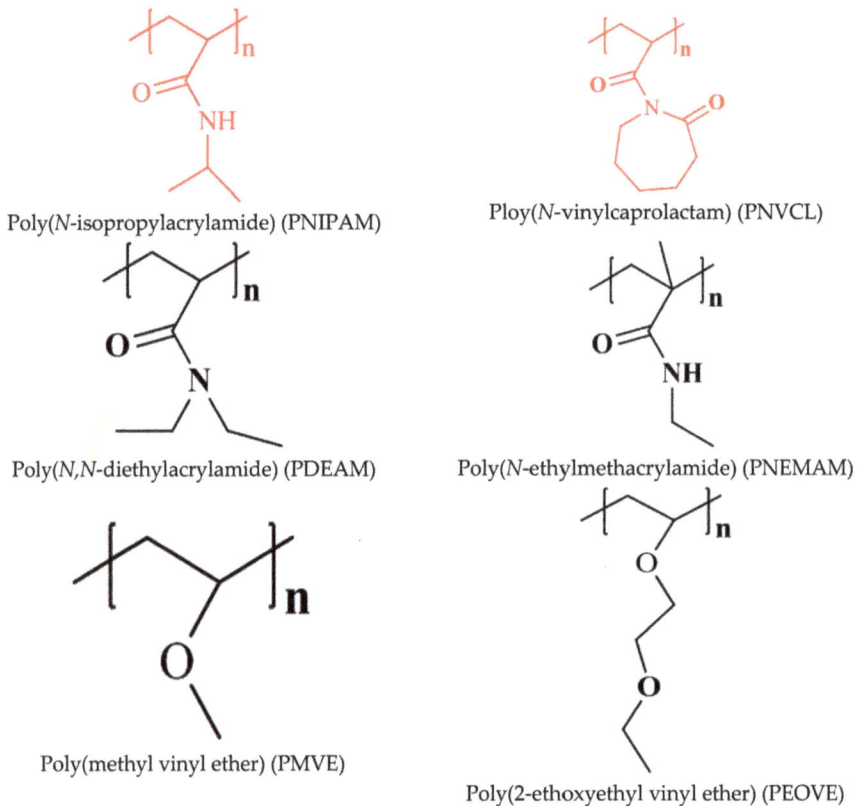

Figure 1. Chemical structures of the thermoresponsive polymers.

2. Stimuli Responsive Poly(Vinyl Caprolactam) Gels

2.1. Macro Hydrogels

Stimuli responsive NVCL gels are formed using physical or chemical crosslinking methods. In the physical crosslinking method, the gelation behavior of a polymer solution can be affected by a range of factors, such as temperature, polymer concentration and type of solvent used. Chemical crosslinking involves covalent bonding through polymeric network chains. Recently, PNVCL-based gels have become important because of their biocompatibility and temperature responsive nature, highlighting their potential in the biomedical field [43]. As an important step towards a fundamental understanding of PNVCL gels and their potential applications, Andy *et al.*, synthesized a high molecular mass of PNVCL-based linear chains by free radical bulk polymerization in the presence of 2,2′-azobisisobutyronitrile (AIBN) as an initiator. The LCST of PNVCL is approximately 31.5 °C, which is close to the shrinking temperature of the PNVCL hydrogel [41]. PNVCL hydrogels were produced by free radical polymerization weakly crosslinked with *N, N′*-methylenebisacrylamide (MBA) as a crosslinker using an AIBN initiator [50]. The resulting gels have thermoresponsive behavior and the collapse is induced approximately at the same physiological temperature as for PNIPAM gels. The PNVCL gel transition depends on the surfactants and additives.

Boyko *et al.*, examined the gelation process on a radical chain crosslinking reaction based on NVCL by dynamic light scattering (DLS) [51]. In this study, free radical polymerization was performed using 3,3'-(ethane-1,1-diyl)bis(1-vinyl-2-pyrrolidone) (BISVP) as a crosslinker in the presence of AIBN as an initiator. The crosslinked gels followed the power law exponent, which is dependent on the concentration of the crosslinker. As the amount of crosslinker increases in the feed ratio of the gel, the network density increases with a higher exponent (μ). The power law behavior of PNVCL gels was found precisely at the sol–gel transition state and not in the gel state [51].

The strength of the PNVCL gels is also very important for applications. Morget *et al.* reported the tensile properties of PVCL gels, which were synthesized using ethylene glycol dimethacrylate and AIBN as the initiator [52]. The thermoresponsive gels showed that the stress-strain behavior was qualitatively different in the collapsed state above the temperature-induced system. At higher temperatures, the gels were stiffer, more ductile and showed greater time dependence.

The combination of thermoresponsive monomer, NVCL, with a pH responsive monomer yielded doubly responsive copolymers. Mamytbekov *et al.*, examined the swelling and mechanical properties of ionized networks of copolymers of NVCL and *N,N*-diallyl-*N,N*-dimethyl ammonium chloride (DMD) using BISVP as a crosslinker in the presence of gamma irradiation [53]. Upon heating, the system exhibited a continuous decrease in swelling with increasing volume phase transition temperature. Unlike other ionic hydrogels, the polyelectrolyte system showed continuous deswelling curves with increasing temperature. The degree of swelling in water and NaCl solutions was analyzed theoretically using the swelling equilibria of the polyelectrolyte network. The incorporation of vinyl pyrrolidine (VP) chains into these ionized networks showed a continuous phase transition from an expanded to collapsed state [54]. The transition temperature increased with an increasing amount of VP chains in the gels because the VP networks helped improve the hydrophilicity of the networks.

Interpenetrating polymer networks (IPNs) are a class of polymer blends, in which two polymeric networks are independently cross-linked. Lenka *et al.*, developed IPN gels of PNVCL and PNIPAM [55]. In this study, the PNVCL network was prepared using an ethylidene-bis-3(*N*-vinyl-2-pyrrolidone) (EBVP) crosslinker. Subsequently, NIPAM and MBA crosslinkers were polymerized in the swollen network structure of PNVCL using photo initiation exposure to UV light. In addition, they examined the volume phase transitions of IPN hydrogels using a range of techniques, such as nuclear magnetic resonance (NMR) spectroscopy, small angle neutron scattering (SANS), DSC, and dynamic mechanical measurements.

In the semi-interpenetrating polymer networks (semi-IPNs), one component of the assembly is crosslinked leaving the other in a linear form. Vyshivannaya *et al.*, developed a semi-IPN hydrogel using PAAm and PNVCL [56]. In their study, scattering experiments were used to examine the mechanism of the turbidity of thermoresponsive PNVCL and non-thermoresponsive poly(acrylamide) (PAAm) polymers and their mixture forming semi-IPNs. The clustered aggregates of PNVCL globules formed in the sols at high temperatures showed an improved contribution of the scattering due to clusters, whereas in the case of semi-IPNs, an increase in both the dynamic and static constituents of the intensity was observed.

The incorporation of cationic and anionic monomers further showed double (pH and thermo) the responsive behavior of the PNVCL hydrogels. Elcin Cakal *et al.*, synthesized a copolymeric hydrogel using a NVCL thermoresponsive monomer with 2-(diethylamino)ethyl methacrylate (DEAEMA) as a cationic monomer in the presence of free radical polymerization at 60 °C [57]. In this system, two important crosslinkers were used and equilibrium swelling was investigated with respect to both the pH and thermoresponsiveness. The pH sensitivity was strongly dependent on the DEAEMA content. Fickian swelling behavior was observed at pH 7.4 when the system contained more than 95 mol % of NVCL in the presence of both ethylene glycol dimethacrylate (EGDMA) and allyl methacrylate (AMA) crosslinkers. In another study, a cationic monomer, *N*-acryloyl-*N*'-ethyl piperazine (AcrNEP), was used for the synthesis of copolymeric hydrogels with NVCL in the presence of free radical polymerization at 75 °C with MBA as the crosslinker. The resulting gels were flexible and responsive to external stimuli with respect to both pH and temperature. They also examined the effects of the crosslinker on the properties of gel stimuli response, water transport mechanism, diffusion, and adsorption of anionic dyes [58].

Selva *et al.*, synthesized poly(*N*-vinylcaprolactam-*co*-itaconic acid) poly(NVCL-*co*-IA) gels in ethanol using the free radical crosslinking polymerization method at 60 °C for 24 h in the presence of AIBN and AMA as the initiator and crosslinking agent, respectively [59]. The incorporation of anionic monomer into the thermoresponsive PNVCL gels exhibited dual pH and temperature responsive behavior. They investigated the swelling behavior with respect to both pH and temperature. Further, swelling kinetics was also performed to know the diffusion mechanism of liquid from the diffusion exponent value (n). In this work, the n values for 5 and 10 mol % of itaconic acid (IA) present in the gels are 0.64 and 1.07, respectively. Based on these n values, the diffusion strongly depends on the IA content. The lower IA gels followed a completely non-Fickian pattern, whereas at higher content of IA, the mechanism changed from non-Fickian to case II. The relaxation controlled transport mechanism is highly dependent on the higher amount of ionizable IA groups for ionic gels. The incorporation of ionic monomers, such as acrylic acid (AAc) and methacrylic acid (MAAc), into physical gels of PNVCL also resulted in pH and temperature responsive behavior. Heba *et al.*, synthesized physically-crosslinked gels based on NVCL and AAc/MAAc via free radical polymerization [60]. The gels of the phase transition temperature was dependent on the solution pH because of the dissociate behavior of –COOH groups in the buffer solution. In addition, the incorporation of *N,N*-dimethylacrylamide (DMAm) into the system further increased the LCST. This increasing behavior of LCST was attributed to the increasing hydrophilicity of the system.

Zavgorodnya, *et al.* studied temperature-responsive properties of PNVCL multilayer hydrogels in the presence of Hofmeister anions [61]. The hydrogels were produced by glutaraldehyde-assisted crosslinking of hydrogen-bonded multilayers of PNVCL-*co*-(aminopropyl)methacrylamide) and poly(methacrylic acid). The authors found that swelling and temperature-induced shrinkage of PNVCL hydrogels were suppressed in the order $SO_4^{2-} > H_2PO_4^- > Cl^-$, following the Hofmeister series. In contrast, I^- increased hydrogel swelling but suppressed thermal response. Further, an optical response is initiated in the presence of anions when a layer of gold nanoparticles that has been stabilized with glutathione is introduced within the PNVCL hydrogel. When the temperature reversibly changed from 20 °C to 50 °C, the signal intensity of $(PNVCL)_{81}$-Au hydrogels and the plasmon band position were remarkably dependent on ion concentration and type. These results are essential to understand the effect of Hofmeister anions on ultrathin non-ionic polymer networks. In addition, a distinct and fast optical monitoring of hydrogel temperature-triggered response depending on ion concentrations can be possible for the $(PNVCL)_{81}$-Au hybrid hydrogels.

Sanna, *et al.* synthesized cellulose nanocrystals reinforced MBA crosslinked with PNVCL hydrogels using free radical polymerization method in the presence of trihexyltetradecylphosphonium persulfate (TETDPPS). The resulting nanocomposite hydrogels showed LCST of 33–34 °C. Rheological studies showed a significant increase of the mechanical properties even at very low CNC concentrations [62].

2.2. Micro or Nanogels

Sometimes, hydrogels can be confined to smaller dimensions, or exist in the form of macroscopic networks, such as microgels, which are crosslinked polymeric particles. When the microgels exist in the submicron range size, they are usually called nanogels [63–65]. Yinbing *et al.* examined the interaction of surfactant and PNVCL microgels [66]. In this study, the authors successfully prepared narrowly-distributed spherical PNVCL microgels by precipitation polymerization. The swelling and shrinkage of microgels were investigated by dynamic laser light scattering (LLS) with both anionic sodium dodecyl sulphate (SDS) and cationic *N*-dodecylpyridinium bromide (DPB) surfactants. The results show that the microgels gradually shrank to a collapsed state when the temperature was increased from 20 to 38 °C. The effects of the surfactant on the swelling and shrinking of the PNVCL microgels was attributed partially to the formation of micelles inside the microgel structure. The combination of hydrophilic monomer VP with NVCL produced a stable dispersion of larger particles by precipitation polymerization [67]. The microgels showed a strong decrease in the radius of gyration, R_g, and hydrodynamic radius, R_h, during the phase transition near 30 °C. The smooth surface of the macrogels indicates the decoration of hydrophilic polymer segments on the surface of the microgels. In another study, amphiphilic macro monomer polyethylene oxide (PEO) segments were used to prepare the thermally responsive microgels with NVCL using a macro monomer technique [65]. In this study, the polymerization yield was low because the potassium persulphate (KPS) initiator can hydrolyze the monomer NVCL. On the other hand, the microgels showed monomodal and narrow particle size distribution. Dispersion polymerization was used for the production of vinyl imadazole (Vim) functionalized poly(N-vinyl caprolactam-*co*-acetoacetoxy methacrylate) P(NVCL-*co*-AAEM) copolymer microgels [68]. A narrow size distribution with a hydrodynamic radius in the range 200–500 nm was obtained. The microgels exhibited both pH and thermo responsive behavior. The swelling property of the microgels depends on the VIm content in the microgel structure.

Crosslinking is also one of the important factors for the formation and their swelling characteristics of PNVCL microgels. Ainara *et al.* examined the formation of PNVCL microgels using different crosslinkers polyethyleneglycol diacrylate (PEGDA) and MBA and different concentrations [69]. The results showed that polymerization was complete within 10 min. In this case, both crosslinkers were consumed faster than the monomer. The crosslinking concentration did not affect the final collapsed diameters. On the other hand, the use of a small amount of crosslinker resulted in unexpected swelling behavior at temperatures below 20 °C. Another study showed similar results in that the MBA crosslinker and 3-*o*-methacryloyl-1,2:5,6-di-o-isopropylidene-α-D-glucofuranose (3MDG) monomers achieved complete conversion, whereas NVCL was not consumed completely [70].

Spherical microgels were produced from the precipitation copolymerization of NVCL and sodium acrylate (NaA) in water at 60 °C [71]. The copolymerization of a few mole-percent of ionic NaA into the P(VCL-*co*-NaA) microgel can raise its volume transition temperature slightly and increase the extent of its swelling. At the transition temperature (32 °C), the PVCL chains become hydrophobic and insoluble. The intra-microgel complexation between Ca^{2+} and these $–COO^-$ groups lead to the shrinkage of microgels but inter microgel complexation induces aggregation. A comparison of the linear copolymer chains and spherical microgels showed that the aggregation of linear chains was much more profound than that of the spherical microgels, presumably because of the competition between the inter- and intra-chain complexation. The same research group also examined Ca^{2+}-induced complexation between the thermally responsive P(NVCL-*co*-NaA) microgels with gelatin linear chains in water. These results suggest that the microgel can form complexes with gelatin in the presence of Ca^{+2} and complexation is induced by both hydrophobic and electrostatic attractions. Gelatin absorbed effectively on the surface of the gel due to the above said mechanism [72].

3. Biomedical Applications

Table 1 summarizes recent significant studies on the stimuli responsive gels of PNVCL for biomedical applications.

3.1. Drug Delivery and Antibacterial Applications

Iskav *et al.* developed thermoresponsive PNVCL hydrogels with sodium itacanote. The gels were used to encapsulate farmazine (Far) as a model release agent [73]. In this study, the phase transition temperature was shifted 2.5 °C to a lower temperature for the drug immobilized gels compared to the pristine gels. The authors attributed this shift to hydrogen bonding of the lactam oxygen atom of the polymer to the amino group of Far. The gels exhibited the controlled release of the drug via diffusion. The half time of drug release was 32 min at 22 °C and 210 min at 37 °C.

Reddy *et al.*, synthesized a dual responsive semi-IPN hydrogels using sodium alginate (SAlg) and poly(acrylamide-*co*-acrylamidoglycolic acid-*co*-N-vinyl caprolactam) P(AAm-*co*-AGA-*co*-NVCL) multi components via free radical redox polymerization [74]. In this study, 5-FU as a model anticancer agent, was used for encapsulation into these semi-IPNs via an equilibrium swelling method. In addition, the authors synthesized silver nanoparticles (Ag NPs) with a size and shape-controlled spherical shape and an average size of approximately 20 nm in the gel networks by simple equilibration of the Ag salts and reduced in the presence of $NaBH_4$ (Figure 2). The authors proved that the gels could control the release of the drug and Ag NPs through stimuli, such as pH and temperature. As shown in Figure 2, the temperature-responsive drug release mainly caused the presence of PNVCL copolymeric chains in the gel networks.

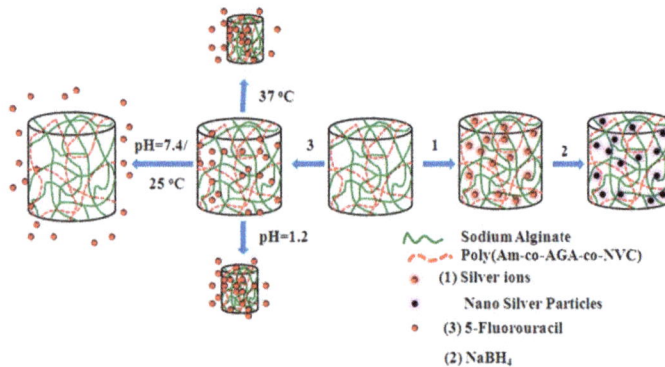

Figure 2. Synthesis and dual responsive-triggered release of the drug from semi-Interpenetrating Polymeric Networks (semi-IPNs) of SAlg-P(AAm-*co*-AGA-*co*-NVCL) multi component gels and the formation of Ag NPs in the hydrogel networks. (Reproduced from Rama Subba Reddy *et al.*, *Macromol. Res.* **2014**, *8*, 832–842. (Ref. 74) copyright 2014 with permission from Springer).

The incorporation of a NVCL thermoresponsive monomer on the backbone of biopolymers, such as chitosan (CS) and SAlg, by grafting could improve the properties without sacrificing the biodegradable nature. As shown in Figure 3, grafting was achieved via the free radical polymerization of NVCL or the coupling reaction of PNVCL onto SAlg or CS. Rao *et al.* prepared microgels using a PNVCL-g-SAlg graft polymer via an ionotropic gelation method. The effects of pH and temperature on the swelling behavior of microgels studied ascertained that they were sensitive to pH and thermoresponsive properties [75]. The mean size of the microgels was approximately 100 μm with a smooth and spherical shape. 5-fluorouracil (5-FU), as the model anticancer drug, was loaded and the encapsulation efficiency was found to be 84%. The authors systematically examined the release of 5-FU as a function of temperature, pH, amount of crosslinker, and % of drug concentration. The gels exhibited controlled release behavior over a period of more than 12 h. In another study, the same graft copolymer was used to produce the gel beads and utilized the encapsulation of the 5-FU model drug using GA as a crosslinker. The mean size of the gel beads was also close to 100 μm. The authors also examined the thermoresponsive behavior and release studies using a range of parameters [76]. Prabaharan, *et al.*, synthesized CS-g-PNVCL by grafting carboxyl-terminated PNVCL chains onto the CS backbone via a DCC coupling reaction and used it as a drug delivery carrier [77]. The MTT (3-(4,5-dimethylthiazol-2-yl)-2,5-diphenyltetrazolium bromide) assay showed no obvious cytotoxicity of CS-g-PNVCL against a human endothelial cell line over a concentration range of 0–400 μg/mL. The beads were crosslinked with TPP and encapsulated Ket as the model drug. Drug release was influenced by both pH and temperature. The authors concluded that the release was very slow at pH 7.4 and 37 °C because the gel beads had a compact structure with a reduced pore size and strong interaction between the drug molecules and polymer chains.

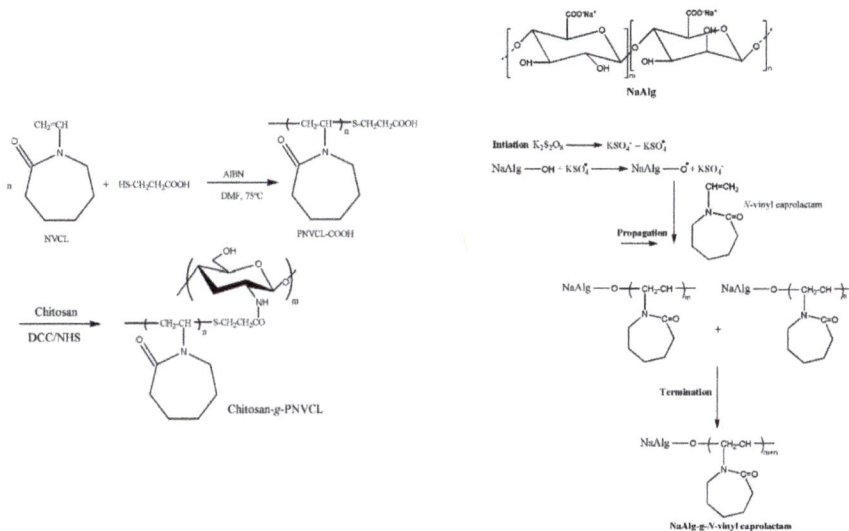

Figure 3. Synthesis of the graft copolymers of CS and SAlg with NVCL [75,77]. (Reproduced from Madhusudana-Rao *et al.*, *J. Appl. Pharm. Sci.* **2013**, *3*, 061–069(Ref. 75) copyright 2013 with permission from Journal and Prabaharan *et al.*, *Macromol. Biosci.* **2008**, *8*, 843–851 (Ref. 77) copyright 2008 with permission from John Wiley & Sons, Inc.).

A graft copolymer of poly(N-vinyl caprolactam-g-PEO macromonomer (PNVCL-g-$C_{11}EO_{12}$) based gels was formed, physically crosslinked with salicylic acid (SA) via H-bonding interactions between PNVCL and SA [78]. In this study, the gel's stability was dependent mainly on the phase transition temperature. At 23 °C, which is below the LCST, the gel stability was poor, whereas more stable hydrogels were formed above the LCST (37 °C). On the other hand, the graft copolymer provided more H-bonding interactions with SA, leading to sustained drug release compared to PNVCL.

The combination of metal nanoparticles with smart polymeric architectures appears to be a promising way of designing new materials. In this context, microgels are promising templates for the synthesis of Ag NPs in the networks. Nadine *et al.* reported that PNVCL based microgels were effective templates for loading of different concentrations (1–12 wt. %). The porous structure of the poly(N-vinyl caprolactam-co-glycedyl methacrylate) P(NVCL-co-GMA) microgels allowed the release of the AgNPs [79]. The Ag NP-embedded microgels exhibited good antibacterial activity against Gram-positive *S. aureus*, and Gram-negative *P. aeruginosa*.

The microgels were developed by the incorporation of undecenoic acid (UA) functionalized PNVCL which showed a double responsive phase transition temperature [80]. In this study, the LCST of the microgels was close to body temperature. Doxorubicin (Dox) as a model anticancer drug was encapsulated via a thermal gelation process. The release of Dox was dependent on the pH and temperature. Cytotoxicity of PNVCL, and their microgels (5% and 10% UA) were performed on 3T3 murine fibroblast cells using CCK-8 assay at concentrations from 0.15 to 10 mg/mL. The results confirmed no obvious toxicity to the cells. These results can be applied to the temperature and pH controlled delivery of anticancer drugs. Based on the intracellular environment of cancer cells, Yang *et al.*, reported intracellular degradable multi responsive microgels composed from NVCL and methacrylic acid (MAAc) from the disulphide methacrylate crosslinker by precipitation polymerization [81]. The microgels were spherical and degradable in response to GSH due to the availability of disulphide bonds in the networks of the microgels. The microgels could effectively encapsulate the Dox drug and exhibited stimuli-triggered drug release in acidic or reducing environments. Further, to confirm the biocompatibility of microgels, the authors tested cytotoxicity of microgels on HK-2 cells using MTT assay. After 48 h incubation, the microgels showed no cytotoxicity in the concentration range 0.1–50 μg/mL. Ninety percent of cells were viable at higher concentrations (100 μg/mL), also proving that the microgels are highly biocompatible. The cytotoxicity of DOX loaded microgels were evaluated against Hela cancer cells. The authors found IC50 (at a concentration 2.1 μg/mL DOX in the microgels) is slightly higher than free DOX (1.4 μg/mL). Further, the DOX loaded microgel studies of intracellular drug release on Hela cancer cells confirmed the high cell uptake and high anticancer activity.

Nanogel platforms are better for the loading and release of bioactive agents than other carriers [82,83] because of the stability, responsiveness and biocompatibility of nanogels. The particle size of the nanogels plays a crucial role in the *in vivo* fate of colloidal drug delivery systems [84]. Therefore, easy and effective control over the particle size is of great importance. Rao *et al.* developed copolymeric nanogels from NVCL and acrylamidoglycolic acid (AGA) crosslinked with MBA by emulsion polymerization [85]. In this method, the particle size of the nanogels produced was approximately 50–100 nm with a spherical shape. In addition, 5-FU model anticancer drug encapsulated and the controlled release of 5-FU was examined at different pH and temperatures. From *in vitro* release, the nanogels showed pH and thermoresponsive behavior (Figure 4). Sudhakar *et al.* used the same emulsion polymerization for the incorporation of hydrophilic 2-hydroxyethyl methacrylate (HEMA) chains into PNVCL nanogels and produced a spherical shape, 150 nm in size [86]. In this study, a hydrophobic curcumin (CUR) model drug was encapsulated successfully during the polymerization process. The nanogels showed high aqueous stability (Figure 5) and were more bioavailable. The *in vitro* release studies also proved that the resulting nanogels were thermoresponsive and useful for targeted drug delivery applications. In another report, Mirian *et al.*, incorporated polyethyleneglycol methacrylate (PEGMA) chains into PNVCL nanogels using surfactant-free emulsion polymerization [87]. In addition, vinyl pyrrolidine (VP), 2-methacryloyloxybenzoic acid (2MBA) monomers were also incorporated to adjust the phase transition temperature. The nanogels sizes are between 120 and 300 nm. The nanogels containing 15.5% 2MBA showed a transition temperature close to 38 °C. In this study, 5-FU was also used as a model bioactive agent. The release of 5-FU was slower at pH 7.4 and 37 °C than under the tumor cellular condition (pH 6.0 and 40 °C). Drug release kinetics studies also proved that the drug release from the nanogels followed Fickian diffusion.

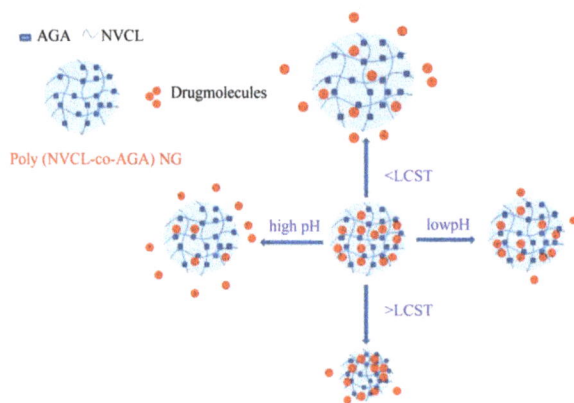

Figure 4. pH and temperature responsive nature of NGs [85]. (Reproduced from Madhusudana Rao *et al.*, Colloids Surf. B **2013**, *102*, 891–897 (Ref. 87) copyright 2013 with permission from Elsevier).

Table 1. Summary of the recent significant studies on the stimuli responsive gels of poly(*N*-vinyl caprolactam) PNVCL for biomedical applications.

Gel Composition	Responsiveness		Release Carriers	Application	Reference
P(NVCL-co-SIA)	thermo	Hydrogel	Far	Controlled release	[73]
P(AAm-co-NVCL-co-AGA)/SAlg	pH/thermo	Composite hydrogel	5-FU/Ag nano	Controlled release/Antibacterial	[74]
SAlg-g-PNVCL	pH/thermo	Microgel	5-FU	Controlled release	[75]
SAlg-g-PNVCL	pH/thermo	Gel beads	5-FU	Controlled release	[76]
CS-g-PNVCL	pH/thermo	Gel beads	Ket	Controlled release	[77]
PNVCL-g-PEO	thermo	Gel particles	Nad/Prop/Ket/SA	Controlled release	[78]
PNVCL-co-GMA	thermo	Hybrid microgel	Ag	antibacterial	[79]
P(NVCL-co-UA)	pH/thermo	Microgel	Dox	Controlled release	[80]
P(NVCL-PMAA)	pH/thermo/reduction	Microgel	Dox	Tumor targeted drug delivery	[81]
P(NVCL-co-AGA)	pH/thermo	Nanogel	5-FU	Controlled release	[85]
PVOH-b-PNVCL	pH/thermo/reduction	Nanogel	NR	Tumor targeted drug delivery	[86]
P(NVCL-co-HEMA)	thermo	Nanogel	CUR	Targeted drug delivery	[87]
P(NVCL-co-PEGMA)/VP/2MBA	pH/thermo	Nanogel	5-FU	Controlled release	[88]
PNVCL/Dex-MA	pH/thermo/Enzyme	Nanogel	dextranase	Enzymatic degradation	[89]
Fib-g-PNVCL	Thermo	Nanogel	5-FU/Meg	Tumor targeted drug delivery	[90]
HPCL-click-PNVCL	Thermo	Films	—	controlled drug delivery	[91]
PVCL-NH$_2$/PMAA	Thermo/pH	Gel cubes, capsules	—	controlled drug delivery	[92]
PNVCL	thermo	Hydrogel	Bacteria and fungi	Medicine	[93]
PNVCL-CaAlg	pH/thermo	Hydrogel	Enzymes and cells	Medicine	[94]
PNVCL-CaAlg	pH/thermo	Hydrogel	Protease	Biotechnology/medicine	[95]

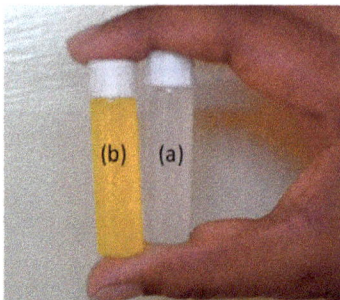

Figure 5. (a) Aqueous dispersions of NGs and (b) curcumin-loaded NGs. (Reproduced from Sudhakar *et al.*, *Des. Monomers Polym.* **2015**, *18*, 705–713. (Ref. 86) copyright 2015 with permission from Taylor and Francis).

Liu *et al.* reported the synthesis and loading/release properties of the reversibly-crosslinked thermo and redox responsive nanogels based on poly(vinyl alcohol)-g-*N*-vinyl caprolactam (PVOH-b-PNVCL) copolymers [88]. The heating and cooling cycles used for the formation of nanogels with 3,3′-dithiodipropionic acid (DPA) as crosslinker. In the presence of dithiothreitol (DTT) (10 mM), the gels exhibited degradation, and reversible crosslinking occurred upon exposure to H_2O_2. The authors used Nile red as the model compound to evaluate the release and cellular uptake. Cytotoxicity of PVOH-b-PNVCL copolymeric nanogels was tested against mouse fibroblast-like L929 cell line using *in vitro* culture via MTS (3-(4,5-dimethylthiazol-2-yl)-5-(3-carboxymethoxyphenyl)-2-(4-sulfophenyl)-2H-tetrazolium) assay. The nanogels exhibited high cell viability above 85%, even after 48 h incubation with 2 mg/mL concentrated nanogels. The results suggested the nanogels have low cytotoxicity. Further, their preliminary studies on cellular uptake of the nanogels within human melanoma (MEL-5) cells confirmed the cellar internalization of nanogels. Two important families of thermo responsive enzymatically degradable nanogels were developed by the batch emulsion polymerization of NVCL with dextran methacrylate (Dx-MA) [89]. The crosslinking nanogels were swollen below the volume phase transition temperature and collapsed above it. On the other hand, the low degree of crosslinked nanogels followed anomalous thermal behavior. The dextranase enzymatic degradation of the nanogels resulted in the release of sugars. Both families can be suitable for drug delivery in tissues or organs.

Like graft copolymeric microgels, Sanoj *et al.*, developed multi-drug loaded thermo responsive fibrinogen-g-*N*-vinyl caprolactam (Fib-g-PNVCL) nanogels and used for breast cancer drug delivery [90]. In this work, 5-FU and Meg multi drugs loaded into these nanogels. The LCST of the nanogels tuned by adjusting the composition of PNVCL/fib was approximately 35 °C. The sizes of the nanogels produced were 150–170 nm.The multi drugs are responsible for enhanced anticancer activity towards the MCF-7 cancer cells. The *in vivo* experiments also examined on Swiss albino mice showed the sustained release of Meg and 5-FU within three days.

Cai *et al.*, synthesized hyperbranched poly(3-caprolactone) with peripheral terminal alkyne groups (HPCL) via thiol–yne click reaction among the AB₂-type α-thiol-ω-alkyne–poly(3-caprolactone) linear precursors [91]. Azide-terminated poly(*N*-vinylcaprolactam) (PNVCL–N₃), prepared *a priori* via xanthate mediated reversible addition–fragmentation chain transfer (RAFT) polymerization of NVCL, was then linked to HPCL chains through Cu(I)-catalyzed alkyne-azide click reaction. Further, the authors fabricated membranes using these polymers with well-defined and uniform pores. The swelling of HPCL-click-PNVCL membranes resulted in the temperature-responsive property. Cell viability studies using MTT assay on 3T3 fibroblast cell line incubated with copolymer membranes for 24 h resulted in more than 90% of cell viability. The cytotoxicity assays thus indicate that the introduction of PNVCL chains on the HPCL membrane and pore surfaces has negligible cytotoxicity effects and the non-cytotoxic nature of HPCL membranes. These stimuli responsive membranes with controllable morphology improved mechanical properties and negligible cytotoxicity, which are useful as biomaterials for controlled drug delivery.

Liang *et al.* reported a novel type of single-component PNVCL multilayer hydrogel films and capsules with a distinct and highly reversible thermoresponsive behavior derived from hydrogen-bonded multilayers of PMAA and PNVCL-NH$_2$ by chemical crosslinking of PNVCL-NH$_2$ copolymer layers and subsequent release of PMAA at basic pH values [92]. Cubical (PNVCL)$_7$ hydrogel capsules retained their cubical shape when temperature was changed from 25 to 50 °C while showing a size decrease of 21% ± 1%. Spherical (PNVCL)$_7$ hydrogel capsules demonstrated the similar shrinkage of 23% ± 1%. The temperature-triggered size changes of both types of capsules were completely reversible. Non crosslinked two component films did not show temperature-responsive behavior because of the presence of PMAA. This work opened up new prospects for developing biocompatible hydrogel-based nanothin coatings and shaped containers for temperature-regulated drug delivery, cellular uptake, sensing, and transport behavior in microfluidic devices.

3.2. Enzyme/Cell Immobilization

The PNVCL gels not only encapsulated the drugs, but also immobilized the cells and enzymes. Marina *et al.* developed bacteria and fungi microbial cells entrapped in thermoresponsive PNVCL gel beads. The results showed that the enzymatic activities of the microorganisms decreased after the entrapment of cells in the gel beads. This study allowed the preparation of gels of different cells as biocatalysts in a single step procedure for the transformation of both hydrophilic and lipophilic substrates [93]. A one step method was used to fabricate macroscopic poly(N-vinyl caprolactam-*co*-calcium alginate (PNVCL-CaAlg hydrogels) for the entrapment of human cells and enzymes, and their characteristics as immobilized enzymes including their thermal stability and storage stabilities were investigated. Three important proteases, *i.e.*, trypsin, α-chymotrypsin, carboxypeptidase B, and thrombin immobilized in hydrogels, could be used at 65–80 °C, whereas the native enzymes were completely inactivated at 50–55 °C. The hydrogel beads with entrapped α-chymotrypsin were used in the enantioselective hydrolysis of Schiff's base of D,L-phenylalanine ethyl ester (SBPH) in an acetonitrile/water medium, thrombin immobilized in PVCL-based hydrogel films used for wound treatment. Hybridoma cell lines producing MAb to interleukin-2 were also cultivated successfully in the hydrogel beads [94]. Another study also used PVCL-CaAlg gels for the entrapment of trypsin and CPB. This method is advantageous for the immobilization of enzymes and their application in biotechnology and medicine [95].

4. Conclusions and Future Outlooks

The number of studies on PNVCL is currently low, at least compared to those on PNIPAM. In addition, PNVCL exhibits good biocompatibility and is attractive to the biomedical field. On the other hand, PNVCL is not yet FDA approved but there are an increasing number of reports highlighting its applicability in drug delivery and tissue-engineering fields. The limitation in the popularity of PNVCL polymerization was not controllable. Recently, however, controlled living polymerizations resolved this problem through the use of reversible addition–fragmentation chain transfer (RAFT) and macromolecular design via the interchange of xanthates (MADIX) polymerization (RAFT/MADIX) and cobalt-mediated polymerization methods [91,96–98]. This review summarized the synthesis and stimuli responsive behavior of PNVCL gels and their recent progress in biomedical fields, such as drug delivery, encapsulation of cells and the immobilization of enzymes. The incorporation of ionic moieties, such as cationic and anionic functional monomers, into thermoresponsive PNVCL gels made them pH and temperature responsive, making them suitable for drug delivery applications. In addition, pH, thermoresponsive and intracellular degradable micro/nanogels were also studied for cancer drug delivery. In future, the research based on PNVCL gels is expected to grow because, already, a few studies have confirmed the biocompatibility of PNVCL alone or in combination with other polymers.

At present, a number of studies on micro/nano gels have been successfully applied in biomedical fields, especially in drug delivery. However, PNVCL-based hydrogels in tissue engineering has not yet been studied. In future, there is a need to develop methods and modifications with other polymers in order to improve the cytocompatability. Although stimuli responsive nanogels of PNVCL have already been reported, it will be necessary to develop new synthetic strategies to modify them with other functional groups, targeting ligands in the biomedical field, particularly cancer targeted drug delivery.

Acknowledgments: The work was supported by the National Research Foundation of Korea (NRF) Grant funded by the Ministry of Science, ICT & Future Planning, Korea (Acceleration Research Program (2014 R1AZA1110054584) and Brain Korea 21 Plus Program (21A2013800002)). Author Kummari Subba Venkata Krishna Rao thanks to University Grants Commission (UGC), New Delhi, India, for financial support under UGC-RAMAN Postdoctoral Fellowship program (No. F 5-84/ 2014 (IC)).

Author Contributions: Kummara Madhusudana Rao drafted the manuscript, which was revised by all authors. All authors read and approved the final manuscript.

Conflicts of Interest: The authors declare no conflict of interest.

Abbreviations

Monomers	Crosslinkrs
NVCL: *N*-vinyl caprolactam	BISVP: 3,3′-(ethane-1,1-diyl)bis(1-vinyl-2-pyrrolidone)
NIPAM: *N*-isopropyl acrylamide	DMD: *N,N*-diallyl-*N,N*-dimethyl ammonium chloride
DEAM: *N,N*-diethyl acrylamide	MBA: *N, N′*-Methylenebisacrylamide
MVE: methyl vinyl ether	EBVP: Ethylidene-bis-3(*N*-vinyl-2-pyrrolidone)
NEMAM: *N*-ethylmethacrylamide	EGDMA: Ethylene glycol dimethacrylate
EOVE: 2-ethoxyethyl vinyl ether	GA: Glutaraldehyde
NVIBAM: *N*-vinylisobutyramide	TPP: Tripolyphosphate
VP: vinyl pyrrolidine	**Polymers**
AAm: acrylamide	PAA: Poly(acrylamide)
DEAEMA: 2-(diethylamino)ethyl methacrylate	PEO: polyethylene oxide
AcrNEP: *N*-acryloyl-*N′*-ethyl piperazine	SAlg: Sodium Alginate
AMA: Allyl methacrylate	CS: Chitosan
IA: Itaconic acid	PVOH: Polyvinyl alcohol
AAc:Acrylic acid	Fib: Fibrinogen
MAAc: Methacrylic acid	PNVCL: Poly(*N*-vinyl caprolactam)
DMAm: *N,N*-dimethylacrylamide	PNIPAM: poly(*N*-isopropylacrylamide)
Vim: Vinyl imadazole	**Intiators**
AAEM: Acetoacetoxy metthacrylate	AIBN: Azobis(isobutyronitrile)
3MDG: 3-*o*-methacryloyl-1,2:5,6-di-*o*-isopropylidene-α-D-glucofuranose	ABIN: 2,2′-azobis(2-methylpropoionate)
AGA: Acrylamidoglycolic acid	KPS: Potassium persulphate
NaA: Sodium acrylate	APS: Ammonium persulphate
$C_{11}EO_{12}$:PEO macromonomer	**Surfactants**
UA: Undecenoic acid	SDS: Sodium dodecyl sulphate
2MBA: 2-methacryloyloxybenzoic acid	DPB: *N*-dodecylpyridinium bromide
Dx-MA: Dextran mehacrylate	
Drugs	**Others**
5-FU: 5-Fluorouracil	NaBH$_4$: Sodium borohydride
Far: Farmazine	GSH: Glutathione
Dox: Doxorubicin	LCST: Lowest critical Solution temperature
Ket: Ketoprofen	LLS: Laser light scattering studies
CUR: Curcumin	DLS: Dynamic light scattering studies
Meg:Megestrol acetate	IPN: Interpenetrating Polymer Network
Nad: Nadolol	MTT: 3-(4,5-dimethylthiazol-2-yl)-2,5-diphenyltetrazolium bromide
Prop: Propranolol	RAFT: Reversible addition–fragmentation chain transfer
SA: Salicylic acid	

References

1. Langer, R.; Peppas, N.A. Advances in biomaterials, drug delivery, and bionanotechnology. *AIChE J.* **2003**, *49*, 2990–3006. [CrossRef]
2. Gil, E.S.; Hudson, S.M. Stimuli responsive polymers and their conjugates. *Prog. Polym. Sci.* **2004**, *9*, 1173–1222. [CrossRef]
3. Enas, M.A. Hydrogel: Preparation, characterization, and applications: A review. *J. Adv. Res.* **2015**, *6*, 105–121.
4. Tanaka, T. Collapse of gels and the critical end point. *Phys. Rev. Lett.* **1978**, *40*, 820–824. [CrossRef]
5. Ashley, S. Artificial muscles. *Sci. Am.* **2003**, *289*, 52–59. [CrossRef] [PubMed]
6. Mather, P.T. Responsive materials: Soft answers for hard problems. *Nat. Mater.* **2007**, *6*, 93–94. [CrossRef] [PubMed]
7. Peteu, S.F. Responsive Materials Configured for Micro and Nano actuation. *J. Intell. Mater. Syst. Struct.* **2007**, *18*, 147–152. [CrossRef]
8. Kashyap, N.; Kumar, N.; Kumar, M. Hydrogels for pharmaceutical and biomedical applications. *Crit. Rev. Ther. Drug Carr. Syst.* **2005**, *22*, 107–149. [CrossRef]
9. Stamatialis, D.F.; Papenburg, B.J.; Gironés, M.; Saiful, S.; Bettahalli, S.N.M. Wessling, Medical applications of membranes: Drug delivery, artificial organs and tissue engineering. *J. Membr. Sci.* **2008**, *308*, 1–34. [CrossRef]
10. Roy, D.; Brooks, W.L.A.; Sumerlin, B.S. New directions in thermoresponsive polymers. *Chem. Soc. Rev.* **2013**, *42*, 7214–7243. [CrossRef] [PubMed]
11. Gibson, M.I.; Reilly, R.K. To aggregate, or not to aggregate? Considerations in the design and application of polymeric thermally-responsive nanoparticles. *Chem. Soc. Rev.* **2013**, *42*, 7204–7213. [CrossRef] [PubMed]
12. Ward, M.A.; Georgiou, T.K. Thermoresponsive terpolymers based on methacrylate monomers: Effect of architecture and composition. *J. Polym. Sci. Part A Polym. Chem.* **2010**, *48*, 775–783. [CrossRef]
13. Li, Z.; Wang, F.; Roy, S.; Sen, C.K.; Guan, J. Injectable, highly flexible, and thermosensitive hydrogels capable of delivering superoxide dismutase. *Biomacromolecules* **2009**, *10*, 3306–3316. [CrossRef] [PubMed]
14. Hacker, M.C.; Klouda, L.; Ma, B.B.; Kretlow, J.D.; Mikos, A.G. Synthesis and characterization of injectable, thermally and chemically gelable, amphiphilic poly(N-isopropylacrylamide)-based macromers. *Biomacromolecules* **2008**, *9*, 1558–1570. [CrossRef] [PubMed]
15. Ma, X.; Xi, J.; Zhao, X.; Tang, X. Deswelling Comparison of Temperature-Sensitive Poly(*N*-isopropylacrylamide) Microgels Containing Functional -OH Groups with Different Hydrophilic Long Side Chains. *J. Polym. Sci. Part A Polym. Chem.* **2005**, *43*, 3575–3583. [CrossRef]
16. Ito, D.; Kubota, K. Solution properties and thermal behavior of poly(N-n-propylacrylamide) in water. *Macromolecules* **1997**, *30*, 7828–7834. [CrossRef]
17. Ono, Y.; Shikata, T. Hydration and Dynamic Behavior of Poly(*N*-isopropylacrylamide)s in Aqueous Solution: A Sharp Phase Transition at the Lower Critical Solution Temperature. *J. Am. Chem. Soc.* **2006**, *128*, 10030–10031. [CrossRef] [PubMed]
18. Wei, H.; Cheng, S.-X.; Zhang, X.-Z.; Zhuo, R.-X. Thermo-sensitive polymeric micelles based on poly(*N*-isopropylacrylamide) as drug carriers. *Prog. Polym. Sci.* **2009**, *34*, 893–910. [CrossRef]
19. Rzaev, Z.M.O.; Dinçer, S.; Pişkin, E. Functional copolymers of *N*-isopropylacrylamide for bioengineering applications. *Prog. Polym. Sci.* **2007**, *32*, 534–595. [CrossRef]
20. Qin, S.; Geng, Y.; Discher, D.E.; Yang, S. Temperature-Controlled Assembly and Release from Polymer Vesicles of Poly(ethylene oxide)-blockpoly(*N*-isopropylacrylamide). *Adv. Mater.* **2006**, *18*, 2905–2909. [CrossRef]
21. Li, W.; Zhao, H.; Qian, W.; Li, H.; Zhang, L.; Ye, Z.; Zhang, G.; Xia, M.; Li, J.; Gao, J.; *et al.* Chemotherapy for gastric cancer by finely tailoring anti-Her2 anchored dual targeting immunomicelles. *Biomaterials* **2012**, *33*, 5349–5362. [CrossRef] [PubMed]
22. Zhou, C.; Hillmyer, M.A.; Lodge, T.P. Micellization and Micelliar Aggregation of Poly(ethylene-alt-propylene)-b-poly(ethyleneoxide)-b-poly(*N*-isopropylacrylamide) Triblock Terpolymers in Water. *Macromolecules* **2011**, *44*, 1635–1641. [CrossRef]
23. Zhou, C.; Hillmyer, M.A.; Lodge, T.P. Efficient Formation of Multicompartment Hydrogels by Stepwise Self-Assembly of Thermoresponsive ABC Triblock Terpolymers. *J. Am. Chem. Soc.* **2012**, *134*, 10365–10368. [CrossRef] [PubMed]
24. Mi, L.; Xue, H.; Li, Y.; Jiang, S.A. Thermoresponsive Antimicrobial Wound Dressing Hydrogel Based on a Cationic Betaine Ester. *Adv. Funct. Mater.* **2011**, *21*, 4028–4034. [CrossRef]

25. Patenaude, M.; Hoare, T. Injectable, Mixed Natural-Synthetic Polymer Hydrogels with Modular Properties. *Biomacromolecules* **2012**, *13*, 369–378. [CrossRef] [PubMed]

26. Gehan, H.; Fillaud, L.; Chehimi, M.M.; Aubard, J.; Hohenau, A.; Felidj, N.; Mangeney, C. Thermo-induced Electromagnetic Coupling in Gold/Polymer Hybrid Plasmonic Structures Probed by Surface-Enhanced Raman Scattering. *ACS Nano* **2010**, *4*, 6491–6500. [CrossRef] [PubMed]

27. Zhu, Z.; Sukhishvili, S.A. Temperature-Induced Swelling and Small Molecule Release with Hydrogen-Bonded Multilayers of Block Copolymer Micelles. *ACS Nano* **2009**, *3*, 3595–3605. [CrossRef] [PubMed]

28. Roy, D.; Cambre, J.N.; Sumerlin, B.S. Future perspectives and recent advances in stimuli-responsive materials. *Prog. Polym. Sci.* **2010**, *35*, 278–301. [CrossRef]

29. Lackey, C.A.; Press, O.W.; Hoffman, A.S.; Stayton, P.S. A Biomimetic pH-Responsive Polymer Directs Endosomal Release and Intracellular Delivery of an Endocytosed Antibody Complex. *Bioconjugate Chem.* **2002**, *13*, 996–1001. [CrossRef]

30. Sugihara, S.; Yamashita, K.; Matsuzuka, K.; Ikeda, I.; Maeda, Y. Transformation of Living Cationic Polymerization of Vinyl Ethers to RAFT Polymerization Mediated by a Carboxylic RAFT Agent. *Macromolecules (Washington DC USA)* **2012**, *45*, 794–804. [CrossRef]

31. Meeussen, F.; Nies, E.; Bergmans, H.; Verbrugghe, S.; Goethals, E.; Du Prez, F. Phase behaviour of poly (N-vinyl caprolactam) in water. *Polymer* **2000**, *41*, 8597–8602. [CrossRef]

32. Sun, S.; Wu, P. Infrared Spectroscopic Insight into Hydration Behavior of Poly(N-vinylcaprolactam) in Water. *J. Phys. Chem. B* **2011**, *115*, 11609–11618. [CrossRef] [PubMed]

33. Jiří, S. NMR investigations of phase transition in aqueous polymer solutions and gels. *Curr. Opin. Colloid Interface Sci.* **2009**, *14*, 184–191.

34. Kermagoret, A.; Fustin, C.-A.; Bourguignon, M.; Detrembleur, C.; Jerome, C.; Debuigne, A. One-pot controlled synthesis of double thermoresponsive N-vinylcaprolactam-based copolymers with tunable LCSTs. *Polym. Chem.* **2013**, *4*, 2575–2583. [CrossRef]

35. Schild, H.G.; Tirrell, D.A. Microcalorimetric determination of lower critical solution temperature in aqueous polymer solutions. *J. Phys. Chem.* **1990**, *94*, 4352–4356. [CrossRef]

36. Kirsh, Y.E. *Water Soluble Poly-N-Vinylamides*; John Wiley & Sons: Chichester, UK, 1998.

37. Chilkoti, A.; Dreher, M.R.; Meyer, D.E.; Raucher, D. Targeted Drug Delivery by Thermally Responsive Polymers. *Adv. Drug Deliv.* **2002**, *54*, 613–630. [CrossRef]

38. Vihola, H. *Studies on Thermosensitive Poly(N-vinylcaprolactam) Based Polymers for Pharmaceutical Applications*; University of Helsinki: Helsinki, Finland, 2007.

39. Shtanko, N.I.; Lequieu, W.; Goethals, E.J.; Prez, F.E. pH- and Thermoresponsive Properties of Poly(N-vinylcaprolactam-co-acrylic acid) Copolymers. *Polym. Int.* **2003**, *52*, 1605–1610. [CrossRef]

40. Makhaeva, E.E.; Tenhu, H.; Khoklov, A.R. Conformational Changes of Poly(vinylcaprolactam) Macromolecules and Their Complexes with Ionic Surfactants in Aqueous Solution. *Macromolecules* **1998**, *31*, 6112–6118. [CrossRef]

41. Lau, A.C.W.; Wu, C. Thermally Sensitive and Biocompatible Poly(Nvinylcaprolactam): Synthesis and Characterization of High Molar Mass Linear Chains. *Macromolecules* **1999**, *32*, 581–584. [CrossRef]

42. Tager, A.A.; Safronov, A.P.; Berezyuk, E.A.; Galaev, I.Yu. Lower Critical Solution Temperature and Hydrophobic Hydration in Aqueous Polymer Solutions. *Colloid Polym. Sci.* **1994**, *272*, 1234–1239. [CrossRef]

43. Vihola, H.; Laukkanen, A.; Valtola, L.; Tenhu, H.; Hirvonen. Cytotoxicity of thermosensitive polymers poly(N-isopropyl acrylamide), poly(N-vinyl caprolactam) and amphiphilically modified poly(N-vinyl caprolactam). *J. Biomater.* **2005**, *26*, 3055–3064. [CrossRef] [PubMed]

44. Liang, X.; Liu, F.; Kozlovskaya, V.; Palchak, Z.; Kharlampieva, E. Thermoresponsive Micelles from Double LCST-Poly(3- methyl-N-vinylcaprolactam) Block Copolymers for Cancer Therapy. *ACS Macro Lett.* **2015**, *4*, 308–311. [CrossRef]

45. Nakayama, M.; Okano, T.; Miyazaki, T.; Kohori, F.; Sakai, K.; Yokoyama, M. Molecular design of biodegradable polymeric micelles for temperature-responsive drug release. *J. Controll. Release* **2006**, *115*, 46–56. [CrossRef] [PubMed]

46. Akimoto, J.; Nakayama, M.; Okano, T. Temperature-responsive polymeric micelles for optimizing drug targeting to solid tumors. *J. Controlled Release* **2014**, *193*, 2–8. [CrossRef] [PubMed]

47. Liu, F.; Kozlovskaya, V.; Medipelli, S.; Xue, B.; Ahmad, F.; Saeed, M.; Cropek, D.; Kharlampieva, E. Temperature-Sensitive Polymersomes for Controlled Delivery of Anticancer Drugs. *Chem. Mater.* **2015**, *27*, 7945–7956. [CrossRef]

48. Prabaharan, M.; Grailer, J.J.; Steeber, D.A.; Gong, S. Thermosensitive micelles based on folate-conjugated poly(N-vinylcaprolactam)-block-poly(ethylene glycol) for tumor targeted drug delivery. *Macromol. Biosci.* **2009**, *9*, 744–753. [CrossRef] [PubMed]

49. Liang, X.; Kozlovskaya, V.; Cox, C.P.; Wang, Y.; Saeed, M.; Kharlampieva, E. Synthesis and self-assembly of thermosensitive double-hydrophilic poly(Nvinylcaprolactam)-*b*-poly(N-vinyl-2-pyrrolidone) diblock copolymers. *J. Polym. Sci. Part A Polym. Chem.* **2014**, *52*, 2725–2737. [CrossRef]

50. Elenu, E.M.; Le, T.M.T.; Sergei, G.S.; Alexei, R.K. Thermoshrinking behavior of poly(vinylcaprolactam) gels in aqueous solution. *Macromol. Chem. Phys.* **1996**, *197*, 1973–1982.

51. Volodymyr, B.; Sven, R. Monitoring of the Gelation Process on a Radical Chain Crosslinking Reaction Based on N-Vinylcaprolactam by Using Dynamic Light Scattering, Macromol. *Chem. Phys.* **2004**, *205*, 724–730.

52. Leslie, D.M.; Jeffrey, A.H. *Tensile Properties of Poly(N-vinyl caprolactam) Gels*; National Aeronautics and Space Administration Langley Research Center Hampton: Virginia, USA, 2004; pp. 1–10.

53. Mamytbekov, G.; Bouchal, K.; Ilavsky, M. Phase transition in swollen gels 26. Effect of charge concentration on temperature dependence of swelling and mechanical behavior of poly(N-vinlycaprolactam) gels. *Eur. Polym. J.* **1999**, *35*, 1925–1933. [CrossRef]

54. Ilavský, M.; Mamytbekov, G.; Sedláková, Z.; Hanyková, L.; Dušek, K. Phase transition in swollen gels 29. Temperature dependences of swelling and mechanical behaviour of poly(N-vinylcaprolactam-co-1-vinyl-2-pyrrolidone) gels in water. *Polym. Bull.* **2001**, *46*, 99–106.

55. Váček, J.; Radecki, M.; Zhigunov, A.; Šťastná, J.; Valentová, H.; Sedláková, Z. Structures and interactions in collapsed hydrogels of thermoresponsive interpenetrating polymer networks. *Colloids Polym. Sci.* **2015**, *293*, 709–720.

56. Vyshivannaya, O.V.; Laptinskaya, T.V.; Makhaeva, E.E.; Khokhlov, A.R. Dynamic Light Scattering in Semi_Interpenetrating Polymer Networks Based on Polyacrylamide and Poly(N_vinylcaprolactam). *Polym. Sci. Ser. A* **2012**, *54*, 693–706. [CrossRef]

57. Cakal, E.; Cavus, S. Novel Poly(N-vinylcaprolactam-co-2-(diethylamino)ethyl methacrylate) Gels: Characterization and Detailed Investigation on Their Stimuli-Sensitive Behaviors and Network Structure. *Ind. Eng. Chem. Res.* **2010**, *49*, 1741–11751. [CrossRef]

58. Roshan Deen, G.; Lim, E.K.; Hao Mah, C.; Heng, K.M. New Cationic Linear Copolymers and Hydrogels of N-Vinyl Caprolactam and N-Acryloyl-N'-ethyl Piperazine: Synthesis, Reactivity, Influence of External Stimuli on the LCST and Swelling Properties. *Ind. Eng. Chem. Res.* **2012**, *51*, 13354–13365. [CrossRef]

59. Selva, C.; Elc, C. Synthesis and Characterization of Novel Poly(N-vinylcaprolactam-coitaconic Acid) Gels and Analysis of pH and Temperature Sensitivity. *Ind. Eng. Chem. Res.* **2012**, *51*, 1218–1226.

60. Gaballa, H.A.; Geever, L.M.; Killion, J.A.; Higginbotham, C.L. Synthesis and characterization of physically crosslinked N-vinylcaprolactam, acrylic acid, methacrylic acid, and N,N-dimethylacrylamide hydrogels. *J. Polym. Sci. Part B Polym. Phys.* **2013**, *51*, 1555–1564. [CrossRef]

61. Zavgorodnya, O.; Kozlovskaya, V.; Liang, X.; Kothalawala, N.; Catledge, S.A.; Dass, A.; Kharlampieva, E. Temperature-responsive properties of poly(N-vinylcaprolactam) multilayer hydrogels in the presence of Hofmeister anions. *Mater. Res. Express* **2014**, *1*. [CrossRef]

62. Sanna, R.; Fortunati, E.; Alzari, V.; Nuvoli, D.; Terenzi, A.; Casula, M.F.; Kenny, J.M.; Mariani, A. Poly(N-vinylcaprolactam) nanocomposites containing nanocrystalline cellulose: A green approach to thermoresponsive hydrogels. *Cellulose* **2013**, *20*, 2393–2402. [CrossRef]

63. Gao, Y.; Au-Yeung, S.C.F.; Wu, C. Interaction between Surfactant and Poly(N-vinylcaprolactam) Microgels. *Macromolecules* **1999**, *32*, 674–3677. [CrossRef]

64. Boyko, V.; Richter, S.; Grillo, I.; Geissler, E. Structure of Thermosensitive Poly(Nvinylcaprolactam-co-N-vinylpyrrolidone) Microgels. *Macromolecules* **2005**, *38*, 5266–5270. [CrossRef]

65. Laukkanen, A.; Hietala, S.; Maunu, S.L.; Tenhu, H. Poly(N-vinylcaprolactam) Microgel Particles Grafted with Amphiphilic Chains. *Macromolecules* **2000**, *33*, 8703–8708. [CrossRef]

66. Yibing, G.; Steve, C.F.A.Y.; Chi, W. Interaction between surfactant and poly(N-vinyl caprolactam) microgels. *Macromolecules* **1999**, *32*, 3674–3677.

67. Imaz, A.; Forcada, J. *N*-Vinylcaprolactam-Based Microgels: Synthesis and Characterization. *J. Polym. Sci. Part A Polym. Chem.* **2008**, *46*, 2510–2524. [CrossRef]

68. Pich, A.; Tessier, A.; Boyko, V.; Lu, Y.; Adler, H.P. Synthesis and Characterization of Poly(vinylcaprolactam)-Based Microgels Exhibiting Temperature and pH-Sensitive Properties. *Macromolecules* **2006**, *39*, 7701–7707. [CrossRef]

69. Imaz, A.; Forcada, J. N-Vinylcaprolactam-Based Microgels: Effect of the Concentration and Type of Cross-linker. *J. Polym. Sci. A Polym. Chem.* **2008**, *46*, 2766–2775. [CrossRef]

70. Imaz, A.; Forcada, J. Optimized buffered polymerizations to produce *N*-vinylcaprolactam-based Microgels. *Eur. Polym. J.* **2009**, *45*, 3164–3175. [CrossRef]

71. Peng, S.; Wu, C. Ca^{2+}-induced Thermoreversible and Controllable Complexation of Poly(*N*-vinylcaprolactam-co-sodium acrylate) Microgels in Water. *J. Phys. Chem. B* **2001**, *105*, 2331–2335. [CrossRef]

72. Peng, S.; Wu, C. Ca^{+2}-induced complexation between thermally sensitive spherical poly(*N*-vinyl-caprolactam-co-sodium acrylate) microgels and linear gelatin chains in water. *Polymer* **2001**, *42*, 7343–7347. [CrossRef]

73. Iskakov, R.M.; Sedinkin, S.V.; Mamytbekov, G.K.; Batyrbekov, E.O.; Bekturov, E.A.; Zhubanov, B.A. Controlled Release of Farmazin from Thermosensitive Gels Based on Poly(*N*-vinylcaprolactam). *Russ J. Appl. Chem.* **2004**, *7*, 339–341. [CrossRef]

74. Rama Subba Reddy, P.; Madhusudana Rao, K.; Krishna Rao, K.S.V.; Schichpanov, Y.; Ha, C.S. Synthesis of Alginate Based Silver Nanocomposite Hydrogels for Biomedical Applications. *Macromol. Res.* **2014**, *8*, 832–842. [CrossRef]

75. Madhusudana-Rao, K.; Krishna-Rao, K.S.V.; Sudhakar, P.; Chowdoji-Rao, K.; Subha, M.C.S. Synthesis and Characterization of biodegradable Poly (Vinyl caprolactam) grafted on to sodium alginate and its microgels for controlled release studies of an anticancer drug. *J. Appl. Pharm. Sci.* **2013**, *3*, 061–069.

76. Swamy, B.Y.; Chang, J.H.; Ahn, H.; Lee, W.K.; Chung, I. Thermoresponsive *N*-vinyl caprolactam grafted sodium alginate hydrogel beads for the controlled release of an anticancer drug. *Cellulose* **2013**, *20*, 1261–1273. [CrossRef]

77. Prabaharan, M.; Grailer, J.J.; Steeber, D.A.; Gong, S. Stimuli-responsive chitosan-graftpoly(*N*-vinylcaprolactam) as promising material for controlled hydrophobic drug delivery. *Macromol.Biosci.* **2008**, *8*, 843–851. [CrossRef] [PubMed]

78. Vihola, H.; Laukkanen, A.; Tenhu, H.; Hirvonen, J. Drug Release Characteristics of Physically Cross-Linked Thermosensitive Poly(*N*-vinylcaprolactam) Hydrogel Particles. *J. Pharm. Sci.* **2008**, *97*, 4783–4793. [CrossRef] [PubMed]

79. Hantzschel, N.; Hund, R.D.; Hund, H.; Schrinner, M.; Lück, C.; Pich, A. Hybrid Microgels with Antibacterial Properties. *Macromol. Biosci.* **2009**, *9*, 444–449. [CrossRef] [PubMed]

80. Lou, S.; Gao, S.; Wang, W.; Zhang, M.; Zhang, Q.; Wang, C.; Li, C.; Kong, D. Temperature/pH dual responsive microgels of crosslinked poly(N-vinylcaprolactam-co-undecenoic acid) as biocompatible materials for controlled release of doxorubicin. *J. Appl. Polym. Sci.* **2014**, *131*. [CrossRef]

81. Wang, Y.; Nie, J.; Chang, B.; Sun, Y.; Yang, W. Poly(N-vinylcaprolactam)-Based Biodegradable Multiresponsive Microgels for Drug Delivery. *Biomacromolecules* **2013**, *14*, 3034–3046. [CrossRef] [PubMed]

82. Malmsten, M. Soft drug delivery systems. *Soft Matter* **2006**, *2*, 760–769. [CrossRef]

83. Napier, M.E.; Desimone, J.M. Nanoparticle drug delivery platform. *Polym. Rev.* **2007**, *47*, 321–327. [CrossRef]

84. Moghimi, S.M.; Hunter, A.C.; Murray, J.C. Long-circulating and target-specific nanoparticles: Theory to practice. *Pharmacol. Rev.* **2001**, *53*, 283–318. [PubMed]

85. Madhusudana Rao, K.; Mallikarjuna, B.; Krishna Rao, K.S.V.; Siraj, S.; Chowdoji Rao, K.; Subha, M.C.S. Novel thermo/pH sensitive nanogels composed from poly(*N*-vinylcaprolactam) for controlled release of an anticancer drug. *Colloids Surf. B* **2013**, *102*, 891–897. [CrossRef] [PubMed]

86. Sudhakara, K.; Madhusudana Rao, K.; Subha, M.C.S.; Chowdoji Rao, K.; RotimiSadiku, E. Temperature-responsive poly(*N*-vinylcaprolactam-co-hydroxyethyl methacrylate) nanogels for controlled release studies of curcumin. *Des. Monomers Polym.* **2015**, *18*, 705–713. [CrossRef]

87. Gonalez-Ayon, M.A.; Adriana Sanudo-Barajas, J.; Picos-Corrales, L.A.; Licea-Claverie, A. PNVCL-PEGMA nanohydrogels with tailored transition temperature for controlled release of 5-Fluorouracil. *J. Polym. Sci. Part A Polym. Chem.* **2015**, *53*, 2662–2672. [CrossRef]

88. Liu, J.; Detrembleur, C.; Hurtgen, M.; Debuigne, A.; Pauw-Gillet, M.C.; Mornet, S.; Duguet, E.; Jérome, C. Reversibly-crosslinked Thermo- and Redox-responsive Nanogels for Controlled Drug Release. *Polym. Chem.* **2014**, *5*, 77–88. [CrossRef]

89. Aguirre, G.; Ramos, J.; Forcada, J. Synthesis of new enzymatically degradable thermo-responsive nanogels. *Soft Matter* **2013**, *9*, 261–270. [CrossRef]

90. Rejinold, N.S.; Baby, T.; Chennazhi, K.P.; Jayakumar, R. Multi Drug Loaded Thermo-Responsive Fibrinogen-graft-Poly(N-vinyl Caprolactam) Nanogels for Breast Cancer Drug Delivery. *J. Biomed. Nanotechnol.* **2015**, *11*, 392–402. [CrossRef] [PubMed]

91. Cai, T.; Li, M.; Zhang, B.; Neoh, K.G.; Kang, E.T. Hyperbranched polycaprolactone-clickpoly (N-vinylcaprolactam) amphiphilic copolymers and their applications as temperature-responsive membranes. *J. Mater. Chem. B* **2014**, *2*, 814–825. [CrossRef]

92. Liang, X.; Kozlovskaya, V.; Chen, Y.; Zavgorodnya, O.; Kharlampieva, E. Thermosensitive Multilayer Hydrogels of Poly(N-vinylcaprolactam) as Nanothin Films and Shaped Capsules. *Chem. Mater.* **2012**, *24*, 3707–3719. [CrossRef] [PubMed]

93. Donova, M.V.; Kuzkina, I.F.; Arinbasarova, A.Y.; Pashkin, I.I.; Markvicheva, E.E.; Baklashova, T.G.; Sukhodolskaya, G.V.; Fokina, V.V.; Kirsh, Y.E.; Koshcheyenko, K.A.; *et al.* Poly(N-vinylcaprolactam) gel. A novel matrix for entrapment of microorganism. *Biotechnol. Tech.* **1993**, *7*, 415–422. [CrossRef]

94. Kuptsova, S.V.; Mareeva, T.Y.; Vikhrov, A.A.; Dugina, T.N.; Strukova, S.M.; Belokon, Y.N.; Kochetkov, K.A.; Baranova, E.N.; Zubov, V.P.; Poncelet, D.; *et al.* Immobilized enzymes and cells in poly(N-vinyl caprolactam)-based hydrogels. *Appl. Biochem. Biotechnol.* **2000**, *88*, 145–157.

95. Markvicheva, E.A.; Bronin, A.S.; Kudryavtseva, N.E.; Rumsh, L.D.; Kirsh, Y.E.; Zubov, V.P. Immobilization of proteases in composite hydrogel based in poly(N-vinylcaprolactam). *Biotechnol. Tech.* **1994**, *8*, 143–148. [CrossRef]

96. Wan, D.; Zhou, Q.; Pu, H.; Yang, G. Controlled radical polymerization of N-vinylcaprolactam mediated by xanthate or dithiocarbamate. *J. Polym. Sci. Part A Polym. Chem.* **2008**, *46*, 3756–3765. [CrossRef]

97. Beija, M.; Marty, J.-D.; Destarac, M. Thermoresponsive poly(N-vinyl caprolactam)-coated gold nanoparticles: Sharp reversible response and easy tenability. *Chem. Commun.* **2011**, *47*, 2826–2828. [CrossRef] [PubMed]

98. Marie, H.; Ji, L.; Antoine, D.; Christine, J.; Christophe, D. Synthesis of thermo-responsive poly(N-vinylcaprolactam)-containing block copolymers by cobalt-mediated radical polymerization. *J. Polym. Sci. Part A Polym. Chem.* **2012**, *50*, 400–408.

gels

MDPI

Article

Development of Novel *N*-isopropylacrylamide (NIPAAm) Based Hydrogels with Varying Content of Chrysin Multiacrylate

Shuo Tang [1,3], Martha Floy [2], Rohit Bhandari [1,3], Thomas Dziubla [1,3] and J. Zach Hilt [1,3,*]

[1] Department of Chemical and Materials Engineering, University of Kentucky, Lexington, KY 40506, USA; shuo.tang@uky.edu (S.T.); rohit.bhandari@uky.edu (R.B.); thomas.dziubla@uky.edu (T.D.)
[2] Department of Chemical Engineering, Kansas State University, Manhattan, KS 66506, USA; mfloy@ksu.edu
[3] Superfund Research Center, University of Kentucky, Lexington, KY 40506, USA
* Correspondence: zach.hilt@uky.edu; Tel.: +1-859-257-9844; Fax: +1-859-323-1929

Received: 20 August 2017; Accepted: 11 October 2017; Published: 22 October 2017

Abstract: A series of novel temperature responsive hydrogels were synthesized by free radical polymerization with varying content of chrysin multiacrylate (ChryMA). The goal was to study the impact of this novel polyphenolic-based multiacrylate on the properties of *N*-isopropylacrylamide (NIPAAm) hydrogels. The temperature responsive behavior of the copolymerized gels was characterized by swelling studies, and their lower critical solution temperature (LCST) was characterized through differential scanning calorimetry (DSC). It was shown that the incorporation of ChryMA decreased the swelling ratios of the hydrogels and shifted their LCSTs to a lower temperature. Gels with different ChryMA content showed different levels of response to temperature change. Higher content gels had a broader phase transition and smaller temperature response, which could be attributed to the increased hydrophobicity being introduced by the ChryMA.

Keywords: hydrogel; *N*-isopropylacrylamide; LCST

1. Introduction

Stimuli responsive hydrogels, which are often referred to as intelligent hydrogels, are a type of hydrogel where swelling behavior changes in response to environmental factors such as pH, salt concentration, ionic strength, and temperature, or a combination therein [1–4]. Among these, pH and temperature responsive gels have gained the greatest attention, and they have been shown to be easily tunable [5]. *N*-isopropylacrylamide (NIPAAm)-based polymers, as one of the most widely studied temperature responsive polymers, have shown great potentials in various fields. NIPAAm-based polymers exhibit a lower critical solution temperature (LCST) at ~32–33 °C in aqueous solution, which can be easily adjusted to physiological temperature through the modification of the hydrophilic/hydrophobic balance in the polymer with comonomers [6–10]. There are numerous applications of the sharp phase transition, which can be a reversible swelling change for crosslinked systems, especially in the designing of a controlled release system e.g., drug delivery, analytical separation and detection [11–15].

Chrysin (5,7-dihydroxyflavone), a naturally occurring flavonoid presents at high levels in honey and propolis, has shown potential pharmacological effects in inhibiting coronary heart disease, stroke, and cancer [16]. As a flavonoid, chrysin has been widely studied both in vitro and in vivo for its anti-inflammatory, antioxidant, antiviral, and immunomodulatory effects [17]. Multiple researchers have shown that chrysin stimulates or inhibits a wide variety of enzyme systems [17–20]. In this work, chrysin is not studied as biomedical agent, but it is acrylated and used as a hydrophobic comonomer/crosslinker that is incorporated in an NIPAAm-based temperature responsive hydrogel system.

In the present study, a series of temperature responsive hydrogels consisting of NIPAAm and varying amount of ChryMA were developed. The LCST of NIPAAm gel can be easily altered by adjusting the hydrophilic and hydrophobic balance of the network [7,8]. The use of comonomers to adjust the LCST of NIPAAm gels have been well studied, although such studies have focused mostly on the hydrophilic modification of the polymer network. Some of the well-known hydrophilic crosslinkers/comonomers such as acrylic acid (AA) [21–26] and methacrylic acid (MA) [3,27–29] have been extensively reported in previous publications. However, few have reported on the hydrophobic crosslinkers/comonomers. Most recently, our group reported two sets of NIPAAm gels using hydrophobic crosslinkers curcumin multiacrylate (CMA) and quercetin multiacrylate (QMA), which have shifted the LCST of NIPAAm gel from 33 °C to 30.4 °C and 28.9 °C, respectively [30]. Other groups, such as Gan et al., investigated the phase transition behavior of hydrophobically modified biodegradable hydrogel using poly(ε-caprolactone) dimethacrylate (PCLDMA) and bisacryloylcysatamine (BACy) as the crosslinkers. The LCST was shifted from 32.6 to 30.68 °C, yet the swelling ratio remained high (~20) [31]. In addition, butyl methacrylate (BMA) [32], di-n-propyl acrylamide (DPAM) [33], polystyrene [34], benzo-12-crown-4-acrylamide (PNB12C4) [35], and others have been reported in recent years on the hydrophobic modification of NIPAAm hydrogel. In this work, we present novel NIPAAm hydrogels with varying content of hydrophobic comonomer/crosslinker ChryMA, which acts as a model compound for the successful synthesis of similar hydrophobic/polyphenolic materials for various fields of application.

2. Results and Discussion

2.1. Characterization of ChryMA

HPLC chromatograms for chrysin and ChryMA are shown in Figure 1. The red and blue lines represent chrysin and ChryMA, respectively. A single and distinctive peak was observed at 7.2 min in the red line, which corresponds to the chrysin with a high purity. Two peaks are shown around 10 and 11 min in blue line, which are the different acrylates of chrysin. On the basis of an increase in the inherent hydrophobicity of the acrylated chrysin, the peak at 10 min was identified as the monoacylate, and the peak at 11 min was identified as the diacrylate. No peak was observed at 7.2 min for the case of ChryMA, indicating that all of the precursor was converted to acrylated product. The composition of mono- and diacrylate was characterized in liquid chromatography time-of-flight (LC-TOF), and these two forms were determined to be present in molar amounts of 36.9% and 63.1%, respectively, which are similar to the observed ratios from HPLC absorbance peak areas. The average molecular weight of ChryMA was calculated to be 365.3 g/mol based on the composition of mono- and diacrylates in the product. The chemical structure of precursor, monoacrylate, and diacrylate are shown in Figure 2.

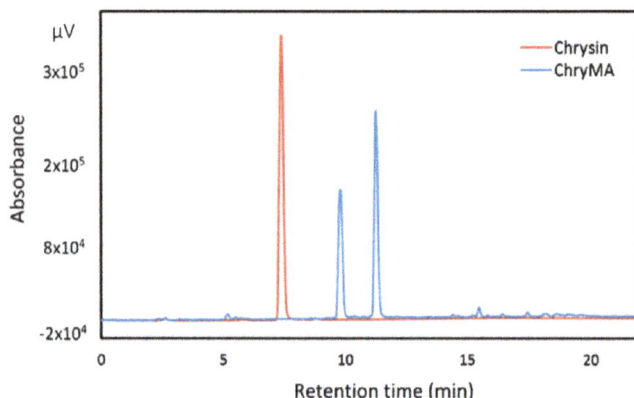

Figure 1. HPLC chromatograms for Chrysin and chrysin multiacrylate (ChryMA).

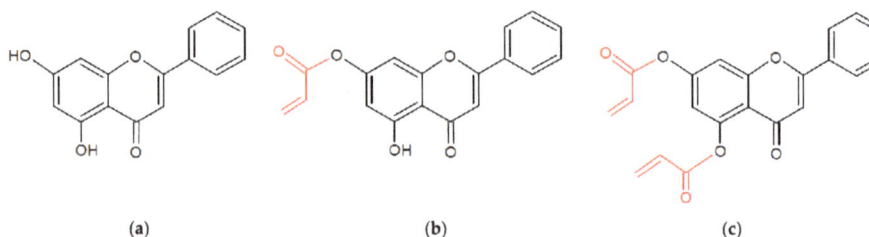

Figure 2. Chemical structure of (**a**) Chrysin; (**b**) Chrysin-monoacrylate; (**c**) Chrysin-diacrylate.

2.2. Synthesis of NIPAAm-co-ChryMA Gels

NIPAAm-*co*-ChryMA gels were synthesized by free radical polymerization using ammonium persulfate (APS) as a thermal initiator. The crosslinker poly(ethylene glycol) 400 dimethacrylate (PEG400DMA) was kept at 5 mol %, while the ratio of NIPAAm and ChryMA was varied to determine the influence of the hydrophobic crosslinker/comonomer in swelling behaviors and phase transition properties. The control group was synthesized using 5 mol % of PEG400DMA with the rest of NIPAAm, which is referred to as ChryMA 0.0. Three other hydrogel systems were synthesized with 2, 4, or 6 mol % of ChryMA, referred to as ChryMA 2.0, ChryMA 4.0, and ChryMA 6.0, respectively. With these reaction conditions, hydrogels could not be synthesized with more than 6 mol % of ChryMA, which is potentially due to the larger ChryMA molecule sterically hinder gel synthesis at the double bond reactive site. The texture of swollen gel became stiff and rubbery as the amount of ChryMA increased, while low ChryMA content gels were soft and flexible. The polymerization schematic of synthesizing NIPAAm-*co*-ChryMA gel is shown in Figure 3.

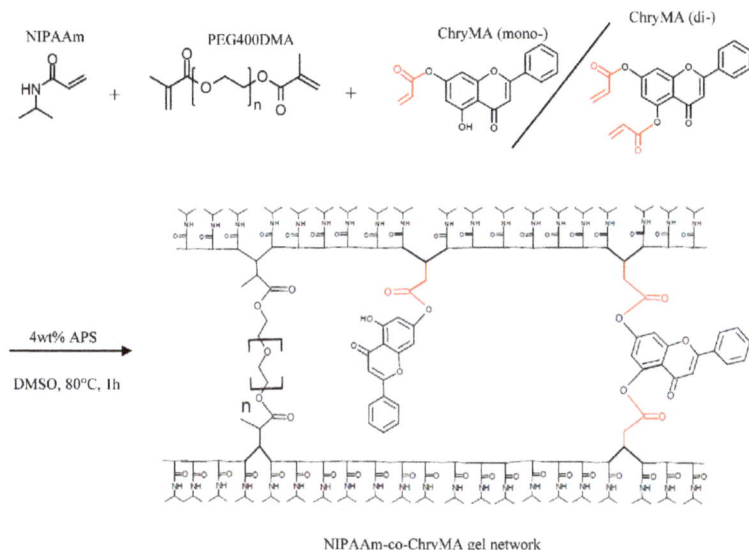

Figure 3. Example polymerization scheme for synthesizing *N*-isopropylacrylamide (NIPAAm)-*co*-ChryMA gels.

2.3. Kinetic Swelling Study

The kinetic swelling behavior of NIPAAm-co-ChryMA gels was studied at 25 °C in a water bath for up to 48 h. The mass swelling ratio "*q*" was defined as the mass at the swollen state divided by the mass at the dry state, and the q as a function of time is shown in Figure 4. From the result, it can be noticed that all gels reached equilibrium swelling by 24 h, and the equilibrium swelling ratios decreased with ChryMA content. The addition of ChryMA increases the hydrophobicity of copolymer gels, and thus, the hydrogels have less affinity for water. Furthermore, the increasing ChryMA content in the gel network increases the degree of crosslinking. Both of these factors can be attributed to the lower swelling ratio of the hydrogel with a higher content of ChryMA.

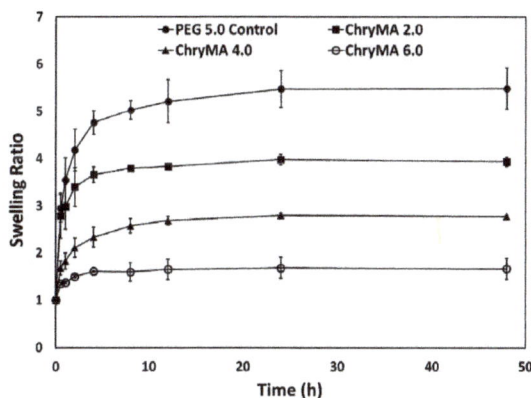

Figure 4. Kinetic swelling behavior of NIPAAm-*co*-ChryMA gels.

2.4. Temperature Dependent Swelling Study

To study the impact of comonomer content on thermoresponsive swelling behaviors, two temperature dependent swelling studies were conducted. The first study aimed to generate a temperature swelling profile. In this study, copolymer gels were allowed to swell at different temperatures, and their mass swelling ratios were measured after 24 h. As shown in Figure 5, higher temperature caused a lower swelling ratio; and the higher ChryMA content led to a lower transition temperature. Meanwhile, higher ChryMA content gels showed lower swelling and broader phase transition due to more crosslinking in the structure and the addition of hydrophobic content. As temperature increased, all copolymer gels almost completely collapsed by 37.5 °C.

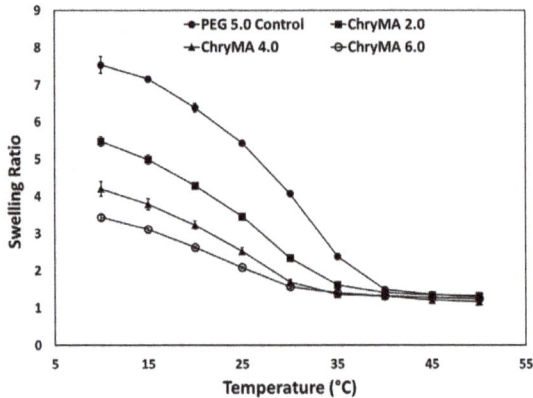

Figure 5. Temperature dependent swelling profile of ChryMA hydrogels.

A second reversible swelling study was also conducted, as shown in Figure 6. The capability of the NIPAAm-*co*-ChryMA hydrogels to swell and deswell repeatedly is important for applications. Reversible temperature changes in this study were designed to span the LCST of the gels going from 10 to 50 °C for ChryMA gels. All gels showed good reversible swelling behavior, retaining swelling within 5%. One possible reason for the slight decrease in the reswelling ratio is the additional crosslink formation through repeated heating. Another reason could be that small fragments of the hydrogel samples were lost during sample handling (especially when they were swollen and fragile), as this could also contribute to the slight decrease in the swelling ratios. Additionally, bubbles formed in ChryMA 6.0 content gels during the heating cycle of the reversible swelling test, and this is likely because the outside thin "skin" barrier is more permeable to water and collapses quickly when undergoing a fast temperature transition. After this outside network collapses, trapped water inside cannot leave the hydrogel and forms a bubble. Zhang et al. reported a similar deswelling and reswelling phenomenon for higher comonomer content gels [36].

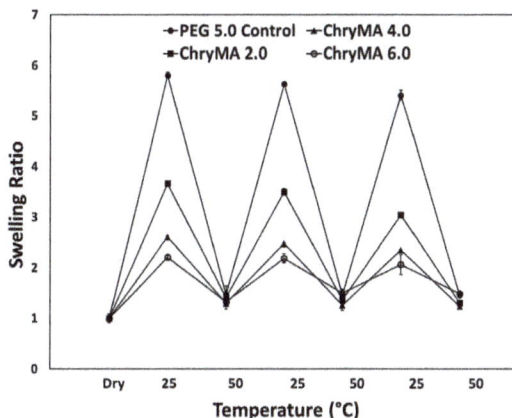

Figure 6. Reversible swelling behavior of NIPAAm-*co*-ChryMA hydrogels.

2.5. LCST Measurements

Various methods are known for determining the LCST of NIPAAm hydrogels, but the most widely used methods are differential scanning calorimetry (DSC) and the equilibrium swelling ratio method. The DSC technique is often more precise since it gives information on the heat released from the cleavage of hydrogen bonds between water and the polymer chain [37,38]. In a typical DSC thermogram, the endothermic peak is referred to as the LCST of the hydrogel, where intramolecular hydrogen bonds between water and NH-break.

Homopolymer NIPAAm is known to have an LCST around 32–33 °C. PEG 5.0 control gel showed an upward shift in LCST to 34.5 °C, which is likely due to the hydrophilicity of PEG400DMA. The incorporation of hydrophilic groups increases the amount of intermolecular hydrogen bonding, so that additional heat is necessary to break the hydrogen bonds, resulting in an increase in the LCST.

The NIPAAm-*co*-ChryMA gels exhibit an LCST that can be tuned from 34.5 to 27.4 °C by altering the ChryMA composition, as shown in Figure 7. The addition of hydrophobic ChryMA shifted the LCST to lower temperatures due to the fewer hydrogen bonds, meaning that less energy was needed to beak the bonds, thus the transition temperature decreased. In addition, broader peaks were observed for high ChryMA content gels. Important characteristics of synthesized gels are summarized in Table 1.

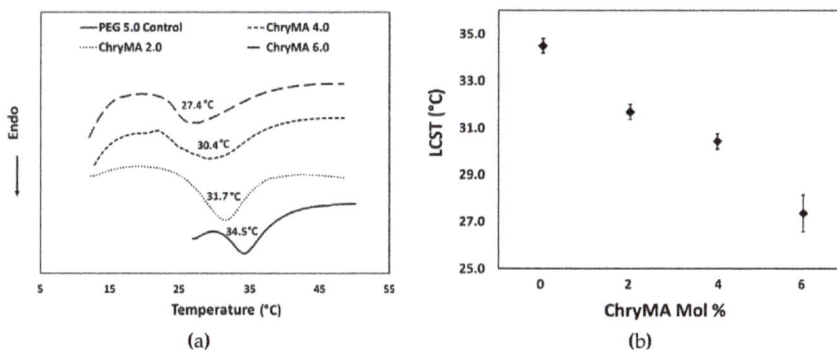

(a)

(b)

Figure 7. (**a**) Lower critical solution temperature (LCST) measurement of NIPAAm-*co*-ChryMA gels; (**b**) LCST as a function of ChryMA content.

3. Conclusions

Hydrophobically modified NIPAAm-*co*-ChryMA hydrogels were successfully developed through free radical polymerization. Gels with four different compositions were synthesized to study the effect of comonomer content on their swelling and phase transition behaviors. The swelling ratios of copolymeric gels decreased with an increase in ChryMA content, and the addition of ChryMA shifted the LCST to lower temperatures. As temperature increased, swelling decreased for all gels, displaying strong temperature responsive behavior, and this response was shown to be reversible. These novel hydrogels have various potential applications in biomedical, environmental, and other fields.

4. Materials and Methods

4.1. Materials

N-isopropylacrylamide (NIPAAm, 97%), initiator ammonia persulfate (APS, $\geq$98%) triethyl amine (TEA), acryloyl chloride (AC), and chrysin were purchased from Sigma-Aldrich Corporation (St. Louis, MO, USA). Poly(ethyleneglycol) 400 dimethacrylate (PEG400DMA) was purchased from Polysciences, Inc (Warrington, FL, USA). All solvents were purchased from VWR International (Radnor, PA, USA). Molecular sieves (3 Å) were added to the solvents to remove any moisture present and to maintain their anhydrous state.

4.2. Synthesis of ChryMA

Comonomer chrysin multiacrylates (ChryMA) were prepared in accordance with the protocols described previously [39,40]. Briefly, chrysin was dissolved in an excess amount of THF, followed by the addition of AC with the ratio to chrysin of 3:1. TEA was added in the same molar ratio as AC to capture the byproduct hydrogen chloride by forming a chloride salt with the progression of the reaction. AC was added dropwise under continuous stirring in an ice bath to prevent the reaction from overheating. Since Chrysin is light-sensitive, the acrylation process was conducted under dark conditions for 16 h. Then, the obtained mixture was filtered from salts and evaporated under vacuum using a liquid N_2 trap. Further purification was conducted by multiple washes with 0.1 M K_2CO_3 and then 0.1 M HCl to remove any unreacted AC and TEA. The final product was filtrated once more and vacuum dried to obtain powdered ChryMA. The product was kept in a freezer at -20 °C until use.

4.3. Characterization of ChryMA Using High Performance Liquid Chromatography (HPLC)

The obtained ChryMA was analyzed through reverse-phase HPLC (Waters Phenomenex C18 column, 5 μm, 250 mm (length) × 4.6 mm (inside diameter, I.D.) on a Shimadzu Prominence LC-20 AB HPLC system) to verify product quality. Samples were dissolved in acetonitrile (ACN) at 100 μg/mL. A gradient from 50/50 ACN/water to 100/0 ACN/water over 24 min at 1 mL/min was used with the column chamber set at 40 °C. The absorbance was measured from 260 nm to 370 nm.

4.4. Synthesis of NIPAAm-co-ChryMA Gels

NIPAAm-*co*-ChryMA hydrogels were synthesized using the free radical polymerization approach. ChryMA and NIPAAm were dissolved in DMSO with the feed ratio of 2/93, 4/91, or 6/89 mol %, and PEG400DMA (crosslinker) was kept at 5 mol %. The initiator, APS, was dissolved in deionized (DI) water to the specified concentration of 0.5 mg/mL and added at 4 wt % combined weight of NIPAAm and ChryMA. To increase solubility, the reaction mixture was preheated at 80 °C before adding the initiator, APS. Instant polymerization occurred, but gels were allowed to continue the reaction for 1 h to ensure high conversion. To remove any unreacted monomers, gels were washed with excess acetone and DI water three times each, for 30 min per wash. Then, gels were cut into small pieces (5 mm in diameter) and freeze-dried overnight until no further mass change occurred. Reaction components for all gels are summarized in Table 1.

Table 1. The compositions, equilibrium swelling ratios, and LCSTs of the NIPAAm-*co*-ChryMA gels studied.

Sample	NIPAAm mol %	ChryMA mol %	PEG400DMA mol %	Equilibrium q (25 °C)	LCST (°C)
Control	95.0	0.0	5.0	5.5	34.5
ChryMA 2.0	93.0	2.0	5.0	3.8	31.7
ChryMA 4.0	91.0	4.0	5.0	2.6	30.4
ChryMA 6.0	89.0	6.0	5.0	1.7	27.4

4.5. Kinetic Swelling Study

Dried gels were swelled in 5 mL of DI water at 25 °C in an isothermal water bath to study equilibrium swelling kinetics. Mass measurements were taken at time points of 0, 0.5, 1, 2, 4, 8, 12, 24, and 48 h. Each sample was removed from the water bath, dabbed dry with Kimwipe to remove excess surface water, and weighed. The mass swelling ratio was defined as the swollen mass divided by the dry mass, as shown in Equation (1), where $M_{swollen}$ is the wet weight measured at each time point, and M_{dry} is the dry mass after freeze-drying.

$$q = M_{swollen} / M_{dry} \tag{1}$$

4.6. Temperature Dependent Swelling Study

To determine the temperature responsiveness of the gels, gels discs were swelled in 5 mL of DI water at different temperatures for 24 h to reach equilibrium swelling. Swelling ratios were measured at temperature increments of 5 °C from 10 to 50 °C. A reversible swelling study was conducted to ensure that the swelling response of the gels was repeatable. Gels were cycled for three times in an isothermal water bath from swollen at 25 °C to collapsed at 50 °C. The mass of the gel was recorded after reaching the swelling equilibrium by 24 h.

4.7. LCST Measurements

The LCSTs of the hydrogels were measured using differential scanning calorimetry (DSC Q200, TA instruments Inc., New Castle, DE, USA). Hydrogels were allowed to swell for at least 24 h in DI water for equilibrium. A small piece of gel was gently dabbed dry, and its mass was carefully measured and recorded. The sample was then hermetically sealed in the T-zero pan in order to eliminate the possibility of water evaporation, and placed along with a reference pan on heaters. Samples were heated from 10 to 50 °C at a rate of 2 °C/min under a dry nitrogen atmosphere at a flow rate of 50 mL/min.

Acknowledgments: This research has been generously supported by the National Institute of Environmental Health Sciences (NIEHS) (Project No: P42ES007380) and the National Science Foundation REU Program (Project No: EEC-1460486).

Author Contributions: Shuo Tang, Martha Floy, Rohit Bhandari, Thomas Dziubla, J. Zach Hilt conceived and designed the experiments. Shuo Tang, Martha Floy, and Rohit Bhandari performed the experiment. Shuo Tang and Martha Floy analyzed the data. Shuo Tang, Martha Floy, Rohit Bhandari, Thomas Dziubla, J. Zach Hilt wrote the paper.

Conflicts of Interest: The authors declare no conflict of interest.

References

1. Lue, S.J.; Chen, C.-H.; Shih, C.-M. Tuning of lower critical solution temperature (LCST) of poly (*N*-isopropylacrylamide-*co*-acrylic acid) hydrogels. *J. Macromol. Sci. Part B* **2011**, *50*, 563–579. [CrossRef]
2. Lee, W.-F.; Hsu, C.-H. Thermoreversible hydrogels: 3. Synthesis and swelling behavior of the (*N*-isopropylacrylamide-*co*-trimethylacrylamidopropyl ammonium iodide) copolymeric hydrogels. *Polymer* **1998**, *39*, 5393–5403. [CrossRef]

3. Brazel, C.S.; Peppas, N.A. Synthesis and Characterization of Thermo-and Chemomechanically Responsive Poly (N-isopropylacrylamide-co-methacrylic acid) Hydrogels. *Macromolecules* **1995**, *28*, 8016–8020. [CrossRef]

4. Kim, J.H.; Lee, S.B.; Kim, S.J.; Lee, Y.M. Rapid temperature/pH response of porous alginate-g-poly (N-isopropylacrylamide) hydrogels. *Polymer* **2002**, *43*, 7549–7558. [CrossRef]

5. Haq, M.A.; Su, Y.; Wang, D. Mechanical properties of PNIPAM based hydrogels: A review. *Mater. Sci. Eng. C* **2017**, *70*, 842–855. [CrossRef] [PubMed]

6. Hirokawa, Y.; Tanaka, T. Volume phase transition in a non-ionic gel. In Proceedings of the AIP Conference, Steamboat Springs, CO, USA, 23–30 May 1984.

7. Ogata, T.; Nonaka, T.; Kurihara, S. Permeation of solutes with different molecular size and hydrophobicity through the poly (vinyl alcohol)-graft-N-isopropylacrylamide copolymer membrane. *J. Membr. Sci.* **1995**, *103*, 159–165. [CrossRef]

8. Hoffman, A.S. Environmentally sensitive polymers and hydrogels. *MRS Bull.* **1991**, *16*, 42–46. [CrossRef]

9. Schild, H.G. Poly (N-isopropylacrylamide): Experiment, theory and application. *Prog. Polym. Sci.* **1992**, *17*, 163–249. [CrossRef]

10. Kuckling, D.; Wohlrab, S. Synthesis and characterization of biresponsive graft copolymer gels. *Polymer* **2002**, *43*, 1533–1536. [CrossRef]

11. Safont, B.; Vitas, A.; Peñas, F. Biodegradation of phenol in a draft-tube spouted bed bioreactor with biomass attached to hydrogel particles. In *Water Resources Management V*; WIT Press: Southampton, UK, 2009; pp. 147–156.

12. Ali, W.; Gebert, B.; Hennecke, T.; Graf, K.; Ulbricht, M.; Gutmann, J.S. Design of thermally responsive polymeric hydrogels for brackish water desalination: Effect of architecture on swelling, deswelling, and salt rejection. *ACS Appl. Mater. Interfaces* **2015**, *7*, 15696–15706. [CrossRef] [PubMed]

13. Ashraf, S.; Park, H.-K.; Park, H.; Lee, S.-H. Snapshot of phase transition in thermoresponsive hydrogel PNIPAM: Role in drug delivery and tissue engineering. *Macromol. Res.* **2016**, *24*, 297–304. [CrossRef]

14. Klouda, L. Thermoresponsive hydrogels in biomedical applications: A seven-year update. *Eur. J. Pharm. Biopharm.* **2015**, *97*, 338–349. [CrossRef] [PubMed]

15. Díez-Peña, E.; Frutos, P.; Frutos, G.; Quijada-Garrido, I.; Barrales-Rienda, J.M. The influence of the copolymer composition on the diltiazem hydrochloride release from a series of pH-sensitive poly [(N-isopropylacrylamide)-co-(methacrylic acid)] hydrogels. *AAPS PharmSciTech* **2004**, *5*, 69–76. [CrossRef] [PubMed]

16. Walle, U.K.; Galijatovic, A.; Walle, T. Transport of the flavonoid chrysin and its conjugated metabolites by the human intestinal cell line Caco-2. *Biochem. Pharmacol.* **1999**, *58*, 431–438. [CrossRef]

17. Comalada, M.; Ballester, I.; Bailón, E.; Sierra, S.; Xaus, J.; Gálvez, J.; de Medina, F.S.; Zarzuelo, A. Inhibition of pro-inflammatory markers in primary bone marrow-derived mouse macrophages by naturally occurring flavonoids: Analysis of the structure–activity relationship. *Biochem. Pharmacol.* **2006**, *72*, 1010–1021. [CrossRef] [PubMed]

18. Wolfman, C.; Viola, H.; Paladini, A.; Dajas, F.; Medina, J.H. Possible anxiolytic effects of chrysin, a central benzodiazepine receptor ligand isolated from Passiflora coerulea. *Pharmacol. Biochem. Behav.* **1994**, *47*, 1–4. [CrossRef]

19. Zeng, Y.-B.; Yang, N.; Liu, W.S.; Tang, N. Synthesis, characterization and DNA-binding properties of La (III) complex of chrysin. *J. Inorg. Biochem.* **2003**, *97*, 258–264. [CrossRef]

20. Nielsen, S.; Breinholt, V.; Justesen, U.; Cornett, C.; Dragsted, L.O. In vitro biotransformation of flavonoids by rat liver microsomes. *Xenobiotica* **1998**, *28*, 389–401. [CrossRef] [PubMed]

21. Xue, W.; Champ, S.; Huglin, M.B. Network and swelling parameters of chemically crosslinked thermoreversible hydrogels. *Polymer* **2001**, *42*, 3665–3669. [CrossRef]

22. Velada, J.L.; Liu, Y.; Huglin, M.B. Effect of pH on the swelling behaviour of hydrogels based on N-isopropylacrylamide with acidic comonomers. *Macromol. Chem. Phys.* **1998**, *199*, 1127–1134. [CrossRef]

23. Yoo, M.; Sung, Y.K.; Lee, Y.M.; Cho, C.S. Effect of polyelectrolyte on the lower critical solution temperature of poly (N-isopropyl acrylamide) in the poly (NIPAAm-co-acrylic acid) hydrogel. *Polymer* **2000**, *41*, 5713–5719. [CrossRef]

24. Kim, S.; Healy, K.E. Synthesis and characterization of injectable poly (N-isopropylacrylamide-co-acrylic acid) hydrogels with proteolytically degradable cross-links. *Biomacromolecules* **2003**, *4*, 1214–1223. [CrossRef] [PubMed]

25. Chen, G.; Hoffman, A.S. Temperature-induced phase transition behaviors of random vs. graft copolymers of *N*-isopropylacrylamide and acrylic acid. *Macromol. Rapid Commun.* **1995**, *16*, 175–182. [CrossRef]

26. Jones, M. Effect of pH on the lower critical solution temperatures of random copolymers of *N*-isopropylacrylamide and acrylic acid. *Eur. Polym. J.* **1999**, *35*, 795–801. [CrossRef]

27. Zhang, J.; Peppas, N.A. Synthesis and characterization of pH-and temperature-sensitive poly (methacrylic acid)/poly (*N*-isopropylacrylamide) interpenetrating polymeric networks. *Macromolecules* **2000**, *33*, 102–107. [CrossRef]

28. Qiu, Y.; Park, K. Environment-sensitive hydrogels for drug delivery. *Adv. Drug Deliv. Rev.* **2001**, *53*, 321–339. [CrossRef]

29. Brazel, C.S.; Peppas, N.A. Pulsatile local delivery of thrombolytic and antithrombotic agents using poly (*N*-isopropylacrylamide-co-methacrylic acid) hydrogels. *J. Control. Release* **1996**, *39*, 57–64. [CrossRef]

30. Tang, S.; Bhandari, R.; Delaney, S.P.; Munson, E.J.; Dziubla, T.D.; Hilt, J.Z. Synthesis and characterization of thermally responsive *N*-isopropylacrylamide hydrogels copolymerized with novel hydrophobic polyphenolic crosslinkers. *Mater. Today Commun.* **2017**, *10*, 46–53. [CrossRef] [PubMed]

31. Gan, J.; Guan, X.X.; Zheng, J.; Guo, H.; Wu, K.; Liang, L.; Lu, M. Biodegradable, thermoresponsive PNIPAM-based hydrogel scaffolds for the sustained release of levofloxacin. *RSC Adv.* **2016**, *6*, 32967–32978. [CrossRef]

32. Feil, H.; Bae, Y.H.; Feijen, J.; Kim, S.W. Effect of comonomer hydrophilicity and ionization on the lower critical solution temperature of *N*-isopropylacrylamide copolymers. *Macromolecules* **1993**, *26*, 2496–2500. [CrossRef]

33. Xue, W.; Hamley, I.W. Thermoreversible swelling behaviour of hydrogels based on *N*-isopropylacrylamide with a hydrophobic comonomer. *Polymer* **2002**, *43*, 3069–3077. [CrossRef]

34. Singh, R.; Deshmukh, S.A.; Kamath, G.; Sankaranarayanan, S.K.R.S.; Balasubramanian, G. Controlling the aqueous solubility of PNIPAM with hydrophobic molecular units. *Comput. Mater. Sci.* **2017**, *126*, 191–203. [CrossRef]

35. Wei, Y.Y.; Liu, Z.; Ju, X.; Shi, K.; Xie, R.; Wang, W.; Cheng, Z.; Chu, L. Gamma-Cyclodextrin-Recognition-Responsive Characteristics of Poly (N-isopropylacrylamide)-Based Hydrogels with Benzo-12-crown-4 Units as Signal Receptors. *Macromol. Chem. Phys.* **2016**, *218*. [CrossRef]

36. Zhang, X.-Z.; Sun, G.M.; Wu, D.Q.; Chu, C.C. Synthesis and characterization of partially biodegradable and thermosensitive hydrogel. *J. Mater. Sci. Mater. Med.* **2004**, *15*, 865–875. [CrossRef] [PubMed]

37. Constantin, M.; Cristea, M.; Ascenzi, P.; Fundueanu, G. Lower critical solution temperature versus volume phase transition temperature in thermoresponsive drug delivery systems. *Express Polym. Lett.* **2011**, *5*, 839–848. [CrossRef]

38. Boutris, C.; Chatzi, E.; Kiparissides, C. Characterization of the LCST behaviour of aqueous poly (*N*-isopropylacrylamide) solutions by thermal and cloud point techniques. *Polymer* **1997**, *38*, 2567–2570. [CrossRef]

39. Gupta, P.; Jordan, C.T.; Mitov, M.I.; Butterfield, D.A.; Hilt, J.Z.; Dziubla, T.D. Controlled curcumin release via conjugation into PBAE nanogels enhances mitochondrial protection against oxidative stress. *Int. J. Pharm.* **2016**, *511*, 1012–1021. [CrossRef] [PubMed]

40. Gupta, P.; Authimoolam, S.P.; Hilt, J.Z.; Dziubla, T.D. Quercetin conjugated poly (β-amino esters) nanogels for the treatment of cellular oxidative stress. *Acta Biomater.* **2015**, *27*, 194–204. [CrossRef] [PubMed]

![gels logo] *gels*

MDPI

Article

Synthesis and Characterization of New Functional Photo Cross-Linkable Smart Polymers Containing Vanillin Derivatives

Momen S.A. Abdelaty [1,2] and Dirk Kuckling [1,*]

[1] Chemistry Department, Paderborn University, Warburger Straße 100, D-33098 Paderborn, Germany
[2] Polymer Lap, Chemistry Department, Faculty of Science, Al-Azhar University, Assiut 71524, Egypt; momensayed2007@yahoo.com
* Correspondence: dirk.kuckling@uni-paderborn.de; Tel.: +49-5251-602-171; Fax: +49-5251-603-245

Academic Editor: David Díaz Díaz
Received: 2 October 2015; Accepted: 18 December 2015; Published: 14 January 2016

Abstract: The synthesis of new functional monomers based on vanillin is reported. The monomers further were used in the synthesis of different temperature-responsive photo cross-linkable polymers via free radical polymerization with N-isopropyl acrylamide and a maleimide photo cross-linker. These polymers were characterized by NMR, FTIR and UV spectroscopy, as well as gel permeation chromatography (GPC) and differential scanning calorimetry (DSC). Critical solution temperatures were determined by UV spectroscopy. Hydrogel thin films were formed by spin coating of a polymer solution over gold with adhesion promotor followed by cross-linking by UV irradiation. The swelling properties were determined by surface plasmon resonance coupled with optical waveguide spectroscopy. The swelling behavior of the hydrogel films was determined as a function of temperature. The incorporation of a dialkyl amino group compensated the hydrophobic effect of the vanillin monomer. Transition temperatures in the physiological range could be obtained.

Keywords: vanillin; N-isopropyl acrylamide; photo cross-linking; hydrogel films; temperature responsive; swelling; lower critical solution temperature; surface plasmon resonance

---

1. Introduction

Smart polymers or stimuli-responsive polymers have been widely used due to their practical interests, and several types of responsiveness, including temperature, pH, light, pressure, magnetic and electrical fields, have been reported [1]. These kinds of materials were known in nature in some examples, like the leaves of Mimosa pudica and Venus flytrap [2]. These polymers and hydrogels are responsive towards more than one stimulus, but the majority of studies have focused on pH and temperature variation [1,3–7]. This has been achieved by a combination of ionizable and hydrophobic functional groups [8–11]. Several authors have recently presented their advances in this field [12–26]. The lower critical solution temperature (LCST) behavior is associated with a critical temperature (T_c) at which the polymer solution undergoes phase separation from one phase (isotropic state) to two phases (anisotropic state), rich and poor in polymer, respectively [27]. Below T_c, the enthalpy term of the free energy of mixing, related to the hydrogen bonding between the polymer and the water molecules, is responsible for the polymer dissolution. When raising the temperature above T_c, the entropy term (hydrophobic interactions) dominates, leading to polymer precipitation. T_c of polymers in aqueous solutions can be modulated by incorporating hydrophilic or hydrophobic moieties [27–31]. For example, when N-isopropyl acrylamide (NIPAAm) is copolymerized with hydrophilic monomers, such as acryl amide (AAm), the T_c increases up to about 45 °C when 18 mol %

of AAm is incorporated with the polymer; whereas T_c decreases to about 10 °C when 40 mol % of hydrophobic *N*-tert-butylacrylamide (NtBAAm) is added to the polymer [29].

Functional hydrogels consisting of at least one polymer segment that enables polymer analogous reactions are of increasing importance. Possible routes utilize active ester chemistry, click chemistry, as well as reactions with polymeric anhydrides, epoxides, aldehydes and ketones. Including Michael-type and Friedel–Crafts reactions, as well as methylations, polymer analogous reactions involve almost all high yield organic reactions [32–35]. The reactive sites allow the modification with a variety of biomolecules for biosensor applications and chromatography based on affinity binding [36–40]; additionally, attaching molecular recognition sites diversified cell culturing, and polymer-assisted drug delivery advanced medication [41]. As mentioned above, responsive behavior can be implemented to create a functional responsive hydrogel or to diversify the responsiveness, *i.e.*, to a magnetic field by incorporating magnetite nanoparticles [42,43]. Recent advances in nanotechnology have increased interest in hydrogel thin films. The advantages of hydrogel thin films have been explored for the fabrication of miniaturized devices with fast response times. Hydrogel thin films have also attracted interest as an approach to responsive surfaces and interfaces, where they compete with grafted polymer layers [44,45]. A 3D polymer network is much more stable at interfaces when compared to polymer brushes, where polymer chains are grafted to the surface via only one functional group, while the polymer network is linked to the surface by multiple anchoring points [2,46]. Surface plasmon resonance spectroscopy coupled with optical waveguide spectroscopy (SPR/OWS) has been widely used in the determination of the LCST behavior of hydrogels [47–51]. Many applications of stimuli-responsive hydrogels have been used in biotechnology, like purification of biomolecules [52–56], switchable wettability [57–61] and sensors (biosensors) [62,63].

Vanillin is a widely-used flavoring, e.g., for food, fragrances and beverages. For sustainability purposes, vanillin can be extracted from natural resources [64] or mass-produced from lignin as a biosource [65–68]. Vanillin has hydroxyl and aldehyde reactive sites that can be used for chemical modification to produce valuable monomers that can be polymerized into materials with different mechanical and thermal properties [69,70]. Furthermore, vanillin has been converted into vinyl monomers for vinyl ester resins [69,71,72]. By this approach, the resulting vanillin-based monomers possess a reactive aldehyde functionality. Here, we describe the synthesis of different vanillin-based monomers and evaluate their influence on the thermal transition of temperature-responsive polymers. The utilization of the reactive aldehyde group will be reported in a future work.

2. Results and Discussion

2.1. Synthesis of Functional Monomers and Polymers

The new monomers 4-formyl-2-methoxyphenylacrylate (VA) **2**, 2-((diethylamino)methyl)-4-formyl-6-methoxyphenyl acrylate (DEAMVA) **4a** and 2-((dimethylamino)methyl)-4-formyl-6-methoxyphenyl acrylate (DMAMVA) **4b** were synthesized in one or two steps, as shown in Scheme 1. The first step was the reaction of vanillin with diethylamine for DEAMVA or dimethylamine for DMAMVA and formaldehyde according to a Mannich reaction mechanism. In this reaction, we did not use any catalysis, especially acid catalysis, which are normally used in Mannich reactions. The second step was the reaction with acryloyl chloride in the presence of triethylamine (TEA) or NaOH to form the respective esters. All compounds have been analyzed by ^{1}H and ^{13}C NMR, as well as IR spectroscopy, and they were in a good agreement with the chemical structures. Details are presented in the experimental part.

The free radical polymerization of NIPAAm and VA, DEAMVA or DMAMVA with dimethylmaleimidoacrylate (DMIA) as a photo-reactive moiety was used to synthesize functional temperature-responsive polymers. Different mole ratios of VA, DEAMVA or DMAMVA were used to probe the influence of the relative content. A scheme of the synthesis of the respective photo cross-linkable polymers with vanillin-based monomer DEAMVA is shown in Scheme 2. The polymers are denominated as VA-05-10, whereas VA is the type of vanillin monomer (DE: DEAMVA; DM: DMAMVA); the first number denominates the feed content of the photo cross-linker, and the second number denominates the feed content of the vanillin derivative.

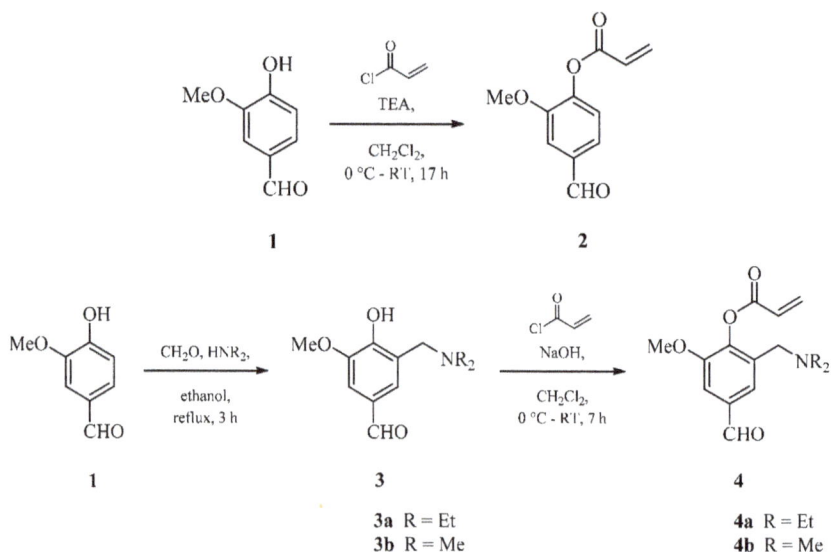

Scheme 1. Scheme of the synthesis of vanillin-based monomers (4-formyl-2-methoxyphenylacrylate (VA) **2**, 2-((diethylamino)methyl)-4-formyl-6-methoxyphenyl acrylate (DEAMVA) **4a** and 2-((dimethylamino)methyl)-4-formyl-6-methoxyphenyl acrylate (DMAMVA) **4b**).

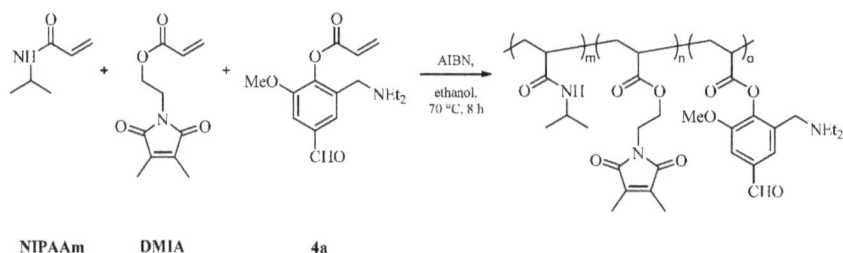

Scheme 2. Scheme of the synthesis of the respective photo cross-linkable polymers with vanillin-based monomer DEAMVA. NIPAAm, *N*-isopropyl acrylamide; DMIA, dimethylmaleimidoacrylate.

The ^{1}H NMR was performed to determine the actual amounts of DMIA and VA, DEAMVA or DMAMVA in the polymer chain. For this, the integrals of two methyl groups of NIPAAm at 0.76–1.30 ppm, of the two methyl groups of DMIA at 1.89–2.02 ppm and of one hydrogen of the aldehyde group at 9.79–10.10 ppm were used. Representative ^{1}H NMR spectra are presented in Figure 1. It can clearly be seen that the aldehyde functional group is still present after copolymerization. In addition, comparable copolymers could be modified with primary amines by imine formation. However, it cannot be excluded that some of the aldehyde functions are destroyed, resulting in relatively low aldehyde content in the polymer. Oppositely, although copolymer parameters are not known, from the comparison of the copolymerization parameters of related structures [73], one can conclude that the phenyl acrylates are less reactive than acrylamides. However, due to the overlap of the signals in the ^{1}H NMR spectra, exact calculations were very difficult. The DMIA content was determined with UV spectroscopy, as well. All calculations were in a logic case with the feeding calculation, as shown in Table 1.

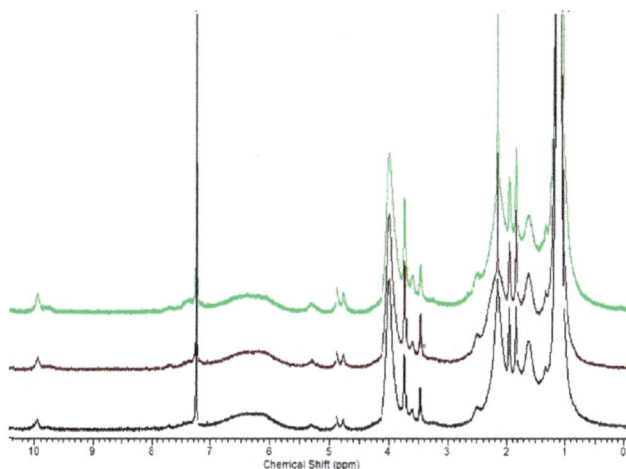

Figure 1. ^{1}H NMR spectra of DEAMVA (DE)-05-10 (black), DE-05-15 (brown) and DE-05-20 (green).

Table 1. Composition of functional photo cross-linkable smart polymers based on vanillin derivatives VA, DEAMVA and DMAMVA (DM).

Polymer	DMIA (mol %)		2 (mol %)	4a (mol %)	4b (mol %)	Yield (%)
	^{1}H NMR	UV				
VA-05-10	5.9	6.0	11.1			83
DE-05-10	4.3	4.2		3.0		60
DE-05-15	4.5	4.7		4.8		57
DE-05-20	4.8	4.8		6.5		58
DM-05-10	4.6	4.7			3.6	66
DM-05-15	4.3	4.4			4.2	56
DM-05-20	4.5	4.5			4.7	43

The number average molecular weight and polydispersity were measured by gel permeation chromatography (GPC). The glass transition temperature T_g, which is important to form a rigid film for photo cross-linking, increased with increasing DEAMVA or DMAMVA composition. The phase transition temperature T_c was determined for all polymers in water. The phase separation for each polymer was studied by UV-VIS spectroscopy. The results are summarized in Table 2.

The T_c of poly(N-isopropyl acrylamide) (PNIPAAm) is readily influenced by hydrophilic or hydrophobic comonomers [11]. Hence, each modification of PNIPAAm might lead to changed T_c values. Here, the introduction of 11 mol % of the simple vanillin-based monomer **2** led to a decrease of T_c of about 14 °C compared to the non-modified photo cross-linkable polymer [48]. In order to overcome this shift, vanillin was modified with a methyl amino group having two alkyl groups of different hydrophobicity. The amino group will be slightly charged at weak acid or neutral pH values, which leads to a compensation of the hydrophobic effect. Indeed, the copolymers with DEAMVA and DMAMVA, respectively, showed increased T_c values in water. No clear difference can be observed for diethyl amino or dimethyl amino modification. The turbidity curve in water showed quite broad transitions, making the determination of T_c more difficult (Figure 2). This can be attributed to the chemical composition distribution of the polymer chains. Nevertheless, the incorporation of the methyl amino groups increased T_c by approximately 20 °C. The strong increase might be also due to some additional carboxylic groups resulting from the oxidation of aldehyde groups. Additionally, the copolymers showed quite complex pH behavior. This fact is still under investigation. As a conclusion, these photo cross-linkable polymers should be used at neutral to acidic pH only.

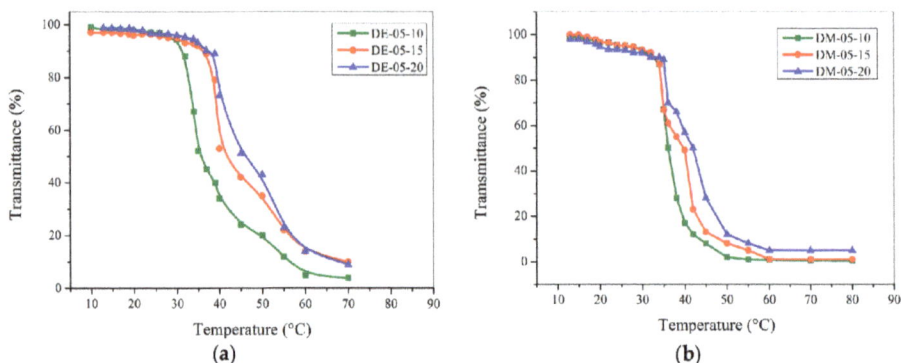

Figure 2. Turbidity measurements of functional thermo-responsive polymers for: (**a**) DEAMVA in water; (**b**) DMAMVA in water (1 wt % of polymer solution).

Table 2. Characterization of functional photo cross-linkable smart polymers based on vanillin derivatives VA, DEAMVA and DMAMVA.

Polymer	M_n (g/mol)	PD	T_g (°C)	T_c (°C)
VA-05-10	20,700	3.1	133.5	13.8
DE-05-10	7600	1.7	131.0	35.0
DE-05-15	5600	1.6	145.0	42.5
DE-05-20	6100	1.5	146.0	46.6
DM-05-10	6500	2.0	128.5	36.0
DM-05-15	5500	1.8	134.0	39.0
DM-05-20	4900	1.9	135.8	35.0

2.2. Photo Cross-Linking and Formation of Hydrogel Thin Film

The formation of a hydrogel layer of vanillin-based copolymers was obtained by photo cross-linking through [2 + 2] cyclodimerization of the dimethyl maleimide moieties. In order to investigate the swelling properties by surface plasmon resonance (SPR) spectroscopy, gold-coated LaSFN9 glass with an approximately 50-nm Au film thickness was used as the support. The gold slide was immersed in a solution of 5 mM thioacetic acid 3-(3,4-dDimethyl-2,5-dioxo-2,5-dihydro-pyrrol-yl)-Ppropyl ester (DMITAc) adhesion promoter. Then, the polymer solution was spin coated. It was observed that the used solvent affects the homogeneity of the surface. Solvents like cyclohexanone worked fine, whereas DMSO or DMF failed to form homogeneous films. Films were prepared in a manner such that the dry films were thick enough (approximately 300–500 nm) to show the first waveguide mode (Figure 3a). This is important, since the curve can be fitted to the plasmon minimum, as well as waveguide mode to determine the film thickness, as well as the refractive index of the dry layer. The photo cross-linking has been done in the near UV (300–430 nm). For this reason, the photosensitizer thioxanthone was always used. It transfers the energy of absorbed light to the polymer's photo-reactive moieties, causing them to enter the excited state and perform the cross-linking reaction. Factors that influenced the photo cross-linking are irradiation wavelength and intensity, reaction time, composition of the photo-reactive polymers and the presence of the photo sensitizer. The presence of the adhesion promoter resulted in surface attachment of the hydrogel to the substrate. The maleimide group in the adhesion promoter reacted with a similar group present in the spin-coated photo cross-linkable polymer through the above-mentioned [2 + 2] cyclodimerization. The Au–S bond, on the other hand, resulted in the coordination covalent bond attachment of the hydrogel to the substrate.

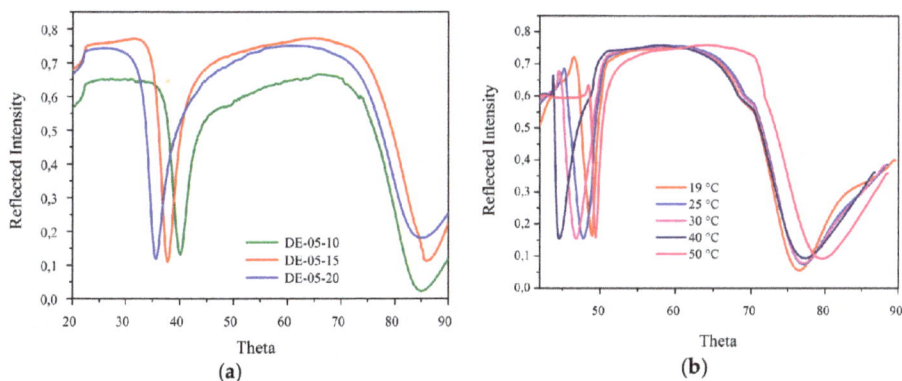

Figure 3. SPR scans of photo cross-linked polymer layers containing DEAMVA: (**a**) in the dry state for different DEAMVA contents; (**b**) in the swollen state for DE-05-15 at different temperatures.

The surface plasmon resonance measurements were carried out according to the Kretschmann configuration. The LaSFN9 glass was optically matched to the base of the glass prism. Monochromatic light from a He/Ne laser at 633 nm was directed through the prism. The external angle of incidence (θ) was varied with a goniometer, and then, the light was collected with a photodiode. The reflected intensity *versus* angle was recorded (Figure 3). The angle-dependent reflectivity was described by the Fresnel equation for a multilayer system [47–49]. The base system consisted of LaSFN9 glass, gold, adhesion promoter (DMITAc) and air. In the case of dry polymer, an additional layer for the thin film was added. For the swelling case, the refractive index of deionized water and buffer solution, respectively, was used to model the continuous layer. During swelling and collapse, the refractive index and thickness of the gel layers changed simultaneously. This can be seen by a shift of the plasmon minimum, as well as the waveguide mode still present in the swollen state. Representative spectra are shown in Figure 3b. Waveguide modes can be guided within layers thicker than 200 nm in the dry state and 500 nm in the swollen state. Hence, the combination of SPR and OWS provided an additional feature to determine the layer thickness, as well as the refractive index of the hydrogel simultaneously. From the refractive index, the volume degree of swelling can be estimated [48]. Temperature control was achieved by placing the sample on a hot stage, on which the stage was heated and cooled, resistively, by a Peltier element to adjust the temperature within 15–55 °C.

Figure 4 shows the swelling behavior of different hydrogel layers. A decrease in the volume degree of swelling with rising temperature occurred. As already seen for the soluble polymers, the temperature-induced phase transition was more pronounced for copolymers with vanillin derivatives containing the dimethyl amino groups. With higher contents of DEAMVA in the hydrogel layer, the swelling was decreased. It turned out that the chemical composition distribution caused a broad transition range. This might be due to π–π-stacking of vanillin moieties, resulting in additional dynamic cross-links. Furthermore, carboxylic groups resulting from the accidental decomposition of the aldehyde group can interact with the dialkyl amino groups by Coulomb interactions, again forming additional junction points. With the increase of the DEAMVA content, the possibility of those interactions increases, leading to a decreased swelling. In the case of DMAMVA, the oxidation did not take part to such an extent. Hence, the swelling transition is more pronounced. To prove this hypothesis, a different synthetic approach had to be developed to prevent decomposition of the aldehyde group.

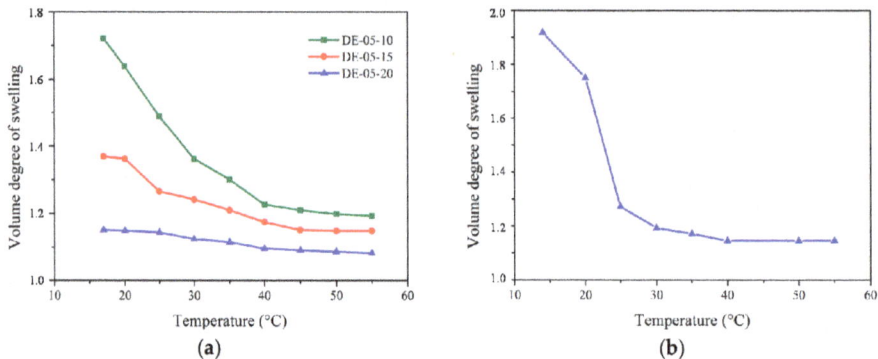

Figure 4. Volume degree of swelling *vs.* temperature of photo cross-linked hydrogels in water; (a) DE-05-10, DE-05-15 and DE-05-20; (b) DM-05-20.

3. Conclusions

In this work, new monomers based on vanillin were copolymerized with NIPAAm and DMIA to obtain photo cross-linkable polymers. Surface-attached photo cross-linked thin gel layers were prepared. Their transition temperature T_c were determined for both soluble polymers and cross-linked gels. The swelling behavior of the thin polymer gel layers was investigated by a combination of surface plasmon resonance (SPR) and optical waveguide spectroscopy (OWS). The incorporation of methyl dialkyl amino functionality could compensate the hydrophobic character of the vanillin backbone with respect to the influence of T_c. Transition temperatures in the physiological range could be obtained. The polymers possess aldehyde functionalities that in the future will be used to bind reversibly-active compounds.

4. Experimental Section

4.1. Instrumentation

^{1}H and ^{13}C NMR spectra were recorded on a Bruker AV 500 spectrometer in CDCl$_3$ at 500 MHz and 125 MHz, respectively. The solvent signal was used as the internal standard. IR spectra were recorded on the Vertex 70 Fourier transform infrared instrument. The samples were milled with KBr and pressed into pellets.

Molecular weights and polydispersity (PD) were analyzed employing size exclusion chromatography (SEC). As eluent, chloroform (containing 0.1 vol % triethylamine) with a flow rate of 0.75 mL/min (Jasco 880-PU pump) was used with a RI-Detector (Waters, Milford, MA, USA) and toluene as the internal standard at 30 °C. The samples (15 mg/mL) were injected by hand via a 20-μL loop. PSS-SDV columns (Polymer Standards Service, Germany) filled with 5-μm gel particles with a defined porosity of 10^6 Å (guard), 10^5 Å, 10^3 Å and 10^2 Å, respectively, were used. Molecular weight determination was based on narrow polystyrene standards. The Differential Scanning Calorimeter Pyris 1 (Perkin Elmer, Waltham, MA, USA) was used for the determination of T_g of solid polymers. The thermogram was recorded at a heating and cooling rate of 5 °C/min. UV-VIS spectra were recorded on a Lambda 45 Spectrometer (Perkin Elmer). For the T_c determination, a polymer solution of 1 wt % in water was used. As T_c, the value at 50% transmission was used. The spin coater model G3P-8 SpinCoat (Speciality Coating System, Inc., Amherst, NH, USA) was used to prepare thin polymer films on the substrates with 250 rpm for 30 s and 1000 rpm for 140 s (polymer concentration 10 wt %). A UV Hg-lamp (OSRAM, Munich, Germany, 100 W) equipped with an optical lens and a mirror was used for the photo cross-linking process. An SPR along with OWS setup (Restech, Munich, Germany) was used to determine the film thickness and swelling ratios of photo cross-linked thin films. A He–Ne laser beam with a wavelength of 632.8 nm was used for excitation of SPs in the Kretschmann configuration. The substrate was LaSFN9 glass slides with an approximately 50-nm gold film. The Au film was deposited by PVD (physical vapor deposition, tectra GmbH, Frankfurt am Main, Germany). After adsorption of the DMITAc adhesion promoter on Au, photo cross-linkable polymer solution was spin coated and irradiated with UV light. For measuring temperature-dependent swelling, 3–4 mL of distilled water were injected manually into the SPR cell (the SPR cell consists of a sample holder, LaSFN9 glass with gold coating, a prism and a Peltier element to maintain the temperature). The temperature inside the cell was measured with a thermocouple of 0.1 °C accuracy.

4.2. Materials

N-isopropyl acrylamide (NIPAAm; Acrōs, Geel, Belgium) was recrystallized from distilled hexane. Vanillin (99% Acrōs), dimethylamine and diethylamine (Merck, Darmstadt, Germany), triethylamine (Merck), 2-hydroxyethylamine (Acrōs), dimethyl maleic anhydride (98%, Aldrich, St. Louis, USA) and acryloyl chloride (AcrCl, Merck) were used as received. 2,2′-Azobis(isobutyronitrile) (AIBN) was recrystallized from methanol. Dioxane, tetrahydrofuran (THF) and diethyl ether were distilled over potassium hydroxide.

4.2.1. Synthesis of the Dimethylmaleimidoacrylate Photo Cross-Linker

The dimethylmaleimidoacrylate (DMIA) monomer was prepared according to the literature [74].

4.2.2. Synthesis of Thioacetic Acid 3-(3,4-Dimethyl-2,5-dioxo-2,5-dihydro-pyrrol-yl)-Propyl Ester Adhesion Promoter

The DMITAc was prepared according to the literature [47].

4.2.3. Synthesis of 4-Formyl-2-Methoxyphenylacrylate

In a two neck flask fitted with an argon balloon, vanillin (4-hydroxy-3-methoxy benzaldehyde) (8 g, 0.052 mol) was dissolved in dry CH_2Cl_2 (100 mL). Under vigorous stirring, TEA (10.52 g, 0.1 mol) was added. The reaction mixture was cooled in an ice bath to 0–5 °C. Acryloyl chloride (5.4 g, 0.059 mol) was added drop wise. The yellowish suspension was stirred at 5 °C for 1 h, then allowed to stir at RT overnight. The precipitate was filtered, and the solvent was evaporated under reduced pressure. The product was extracted by CH_2Cl_2 and washed three times with distilled water, one time with sodium carbonate and one time with 0.1 M HCl. The organic phase was dried with $MgSO_4$ overnight, then filtered, and the product was distilled using an oil pump; yield 85%; the colorless oil changed to a white, pasty solid after cooling overnight in a refrigerator.

^{1}H NMR (500 MHz, CDCl$_3$): δ (ppm) = 3.73 (s, 3 H, OCH$_3$), 5.92 (dd, ^2J = 0.7 Hz, ^3J = 10.5 Hz, 1 H, =CH$_2$), 6.23 (dd, ^3J = 10.5 Hz, ^3J = 17.3 Hz, 1 H, =CH), 6.49 (dd, ^2J =0.7 Hz, ^3J = 17.3 Hz, 1 H, =CH$_2$), 7.12 (d, ^3J = 8.0 Hz, 1 H, 6-Ar–CH), 7.34 (dd, ^3J = 8.0 Hz, ^4J = 1.5 Hz, 1 H, 5-Ar–CH), 7.37 (d, ^4J = 1.3 Hz, 1 H, 3-Ar–CH), 9.96 (s, 1 H, CHO).

^{13}C NMR (125 MHz, CDCl$_3$): δ (ppm) = 55.91 (1 C, OCH$_3$), 111.09 (1 C, 3-Ar–CH), 123.36 (1 C, 5-Ar–CH), 124.24 (1 C, 6-Ar–CH), 127.08 (1 C, =CH), 133.16 (1 C, =CH$_2$), 135.26 (1 C, 4-Ar–C), 144.62 (1 C, 1-Ar–CH), 151.09 (1 C, 2-Ar–C), 163.22 (1 C, COO), 190.99 (1 C, CHO).

IR (KBr): v (cm^{-1}) = 2970–2950 (s) (CH$_2$, CH$_3$), 2840 (m) (OCH$_3$), 1745 (s) (COO), 1695 (s) (CHO), 1600 (s) (C=C), 885–784 (m) (Ar–CH).

4.2.4. Synthesis of 2-((Diethylamino)methyl)-4-Formyl-6-Methoxyphenyl Acrylate (**2**)

Step 1: Synthesis of 3-((diethylamino)methyl)-4-hydroxy-5-methoxy-benzaldehyde (**3a**).

In a 250-mL round-bottomed flask fitted with a reflux condenser, vanillin (4-hydroxy-3-methoxy benzaldehyde) (10.0 g, 0.065 mol), formaldehyde (10.0 g, 0.33 mol) and diethylamine (10.0 g, 0.136 mol) were dissolved in ethanol (150 mL). The mixture was refluxed for 3 h. Then, the mixture was allowed to cool to room temperature. The solvent was removed under reduced pressure to collect the product; yield 97%; yellowish white solid.

^{1}H NMR (500 MHz, CDCl$_3$): δ (ppm) = 1.18 (t, ^3J = 7.2 Hz, 6 H, CH$_3$), 2.73 (q, ^3J = 7.2 Hz, 4 H, C̲H$_2$CH$_3$), 3.92 (s, 2H, NCH$_2$), 3.94 (s, 3 H, OCH$_3$), 7.25 (br, 1 H, 2- or 6-Ar–CH), 7.34 (d, ^4J = 1.6 Hz, 1 H, 2- or 6-Ar–CH), 9.78 (s, 1 H, CHO).

^{13}C NMR (125 MHz, CDCl$_3$): δ (ppm) = 10.82 (2 C, CH$_3$), 46.35 (2 C, C̲H$_2$CH$_3$), 55.85 (1 C, NCH$_2$), 56.01 (1 C, OCH$_3$),109.68 (1 C, 6-Ar–CH), 120.84 (1 C, 2-Ar–CH), 125.75 (1 C, 3-Ar–C), 127.99 (1 C, 1-Ar–C), 148.65 (1 C, 4-Ar–C), 154.87 (1 C, 5-Ar–C), 191.65 (1 C, CHO).

IR (KBr): ν (cm^{-1}) = 2987 (s) (CH$_2$, CH$_3$), 1706 (s) (C=O), 1650 (s) (C=C), 868–820 (m) (Ar–CH).

Step 2: Synthesis of 2-((diethylamino)methyl)-4-formyl-6-methoxyphenyl acrylate (**4a**) (DEMAVA).

In a two-neck flask fitted with an argon balloon, **1** (13.9 g, 0.058 mol) was dissolved in dry CH$_2$Cl$_2$ (200 mL) and stirred strongly, and sodium hydroxide (10.0 g, 0.25 mol) was added. The reaction mixture was allowed to cool in an ice bath to 0–5 °C. Then acryloyl chloride (5.4 g, 0.059 mol) was added drop wise. The yellowish suspension was stirred at 5 °C for 1 h and then allowed stirred at RT for 6 h. The precipitate was filtered, and the solvent was evaporated under reduced pressure. The product was dissolved in CH$_2$Cl$_2$ and washed three times with DI water, one time with 0.1 M Na$_2$CO$_3$ and again with DI water. After drying over MgSO$_4$ overnight, the solvent was removed under reduced pressure to collect the product; yield 75%; orange viscous liquid.

^{1}H NMR (500 MHz, CDCl$_3$): δ (ppm) = 1.01 (t, ^3J = 7.1 Hz, 6 H, CH$_3$), 2.50 (q, ^3J = 7.1 Hz, 4 H, C̲H$_2$CH$_3$), 3,52 (s, 2 H, NCH$_2$), 3.87 (s, 3 H, OCH$_3$), 6.05 (dd, ^2J = 1.2 Hz, ^3J = 10.5 Hz, 1 H, =CH$_2$), 6.36 (dd, ^3J = 10.5 Hz, ^3J = 17.3 Hz, 1 H, =CH), 6.64 (dd, ^2J = 1.2 Hz, ^3J = 17.3 Hz, 1 H, =CH$_2$), 7.39 (d, ^4J = 1.7 Hz, 1 H, 3- or 5-Ar–CH), 7.66 (d, ^4J = 1.7 Hz, 1 H, 3- or 5-Ar–CH), 9.95 (s,1H, CHO).

^{13}C NMR (125 MHz, CDCl$_3$): δ (ppm) = 11.64 (2 C, CH$_3$), 47.00 (2 C, C̲H$_2$CH$_3$), 51.58 (1 C, NCH$_2$), 56.20 (1 C, OCH$_3$), 108.60 (1 C, 5-Ar–CH), 126.45 (1 C, 3-Ar–CH), 127.19 (1 C, =CH), 127.19 (1 C, 2-Ar–C), 130.52 (1 C, 4-Ar–C), 134.52 (1 C, =CH$_2$), 143.49 (1 C, 1-Ar–C), 152.11 (1 C, 6-Ar–C), 163.03 (1 C, COO), 191.52 (1 C, CHO).

IR (KBr): ν(cm^{-1}) = 2915, 2834 (s) (CH$_2$, CH$_3$), 1639 (s) (C=O), 1600 (s) (C=C), 866–820 (m) (Ar–CH).

4.2.5. Synthesis of 2-((Dimethylamino)methyl)-4-Formyl-6-Methoxyphenyl Acrylate (4)

Step 1: Synthesis of 3-((dimethylamino)methyl)-4-hydroxy-5-methoxy-benzaldehyde (**3b**).

In this experiment, dimethylamine was used. The same conditions were used as discussed earlier for DEAMVA Step 1.

^{1}H NMR (500 MHz, CDCl$_3$): δ (ppm) = 2.37 (s, 6 H, NCH$_3$), 3.75 (s, 2 H, NCH$_2$), 3.93 (s, 3 H, OCH$_3$), 6.37 (br, s, 1 H, OH), 7.15 (d, 1 H, ^4J = 1.8 Hz, 2- or 6-Ar–CH), 7.33 (d, 1 H, ^4J = 1.8 Hz, 2- or 6-Ar–CH), 9.76 (s, 1 H, CHO).

^{13}C NMR (125 MHz, CDCl$_3$): δ (ppm) = 44.32 (2 C, NCH$_3$), 56.01 (1 C, NCH$_2$), 62.21 (1 C, OCH$_3$), 109.97 (1 C, 6-Ar–CH), 123.70 (1 C, 2-Ar–CH), 125.21 (1 C, 3-Ar–C), 128.09 (1 C, 1-Ar–C), 148.68 (1 C, 4-Ar–C), 154.54 (1 C,5-Ar–C), 190.67 (1 C, CHO).

IR (KBr): ν (cm^{-1}) = 2977 (s) (CH$_2$, CH$_3$), 1742 (s) (C=O), 1646 (s) (C=C), 865–818 (m) (Ar–CH).

Step 2: Synthesis of 2-((dimethylamino)methyl)-4-formyl-6-methoxyphenyl acrylate (**4b**) (DMAMVA).

The same conditions were used as discussed earlier for DEAMVA Step 2; yield 73%; orange viscous liquid.

^{1}H NMR (500 MHz, CDCl$_3$): δ (ppm) = 2.19 (s, 6 H, CH$_3$), 3.37 (s, 2 H, NCH$_2$), 3.84 (s, 3 H, OCH$_3$), 6.03 (dd, ^2J = 1.1 Hz, ^3J = 10.5 Hz, 1 H, =CH$_2$), 6.34 (dd, ^3J = 10.5 Hz, ^3J = 17.3 Hz, 1 H, =CH), 6.64 (dd, ^2J = 1.1 Hz, ^3J = 17.3 Hz, =CH$_2$), 7.37 (d, 1 H, ^4J = 1.7 Hz, 3- or 5-Ar–CH, 7.34 (d, 1 H, ^4J = 1.7 Hz, 3- or 5-Ar–CH), 9.91 (s,1 H, CHO).

^{13}C NMR (125 MHz, CDCl$_3$): δ (ppm) = 45.42 (2 C, CH$_3$), 55.72 (1 C, NCH$_2$), 62.41 (1 C, OCH$_3$), 108.75 (1 C, 5-Ar–CH), 122.37 (1 C, 3-Ar–CH), 126.25 (1 C, =CH), 127.30 (1 C, 2-Ar–C), 132.22 (1 C, 4-Ar–C), 134.39 (1 C, =CH$_2$), 143.50 (1 C, 1-Ar–C), 152.51 (1 C, 6-Ar–C), 162.81 (1 C, COO), 191.64 (1 C, CHO).

IR (KBr): ν (cm^{-1}) = 2935 (s) (CH$_2$, CH$_3$), 1665 (s) (C=O), 1610 (s) (C=C), 873–826 (m) (Ar–CH).

4.2.6. Synthesis of Photo Cross-Linkable Poly(NIPAAm-*co*-DEAMVA-*co*-DMIA) (5) and Poly(NIPAAm-*co*-DMAMVA-*co*-DMIA) (6)

General procedure: In a 100-mL round-bottomed flask DMIA (5 mol %) was added 2-((diethylamino)methyl)-4-formyl-6-methoxyphenyl acrylate or 2-((dimethylamino)methyl)-4-formyl-6-methoxyphenyl acrylate, respectively (10, 15 and 20 mol %) and NIPAAm (2.00 g, 17.6 mmol) in ethanol (50 mL), and AIBN (ca. 10 mg) was also added. The reaction mixture was purged in argon for 20 min and then heated in an oil bath to 70 °C for 8 h. After cooling to room temperature, the polymer was precipitated in diethyl ether at −50 °C. The precipitate was collected, redissolved in THF and reprecipitated in diethyl ether to remove the unreacted monomers and impurities. Polymerization has also been done in 1,4-dioxane.

Acknowledgments: The authors are thankful to the Egyptian Government and Deutscher Akademischer Austauschdienst (DAAD) for providing a scholarship to Momen S.A. Abdelaty.

Author Contributions: Momen S.A. Abdelaty performed the overall experimental work and contributed to the writing of the manuscript. Dirk Kuckling coordinated the study and contributed to the writing of the manuscript.

Conflicts of Interest: The authors declare no conflict of interest.

References

1. Sato, E.; Masuda, Y.; Kadota, J.; Nishiyama, T.; Horibe, H. Dual stimuli-responsive homopolymers: Thermo- and photo-responsive properties of coumarin-containing polymers in organic solvents. *Eur. Polym. J.* **2015**, *69*, 605–615.
2. Chen, J.-K.; Chang, C.-J. Fabrications and applications of stimulus-responsive polymer films and patterns on surfaces: A review. *Materials* **2014**, *7*, 805–875. [CrossRef]
3. Guenther, M.; Kuckling, D.; Corten, C.; Gerlach, G.; Sorber, J.; Suchaneck, G.; Arndt, K. Chemical sensors based on multiresponsive block copolymer hydrogels. *Sens. Actuat. B Chem.* **2007**, *126*, 97–106. [CrossRef]
4. Matsukuma, D.; Yamamoto, K.; Aoyagi, T. Stimuli-responsive properties of N-isopropylacrylamide-based ultrathin hydrogel films prepared by photo-cross-linking. *Langmuir* **2006**, *22*, 5911–5915. [CrossRef] [PubMed]
5. Chen, Y.; Pang, X.-H.; Dong, C.-M. Dual stimuli-responsive supramolecular polypeptide-based hydrogel and reverse micellar hydrogel mediated by host–guest chemistry. *Adv. Funct. Mater.* **2010**, *20*, 579–586. [CrossRef]
6. Schattling, P.; Jochum, F.D.; Theato, P. Multi-stimuli responsive polymers—The all-in-one talents. *Polym. Chem.* **2014**, *5*, 25–36. [CrossRef]
7. Li, Y.; Zhang, C.; Zhou, Y.; Dong, Y.; Chen, W. Novel multi-responsive polymer materials: When ionic liquids step in. *Eur. Polym. J.* **2015**, *69*, 441–448. [CrossRef]
8. Bulmus, V.; Ding, Z.; Long, C.J.; Stayton, P.S.; Hoffman, A.S. Site-specific polymer-streptavidin bioconjugate for pH-controlled binding and triggered release of biotin. *Bioconj. Chem.* **2000**, *11*, 78–83. [CrossRef]
9. Brazel, C.S.; Peppas, N.A. Synthesis and characterization of thermo- and chemomechanically responsive poly(N-isopropylacrylamide-*co*-methacrylic acid) hydrogels. *Macromolecules* **1995**, *28*, 8016–8020. [CrossRef]
10. Zareie, H.M.; Volga Bulmus, E.; Gunning, A.P.; Hoffman, A.S.; Piskin, E.; Morris, V.J. Investigation of a stimuli-responsive copolymer by atomic force microscopy. *Polymer* **2000**, *41*, 6723–6727. [CrossRef]
11. Kuckling, D.; Adler, H.-J.; Arndt, K.-F.; Ling, L.; Habicher, W.D. Temperature and pH dependent solubility of novel poly(N-isopropylacrylamide) copolymers. *Macromol. Chem. Phys.* **2000**, *201*, 273–280. [CrossRef]
12. Leung, M.F.; Zhu, J.; Harris, F.W.; Li, P. Novel synthesis and properties of smart core-shell microgels. *Macromol. Symp.* **2005**, *226*, 177–186. [CrossRef]

13. Rodríguez-Cabello, C.J.; Reguera, J.; Girotti, A.; Alonso, M.; Testera, A.M. Developing functionality in elastin-like polymers by increasing their molecular complexity: The power of the genetic engineering approach. *Prog. Polym. Sci.* **2005**, *30*, 1119–1145. [CrossRef]

14. Alonso, M.; Reboto, V.; Guiscardo, L.; San Martín, A.; Rodríguez-Cabello, J.C. Spiropyran derivative of an elastin-like bioelastic polymer: Photoresponsive molecular machine to convert sunlight into mechanical work. *Macromolecules* **2000**, *33*, 9480–9482. [CrossRef]

15. Kurata, K.; Dobashi, A. Novel temperature- and pH-responsive linear polymers and crosslinked hydrogels comprised of acidic l-α-amino acid derivatives. *J. Macromol. Sci. A* **2004**, *41*, 143–164. [CrossRef]

16. Ramkissoon-Ganorkar, C.; Baudys, M.; Kim, S.W. Effect of ionic strength on the loading efficiency of model polypeptide/protein drugs in pH-/temperature-sensitive polymers. *J Biomater. Sci. Polym. Ed.* **2000**, *11*, 45–54. [CrossRef] [PubMed]

17. Ju, H.K.; Kim, S.Y.; Kim, S.J.; Lee, Y.M. pH/temperature-responsive semi-ipn hydrogels composed of alginate and poly(*N*-isopropylacrylamide). *J. Appl. Polym. Sci.* **2002**, *83*, 1128–1139. [CrossRef]

18. Benrebouh, A.; Avoce, D.; Zhu, X.X. Thermo- and pH-sensitive polymers containing cholic acid derivatives. *Polymer* **2001**, *42*, 4031–4038. [CrossRef]

19. Ning, L.; Min, Y.; Maolin, Z.; Jiuqiang, L.; Hongfei, H. Radiation synthesis and characterization of polydmaema hydrogel. *Radiat. Phys. Chem.* **2001**, *61*, 69–73. [CrossRef]

20. Gan, L.H.; Gan, Y.Y.; Deen, G.R. Poly(*N*-acryloyl-*N'*-propylpiperazine): A new stimuli-responsive polymer. *Macromolecules* **2000**, *33*, 7893–7897. [CrossRef]

21. Dumitriu, R.P.; Mitchell, G.R.; Vasile, C. Multi-responsive hydrogels based on *N*-isopropylacrylamide and sodium alginate. *Polym. Int.* **2011**, *60*, 222–233. [CrossRef]

22. Chen, D.; Liu, H.; Kobayashi, T.; Yu, H. Multiresponsive reversible gels based on a carboxylic azo polymer. *J. Mater. Chem.* **2010**, *20*, 3610–3614. [CrossRef]

23. Pasparakis, G.; Vamvakaki, M. Multiresponsive polymers: Nano-sized assemblies, stimuli-sensitive gels and smart surfaces. *Polym. Chem.* **2011**, *2*, 1234–1248. [CrossRef]

24. Beck, J.B.; Rowan, S.J. Multistimuli, multiresponsive metallo-supramolecular polymers. *J. Am. Chem. Soc.* **2003**, *125*, 13922–13923. [CrossRef] [PubMed]

25. Xia, F.; Ge, H.; Hou, Y.; Sun, T.; Chen, L.; Zhang, G.; Jiang, L. Multiresponsive surfaces change between superhydrophilicity and superhydrophobicity. *Adv. Mater.* **2007**, *19*, 2520–2524. [CrossRef]

26. Bousquet, A.; Ibarboure, E.; Teran, F.J.; Ruiz, L.; Garay, M.T.; Laza, J.M.; Vilas, J.L.; Papon, E.; Rodríguez-Hernández, J. pH responsive surfaces with nanoscale topography. *J. Polym. Sci. A Polym. Chem.* **2010**, *48*, 2982–2990. [CrossRef]

27. Shibayama, M.; Tanaka, T. Volume Phase Transition and Related Phenomena of Polymer Gels. In *Responsive Gels: Volume Transitions I*; Dušek, K., Ed.; Springer: Berlin, Germany; Heidelberg, Germany, 1993; Volume 109, pp. 1–62.

28. Chen, G.; Hoffman, A.S. Graft copolymers that exhibit temperature-induced phase transitions over a wide range of pH. *Nature* **1995**, *373*, 49–52. [CrossRef] [PubMed]

29. Hoffman, A.S.; Stayton, P.S.; Bulmus, V.; Chen, G.; Chen, J.; Cheung, C.; Chilkoti, A.; Ding, Z.; Dong, L.; Fong, R.; *et al.* Really smart bioconjugates of smart polymers and receptor proteins. *J. Biomed. Mater. Res.* **2000**, *52*, 577–586. [CrossRef]

30. Costa, E.; Coelho, M.; Ilharco, L.M.; Aguiar-Ricardo, A.; Hammond, P.T. Tannic acid mediated suppression of pnipaam microgels thermoresponsive behavior. *Macromolecules* **2011**, *44*, 612–621. [CrossRef]

31. Yang, H.-W.; Chen, J.-K.; Cheng, C.-C.; Kuo, S.-W. Association of poly(*N*-isopropylacrylamide) containing nucleobase multiple hydrogen bonding of adenine for DNA recognition. *Appl. Surf. Sci.* **2013**, *271*, 60–69. [CrossRef]

32. Gauthier, M.A.; Gibson, M.I.; Klok, H.-A. Synthesis of functional polymers by post-polymerization modification. *Angew. Chem. Int. Ed.* **2009**, *48*, 48–58. [CrossRef] [PubMed]

33. Fuchs, A.D.; Tiller, J.C. Contact-active antimicrobial coatings derived from aqueous suspensions. *Angew. Chem. Int. Ed.* **2006**, *45*, 6759–6762. [CrossRef] [PubMed]

34. Zolotukhin, M.G.; Colquhoun, H.M.; Sestiaa, L.G.; Rueda, D.R.; Flot, D. One-pot synthesis and characterization of soluble poly(aryl ether-ketone)s having pendant carboxyl groups. *Macromolecules* **2003**, *36*, 4766–4771. [CrossRef]

35. Zou, Y.; Brooks, D.E.; Kizhakkedathu, J.N. A novel functional polymer with tunable lcst. *Macromolecules* **2008**, *41*, 5393–5405. [CrossRef]

36. Schneider, B.H.; Dickinson, E.L.; Vach, M.D.; Hoijer, J.V.; Howard, L.V. Highly sensitive optical chip immunoassays in human serum. *Biosens. Bioelectron.* **2000**, *15*, 13–22. [CrossRef]

37. Vaisocherová, H.; Yang, W.; Zhang, Z.; Cao, Z.; Cheng, G.; Piliarik, M.; Homola, J.; Jiang, S. Ultralow fouling and functionalizable surface chemistry based on a zwitterionic polymer enabling sensitive and specific protein detection in undiluted blood plasma. *Anal. Chem.* **2008**, *80*, 7894–7901. [CrossRef] [PubMed]

38. Wark, A.W.; Lee, H.J.; Corn, R.M. Long-range surface plasmon resonance imaging for bioaffinity sensors. *Anal. Chem.* **2005**, *77*, 3904–3907. [CrossRef] [PubMed]

39. Lakhiari, H.; Okano, T.; Nurdin, N.; Luthi, C.; Descouts, P.; Muller, D.; Jozefonvicz, J. Temperature-responsive size-exclusion chromatography using poly(*N*-isopropylacrylamide) grafted silica. *Biochim. Biophys. Acta* **1998**, *1379*, 303–313. [CrossRef]

40. Kanazawa, H.; Yamamoto, K.; Matsushima, Y.; Takai, N.; Kikuchi, A.; Sakurai, Y.; Okano, T. Temperature-responsive chromatography using poly(*N*-isopropylacrylamide)-modified silica. *Anal. Chem.* **1996**, *68*, 100–105. [CrossRef] [PubMed]

41. Qiu, Y.; Park, K. Environment-sensitive hydrogels for drug delivery. *Adv. Drug Deliv. Rev.* **2001**, *53*, 321–339. [CrossRef]

42. Bhattacharya, S.; Eckert, F.; Boyko, V.; Pich, A. Temperature-, pH-, and magnetic-field-sensitive hybrid microgels. *Small* **2007**, *3*, 650–657. [CrossRef] [PubMed]

43. Xulu, P.M.; Filipcsei, G.; Zrínyi, M. Preparation and responsive properties of magnetically soft poly(*N*-isopropylacrylamide) gels. *Macromolecules* **2000**, *33*, 1716–1719. [CrossRef]

44. Mateescu, A.; Wang, Y.; Dostalek, J.; Jonas, U. Thin hydrogel films for optical biosensor applications. *Membranes* **2012**, *2*, 40. [CrossRef] [PubMed]

45. Feng, X.; Wu, H.; Sui, X.; Hempenius, M.A.; Julius Vancso, G. Thin film hydrogels from redox responsive poly(ferrocenylsilanes): Preparation, properties, and applications in electrocatalysis. *Eur. Polym. J.* **2015**, *72*, 535–542. [CrossRef]

46. Tokarev, I.; Minko, S. Stimuli-responsive hydrogel thin films. *Soft Matter* **2009**, *5*, 511–524. [CrossRef]

47. Harmon, M.E.; Kuckling, D.; Pareek, P.; Frank, C.W. Photo-cross-linkable pnipaam copolymers. 4. Effects of copolymerization and cross-linking on the volume-phase transition in constrained hydrogel layers. *Langmuir* **2003**, *19*, 10947–10956. [CrossRef]

48. Kuckling, D.; Harmon, M.E.; Frank, C.W. Photo-cross-linkable pnipaam copolymers. 1. Synthesis and characterization of constrained temperature-responsive hydrogel layers. *Macromolecules* **2002**, *35*, 6377–6383. [CrossRef]

49. Harmon, M.E.; Kuckling, D.; Frank, C.W. Photo-cross-linkable pnipaam copolymers. 2. Effects of constraint on temperature and pH-responsive hydrogel layers. *Macromolecules* **2003**, *36*, 162–172. [CrossRef]

50. Zhang, N.; Knoll, W. Thermally responsive hydrogel films studied by surface plasmon diffraction. *Anal. Chem.* **2009**, *81*, 2611–2617. [CrossRef] [PubMed]

51. Anac, I.; Aulasevich, A.; Junk, M.J.N.; Jakubowicz, P.; Roskamp, R.F.; Menges, B.; Jonas, U.; Knoll, W. Optical characterization of co-nonsolvency effects in thin responsive pnipaam-based gel layers exposed to ethanol/water mixtures. *Macromol. Chem. Phys.* **2010**, *211*, 1018–1025. [CrossRef]

52. Chan, C.-H.; Chen, J.-K.; Chang, F.-C. Specific DNA extraction through fluid channels with immobilization of layered double hydroxides on polycarbonate surface. *Sens. Actuat. B Chem.* **2008**, *133*, 327–332. [CrossRef]

53. Chen, J.-K.; Li, J.-Y. Fabrication of DNA extraction device with tethered poly(*N*-isopropylacrylamide) brushes on silicon surface for a specific DNA detection. *Sens. Actuat. B Chem.* **2010**, *150*, 314–320. [CrossRef]

54. Hosoya, K.; Kubo, T.; Tanaka, N.; Haginaka, J. A possible purification method of DNAs' fragments from humic matters in soil extracts using novel stimulus responsive polymer adsorbent. *J. Pharm. Biomed. Anal.* **2003**, *30*, 1919–1922. [CrossRef]

55. Lu, Y.; Mei, Y.; Drechsler, M.; Ballauff, M. Thermosensitive core-shell particles as carriers for Ag nanoparticles: Modulating the catalytic activity by a phase transition in networks. *Angew. Chem. Int. Ed.* **2006**, *45*, 813–816. [CrossRef] [PubMed]

56. Li, X.; Yu, X.; Han, Y. Intelligent reversible nanoporous antireflection film by solvent-stimuli-responsive phase transformation of amphiphilic block copolymer. *Langmuir* **2012**, *28*, 10584–10591. [CrossRef] [PubMed]

57. Ebara, M.; Yamato, M.; Hirose, M.; Aoyagi, T.; Kikuchi, A.; Sakai, K.; Okano, T. Copolymerization of 2-carboxyisopropylacrylamide with *N*-isopropylacrylamide accelerates cell detachment from grafted surfaces by reducing temperature. *Biomacromolecules* **2003**, *4*, 344–349. [CrossRef] [PubMed]

58. Yakushiji, T.; Sakai, K.; Kikuchi, A.; Aoyagi, T.; Sakurai, Y.; Okano, T. Graft architectural effects on thermoresponsive wettability changes of poly(*N*-isopropylacrylamide)-modified surfaces. *Langmuir* **1998**, *14*, 4657–4662. [CrossRef]

59. Chen, J.-K.; Wang, J.-H.; Chang, J.-Y.; Fan, S.-K. Thermally switchable adhesions of polystyrene-block-poly(*N*-isopropylacrylamide) copolymer pillar array mimicking climb attitude of geckos. *Appl. Phys. Lett.* **2012**, *101*, 123701. [CrossRef]

60. Chang, C.-J.; Kuo, E.-H. Roughness-enhanced thermal-responsive surfaces by surface-initiated polymerization of polymer on ordered zno pore-array films. *Thin Solid Films* **2010**, *519*, 1755–1760. [CrossRef]

61. Chen, J.-K.; Qui, J.-Q. Patterned 3D assembly of Au nanoparticle on silicon substrate by colloid lithography. *J. Nanopart. Res.* **2012**, *14*, 1–14. [CrossRef]

62. Huang, H.L.; Chen, J.-K.; Houng, M.P. Fabrication of two-dimensional periodic relief grating of tethered polystyrene on silicon surface as solvent sensors. *Sens. Actuat. B Chem.* **2013**, *177*, 833–840. [CrossRef]

63. Chen, J.-K.; Bai, B.-J. pH-switchable optical properties of the one-dimensional periodic grating of tethered poly(2-dimethylaminoethyl methacrylate) brushes on a silicon surface. *J. Phys. Chem. C* **2011**, *115*, 21341–21350. [CrossRef]

64. Brazinha, C.; Barbosa, D.S.; Crespo, J.G. Sustainable recovery of pure natural vanillin from fermentation media in a single pervaporation step. *Green Chem.* **2011**, *13*, 2197–2203. [CrossRef]

65. Fries, D.M.; Voitl, T.; von Rohr, P.R. Liquid extraction of vanillin in rectangular microreactors. *Chem. Eng. Technol.* **2008**, *31*, 1182–1187. [CrossRef]

66. Lora, J.H.; Glasser, W.G. Recent industrial applications of lignin: A sustainable alternative to nonrenewable materials. *J. Polym. Environ.* **2002**, *10*, 39–48. [CrossRef]

67. Voitl, T.; Rudolf von Rohr, P. Oxidation of lignin using aqueous polyoxometalates in the presence of alcohols. *ChemSusChem* **2008**, *1*, 763–769. [CrossRef] [PubMed]

68. Mialon, L.; Pemba, A.G.; Miller, S.A. Biorenewable polyethylene terephthalate mimics derived from lignin and acetic acid. *Green Chem.* **2010**, *12*, 1704–1706. [CrossRef]

69. Zhang, C.; Madbouly, S.A.; Kessler, M.R. Renewable polymers prepared from vanillin and its derivatives. *Macromol. Chem. Phys.* **2015**, *216*, 1816–1822. [CrossRef]

70. Meylemans, H.A.; Harvey, B.G.; Reams, J.T.; Guenthner, A.J.; Cambrea, L.R.; Groshens, T.J.; Baldwin, L.C.; Garrison, M.D.; Mabry, J.M. Synthesis, characterization, and cure chemistry of renewable bis(cyanate) esters derived from 2-methoxy-4-methylphenol. *Biomacromolecules* **2013**, *14*, 771–780. [CrossRef] [PubMed]

71. Renbutsu, E.; Okabe, S.; Omura, Y.; Nakatsubo, F.; Minami, S.; Saimoto, H.; Shigemasa, Y. Synthesis of UV-curable chitosan derivatives and palladium (II) adsorption behavior on their UV-exposed films. *Carbohydr. Polym.* **2007**, *69*, 697–706. [CrossRef]

72. Stanzione, J.F.; Sadler, J.M.; la Scala, J.J.; Wool, R.P. Lignin model compounds as bio-based reactive diluents for liquid molding resins. *ChemSusChem* **2012**, *5*, 1291–1297. [CrossRef] [PubMed]

73. Brandrup, J.; Immergut, E.H.; Grulke, E.A. *Polymer Handbook*, 4th ed.; Wiley: New York, NY, USA; Chichester, UK, 2004.

74. Gupta, S.; Kuckling, D.; Kretschmer, K.; Choudhary, V.; Adler, H.-J. Synthesis and characterization of stimuli-sensitive micro- and nanohydrogels based on photocrosslinkable poly(dimethylaminoethyl methacrylate). *J. Polym. Sci. A Polym. Chem.* **2007**, *45*, 669–679. [CrossRef]

gels

MDPI

Review

Stimuli-Responsive Cationic Hydrogels in Drug Delivery Applications

G. Roshan Deen [1],* and Xian Jun Loh [2]

[1] Soft Materials Laboratory, Natural Sciences and Science Education AG, National Institute of Education, Nanyang Technological University, 1-Nanyang Walk, Singapore 637616, Singapore
[2] Institute of Materials Research and Engineering, 2-Fusionopolis Way, Singapore 138634, Singapore; lohxj@imre.a-star.edu.sg
* Correspondence: roshan.gulam@nie.edu.sg; Tel.: +65-6790-3816

Received: 13 December 2017; Accepted: 30 January 2018; Published: 1 February 2018

Abstract: Stimuli-responsive, smart, intelligent, or environmentally sensitive polymers respond to changes in external stimuli such as pH, temperature, ionic strength, surfactants, pressure, light, biomolecules, and magnetic field. These materials are developed in various network architectures such as block copolymers, crosslinked hydrogels, nanogels, inter-penetrating networks, and dendrimers. Stimuli-responsive cationic polymers and hydrogels are an interesting class of "smart" materials that respond reversibly to changes in external pH. These materials have the ability to swell extensively in solutions of acidic pH and de-swell or shrink in solutions of alkaline pH. This reversible swelling-shrinking property brought about by changes in external pH conditions makes these materials useful in a wide range of applications such as drug delivery systems and chemical sensors. This article focuses mainly on the properties of these interesting materials and their applications in drug delivery systems.

Keywords: cationic polymers; hydrogels; stimuli-responsive; heterocyclic; swelling of gels; drug-delivery systems

---

1. Introduction

Polymers that respond to changes in external stimuli are called stimuli-responsive, smart, intelligent, or environmentally sensitive polymers. These materials can be in the form of linear polymers (homo polymers, copolymers), crosslinked polymers in the form of hydrogels, inter-penetrating networks, and micro/nanogels. The external stimuli can be either physical, chemical, or biochemical in nature, in the form of pH, temperature, salts, surfactants, light, pressure, biomolecules, and magnetic field [1–5]. The responses to these external stimuli occur in the form of conformational, optical, and chemical changes. The different types of stimuli that bring about these changes are illustrated in Figure 1. These changes are accompanied by variation in the physical properties of the polymer. On a macroscopic level, the changes are apparent as phase separation from aqueous solution (for linear polymers and linear copolymers) or volume changes (for crosslinked systems) [5–10]. The temperature at which such transitions occur is termed the lower critical solution temperature (LCST) or volume-phase transition temperature (VPT) [11]. This article focuses specifically on cationic polymers in the form of hydrogels and their applications in drug delivery applications.

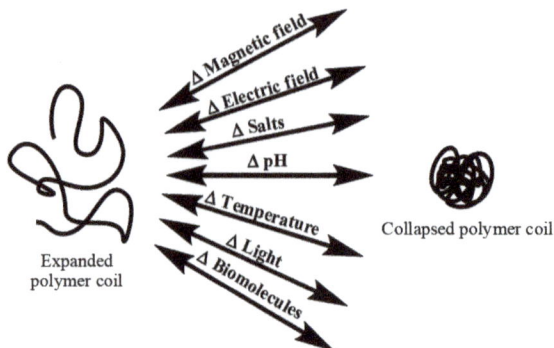

Figure 1. Stimuli-response behavior of polymer chains in response to various external stimuli.

Hydrogels that contain polycations and are sensitive to external pH changes are called cationic hydrogels. The tertiary amine functional groups present in these hydrogels are protonated below their dissociation constant (pK_a) in acidic solutions, causing the hydrogels to swell extensively. This pH-dependent swelling is shown in Figure 2 [12–16]. A few examples of cationic hydrogels are based on poly(N,N-dialkylaminoethyl methacrylate), poly(lysine), chitosan, poly(amido-amine), etc. The chemical structures of some pH-sensitive synthetic cationic polymers such as poly(dimethylaminoethyl methacrylate) (PDMAEMA), poly(diethylaminoethyl methacrylate) (PDEAEMA), poly(ethyl pyrrolidine methacrylate) (PEP), and poly(ethyl piperazine acrylate) (PAcrNEP) are shown in Figure 3. Hydrogels based on these chemical moieties swell extensively in solutions with a low pH due to the protonation of tertiary amine functional groups, which leads to the formation of fixed electric charges on the macromolecule. The degree of swelling of cationic hydrogel depends largely on the pK_a of the ionic group, concentration of the monomer, crosslinking ratio, pH, and ionic strength of the external medium [17,18].

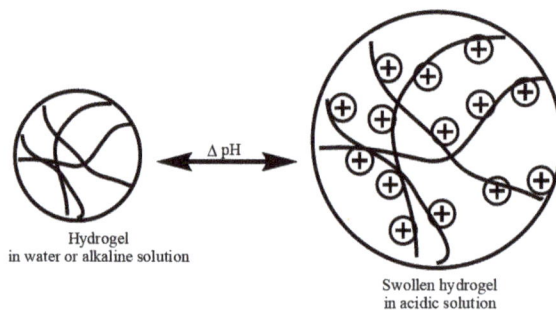

Figure 2. Swelling of hydrogel in acidic solution due to the formation of fixed charges on the polymer network.

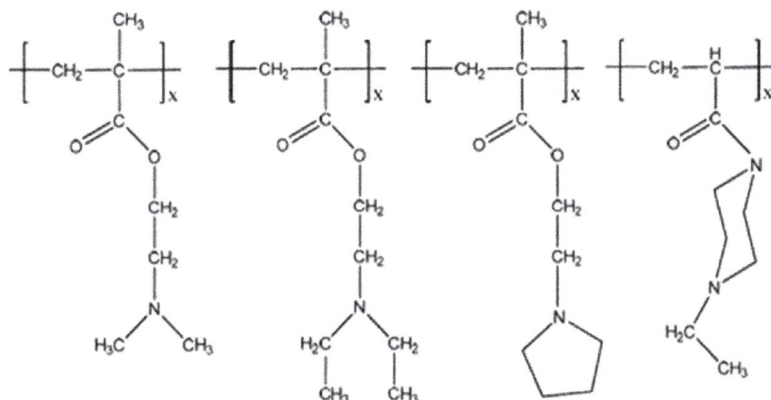

Figure 3. Chemical structure of cationic polymers (1) poly(dimethylaminoethyl methacrylate) (PDMAEMA); (2) poly(diethylaminoethyl methacrylate) (PDEAEMA); (3) poly(ethyl pyrrolidine methacrylate) (PEP); (4) poly(ethyl piperazine acrylate) (PAcrNEP).

In recent years, cationic hydrogels that exhibit response to more than one stimuli, such as pH and temperature, have attracted much research attention in biomedical applications. These polymeric materials are obtained from new synthetic monomers containing tertiary amine functional groups [17–19]. This has been achieved by the following chemical synthesis approaches: (i) the copolymerization of monomers with desired functional groups [20–25], (ii) the copolymerization or combination of thermo-sensitive polymers with polyelectrolytes [26,27], (iii) the formation of inter-penetrating networks through combination of various stimuli-sensitive polymers [28,29], and (iv) the chemical synthesis and polymerization of new functional monomers [30,31].

Cationic polymers and hydrogels are an interesting class due to their pH sensitivity and due to their ability to complex with other systems with anionic character [32]. As a result of these characteristics, cationic polymers and hydrogels have been widely evaluated as alternative vectors to viruses in gene delivery and therapy [33,34]. The nature of cationic groups and their influence on transfection efficiency and biodegradability has been studied for cationic polymers based on poly(phosphoramides) [33–35].

2. Swelling Behavior of Cationic Hydrogels

The most favorable property of hydrogels is their ability to swell when placed in contact with a thermodynamically compatible solvent. Solvent molecules penetrate the glassy surface of the hydrogel and slowly diffuse into the network. The physicochemical models that describe the swelling of cationic hydrogels are usually based on the free energy change that takes place due to the following factors [34–37]: (i) the osmotic pressure of counterions within the gel (Donnan theory); (ii) the mixing of polymer with solvent (Flory-Huggins theory); (iii) the stretching of polymer chains (Flory-Rehner theory).

The total free energy change of a hydrogel at equilibrium swelling is given by the Gibbs free energy equation [38]:

$$\Delta G_{Total} = \Delta G_{Mixing} + \Delta G_{Elastic} \tag{1}$$

where ΔG_{Total} is the change of total free energy in hydrogel, ΔG_{mixing} is the free energy due to the mixing of solvent molecules with polymer chains, and $\Delta G_{elastic}$ is the free energy of the elastic retractive force of the hydrogel.

In the case of ionic hydrogels, the theoretical treatment of equilibrium swelling is more complicated, as swelling depends to a large extent on the degree of ionization of functional groups, as well as the ionic strength of the external solution. Therefore, the ionic groups impart an additional free energy change (ΔG_{ionic}) to the total free energy of the hydrogel. The total free energy change of the ionic hydrogel is given as:

$$\Delta G_{Total} = \Delta G_{Mixing} + \Delta G_{Elastic} + \Delta G_{ionic} \tag{2}$$

The ionic gel is subjected to a swelling pressure, π, which comprises three components similar to the total free energy:

$$\pi_{Total} = \pi_{mixing} + \pi_{elastic} + \pi_{ionic} \tag{3}$$

where π_{mix} is the osmotic pressure due to the mixing of solvent with polymer, $\pi_{elastic}$ is the osmotic pressure due to the elastic force of the gel, and π_{ionic} is the osmotic pressure due to the ionic contribution. The equilibrium swelling is obtained when π_{Total} is set equal to zero.

The osmotic pressure due to mixing, π_{mix}, is given by the Flory-Huggins theory [36]:

$$\pi_{mixing} = -\frac{RT}{V_1}\left[\ln\left(1-v\right)+v+\chi v^2\right] \tag{4}$$

where v is the polymer volume fraction, V_1 is the molar volume of solvent, and χ is the Flory-Huggins interaction parameter.

The osmotic pressure due to elastic or configurational contribution ($\pi_{elastic}$) is obtained from $\Delta G_{elastic}$, during the swelling of hydrogels. Under isotropic swelling, this is obtained by differentiating $\Delta G_{elastic}$ with respect to volume and expanding the inverse Langevin function in a power series, as given by the following expression [39]:

$$\pi_{elastic} = -v_0 RT\left[\left(\frac{v}{v_0}\right)^{1/3} - \frac{1}{2}\left(\frac{v}{v_0}\right)\right] - v_0 RT$$
$$\left[\frac{3}{5}\left(\frac{v_0}{v}\right)^{1/3}\times\frac{1}{n} = \frac{99}{175}\left(\frac{v_0}{v}\right)\frac{1}{n^2}+\frac{513}{875}\left(\frac{v_0}{v}\right)^{5/3}\frac{1}{n^3}+...\right] \tag{5}$$

where $v_0 = v_0\, v_d$ is the concentration of polymer chains when the gel is formed.

The ionic contribution to the osmotic pressure (π_{ion}) arises due to the difference between the osmotic pressure of mobile ions in the gel and in the external solution. This is given by [39]:

$$\pi_{ionic} = RT\left[\Phi\sum_i\overline{C_i}-\phi\sum_i C_i\right] \tag{6}$$

where C_i and $\overline{C_i}$ are the concentrations of mobile ions in the external solution and gel; ϕ and Φ are the corresponding osmotic coefficients, respectively. This implies that charged or ionizable groups present in gels play an important role in swelling behavior. The following equation describes the swelling of cationic hydrogels in the presence of a solvent and their abovementioned dependencies [37–39]. Using this expression, the network structure of cationic gels can be characterized.

$$\frac{V_1}{4I}\left(\frac{v_{2,s}^2}{\overline{v}}\right)\left(\frac{K_b}{10^{pH-14}-K_a}\right)^2 = \left[\ln(1-v_{2,s})+v_{2,s}+\chi v_{2,s}^2\right]+$$
$$\left(\frac{V_1}{v M_c}\right)\times\left(1-\frac{2M_c}{M_n}\right)v_{2,r}\left[\left(\frac{v_{2,s}}{v_{2,r}}\right)^{1/3}-\left(\frac{v_{2,s}}{2v_{2,r}}\right)\right] \tag{7}$$

where I is the ionic strength, K_a is the acid dissociation constant, K_b is the base dissociation constant, M_c is the molecular weight between crosslinks, M_n is the molecular weight of polymer chains without crosslinks, V_1 is the molar volume of water, v_2, r is the volume fraction of the polymer in the relaxed state, $v_{2,s}$ is the volume fraction of the polymer in the swollen state, and χ is the polymer-solvent interaction parameter.

3. Applications of Cationic Hydrogels

Cationic hydrogels have found several applications in the biomedical industry such as targeted drug delivery, gene delivery, and tissue engineering.

3.1. Drug Delivery Systems

The process of administering a pharmaceutical compound to a targeted organ to achieve a therapeutic effect in humans or animals is the principle of targeted drug delivery systems. Due to the variation of pH (between 2 in the stomach to neutral in the small intestine) along the gastrointestinal tract (GI), it is still the most sought after route for drug delivery [40]. It is also the most complex route and therefore various approaches are required for the effective delivery of drugs. A few of the obstacles that need to be overcome for efficient drug delivery include the degradation of drugs and the carrier (gels) by enzymes, rapid removal of the carrier from the body, non-specific toxicity of carrier, etc. In this regard, stimuli-responsive systems allow the advantage of delivering the right amount of drug at the right time in response to changes in external stimuli. Cationic gels, for example, can expand as a result of the ionization of functional moieties present along the macromolecular chain in an acidic condition, thus promoting drug diffusion and release in the stomach [40].

Several cationic hydrogels based on heterocyclic compounds, such as morpholine and pyrolidinone, have been widely studied by San Román and co-workers [27,28,30]. The same group also developed polymers in which an anti-aggregant drug called Triflusal was covalently attached (polymeric prodrugs) [41]. A novel poly(vinylpyrrolidone-co-dimethylmaleic anhydride) (PVD) carried drug was synthesized by Kamada and co-workers [42]. The incorporated pH-sensitive vinylpyrrolidone cationic monomers allowed the conjugation of the drug Adriamycin at pH 8.5, with gradual release at pH 6 to 7. The PVD-Adriamycin conjugate has shown anti-tumor activity against Sarcoma-180 solid tumor in mice.

An injectable hydrogel based on vinylpyrrolidine, NIPAM, and acrylamide was reported by You and co-workers [43]. This hydrogel swelled in a solution of pH 6.5 at 25 °C and shrunk in a solution of pH 8.5 at 37 °C. The combined pH and temperature sensitivity facilitated the extended delivery of an opioid receptor antagonist, naltrexone, over a period of 28 days. Cationic graft copolymers in the form of nanoparticles based on PDMAEMA and polycaprolactone (PCL) were prepared by Gua and co-workers [44] for the encapsulation of paclitaxel and hydrophilic biomolecules. A faster release of the drug was achieved in solutions with a low pH.

Cationic polymers have also played a role in the formulation of sustained release matrix tablets. Matrix tablets were prepared from a combination of hydrophobic ethyl cellulose and hydrophilic sodium carboxymethyl cellulose polymers. The in vitro release of losartan potassium (drug for hypertension treatment) from the matrix tablet was studied. The results showed that the formulation produced sustained drug release over a period of 12 h. Cationic polymers in the form of biodegradable micelles have also been reported as drug carriers [45–47]. The polymer micelles were based on PDMAEMA-PCL-PDMAEMA triblocks. Successful delivery of paclitaxel into tumor cells was achieved, thus optimizing the treatment of tumor cells using cationic polymer micelles. Cationic polymers containing disulfide links prepared through a Michael addition reaction was found to conjugate doxorubicine. The polymer-drug conjugate displayed good stability in physiological pH conditions [48].

Amphiphilic cellulose cationic micelles were prepared by Song and co-workers [49]. This polymer self-assembled into spherical micelles in water which was used as a carrier for the lipophilic drug, prednisolone. Micelles based on PNIPAM, DMAM, and poly (lactic acid) (PLA) were prepared by Akimoto and co-workers [50]. These micelles were able to diffuse into the cells above the LCST of the polymer, due to increased interaction between the solvated micelles and cells.

ABA type triblock cationic copolymers based on PDMAEMA (block A) and PVCL (block B) were prepared by San Miguel and co-workers [51]. The formation of pH- and temperature-responsive micelles as well as in vitro sustained drug release were demonstrated in this study.

3.2. Gene Delivery Systems

The process of administering a gene (DNA) for correcting defective genes to achieve the treatment of many genetic diseases is the principle of gene delivery/gene therapy systems. The delivery of the appropriate therapeutic gene into the cell to replace or regulate the defective gene is a vital step in gene delivery. The gene is transported in gene delivery carriers called gene carriers or vectors or vehicles. DNA is a negatively charged hydrophilic molecule that has a large size at physiological conditions and is therefore very difficult to incorporate into cells. Liposomes and polycations are two important classes of non-viral chemical gene delivery methods to condense DNA that can be transported into cell to replace the defective gene [52].

This gene delivery process using a cationic polymer or hydrogel is also called transfection. This process involves four main steps:

(1) Complexation for DNA with the cationic polymer/hydrogel. The DNA-cationic polymer/hydrogel is termed as a polyplex.
(2) Addition of the polyplex to the cell containing the defective DNA for a certain period of time.
(3) Release of DNA into the cytoplasm and removal of cationic polymer/hydrogel.
(4) Transfer of DNA into nucleus. This step involves incubation for a period of time until the desired results are obtained.

Cationic polymers can either complex with and condense DNA into small particles, or combine with DNA through conjugation, in which the covalent bond will be cleaved in order to release the DNA. To note, a major demerit of many polymer-based non-viral vectors, such as poly(ethyleneimine), PDMAEMA, and dendrimers based on poly(amido-amine), is that they show significant cytotoxicity and lower transfection efficiency [45]. Interestingly, polyplexes based on biodegradable cationic polymers show lower cytotoxicity and improved transfection efficiencies. Such polymers include poly(4-hydroxy-L-proline ester), poly[α-(4-aminobutyl)-L-glycolic acid] (PAGA), cationic polyphosphazenes, and linear or branched poly(amino ester)s [53–55].

Cationic polymers based on poly(amido amine)s show improved stability against hydrolysis compared to poly(amino ester)s, as the amide group is less sensitive to hydrolysis than the ester group. Ferruti and co-workers [56–60] synthesized and studied the properties and application of a wide variety of poly(amido amine)s. For these polymers, the endosomal escape of the polyplexes was attributed to the protonation of the tertiary amino groups, which induces a conformation change.

A few major barriers for non-virial gene delivery are: (i) the inefficient endosomal escape of the polyplexes to avoid the lysosome degradation pathway, and (ii) the unpacking of DNA from the polyplexes to allow transcription [61]. Lin and co-workers [62,63] reported the synthesis of novel poly(amido-amine) linear and copolymers containing disulfide linkages along the polymer backbone (Figure 4).

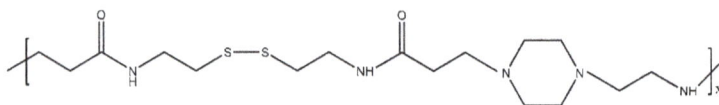

Figure 4. Chemical structure of poly(amido-amine) with disulfide linkage along the polymer backbone.

These polymers were biodegradable and showed improved biophysical properties. These polymers effectively condensed DNA into polyplexes with a size of less than 150 nm with a positive surface charge. These novel polymers containing disulfide linkages also facilitated polyplex unpacking, leading to enhanced gene expressions and lower levels of cytotoxicity. Hoffmann and co-workers [64] designed and functionalized a new monomer containing a disulfide linkage as a pendant group viz. pyridyl disulfate acrylate (PDSA). This polymer allowed efficient conjugation through disulfide linkages for effective endosomal translocation of therapeutics.

Hydrophobically modified cationic polymers with small or bulky lipids, such as cholesterol, have facilitated the translocation of DNA/s-RNA complexes through the cell membrane [65]. A biodegradable non-toxic poly[α-(4-aminobutyl)-L-glycolic acid] (PAGA) used as gene carrier was reported by Lim and co-workers [66]. The polymer condensed with DNA and exhibited fast degradation. The transfection efficiency of this polymer with DNA was found to be higher than the poly(L-lysine) analogues. Forrest and co-workers [67] showed that an increase in the hydrophobicity of acylated branched poly(ethylene amine) caused a fourfold uptake of polyplexes and better transfection. Among poly-L-lysine, the cholesterol-modified analogues showed improved transfection. Further, it was also shown that the incorporation of a hydrophobic cholanic acid moiety in glycol chitosan (Figure 5) was essential for the formation of polyplexes of hydrophobic p-DNA [66,67].

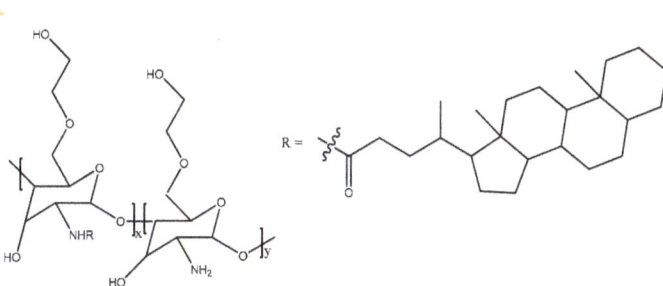

Figure 5. Chemical structure of glycol chitosan modified with cholanic acid.

3.3. Tissue Engineering

Natural and synthetic cationic polymers in particular porous hydrogels have been used as scaffolds to engineer various forms of new tissues. The role of most polymer scaffolds is to provide a modified surface of suitable porosity for seeded cell adhesion and interaction, very much like the extracellular matrix. The volume of tissue developed depends on the crosslinking density and the pore size of the polymer scaffolds [68]. Biocompatible and biodegradable polymers, such as cationic chitosan, poly-L-lysine, poly(ethylene imine) (PEI), etc., have been widely used for this purpose. Various PEI polymers [69,70] have been used to fabricate scaffolds for cultivating bovine chondrocytes and normal human fibroblast cells.

The porous scaffolds of chitosan reinforced with calcium phosphate for improved osteo-retention have been used in bone tissue engineering. This reinforced cationic polymer scaffold was used to regenerate new bone tissue of increased fracture and fatigue resistance [71]. Fiber mesh scaffolds and microspheres based on chitosan have also been studied for bone generation, cell adhesion, and viability [72]. Using chitosan and chitin scaffolds grafted with poly(L-lysine) (PLL), the efficient generation of cartilaginous components was achieved, thus showing the potential application of cationic polymers in articular cartilage engineering.

3.4. Nanoparticles and Microparticles Based on Cationic Polymers

Polymeric nanoparticles and microparticles are of special interest because of their properties such as nontoxicity, biocompatibilty, and stimuli-sensitivity. Nano/microparticles based on cationic polymers may enhance the cellular uptake and endosomal escape of the particles [73]. These materials have been synthesized by methods such as solvent evaporation [74], spray drying [75], emulsification [76], ionotropic gelation [77], and controlled polymerization methods such as atom transfer radical polymerization (ATRP) [78]. Nano or microsheres of chitosan have been widely used for the delivery of insulin [79], heparin [80], cyclosporin A [81], and a variety of proteins [82].

An important application of cationic polymers in the form of nanoparticles is in the deliverly of hydrophobic drugs. Almost one-thrid of newly deiscovered drug molecules are sparingly soluble in water [83]. Cationic copolymers based on 5-Z-amino-δ-valerolactone and ε-caprolactone assemble into nanoaggregates at concentrations above 0.5 mg mL^{-1}, thereby increasing the water solubility of hydrophobic drugs by about 100–1000 times [84].

3.5. Multilayer Films or Coatings

Multilayer films of polymers are usually obained by the layer-by-layer method, usually from oppositely charged polyelectrolytes, neutral polymers, cationic dendrimers [85], or polycations. The films may be deposited onto nano/microspheres, forming the desired capsules for drug encapsulation [86]. A schematic represenation of the various methods used to obtain coated multilayer films for drug delivery is shown in Figure 6. The drug can also be introduced into the nano/microshperes before coating them with multilayer film, or the core of the capsule can be selectively removed before encapsulating the drug. In this case, the capsules serve as sacrifical templates [87].

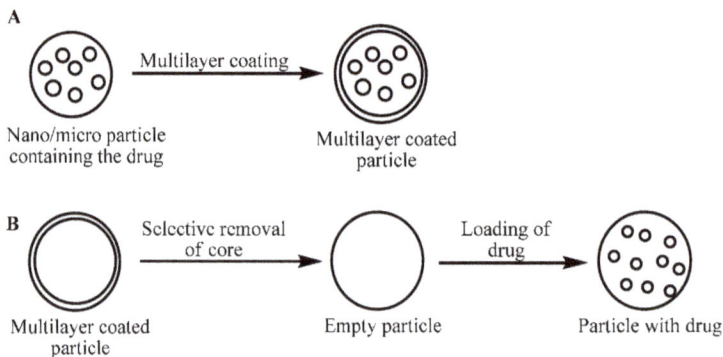

Figure 6. Methods to obtain multilayer polymer films of drug delivery systems. (**A**) Multilayer coating on particle with encapsualted drug; (**B**) sacrificial removal of core followed by drug loading.

The main disadvantage of hollow capsules is the low efficiency of drug loading due to the adsorption of the drug on the walls of the capsule rather than in the core. Various cationic polymers, such as chitosan, protamine, poly(4-vinyl pyridine), poly(diallyl dimethylammonium chloride), poly(L-arginine), poly(dimethyl aminoethyl methacrylate) [88–91], etc., have been used in the preparation of multilayer films. The number of layers or thickness of the capsules is related to the permeability and the release of the active substances. It has been found that the permeability decreases with an increase in the thickness of the layers [91]. The rate of release of a drug can be greatly controlled by coating the drug directly with a multilayer film. The release rate of furosemide microcrystals coated with cationic polymers with a thickness of 150 nm was found to reduce the rate up to 300 times compared to the uncoated drug [92].

4. Conclusions

In summary, stimuli-responsive cationic polymers show interesting properties in response to changes in the external pH of the medium. In this paper, a brief introduction to cationic polymers/hydrogels in terms of design, theory of swelling, and recent biomedical applications such as drug and gene delivery and tissue engineering has been compiled. Based on the variety of cationic polymers and their interesting properties, these materials show promise and will have a definite impact on the formation of polymer-based biomaterials and chemical sensors.

Author Contributions: G. Roshan Deen and Xian Jun Loh jointly wrote the paper.

Conflicts of Interest: The authors declare no conflict of interest.

References

1. Roshan Deen, G.; Mah, C.H. Influence of external stimuli on the network properties of cationic poly(*N*-acryloyl-*N'*-propyl piperazine) hydrogels. *Polymer* **2016**, *89*, 55–68. [CrossRef]
2. Tang, S.; Floy, M.; Bhandari, R.; Dziubla, T.; Zach Hilt, J. Development of novel *N*-isopropylacrylamide (NIPAAm) based hydrogels with varying content of chrysin multiacrylate. *Gels* **2017**, *3*, 40. [CrossRef]
3. Stayton, P.S.; Shimoboji, T.; Long, C. Control of protein-ligand recognition using a stimuli-responsive polymer. *Nature* **1995**, *378*, 472–474. [CrossRef] [PubMed]
4. Jeong, B.; Bae, Y.H.; Kim, S.W. Biodegradable block copolymers as injectable drug delivery systems. *Nature* **1997**, *388*, 860–862. [CrossRef] [PubMed]
5. Hatefi, A.; Amsden, B. Biodegradable injectable in-situ forming drug delivery systems. *J. Control Release* **2002**, *80*, 9–28. [CrossRef]
6. Pasparakis, G.; Vamvakaki, M. Multiresponsive polymers: Nano-sized assemblies, stimuli-sensitive gels and smart surfaces. *Polym. Chem.* **2011**, *2*, 1234–1248. [CrossRef]
7. Ahn, S.; Kasi, R.M.; Kim, S.; Sharma, N.; Zhou, Y. Stimuli-responsive polymer gels. *Soft Matter* **2008**, *4*, 1151–1157. [CrossRef]
8. Almdal, K.; Dyre, J.; Hvidt, S.; Kramer, O. Towards a phenomenological definition of the term 'gel'. *Polym. Gels Netw.* **1993**, *1*, 5–17. [CrossRef]
9. Tanaka, T. Collapse of Gels and the Critical Endpoint. *Phys. Rev. Lett.* **1978**, *40*, 820–823. [CrossRef]
10. Vihola, H.; Laukkanen, A.; Tenhu, H.; Hirvonen, J. Drug release chracteristics of physically cross-linked thermosensitive poly(*N*-vinylcaprolactam) hydrogel particles. *J. Pharm. Sci.* **2008**, *97*, 4783–4793. [CrossRef] [PubMed]
11. Qiu, Y.; Park, K. Environment-sensitive hydrogels for drug delivery. *Adv. Drug Del. Rev.* **2001**, *53*, 321–339. [CrossRef]
12. Gil, E.S.; Hudosn, S.M. Stimuli-responsive polymers and their bioconjugates. *Prog. Polym. Sci.* **2004**, *29*, 1173–1222. [CrossRef]
13. Aguilar, M.R.; Elvira, C.; Gallardo, A.; Vázquez, B.; Román, J.S. Smart polymers and their applications as biomaterials. *Top. Tissue Eng.* **2007**, *3*, 1–27.
14. Katchalsky, A. Rapid swelling and deswelling of reversible gels of polymeric acids by ionization. *Cell. Mol. Life Sci.* **1949**, *5*, 319–320. [CrossRef]
15. Tanaka, T. Phase transition in gels and a single polymer. *Polymer* **1979**, *20*, 1404–1412. [CrossRef]
16. Firestone, B.A.; Siegel, R.A. Dynamic pH-dependent swelling of a hydrophobic polyelectrolyte gel. *Polym. Commun.* **1988**, *29*, 204–208.
17. Peppas, N.A.; Bures, P.; Leobandung, W.; Ichikawa, H. Hydrogels in pharmaceutical formulations. *Eur. J. Pharm. Biopharm.* **2000**, *50*, 27–46. [CrossRef]
18. Ferruti, P.; Marchisio, M.A.; Duncan, R. Poly(amido-amine)S: Biomedical applications. *Macromol. Rapid Commun.* **2002**, *23*, 332–355. [CrossRef]
19. Gan, L.H.; Gan, Y.Y.; Roshan Deen, G. Poly(*N*-acryloyl-*N'*-propyl piperazine): A new stimuli-responsive polymer. *Macromolecules* **2000**, *33*, 7893–7897. [CrossRef]
20. Bulmus, V.; Ding, Z.; Long, C.J.; Stayton, P.S.; Hoffman, A.S. Site-specific polymer-streptavidin bioconjugate for pH-controlled binding and triggered release of biotin. *Bioconjug. Chem.* **2000**, *11*, 78–83. [CrossRef] [PubMed]
21. Brazel, C.S.; Peppas, N.A. Synthesis and Characterization of thermo- and chemomechanically responsive poly(isopropylacrylamide-co-methacrylic acid) hydrogels. *Macromolecules* **1995**, *28*, 8016–8020. [CrossRef]
22. Roshan Deen, G.; Gan, L.H. Determination of reactivity ratios and swelling characteristics of stimuli-responsive copolymers of *N*-acryloyl-*N'*-ethyl piperazine and MMA. *Polymer* **2006**, *47*, 5025–5034. [CrossRef]
23. Roshan Deen, G.; Chua, V.; Ilyas, U. Synthesis, swelling properties and network structure of new stimuli-responsive poly(*N*-acryloyl-*N'*-ethyl piperazine-co-*N*-isopropylacrylamide) hydrogels. *J. Polym. Sci. Polym. Chem.* **2012**, *50*, 3363–3372. [CrossRef]

24. Roshan Deen, G.; Lim, E.K.; Mah, C.H.; Heng, K.M. New cationic linear copolymers and hydrogels of N-vinyl caprolactam and *N*-acryloyl-*N'*-ethyl piperazine: Synthesis, reactivity, influence of external stimuli on the LCST and swelling properties. *Ind. Eng. Chem. Res.* **2012**, *51*, 13354–13365. [CrossRef]

25. Gan, L.H.; Roshan Deen, G.; Loh, X.J.; Gan, Y.Y. New stimuli-responsive copolymers of *N*-acryloyl-*N'*-alkyl piperazine and methyl methacrylate. *Polymer* **2001**, *42*, 65–69. [CrossRef]

26. Roshan Deen, G.; Lee, T.T. New pH-responsive linear and crosslinked functional copolymers of *N*-acryloyl-*N'*-phenyl piperazine with acrylic acid and hydroxyethyl methcrylate: Synthesis, reactivity, and effect of steric hindrance on swelling. *Polym. Bull.* **2012**, *69*, 827–846. [CrossRef]

27. Verestiuc, L.; Ivanov, C.; Barbu, E.; Tsibouklis, J. Dual-stimuli-responsive hydrogels based on poly(*N*-isopropylacrylamide)/chitosan semi-interpenetrating networks. *Int. J. Pharm.* **2004**, *269*, 185–194. [CrossRef] [PubMed]

28. Gonzalez, N.; Elvira, C.; Roman, S.J. Novel dual-stimuli-responsive polymers derived from ethylpyrrolidine. *Macromolecules* **2005**, *38*, 9298–9303. [CrossRef]

29. Kurata, K.; Dobashi, A. Novel temperature and pH-responsive linear polymers and crosslinked hydrogels comprised of acidic L-α-amino acid derivatives. *J. Macromol. Sci. Part A Pure Appl. Chem.* **2004**, *41*, 143–164. [CrossRef]

30. Roshan Deen, G.; Gan, Y.Y.; Gan, L.H.; Teng, S.H. New functional copolymers of *N*-acryloyl-*N'*-methyl piperazine and 2-hydroxyethyl methacrylate: Synthesis, determination of reactivity ratios and swelling characteristics of gels. *Polym. Bull.* **2011**, *66*, 301–313. [CrossRef]

31. Velasco, D.; Elvira, C.; Román, J.S. New stimuli-responsive polymers derived from morpholine and pyrrolidine. *J. Mater. Sci. Mater. Med.* **2008**, *19*, 1453–1458. [CrossRef] [PubMed]

32. Oupicky, D.; Parker, A.L.; Seymour, L.W. Laterally stabilized complexes of DNA with linear reducible polycations: Strategy for triggered inracellular activation of DNA delivery vectors. *J. Am. Chem. Soc.* **2002**, *124*, 8–9. [CrossRef] [PubMed]

33. Lim, Y.B.; Kim, S.M.; Lee, Y.; Lee, W.K.; Yang, T.G.; Lee, M.J.; Suh, H.; Park, J.S. Cationic hyperbranched poly(amino ester): A novel class of DNA condensing molecule with cationic surface, biodegradable three-dimensional structure, and tertiary amine group in the interior. *J. Am. Chem. Soc.* **2001**, *123*, 2460–2461. [CrossRef] [PubMed]

34. Matsuo, E.S.; Tanaka, T. Kinetic of discontinuous volume-phase transition of gels. *J. Chem. Phys.* **1988**, *89*, 1695–1703. [CrossRef]

35. Dušek, K.; Patterson, D. Transition in swollen polymer network induced by intermolecular condensation. *J. Polym. Sci. Part A-2* **1986**, *6*, 1209–1216. [CrossRef]

36. Flory, P.J. Phase equilibria in polymer systems: Swelling of network structures. In *Principles of Polymer Chemistry*; Cornell University: Ithaca, NY, USA, 1953.

37. Flory, P.J.; Rehner, J. Statistical mechanics of cross-linked polymer networks II. Swelling. *J. Chem. Phys.* **1943**, *11*, 521–546. [CrossRef]

38. Ricka, J.; Tanaka, T. Swelling of ionic gels: Quantitative performance of the Donnan theory. *Macromolecules* **1984**, *17*, 2916–2921. [CrossRef]

39. Vasheghani-Farahani, E.; Vera, J.H.; Cooper, D.G.; Weber, M.E. Swelling of ionic gels in electrolyte solutions. *Ind. Eng. Chem. Res.* **1990**, *29*, 554–560. [CrossRef]

40. Patil, V.R.; AMiji, M.M. Preparation and characterization of freeze-dried chitosan-poly(ethylene oxide) hydrogels for site-specific antibiotic delivery in the stomach. *Pharm. Res.* **1996**, *13*, 588–593. [CrossRef]

41. Gallardo, A.; Rodriguez, G.; Aguilar, M.R.; Fernandez, M.M.; Román, S.J. A kinetic model to explain the zero-order release of drugs from ionic polymer drug conjugates: Application to AMPS-Triflusal-derived polymeric drugs. *Macromolecules* **2003**, *36*, 8876–8880. [CrossRef]

42. Kamada, H.; Tsutsumi, Y.; Yoshioka, Y. Design of a pH-Sensitive Polymeric Carrier for Drug Release and its Application in Cancer Therapy. *Clin. Cancer Res.* **2004**, *10*, 2545–2550. [CrossRef] [PubMed]

43. You, J.; Almeda, D.; Ye, G.J.C.; Auguste, D.T. Bioresponsive matrices in drug delivery. *J. Biol. Eng.* **2010**, *4*, 1–12. [CrossRef] [PubMed]

44. Gua, S.; Qiao, Y.; Wang, W.; He, H.; Deng, L.; Xing, J.; Xu, J.; Liang, X.J.; Dong, A. Poly(ε-caprolactone)-*graft*-poly(2-*N,N*-dimethylamino) ethyl methacrylate) nanoparticles: PH dependent thermo-sensitive multifunctional carriers for gene and drug delivery. *J. Mater. Chem.* **2010**, *20*, 6935–6941. [CrossRef]

45. Zhu, C.; Jung, S.; Luo, S.; Meng, F.; Zhu, X.; Park, T.G.; Zhong, Z. Co-delivery of siRNA and paclitaxel into cancer cells by biodegradable cationic micelles based on PDMAEMA-PCL-PDMAEMA triblock copolymers. *Biomaterials* **2010**, *31*, 2408–2416. [CrossRef] [PubMed]

46. Huang, X.N.; Du, F.S.; Cheng, J.; Dong, Y.Q.; Liang, D.H.; Ji, S.P.; Lin, S.S.; Li, Z.C. Acid-sensitive polymeric micelles based on thermoresponsive block copolymers with pendent cyclic orthoester groups. *Macromolecules* **2009**, *42*, 783–790. [CrossRef]

47. Hruby, M.; Filippov, S.K.; Panke, J.; Novakova, M.; Mackova, H.; Kucka, J.; Vetvicka, D.; Ulbrich, K. Polyoxazoline thermoresponsive micelles as radionuclide delivery systems. *Macromol. Biosci.* **2010**, *10*, 916–924. [CrossRef] [PubMed]

48. Lagvinac, N.; Nichols, J.L.; Ferruti, P.; Duncan, R. Poly(amidoamine) Conjugates Containing Doxorubicin Bound via an Acid-Sensitive Linker. *Macromol. Biosci.* **2009**, *9*, 480–487.

49. Song, Y.; Zhong, L.; Gan, W.; Zhou, J.; Zhang, L. Self-assembled micelles based on hydrophobically modified quaternized cellulose for drug delivery. *Colloids Surf. B* **2011**, *83*, 313–320. [CrossRef] [PubMed]

50. Akimoto, J.; Nakayama, M.; Sakai, K.; Okano, T. Temperature-induced intracellular uptake of thermoresponsive polymeric micelles. *Biomacromolecules* **2009**, *10*, 1331–1336. [CrossRef] [PubMed]

51. San Miguel, V.; Limer, A.J.; Haddleton, D.M.; Catalina, F.; Peinado, C. Biodegradable and thermoresponsive micelles of triblock copolymers based on 2-(N,N-dimethylamino)ethyl methacrylate and epsilon-caprolactam for controlled drug delivery. *Eur. Polym. J.* **2008**, *44*, 3853–3863. [CrossRef]

52. Godbey, W.T.; Mikos, A.G. Recent progress in gene delivery using non-virial transfer complexes. *J. Control Release* **2001**, *72*, 115–125. [CrossRef]

53. Lim, Y.B.; Choi, Y.H.; Park, J.S. A self-destroying polycationic polymer: Biodegradable poly(4-hydroxyl-L-proline ester). *J. Am. Chem. Soc.* **1999**, *121*, 5633–5639. [CrossRef]

54. Putnam, D.; Langer, R. Poly(4-hydroxyl-L-proline ester): Low temperature polycondensation and plasmid DNA complexation. *Macromolecules* **1999**, *32*, 3658–3662. [CrossRef]

55. Lim, Y.B.; Kim, C.H.; Kim, K.; Kim, S.W.; Park, J.S. Development of a safe gene delivery system using biodegrdable polymer, poly[α-(4-aminobutyl)-L-glycolic acid]. *J. Am. Chem. Soc.* **2000**, *122*, 6524–6525. [CrossRef]

56. Lynn, D.M.; Langer, R. Degradable poly(β-amino esters): Synthesis, characterization, and self-assembly with plasmid DNA. *J. Am. Chem. Soc.* **2000**, *122*, 10761–10768. [CrossRef]

57. Emilitri, E.; Ranucci, E.; Ferruti, P. New poly(amidoamine)s containing disulfide linkages in their main chain. *J. Polym. Sci. Polym. Chem.* **2005**, *43*, 1404–1416. [CrossRef]

58. Richardson, S.C.W.; Pattrick, N.G.; Man, Y.K.S.; Ferruti, P.; Duncan, R. Poly(amido-amine)s as potential nonviral vectors: Ability to form interpolyelectrolyte complexes and to mediate transfection in vitro. *Biomacromolecules* **2001**, *2*, 1023–1028. [CrossRef] [PubMed]

59. Ferruti, P.; Manzoni, S.; Richardson, S.C.W.; Duncan, R.; Pattrick, N.G.; Mendichi, R.; Casolaro, M. Amphoteric linear poly(amio-amine)s as endosomolytic polymers: Correlation between physicochemical and biological properties. *Macromolecules* **2000**, *33*, 7793–7800. [CrossRef]

60. Ferruti, P.; Knobloch, S.; Ranucci, E.; Duncan, R.; Gianasi, E. A Novel Modification of poly(L-lysine) leading to a soluble cationic polymer with reduced toxicity and with potential as a transfection agent. *Macromol. Chem. Phys.* **1998**, *199*, 2565–2575. [CrossRef]

61. Boussif, O.; Lezoulach, F.; Zanta, M.A.; Mergny, M.D.; Scherman, D.; Demeneix, B.; Behr, J.P. A versatile vector for gene and oligonucleotide transfer into cells in culture and in vivo: Polyethylenimine. *Proc. Natl. Acad. Sci. USA* **1995**, *92*, 7297–7301. [CrossRef] [PubMed]

62. Lin, C.; Zhong, Z.; Lok, M.C.; Jiang, X.; Hennink, W.E.; Feijen, J.; Engbersen, J.F.J. Linear poly(amido amine)s with secondary and tertiary amino groups and variable amounts of disulfide linkages: Synthesis and in vitro gene transfer properties. *J. Control. Rel.* **2006**, *116*, 130–137. [CrossRef] [PubMed]

63. Jiang, X.L.; van der Horst, A.; van Steenbergen, M.J.; Akeroyd, N.; van Nostrum, C.F.; Schoenmakers, P.J.; Hennink, W.E. Molar-mass characterization of cationic polymers for gene delivery by aqueous size-exclusion chromatrography. *Pharm. Res.* **2006**, *23*, 595–603. [CrossRef] [PubMed]

64. Stayton, P.S.; Hoffman, A.S.; Murthy, N. Molecular engineering of proteins and polymers for targeting and intracellular delivery of therapeutics. *J. Control. Release* **2000**, *65*, 203–220. [CrossRef]

65. Vanessa, I.; Afsaneh, L.; Hasana, U. Lipid and Hydrophobic Modification of Cationic Carriers on Route to Superior Gene Vectors. *Soft Matter* **2010**, *6*, 2124–2138.

66. Lim, Y.B.; Han, S.-A.; Kong, H.-U.; Park, J.-S.; Jeong, B.; Kim, S.W. Biodegradable polyester, Poly[α-(4-Aminobutyl)-I-Glycolic Acid]], as a Non-Toxic Gene Carrier. *Pharm. Res.* **2000**, *17*, 811–816. [CrossRef] [PubMed]

67. Forrest, M.L.; Meister, G.E.; Koerber, J.T.; Pack, D.W. Partial Acetylation of Polyethylenimine Enhances in vitro Gene Delivery. *Pharm. Res.* **2004**, *21*, 365–371. [CrossRef] [PubMed]

68. Lee, K.Y.; David, J.M. Hydrogels for Tissue Engineering. *Chem. Rev.* **2001**, *101*, 1869–1880. [CrossRef] [PubMed]

69. Sangram, K.S.; Mamoni, D.; Sandra, V.V.; David, L.K.; Emo, C.; Clemens, V.B.; Lorenzo, M.; Peter, D. Cationic Polymers and their Therapeutic Potential. *Chem. Soc. Rev.* **2012**, *41*, 7147–7194.

70. Kuo, Y.C.; Ku, N. Appplication of Polyethyleneimine-modified Scaffolds to the regeneration of Cartilaginous Tissue. *Biotechnol. Prog.* **2009**, *25*, 1459–1467. [CrossRef] [PubMed]

71. Zhao, L.; Burguera, E.F.; Xu, H.H.K.; Amin, N.; Ryon, H.; Arola, D.D. Fatigue and Human Umbilical Cord Stem Cell Seeding Characteristics of Calcium Phosphate-chitosan-biodegradable Fiber Scaffolds. *Biomaterials* **2010**, *31*, 840–847. [CrossRef] [PubMed]

72. Tuzlakoglu, K.; Alves, C.M.; Mano, J.F.; Reis, R.L. Production and Characterization of Chitosan Fibers and 3-D Fiber Mesh Scaffolds for Tissue Engineering Applications. *Macromol. Biosci.* **2004**, *4*, 811–819. [CrossRef] [PubMed]

73. Elsabahy, M.; Wooley, K.L. Design of Polymeric Nanoparticles for Biomedical Delivery. *Chem. Soc. Rev.* **2012**, *41*, 2545–2561. [CrossRef] [PubMed]

74. Hoffart, V.; Lamprecht, A.; Maincent, P.; Lecompte, T.; Vigneron, C.; Ubrich, N. Oral bioavailability of a Low Molecular Weight Heparin using a Polymeric Delivery system. *J. Control. Release* **2006**, *113*, 38–42. [PubMed]

75. Zhang, S.; Uludağ, H. Nanoparticle Systems for Growth Factor Delivery. *Pharm. Res.* **2009**, *26*, 1561–1580. [CrossRef] [PubMed]

76. Hoffart, V.; Ubrich, N.; Lamprecht, A.; Bachelier, K.; Vigneron, C.; Lecompte, T.; Hoffman, M.; Maincent, P. Microencapsulation of Low Molecular weight Heparin into polymeric Particles Designed with Biodegradabel and Nonbiodegradable Polycationic Polymers. *Drug Deliv.* **2003**, *10*, 1–7. [CrossRef] [PubMed]

77. Hamidi, M.; Azadi, A.; Mohamdi-Samani, S.; Rafiei, P.; Ashrafi, H. A Pharmokinetic Overveiw of Nanotechnology based Drug Delivery System. *J. Appl. Polym. Sci.* **2012**, *124*, 4686–4693.

78. Simakova, A.; Averick, S.E.; Konkolewicz, D.; Matyjaszewski, K. Aqueous ARGET ATRP. *Macromolecules* **2012**, *45*, 6371–6379. [CrossRef]

79. Mahkam, M. Modified Chitosan Crosslinked Starch Polymers for Oral Insulin Delivery. *J. Bioact. Compat. Polym.* **2010**, *25*, 406–411. [CrossRef]

80. Shao, Y.; Zhu, B.; Li, J.; Liu, X.; Tan, X.; Yang, X. Novel Chitosan Microshphere-Templated Microcapsules. *Mater. Sci. Eng. C* **2009**, *29*, 936–942. [CrossRef]

81. El-Shabouri, M.H. Positively Charged Nanoparticles for Improving the Oral Bioavailability of Cyclosporin-A. *Int. J. Pharm.* **2002**, *249*, 101–108. [CrossRef]

82. Zubareva, A.; Ilyina, A.; Prokhorov, A.; Kurek, D.; Efremov, M.; Varlamov, V.; Senel, S.; Ignatyev, P.; Svirshchevskaya, E. Characterization of Protein and Peptide Binding to Nanogels Form3ed by Differently Charged Chitosan Derivatives. *Molecules* **2013**, *18*, 7848–7864. [CrossRef] [PubMed]

83. Lipinski, C.A. Drug-like Properties and the Causes of Poor Solubility and Poor Permeability. *J. Pharmacol. Toxicol. Methods* **2000**, *44*, 235–249. [CrossRef]

84. Nottelet, B.; Patterer, M.; Francois, B.; Schott, M.; Domurado, M.; Garric, X.; Domurado, D.; Coudane, J. Nanoaggregates of Biodegradable Amphiphilic Randon Polycations for delivering Water-insoluble Drugs. *Biomacromolecules* **2012**, *13*, 1544–1553. [CrossRef] [PubMed]

85. Khopade, A.J.; Caruso, F. Electrostatically Assembled Polyelectrolyte/Dendrimer Multilayer Films as Ultrathin Nanoreservoirs. *Nano Lett.* **2002**, *2*, 415–441. [CrossRef]

86. Tong, W.; Song, X.; Gao, C. Layer-by-Layer Assembly of Micorcapsules and their Biomedical Applications. *Chem. Soc. Rev.* **2012**, *41*, 6103–6124. [CrossRef] [PubMed]

87. Gao, C.; Donath, E.; Mohwald, H.; Shen, J. Spontaneous Deposition of Water-Soluble Substances into Microcapsules: Phenomenon, Mechanism, and Application. *Angew. Chem. Int. Ed.* **2002**, *41*, 3789–3793. [CrossRef]

88. Ye, S.; Wang, C.; Liu, X.; Tong, Z. Chitosan in Nanostructured Thin Films. *J. Control. Release* **2005**, *106*, 319–328. [CrossRef] [PubMed]

89. Qiu, X.; Donath, E.; Mohwald, H. Permeability of Ibuprofen in Various Polyelectrolyte Multilayers. *Macromol. Mater. Eng.* **2001**, *286*, 591–597. [CrossRef]
90. Ai, H.; Fang, M.; Jones, S.A.; Lvov, Y.M. Electrostatic Layer-by-Layer Nanoassembly on Biological Microtemplates: Platelets. *Biomacromolecules* **2002**, *3*, 560–564. [CrossRef] [PubMed]
91. Borodina, T.; Markvicheva, E.; Kunizhev, S.; Mohawld, H.; Sukhorukov, G.B.; Kreft, O. Controlled Release of DNA from Self-degrading Microcapsules. *Macromol. Rapid Commun.* **2007**, *28*, 1894–1899. [CrossRef]
92. Ai, H.; Jones, S.A.; De Villiers, M.M.; Lvov, Y.M. Nano-encapsulation of Furosemide Microcrystals for Controlled Drug Release. *J. Control. Release* **2003**, *86*, 59–68. [CrossRef]

Review

Responsive Hydrogels from Associative Block Copolymers: Physical Gelling through Polyion Complexation

Christine M. Papadakis [1],* and Constantinos Tsitsilianis [2],*

[1] Fachgebiet Physik weicher Materie, Physik-Department, Technische Universität München, James-Franck-Str. 1, 85748 Garching, Germany
[2] Department of Chemical Engineering, University of Patras, 26504 Patras, Greece
* Correspondence: papadakis@tum.de (C.M.P.); ct@chemeng.upatras.gr (C.T.); Tel.: +49-89-289-12-447 (C.M.P.); +30-2610-969531 (C.T.)

Academic Editor: Dirk Kuckling
Received: 14 October 2016; Accepted: 6 December 2016; Published: 1 January 2017

Abstract: The present review article highlights a specific class of responsive polymer-based hydrogels which are formed through association of oppositely charged polyion segments. The underpinning temporary three-dimensional network is constituted of hydrophilic chains (either ionic or neutral) physically crosslinked by ion pair formation arising from intermolecular polyionic complexation of oppositely charged repeating units (polyacid/polybase ionic interactions). Two types of hydrogels are presented: (i) hydrogels formed by triblock copolymers bearing oppositely charged blocks (block copolyampholytes), forming self-assembled networks; and (ii) hydrogels formed by co-assembly of oppositely charged polyelectrolyte segments belonging to different macromolecules (either block copolymers or homopolyelectrolytes). Due to the weak nature of the involved polyions, these hydrogels respond to pH and are sensitive to the presence of salts. Discussing and evaluating their solution, rheological and structural properties in dependence on pH and ionic strength, it comes out that the hydrogel properties are tunable towards potential applications.

Keywords: physical hydrogel; polyelectrolyte; pH-responsive; block polyampholyte; polyion association; ionic interactions; three-dimensional (3D) network

1. Introduction

In the last decades, hydrogels have attracted extraordinary attention from theoretical computational, experimental and application point of view, thanks to their valuable properties that are suitable for a large variety of applications [1–5]. Hydrogels are water-born soft materials based mainly on a three-dimensional (3D) network, formed either by small or large organic molecules, through hierarchical self-assembly and/or crosslinking procedures On the one hand, appropriate macromolecules for the 3D network creation constitute hydrophilic chains bearing functional pendant or end groups capable of undergoing crosslinking reactions (chemical gels). On the other hand, amphiphilic polymers, namely associative polymers [6,7], are the best-suited building elements for the formation of the so-called physical hydrogels. In this case, short hydrophobic sequences, attached to long hydrophilic chains, play the role of stickers which associate intermolecularly, through hydrophobic interactions, forming non-permanent reticulation nodes. Beyond these two initially appearing categories of hydrogels, new developments have emerged in the field, due to the enormous progress of macromolecular chemistry, encompassing supramolecular and "click" chemistry, opening new strategies for designing novel polymeric materials as innovative hydrogelators. For instance, well-designed functional block copolymers, of various topologies, have been involved for the fabrication of physical hydrogels and endowed them with novel functionalities. The so-formed hydrogels are referred also as self-assembling hydrogels [8].

The involved block copolymers are constituted of a hydrophilic major part which stabilizes the hydrogel. It can be either a neutral (in most of the cases) or an ionic chain, bearing a number of ionic groups along the chain. The associative part is responsible for the development of the secondary intermolecular interactions, namely hydrophobic, ionic, H-bonding, π–π stacking, host/guest, etc., which drives the macromolecules to self-assemble in water, creating the 3D network structure. Hydrophobic interactions, exerted from the hydrophobic stickers, are the most widely studied case in hydrogels. Depending on the strength of hydrophobicity, correlated with the exchange dynamics of the stickers from the reticulation nodes, dynamic or "frozen" hydrogels can be formed [9].

"Smart" hydrogels belong to a relatively novel class of hydrogels, arising mostly from the self-assembly of the so-called stimuli-responsive block copolymers, triggered by various stimuli such as temperature, pH, light, etc. [10–14]. The latter strategy endows the hydrogels with some particular properties like self-healing and injectability [15,16] allowing "in situ gelling" [17] that meets the requirements for some specific biomedical applications like tissue engineering [18] and controlled drug delivery [19]. For instance, the hydrophobic association can be triggered by a stimulus like temperature or pH when the stickers are hydrophilic sequences, exhibiting lower critical solution temperature (LCST) behavior or suitable pK_α, respectively [20].

Beyond hydrophobic interactions, other non-covalent intermolecular secondary interactions have recently been used for designing "smart" hydrogelators. Among those, metal–ligand coordination has been utilized to crosslink macromolecules into 3D structures [21]. The so-formed physical gels, referred to as metallo-supramolecular polymer gels, are based on 3D networks where the linkages between the macromolecules are provided by reversible, labile metal–ligand coordination bonds. Recently, a supramolecular stimuli-responsive polymer gel was fabricated by well-designed heterotelechelic block copolymers, one extremity being ended by a short associating sticker and the other bearing a chelating ligand. Through the hydrophobic and coordination terminal moieties, the copolymer was hierarchically associated into a supramolecular network with tunable viscoelastic response and yield behavior [22]. An early review summarizes work up to 2006 [23].

Ionic interactions between oppositely charged repeating units (polyacid–polybase), located in the same or in different macromolecules, constitute another strategy to design "smart" hydrogels, and this is the topic of the present review. The hydrogel formation relies on the oppositely charged polyion association, and it is driven by ion pairing (ionic bonds) and the entropy gain due to the release of counterions and hydration water [24], leading to the formation of polyion complexes which form the physically reversible cross-links of the transient network, also named complex coacervate cores (CCCs) [25,26]. The involved interactions between the oppositely charged segments were referred, in most of the cases in the literature (also by us) as electrostatic, which seems inappropriate according to the thermodynamics of polyion association which is mainly entropic [27–29]. Herein, we use the term "ionic interactions" characterizing the interactions between oppositely charged groups (segments) that lead to electro-neutralization of the formed polyion complexes [24]. Depending on the environmental conditions, e.g., presence of salt, these interactions can lead to the formation of the so-called interpolyelectrolyte complexes, namely: (i) (dense) polyelectrolyte complexes or complex precipitates; and/or (ii) polyelectrolyte coacervates (hydrated domains), which are characterized by loose association and liquid like properties [30,31].

By using weak electrolyte repeating units, the so-formed hydrogels are strongly pH-responsive since pH affects the degree of ionization of the oppositely charged polyions and hence their extent of association, rendering them promising candidates for biomedical applications. Two types of systems, block copolymer polyampholytes and mixtures of two oppositely charged polyions (either in the form of triblock/pure polyelectrolyte or triblock/triblock), will be presented and discussed.

We note here that from the strictly rheological point of view, physical hydrogels are soft solids (elastic response), exhibiting very long relaxation times (i.e., very slow exchange dynamics) much higher than the experimental time, having, hence, immeasurable zero-shear viscosity and yield behavior. However, a broader definition of hydrogels, also referred to as free-standing gels (in the time of observation), comprise systems that feature viscoelasticity with measurable long relaxation times and high viscosities, has been used in the literature. In this article, the broader definition is adopted.

The paper is structured as follows. In the first section, we describe systems based on charge-driven self-assembly, afforded by one block copolymer polyampholyte. Afterwards, systems based on charge-driven co-assembly, comprising two oppositely charged macromolecules, are presented. In each case, the solution properties as well as the rheological and structural properties of the hydrogels are discussed. Finally, we summarize the findings and give an outlook.

2. Systems Based on Self-Assembly

The first system enabling self-assembly in aqueous environment was formed by asymmetric triblock copolymers having negatively charged short end blocks and a positively charged long middle block [32]. In addition to the usual possibilities to vary the network properties by variation of the block lengths, the choice of the nature of the polyelectrolyte (strong versus weak) controls the properties of the hydrogels. Moreover, the pH value (weak polyelectrolyte case) and the ionic strength have a strong influence. The system based on the triblock polyampholyte PAA-*b*-P2VP-*b*-PAA (APA) offers great variability because the degrees of ionization of both blocks are pH-dependent: poly(acrylic acid) (A) has a pK_a of 4.5 and poly(2-vinyl pyridine) (P) a pK_b of 5.0. Thus, varying pH alters the net charge and the anion/cation molar ratio along the polymer. Moreover, the uncharged monomers of both blocks are either hydrophobic or capable of developing H-bonding, and their association contributes to the charge-driven self-assembling behavior and the mechanical properties.

2.1. Dilute Solution Properties

The intermolecular association of the PAA$_{134}$-*b*-P2VP$_{628}$-*b*-PAA$_{134}$ (APA$_1$) block polyampholyte towards a 3D network was revealed exploring salt-free dilute aqueous solutions by electrophoresis (zeta potential), turbidimetry (visible light) and capillary viscometry (reduced viscosity) as a function of pH (Figure 1a,b) [32]. The system exhibited three phases: (i) low pH, clear solution; (ii) intermediate pH, two phases, namely the isoelectric point (*iep*) region; and (iii) high pH, clear solution, where the polyampholyte has been transformed to an amphiphilic polyelectrolyte (charged A and hydrophobic P). At pH values just prior to the insoluble *iep* region, a maximum in the reduced viscosity indicated intermolecular association which leads to a 3D network by increasing concentration (Figure 1c) [33]. The percolation threshold was determined at 2.4 wt % polymer concentration, while above 4.5 wt %, a free-standing gel was observed. The hydrogel exhibited shear thinning properties and viscoelastic response with relaxation times of the order of hundreds of seconds (Figure 1d). The network formation was attributed to the ionic association among the protonated positively charged P and a limited number of deprotonated, negatively charged A.

Recent developments in this system encompass quaternization of the P block resulting in PAA-*b*-QP2VP-*b*-PAA (A(QP)A) with the strong cationic polyelectrolyte quaternized poly(2-vinyl pyridine) (QP) as the long middle block (absence of hydrophobic uncharged P units), which renders it soluble in the entire pH range (cationic P moieties always predominate). In addition, only the A end blocks exhibit a pH-dependent degree of ionization, facilitating the control of the charge imbalance (anion/cation molar ratio) which promotes better understanding of the system.

Parallel investigations of these two triblock polyampholytes (of the same block lengths) in dilute solutions (0.2 wt %) revealed that the quaternized version of the triblock polyampholyte, PAA$_{163}$-*b*-QP2VP$_{1397}$-*b*-PAA$_{163}$ (A(QP)A$_2$), stayed water-soluble in the entire pH region, which was attributed to its high net charge as was corroborated by the positive zeta potential for pH values between 1 and 13 [34]. In contrast, PAA$_{163}$-*b*-P2VP$_{1397}$-*b*-PAA$_{163}$ (APA$_2$) precipitated above pH 5, and its zeta potential changed sign from positive below pH 5, where the ionized P predominates, to negative above pH 7, where the deprotonated A segments prevail. Thus, both polyampholytes had a positive net charge in the low pH region, which was of interest for the formation of hydrogels.

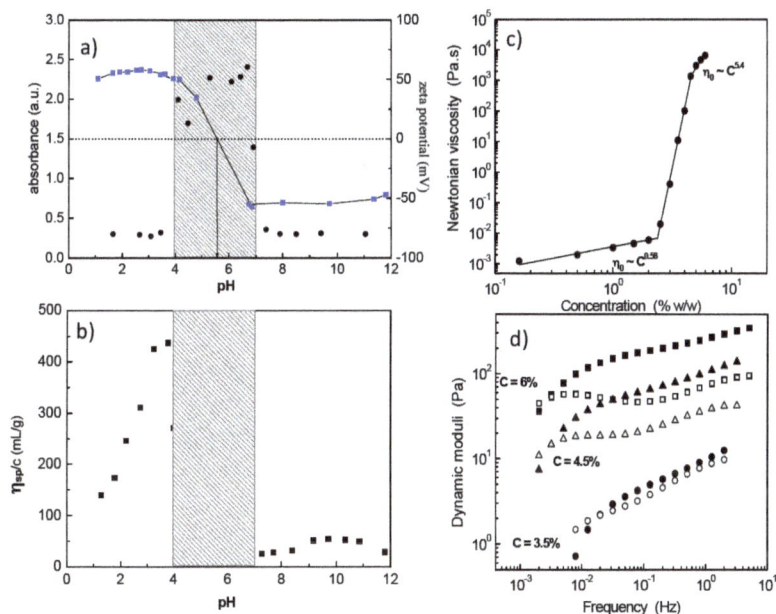

Figure 1. pH dependence of: (**a**) the optical absorbance (black symbols), the zeta potential (blue symbols); and (**b**) of the reduced viscosity, η_{sp}/c, for a APA$_1$ aqueous solution (0.2 wt %) at 25 °C. (**a,b**) Adapted with permission from [32]. Copyright 2003 American Chemical Society. (**c**) Zero-shear viscosity as a function of polymer concentration; (**d**) Storage modulus G' (closed symbols) and loss modulus G'' (open symbols) as a function of frequency at different polymer concentrations: circles (3.5 wt %), triangles (4.5 wt %), squares (6.0 wt %). Adapted with permission from [33]. Copyright 2004 American Chemical Society.

2.2. Rheological Properties of the Hydrogels

Steady-state shear viscosity measurements and tube inversion tests on more concentrated (4 wt %) aqueous solutions of A(QP)A$_2$ at pH values between 2.5 and 7.0 revealed the formation of transparent, free-standing hydrogels (Figure 2a), particularly at pH 3 and 4. Gelation occurred in the same pH region as for the non-quaternized precursor (APA$_2$), and both systems exhibit maximum value in zero shear viscosity close to pH 3 (Figure 2b). Thus, the driving force for the formation of 3D network is of ionic nature (polyion complexation), since, in the quaternized version, all P moieties of the central block are permanently charged, and hence hydrophobic interactions and H-bonding with A are negligible. The drastic effect of the charge imbalance was manifested in the marked decrease of the viscosity at pH 5 and 2.5, which demonstrated that the degree of ionization of the A end blocks, which are weak polyelectrolytes, is at the origin of the pH responsive behavior. This pH dependent behavior was confirmed in oscillatory measurements in the linear viscoelastic regime (Figure 2c). At pH 3 and 4, G' was higher than G'' in the whole frequency range with relaxation times higher than 500 s, which confirmed the appearance of free-supporting hydrogels at these pH values. At pH 5, in contrast, viscoelastic behavior was observed with a terminal relaxation time of ca. 50 s. For the non-quaternized triblock polyampholyte APA$_2$, the zero shear viscosity at 1.2 wt % was also maximum at pH 3, however, precipitation set in above pH 4 (inset of Figure 2b). For this polymer, the viscoelastic behavior appeared at lower pH: At pH 4.35, the terminal relaxation time already decreased to 13 s. Both observations confirmed the important role of the increasing hydrophobicity of the P block with pH.

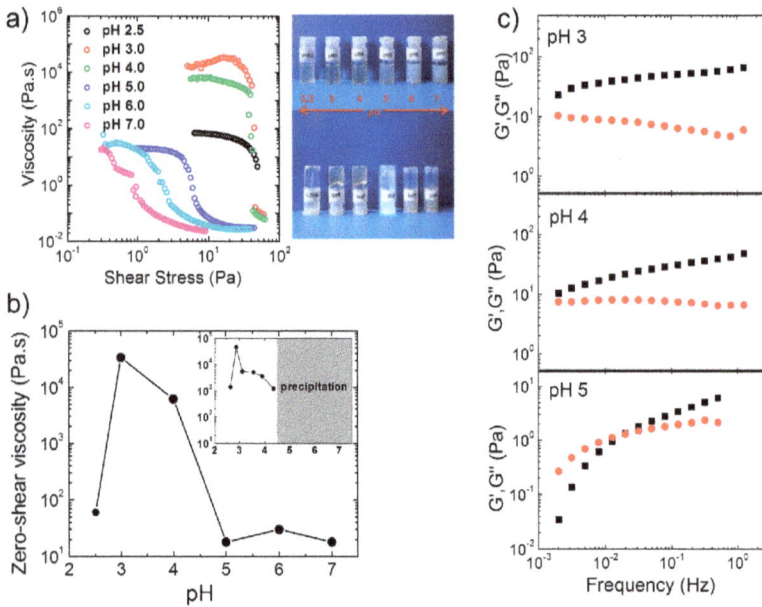

Figure 2. pH-dependent rheological properties of salt-free aqueous solutions of A(QP)A$_2$. (a) **Left**: Apparent viscosity as a function of shear stress for A(QP)A$_2$ at c = 4 wt % and various pH conditions; **right**: photos showing free-standing gels at pH 3 and 4, whereas solution behavior is observed at other pH values; (b) pH dependence of zero-shear viscosity, η_0, at c = 4 wt % and its precursor APA$_2$ at c = 1.2 wt % (**inset**); (c) Dynamic moduli, G' (black symbols) and G'' (red symbols) versus frequency at different pH values of A(QP)A$_2$ at c = 4 wt % at the pH values given in the graphs. Adapted with permission from [34]. Copyright 2014 American Chemical Society.

The concentrations, at which gel formation sets in, were found to be very low for the two polymers and to differ slightly (Figure 3a): For A(QP)A$_2$, the zero-shear viscosity increased by ca. 6 orders of magnitude between a polymer concentration of 1.0 wt % and 1.5 wt %, which pointed to the formation of a transient network above this concentration, critical gel concentration (C_{gel}). For APA$_2$, the same behavior was observed, however, C_{gel} was with ca. 0.4 wt % much lower than in the quaternized copolymer. Thus, C_{gel} also depends remarkably on the ratio of the opposite charges. We should also note that C_{gel} is remarkably lower than that of APA$_1$ (Figure 3a), indicating the influence of the molecular features of the copolymer on the rheological properties and network structure. The nonlinear behavior of the hydrogel from A(QP)A$_2$ was probed in a steady state shear experiment (Figure 3b). At a concentration just above C_{gel}, shear thickening was observed prior a dramatic shear thinning, marked by a drop of the viscosity of five orders of magnitude at a relatively low stress (ca. 8 Pa), implying easy disruption of the network structure. Remarkable hysteresis was observed, pointing to the slow structure recovery, in accordance with the long relaxation time observed by oscillatory measurements (Figure 2c).

Thus, rheological characterization established that both, ionic and hydrophobic (for the non-quaternized polyampholyte) interactions are of importance for the formation of a 3D network. These are accessible by altering pH and thus the degrees of ionization of the oppositely charged moieties.

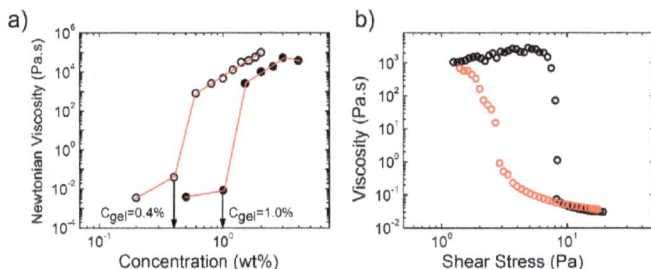

Figure 3. (a) Concentration dependence of the zero shear viscosity at the pH of maximum viscosity for A(QP)A$_2$ (closed symbols) and APA$_2$ (pH 2.9) (open symbols). Lines along the data guide the eyes and arrows indicate the gelation concentration; (b) Apparent viscosity versus shear stress of 1.6 wt % aqueous A(QP)A$_2$ solution at pH 3: increasing stress (black circles) and decreasing stress (red circles). Adapted with permission from [34]. Copyright 2014 American Chemical Society.

2.3. Structural Properties of the Hydrogels in Dependence on pH and Ionic Strength

At a concentration safely above C_{gel}, namely 4 wt % (Figure 3a), a pronounced effect of the pH value on the mechanical properties was observed (Figure 2). At this concentration in heavy water, D$_2$O, the structural characteristics also showed gross differences, as found using small-angle neutron scattering (SANS). The SANS curves from the quaternized polyampholyte A(QP)A$_2$ at pD 7.0, 5.0, and 3.0 as well as the one from APA$_2$ at pD 3.0 are compiled in Figure 4a. As evident from their very different shapes, the morphology depends strongly on the charge imbalance (controlled by pH). The curves at pD 3.0 from A(QP)A$_2$ and APA$_2$ have similar shape, but the features are shifted along the momentum transfer, i.e., the length scales involved differ. Structural models were fitted to describe the morphology and to extract structural parameters.

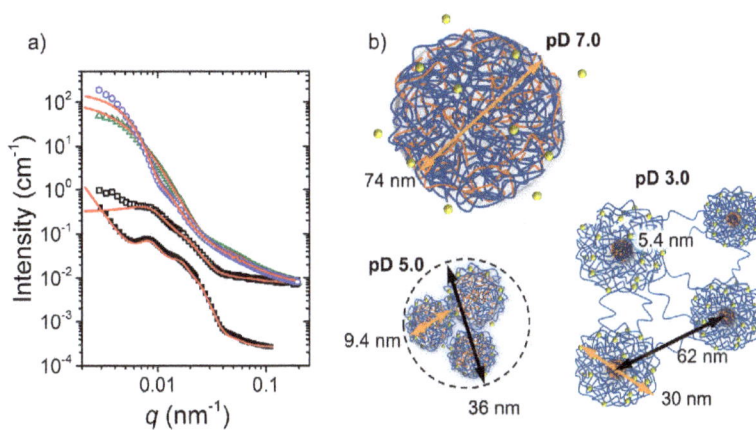

Figure 4. SANS results from A(QP)A$_2$ and APA$_2$. Data from [34]. (a) Scattered intensity, $I(q)$, as a function of the momentum transfer, q, of solutions of A(QP)A$_2$ at pD 7.0 (open blue circles), pD 5.0 (open green triangles) and pD 3.0 (open black squares), and APA$_2$ at pD 3.0 (closed black squares, shifted vertically for clarity). All scattering curves were measured at a concentration of 4 wt % and at 26 °C. The red lines are the model fits; (b) Corresponding nanostructures at the pH values given. Blue lines: QP, orange lines: A, yellow circles: counter ions. Average length scales are given. Adapted with permission from [34]. Copyright 2014 American Chemical Society.

The scattering curve of A(QP)A$_2$ at pD 7.0, where the charge imbalance is minimum (degree of ionization of the A block ca. 80%) and extended ionic interactions take place, was successfully modeled with large spherical particles (radius ~37 nm) together with concentration fluctuations (for details see [34]). These particles had a loose internal structure with an inner correlation length of ~2.6 nm. They were composed of complexed negatively charged groups of the A end blocks and of positively charged groups of the long QP middle block together with about 14 wt % of water (coacervate type structures). The ratio of positively and negatively charged chains within the particle was imbalanced (due to asymmetric blocks), and the excess positively charged QP moieties stabilized the particles. This morphology—charged, unconnected, microgel-like associates of mesoscopic size, as sketched in Figure 4b—is in accordance with the low viscosity at pH 7.0 (Figure 2b).

At pD 5.0, the charge imbalance on the chain was higher due to the smaller fraction of negatively charged A units (~50%). The SANS curve (Figure 4a) was successfully modeled by small spherical core-shell particles having an average core radius of 3.4 nm and a shell thickness of 1.3 nm. The cores consist again of CCCs from negatively charged A segments and positively charged QP segments. The smaller size than the one at pD 7.0 is due to the increasing charge imbalance, resulting in smaller polyion complexes. The shell contains the remaining QP groups. These associates formed larger micellar clusters, containing ca. 50 small micelles. From the viscoelastic behavior described above, it was concluded that these clusters are loosely connected (Figure 4b) and may be disrupted by shear forces.

At pD 3.0, the charge imbalance is maximum with ca. 25% of the A segments being charged (note that the apparent pK_a of A is lowered due to the presence of QP [34]). The SANS curve was fitted using a model including the form factor of polydisperse spherical core-shell particles which are correlated with each other. The particles were found to have an average core radius of ~3 nm and a (rather high) shell thickness of ~12 nm. The average distance between the particle cores was ~60 nm. Thus, small cores were surrounded by a shell of (possibly backfolding) QP blocks and connected with each other by QP bridges. This became possible since less QP segments were involved in the formation of the crosslinked domains than at the higher pD values. At pD 3.0, the sample thus forms a 3D network (Figure 4b) in accordance with the observed gel-like behavior.

The SANS curve of the non-quaternized triblock polyampholyte APA$_2$ at 4 wt % in D$_2$O pD 3.0 looked similar to the one of A(QP)A$_2$, albeit shifted to lower values of the momentum transfer, q, and with more pronounced features. Again, spherical core-shell particles were found, having an average core radius of ~4 nm, which is larger than in A(QP)A$_2$, probably because of the overall higher amount of uncharged, hydrophobic P segments, contributing to the core size augmentation. Accordingly, the shell thickness was found to be smaller than in A(QP)A$_2$, namely only ~6 nm. The higher cross-link functionality is in accordance with the higher zero-shear viscosity observed in the non-quaternized hydrogel (Figure 3a). The hard-sphere radius was higher in APA$_2$ (~36 nm in comparison with ~31 nm in A(QP)A$_2$). Thus, the bridges between the crosslinked cores are longer, resulting in a higher connectivity, in accordance with the lower C_{gel} value for APA$_2$.

The results from the systematic study of the self-assembled, quaternized and non-quaternized triblock polyampholyte system showed that the overall degree of ionization and the charge imbalance together with the hydrophobicity of the uncharged segments result in strongly pH dependent morphologies and, consequently, in vastly different rheological properties.

2.4. Influence of Ionic Strength

The above-described system was also found to be sensitive to ionic strength because the addition of salt reduces the charge density of both type of blocks [35]. It was found that the addition of salt may alter both, the stability of the complex coacervate cores and the bridging ability of the middle blocks. This became evident by comparing the effect of NaCl on two systems: (a) a hydrogel from (PAA$_{109}$-*b*-P2VP$_{819}$-*b*-PAA$_{109}$) APA$_3$ which was investigated at pD 3.0; and (b) a viscoelastic liquid formed by PAA_{109}-*b*-$QP2VP_{819}$-*b*-PAA_{109} (A(QP)A$_3$) which was investigated at pD 5.0, both at a concentration of 3 wt %.

The zero-shear viscosity of APA$_3$ at pH 3.0, extracted from creep measurements, diminished continuously with increasing NaCl concentration (Figure 5a). Starting from a rather high value, which reflects the stiffness of the hydrogel, in salt-free conditions, the viscosity decreased strongly with ionic strength. The ion-pairing dissociation induced by salt addition, weakens the integrity of the physical ionic crosslinks, facilitating therefore the exchange of the A chain-ends from their junctions, decreasing therefore the terminal relaxation times. The stability of these junctions, are affected by the NaCl ions, which are able to penetrate them and to break ionic bonds (through counterion exchange) between oppositely charged A and P blocks, thus leading to their partial disintegration and a softer gel. Due to the high polymer concentration, the number of salt ions was not sufficient, though, to break all polyion pairs, and, therefore, the elastic character of the hydrogels was preserved even at 0.5 M NaCl [35].

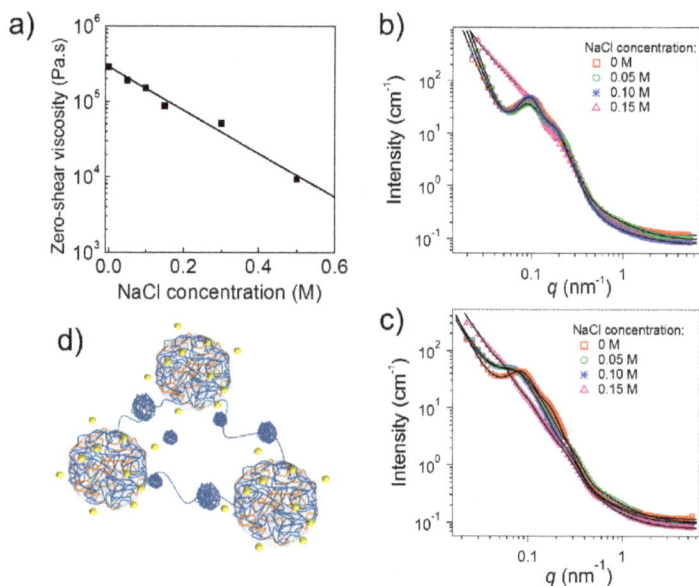

Figure 5. Effect of NaCl on the self-assembled hydrogels from A(QP)A$_3$ and APA$_3$. (**a**) Zero shear viscosity, η_0, as a function of NaCl concentration of 3 wt % APA$_3$ aqueous solutions at pH 3.0; (**b,c**) SANS curves of solutions of APA$_3$ at pD 3.0 (**b**) and A(QP)A$_3$ at at pD 5.0 (**c**), both at 3 wt % in D$_2$O and at 26 °C for different NaCl concentrations (symbols) together with the model curves (lines); and (**d**) sketch of the morphology at a NaCl concentration of 0.15 M. Adapted with permission from [35]. Copyright 2015 American Chemical Society.

The viscosity changes upon salt addition were accompanied by morphological changes, as evidenced by SANS (Figure 5b). For the salt-free solution of APA$_3$ at pD 3 as well as the ones at 0.05 M and 0.10 M, the scattering curves had a shape characteristic of a network formed by interconnected coacervate cores (cf. Figure 4a). Only at a salt concentration of 0.15 M, the shape strongly differed. At NaCl concentrations up to 0.10 M, coacervate domains having radii of ca. 8 nm were present (Figure 5d). The hard-sphere radius was ca. 30 nm. Thus, in this range of NaCl concentrations, SANS shows that there are no dramatic changes in the network architecture, but rather a redistribution of chains between the complexes.

At 0.15 M NaCl, the same structural model could be fitted. The average radius of the coacervate domains was ~17 nm, thus larger than that at lower NaCl concentrations. This was attributed to the increased hydrophobicity of the coacervate domains and to the screening of the charges of the P blocks, and that some of these uncharged P segments joined the coacervate domains. Moreover, additional small spheres were present, having an average radius of ~7 nm which were attributed to small globules of uncharged P (Figure 5d). Moreover, increased scattering at small momentum transfers indicated that smaller and looser aggregates of coacervate domains were present at 0.15 M NaCl, in contrast to the larger aggregates giving rise to a steep increase of intensity with decreasing momentum transfer at 0–0.10 M NaCl (Figure 5b). The addition of salt thus caused a breakup of the large aggregates which is in accordance with the decrease of the zero-shear viscosity discussed above.

The A(QP)A system at pD 5.0 was strongly charged—the QP block was nearly fully charged and the fraction of charged A segments was ca. 55%. Thus, this system offered the possibility to investigate the role of ionic effects upon addition of salt which were expected to dominate over hydrophobic effects due to non-charged segments. The SANS curves again revealed network formation at low NaCl concentrations (Figure 5c), but—in contrast to the APA$_3$ system at pD 3.0—changes were already apparent at 0.05 M NaCl. Model fitting revealed an average radius of the coacervate domains of ~7 nm in the salt-free state with an average hard-sphere radius of ~34 nm. Their radius grew steadily to ~19 nm at 0.15 M NaCl, which was attributed to the fact that the presence of small amounts of salt may enhance complex formation by weakening ionic interactions, favoring more accessible polymer conformations, and enabling chain rearrangement [36]. Computer simulations confirmed the growth of coacervate domains upon increasing ionic strength (Figure 6) [34]. Moreover, QP globules having an average radius of ~8 nm were again detected. The hard-sphere radius follows the same tendency as the radius of the complex coacervate domains.

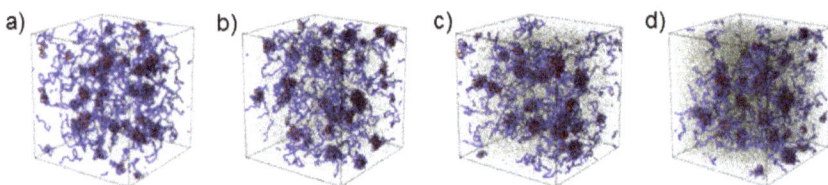

Figure 6. Snapshots from computer simulations of polyampholyte solutions PAA$_{20}$-*b*-QP2VP$_{172}$-*b*-PAA$_{20}$ (A(QP)A$_4$) at a polymer concentration of 2.2 wt % in salt-free solution (**a**); and at salt concentrations of: 0.153 M (**b**); 0.305 M (**c**); and 0.61 M (**d**). A and QP monomers are shown in orange and blue color, respectively. Reprinted with permission from [35]. Copyright 2015 American Chemical Society.

Comparing the two systems, which were studied experimentally, reveals that the hydrogel formed by the non-quaternized APA$_3$ at pD 3.0 is less sensitive to ionic strength than the ones from A(QP)A$_3$ at pD 5.0. In the former system, the impact of screening by the small ions is not as pronounced as in the latter system because of the higher fraction of non-ionized segments. The hydrophobic interactions between non-ionized P segments hamper the effect of the salt addition. In the quaternized system, in contrast, strong effects of the salt on the network structure are observed, because ionic interactions play a dominant role in the highly charged system.

3. Systems Based on Co-Assembly

3.1. Co-Assembly of Triblock Copolymers Having Charged Endblocks and Oppositely Charged Homopolymers

Charge-driven co-assembly is based on the co-dissolution of oppositely charged polymers. These may form multiresponsive reversible gels which are sensitive not only to changes in temperature and concentration but also to ionic strength, cationic/anionic composition and, if weak polyelectrolytes are used, pH value. Such a system was designed by Lemmers et al. [37–39] who investigated the co-assembly of a triblock copolymer with negatively charged end blocks and a neutral, hydrophilic middle block with a chemically different, positively charged homopolymer. The hydrophilic middle blocks prevented phase separation but led instead to microphase separation. At low polymer concentrations, flower-like micelles with CCCs, arising from the polyion complexation of the negatively charged end blocks and the positively charged homopolymers and loops of the middle blocks forming the shell were observed. At higher concentrations, the micelles were bridged by the middle blocks, resulting in the formation of a transient micellar network.

The polymers under study were the triblock copolymer $PSPMA_{28}$-*b*-PEO_{210}-*b*-$PSPMA_{28}$ (SES) (Figure 7a) and the homopolymer poly(allylamine hydrochloride) (AH) (Figure 7b). The appropriate mixing ratio was determined using titration dynamic light scattering on dilute solutions (ca. 1 g·L^{-1} in a 0.2 M KCl aqueous solution). At the composition where the scattering intensity was maximum, the triblock copolymers and the homopolymers formed micelles with a hydrodynamic radius of ca. 20 nm [37]. This occurred at the charge composition variable $f^{+} = [+]/([+]+[-]) \approx 0.5$ (charge stoichiometry) where $[+]$ and $[-]$ denote the concentrations of positive and negative charges, respectively.

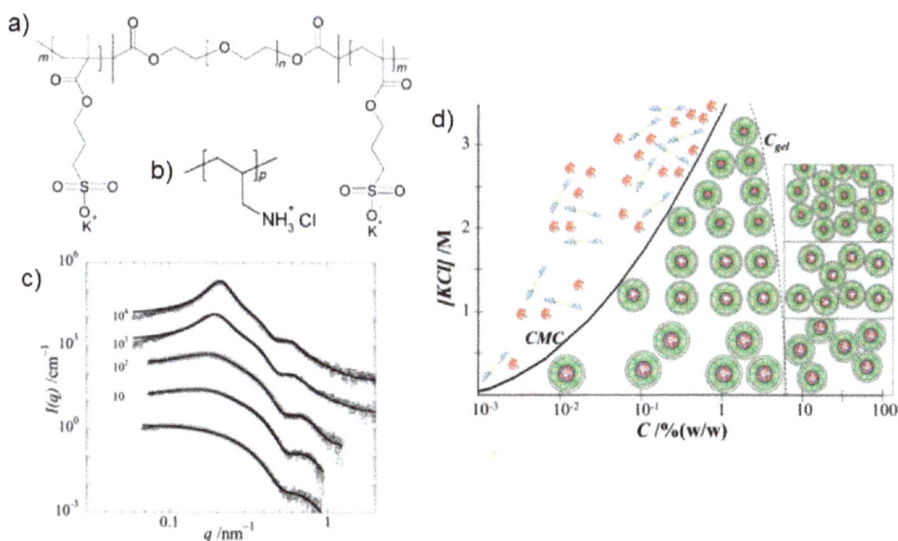

Figure 7. Co-assembled system formed by a triblock copolymer with negatively charged end blocks and: a neutral middle block (SES) (**a**); and AH homopolymer (**b**); (**c**) Small-angle X-ray scattering (SAXS) data for gels from SES and AH at 0.4 M KCl at 20 °C. The polymer concentrations are (from bottom to top) 1%, 4%, 6%, 12% and 20%. Lines are model fits. The curves are shifted vertically as indicated in the graph; (**d**) Schematic morphology diagram for gels from SES and AH in dependence on polymer concentration, *C*, and salt concentration [KCl]. Reproduced with permission from [38]. Copyright 2011 Royal Society of Chemistry.

At higher polymer concentrations, the viscosity increases by several orders of magnitude between 5 wt % and 20 wt %, and the critical gel concentration C_{gel} was determined at 4 wt % [37]. Small-angle X-ray scattering (SAXS) at concentrations of 1–16 wt % revealed spherical particles having a radius of gyration of ca. 8 nm and a distance between them of ca. 30 nm [37]. Addition of KCl led to a decrease of both, the intensity of scattered light and the viscosity. Thus, the almost solid-like gel was transformed into a water-like fluid by adding KCl salt. The gels were also found to be pH responsive because the AH homopolymer is a weak polyelectrolyte. Increasing the pH of a dilute solution of micelles resulted in the decrease in the number of micelles above pH 8. In a gel formed at 18 wt % and 0.4 M KCl, the viscosity decreased drastically upon addition of KOH.

In a later study [38], the same polymers, but having slightly different molar masses, were investigated. A detailed SAXS investigation was carried out in a wide concentration range at 0.4 M (Figure 7c) and 1.0 M KCl. At the lowest polymer concentration, 1 wt %, nearly uncorrelated micelles were observed and could be described as homogeneous spheres having an average radius of 8 nm. With increasing concentration, the spheres became more strongly correlated, as evidenced from the more pronounced peak in the SAXS curves (Figure 7c). The sphere radius resulting from model fitting was nearly independent on concentration. The hard-sphere radius, obtained from the structure factor in the model function, did not depend on concentration either, which may be counterintuitive. However, the volume fraction of the spheres increased nearly linearly with polymer concentration up to 15 wt %, then it leveled off. From the resulting number density of micelles, the authors calculated an average center-to-center distance of the spheres, D. A decrease with increasing polymer concentration following the expected relation $D \propto C^{-1/3}$ was found. The aggregation number of the spheres was found to be ca. 110 at a KCl concentration of 0.4 M and to decrease linearly with [KCl].

Moreover, the mechanical properties were investigated [38]. In frequency sweeps, viscoelastic behavior was demonstrated for a 20 wt % polymer concentration at 0.4 M KCl. The plateau modulus was found to increase sharply between 8 wt % and 10 wt % which the authors attributed to an increasing number of bridges. Based on the structural and rheological characterization, the authors proposed a morphology diagram of the charge-driven, co-assembled system in dependence on polymer concentration and KCl concentration (Figure 7d). The critical micelle concentration (CMC), i.e., the concentration above which (flower-like) micelles form, increased with salt concentration. At high polymer concentration, the micelles were found to pack more closely and to be strongly bridged. The co-assembled charge-driven system thus offers great variability in terms of composition and responsivity to pH, ionic strength and temperature.

In a later investigation, the authors investigated the effect of charge composition, f^+, while keeping the overall polymer concentration constant [39]. For $f^+ < 0.5$, excess negative charges persisted, leading to small, negatively charged aggregates (radius 5–10 nm) from negatively charged homopolymers and positively charged triblock copolymers, called "soluble complexes" (Figure 8). The viscosity was relatively low. At charge stoichiometry, $f^+ = 0.5$, the flowerlike micelles (radius 15–20 nm) were interconnected and tightly packed. At excess positive charge, $f^+ > 0.5$, the micellar size decreased, and a number of dangling positive end blocks stabilized the micelles, maintaining electroneutral complex cores. The authors suggested that bridge formation between these positively charged particles is suppressed, being at the origin of the relatively low viscosity.

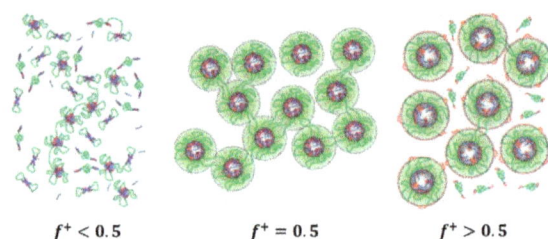

$f^+ < 0.5$ $\qquad\qquad$ $f^+ = 0.5$ $\qquad\qquad$ $f^+ > 0.5$

Figure 8. Schematics of the morphologies of solutions from SES and AH for excess negative charge, charge stoichiometry and excess positive charge for the concentrated regime. Adapted with permission from [39]. Copyright 2012 Royal Society of Chemistry.

Ishii et al. designed a system which is responsive to both, temperature and ionic strength and therefore qualifies as an injectable hydrogel for medical purposes [40]. Triblock copolymers featuring a water-soluble poly(ethylene glycol) (EG) middle block and two cationic poly[4-(2,2,6,6-tetramethylpiperidine-*N*-oxyl)-aminomethylstyrene] (M) end blocks were codissolved with anionic poly(acrylic acid) (A) homopolymers. PMNT-*b*-PEG-*b*-PMNT (M(EG)M) and A were mixed at several molar ratios in a phosphate buffer solution (Figure 9). At low concentrations, they formed polyion complex flower-like micelles at room temperature. Heating a micellar solution having a molar ratio $r = 1:1$ (the molar ratio was defined as the ratio of the molar unit of cationic amine groups of MEM and the molar unit of the anionic carboxyl groups of A and a concentration of 55 mg/mL) resulted in a steep increase of the storage and the loss modulus at 27.6 °C. Interestingly, for $r = 1:1$ and 1:2, the high moduli persisted when cooling back from 45 °C to 15 °C, i.e., the gel formation was irreversible. The authors attributed the gel formation to the formation of ionic cross-links between the cationic MEM and the anionic A, namely by destruction of the flower-like structure and not by their aggregation. This is corroborated by the high modulus in the gel state (>1000 Pa). At low temperatures, the gel formation is prevented by the EG shell of the micelles. Gel formation occurred at concentrations as low as 2 wt %. For $r = 2:1$, the gel formation was found to be reversible. The sol-gel transition temperature decreased upon addition of NaCl, the system is thus responsive to ionic strength. The authors suggested the use of the co-assembled hydrogel as injectable hydrogels: While the micellar solutions at room temperature have low viscosity and can be injected easily into the body, they form gels in the body because of an increase in both, temperature and ionic strength. Since a certain ionic strength is necessary, gelation will not happen in the catheter, but only in the body. It was shown that the system can also incorporate charged macromolecules and therefore may be used as a local delivery carrier of charged drugs.

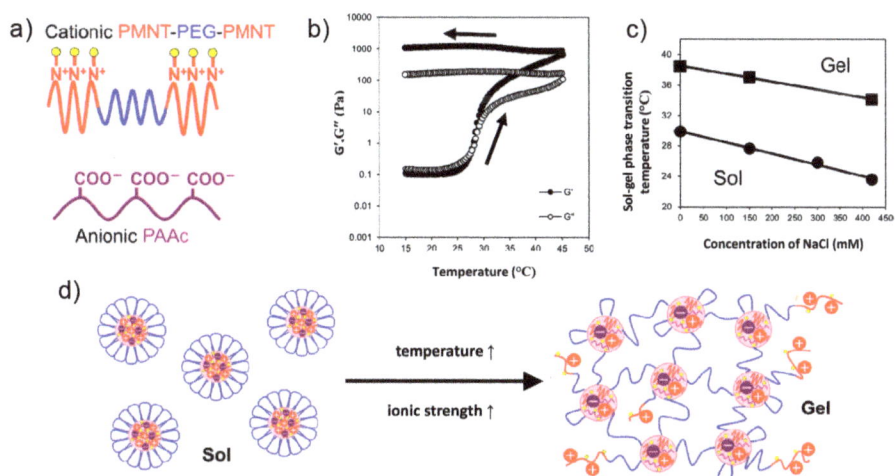

Figure 9. (**a**) Schematics of the polymers MEM and A; (**b**) storage modulus G′ and loss modulus G″ of polyion complexes flower micelles (55 mg/mL, $r = 1 : 1$, 150 mM NaCl, pH 6.2, 550 mM phosphate buffer) with increasing temperature and decreasing temperature, as indicated by the arrows; (**c**) ionic strength dependence (55 mg/mL, $r = 1 : 1$, pH 6.2): concentration of phosphate buffer 330 mM (closed squares) and 550 mM (closed circles); and (**d**) schematics of the solution of the polyion complexes micelles and the irreversible gel formation upon increasing temperature and ionic strength. Adapted with permission from [40]. Copyright 2015 American Chemical Society.

3.2. Co-Assembly of Triblock Copolymers Having Oppositely Charged Endblocks

Several investigations have addressed responsive, co-assembled hydrogels formed by symmetrical triblock copolymers having the same hydrophilic middle block and oppositely charged end blocks [41,42]. Theoretical work [43] addressed the influence of the interaction parameters of the various segments with each other and with the solvents.

Hunt et al. [41] and Krogstad et al. [42] designed high performance hydrogels by exploiting the complex coacervate formation of oppositely charged end blocks of two triblock copolymers which have the same long and hydrophilic middle block (Figure 10a). These are both water-soluble, facilitating gel preparation by mixing. The coacervate domains formed by the end blocks are sensitive to pH and ionic strength, and the gels are thus responsive materials. The charged triblock copolymers were synthesized from a common uncharged precursor triblock copolymer, namely PAGE-*b*-PEG-*b*-PAGE ((GE)(EG)(GE)). The alkene groups in the poly(allyl glycidyl ether) (GE) end blocks were used to introduce ionic groups, namely sulfonate, carboxylate, ammonium or guanidinium. At this, the molar mass of the EG middle block was varied between 10 and 35 kg/mol and the number of charged groups in the end blocks between 22 and 53 per end block. Following this concept, charge imbalance could be avoided.

Mixing triblock copolymers having carboxylate and guadinium groups at 10 wt % resulted in transparent and mechanically robust hydrogels. In dynamic mechanical measurements, the shear and the loss modulus showed a crossover at 15.8 rad/s, i.e., they were viscoelastic. Hydrogels were also formed upon mixing of triblock copolymers with sulfonate and guadinium groups. These were more elastic with a longer relaxation time. When the number of charged groups was sufficiently high, the coacervate domains formed a body-centered cubic lattice, as evidenced using SAXS. The modular approach thus resulted in a versatile system which allows for tuning of the properties.

In a later work, the same group systematically investigated hydrogels formed by mixtures of triblock copolymers with sulfonate and guadinium groups (Figure 10b, [42]). They focused on the effects of polymer concentration and salt (NaCl) concentration as well as pH and stoichiometry of the charged moieties on the mechanical properties and the morphologies. The block lengths were chosen at 31 for the end blocks and 455 for the middle block. Using dynamic mechanical spectroscopy and SAXS, a phase diagram was mapped out (Figure 10c). Increasing the polymer concentration or decreasing the salt concentration resulted in more ordered morphologies. At low salt concentration and intermediate polymer concentrations, a body-centered cubic structure was observed, which transformed into a hexagonal structure at high polymer concentration.

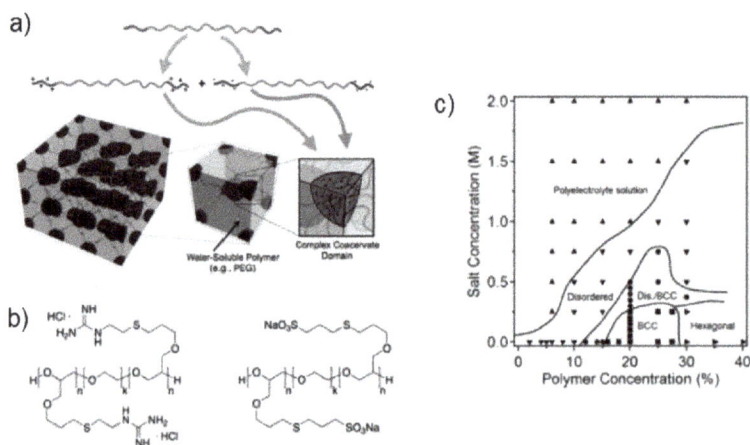

Figure 10. (**a**) Co-assembled system formed by charged triblock copolymers, which were synthesized from the same precursor having a hydrophilic middle block and uncharged end blocks. Schematics of co-assembly of the triblock copolymers having oppositely charged end blocks; (**b**) chemical structures of the charged triblock copolymers having guadinium and sulfonate groups; and (**c**) phase diagram of mixtures of these triblock copolymers in dependence on polymer and NaCl concentration. BCC stands for body-centered cubic. Reprinted with permission from [42]. Copyright 2013 American Chemical Society.

In comparison with the system described by Lemmers et al. [37–39] described above, the authors concluded that the higher amount of bridges in the triblock copolymer mixtures promoted the formation of ordered morphologies which had not been observed in the triblock copolymer/homopolymer system. The system presented [42] was proposed for the use as membranes, for injectable drug delivery or as tissue growth scaffolds.

Theoretical work validated by computer simulations elucidated the role of various interaction parameters on the phase behavior of mixtures of triblock copolymers with oppositely charged end blocks [43]. The experimentally observed behavior (Figure 10c) could be qualitatively reproduced. Phase diagrams were presented in dependence on polymer concentration and endblock fraction (Figure 11). The following interaction parameters were considered: B_{em}, which describes the strength of the (solvent-mediated) repulsive interaction between the middle block and the endblocks; B_{mm}, which is a measure of the solvent quality for the middle blocks; and B_{ee}, which is a measure of the endblock solubility in the solvent (higher values of B_{mm} and B_{ee} mean better solubility of the middle or the endblock, respectively).

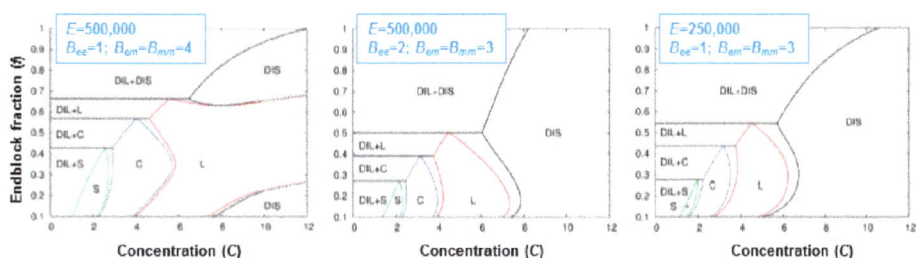

Figure 11. Phase diagrams of mixtures of triblock copolymers with a hydrophilic middle block and oppositely charged end blocks, calculated using the embedded fluctuation model. The morphologies are given in dependence on polymer number concentration and on the endblock fraction for different sets of parameters. *E* is the electrostatic strength parameter. DIS denotes a polymer rich homogeneous phase, DIL a nearly pure water phase, L lamellae, C hexagonally packed cylinders and S cubically packed spheres (body-centered or face-centered). Adapted with permission from [43]. Copyright 2015 Royal Society of Chemistry.

The phase diagrams showed that ordered morphologies could be obtained for low endblock fractions. In this regime, the morphology may be altered by hydration or dehydration. Comparing the phase diagrams obtained for different parameter sets, it was found that increasing B_{em} or B_{mm} promoted microphase separation. Increasing the value of B_{ee}, i.e., improving the end block solubility in the solvent, reduces the microphase-separated regions in the phase diagram. Decreasing the electrostatic strength parameter, *E*, decreased the driving force for microphase separation. This could be achieved by reducing the charge density on the endblocks. These results allow identifying the polymer architecture needed to obtain a desired morphology and to alter this morphology, e.g., by hydration or dehydration.

Cui et al. designed biodegradable hydrogels by co-assembly of triblock copolymers having oppositely charged polypeptides as endblocks and water-soluble EG as the middle block [44]. These hydrogels from PGA-*b*-PEG-*b*-PGA (G(EG)G, where PGA stands for poly(L-glutamic acid)) and PLL-*b*-PEG-*b*-PLL (L(EG)L, where PLL stands for poly(L-lysine)) could be reversibly assembled and disassembled through a change of pH. In more detail, when equimolar solutions of the triblock copolymers in phosphate-buffered saline (PBS) buffer were mixed, gels formed within a few seconds even at polymer concentrations as low as 3–5 wt %. Since both types of endblocks are weak polyelectrolytes, reversible gelation occurs at intermediate pH values only, whereas the gels disassembled at pH < 3 or pH > 11. In contrast to other systems [42], also described above, the addition of NaCl at 1 M led to an increase of the storage modulus of the hydrogels. The authors ascribed this phenomenon to the hydrophobic interaction of the polypeptides which maintained the stability of the coacervate domains, despite the decreasing polyionic interactions. The authors demonstrated furthermore the controlled release of the charged protein bovine serum albumin, the cell viability and the in vivo gel formation and maintenance in rats.

4. Summary and Outlook

The aim of this mini review is to highlight a particular class of physically crosslinked hydrogels, arising from the intermolecular association of hydrophilic triblock copolymers bearing polyion sequences. The hydrogel formation relies on the ionic interactions between the oppositely charged repeating units that lead to the formation of hydrated polyion complexes (complex coacervate nanodomains) which constitute the cross-links of the formed transient network. The driving force of the intermolecular association of the participating oppositely charged polyions in the aqueous media is mainly entropic due to ion pairing and the release of the counterions along with the water molecules.

Two systems are distinguished, depending on the polymer constituents involved in the network formation: (i) self-assembly of a highly asymmetric triblock polyampholyte (bearing cationic and anionic blocks); and (ii) co-assembly of mixtures of triblock polyelectrolytes (having polyionic end blocks) with oppositely charged polyions. In the first system, the remarkable charge asymmetry (i.e., a high charge imbalance) favors the network formation and its mechanical response. On the contrary, in the second system, a charge stoichiometry (charge balance) is needed for the best mechanical performance of the gel. In both cases, the charge ratio of the oppositely charged blocks can be tuned at will through controlled polymerization methods available for the synthesis of the copolymers (macromolecular engineering). A worthy of mention difference between the two systems is that the bridging chains between the crosslinked polyionic complexes of the network are ionic for the first one (arising from the excess part of the larger ionic middle block) and neutral for the second one (arising from the middle block of the copolymer). This might be the reason why the percolation concentration is considerably lower in the first system. Moreover, the length of the bridging chain in the second case is defined at will from the degree of polymerization of the middle block resulting in well-defined network structures.

Another classification can also be seen in the co-assembly systems, depending on the employed type of the polyions: (i) mixtures of a triblock copolymer with ionic end blocks and an oppositely charged homopolymer; and (ii) mixtures of two triblock copolymers with the same neutral central block and oppositely charged end blocks of the same degree of polymerization. In the latter system, equal concentrations of the different triblock copolymers provide exact charge stoichiometry. Moreover, ordered 3D structures can be achieved at elevated concentrations.

By using weak electrolyte repeat units in the copolymers, the hydrogel formation depends strongly on the pH of the medium, in all cases, since it affects the degree of ionization of the polyelectrolyte blocks and, in turn, the charge imbalance of the system. Provided that the network formation is based on ionic complexations, the presence of salt is critical as well. Thus, two external stimuli (pH, ionic strength) will affect the network structure and the mechanical response of the hydrogels, rendering them stimuli responsive.

It is well known that the pH responsive hydrogels are good candidates as injectable carriers of therapeutic means in biomedical applications. A novel property provided by these new polyion complex-based hydrogels is that the coacervate cores of the network are hydrated. This may, for certain drugs, favor their encapsulation with respect to the hydrophobically associated systems. Inspired by the unique, controlled and responsive properties of the hydrogels, demonstrated herein, novel hydrogel systems can be designed, meeting the requirements for ideal injectable hydrogels, exhibiting biocompatibility and biodegradability, as well as proper elasticity and rapid self-healing. Theoretical modeling and computer simulations may help in identifying relevant molecular architectures and conditions. Some recent examples towards this direction have already been presented in this review.

Acknowledgments: We thank the Deutscher Akademischer Austauschdienst (DAAD) and the Greek State Scholarship Foundation (IKY) for financial support of mutual visits in the framework of the program for the promotion of the exchange and scientific cooperation between Germany and Greece, IKYDA 2013. This work is based upon experiments performed at the KWS-2 instrument operated by Jülich Centre for Neutron Science (JCNS) at the Heinz Maier-Leibnitz Zentrum (MLZ), Garching, Germany, and at D22 at Institut Laue-Langevin (ILL), Grenoble, France. We thank these facilities for providing beamtime and excellent equipment. We thank (in alphabetical order) Anatoly V. Berezkin, Zhenyu Di, Margarita A. Dyakonova, Sergey K. Filippov, Sandra Gkermpoura, Isabelle Grillo, Sebastian Jaksch, Konstantinos Kyriakos, Martine Philippe, Maria T. Popescu, Nicoletta Stavrouli and Petr Štěpánek for help with the experiments and computer simulations.

Conflicts of Interest: The authors declare no conflict of interest.

Abbreviations

3D	three-dimensional
A	poly(acrylic acid), PAA
AH	poly(allylamine hydrochlorid)
APA	PAA-*b*-P2VP-*b*-PAA
APA_1	PAA_{134}-*b*-$P2VP_{628}$-*b*-PAA_{134}
APA_2	PAA_{163}-*b*-$P2VP_{1397}$-*b*-PAA_{163}
APA_3	PAA_{109}-*b*-$P2VP_{819}$-*b*-PAA_{109}
A(QP)A	PAA-*b*-QP2VP-*b*-PAA
$A(QP)A_2$	PAA_{163}-*b*-$QP2VP_{1397}$-*b*-PAA_{163}
$A(QP)A_3$	PAA_{109}-*b*-$QP2VP_{819}$-*b*-PAA_{109}
$A(QP)A_4$	PAA_{20}-*b*-$QP2VP_{172}$-*b*-PAA_{20}
CMC	critical micelle concentration
CCC	complex coacervate core
C_{gel}	critical gel concentration
E	poly(ethylene oxide), PEO
EG	poly(ethylene glycol), PEG
G	poly(L-glutamic acid), PGA
GE	poly(allyl glycidyl ether), PAGE
(GE)(EG)(GE)	PAGE-*b*-PEG-*b*-PAGE
G(EG)G	PGA-*b*-PEG-*b*-PGA
iep	isoelectric point
L	poly(L-lysine), PLL
LCST	lower critical solution temperature
L(EG)L	PLL-*b*-PEG-*b*-PLL
M	poly[4-(2,2,6,6-tetramethylpiperidine-N-oxyl)-aminomethylstyrene], PMNT
M(EG)M	PMNT-*b*-PEG-*b*-PMNT
P	poly(2-vinyl pyridine), P2VP
PBS	phosphate-buffered saline
QP	quaternized poly(2-vinyl pyridine), QP2VP
S	poly(3-sulfopropylmethacrylic acid), PSPMA
SANS	small-angle neutron scattering
SAXS	small-angle X-ray scattering
SES	$PSPMA_{28}$-*b*-PEO_{210}-*b*-$PSPMA_{28}$

References

1. Bhatia, A.; Mourchid, S.R.; Joanicot, M. Block copolymer assembly to control fluid rheology. *Curr. Opin. Colloid Interface Sci.* **2001**, *6*, 471–478. [CrossRef]

2. Van de Manakker, F.; Vermonden, T.; van Nostrum, C.F.; Hennink, W.E. Cyclodextrin-based polymeric materials: Synthesis, properties, and pharmaceutical/biomedical applications. *Biomacromolecules* **2009**, *10*, 3157–3175. [CrossRef] [PubMed]

3. Nyström, B.; Kjøniksen, A.-L.; Beheshti, N.; Zhu, K.; Knudsen, K.D. Rheological and structural aspects on association of hydrophobically modified polysaccharides. *Soft Matter* **2009**, *5*, 1328–1339. [CrossRef]

4.   Madsen, J.; Armes, S.P. (Meth)acrylic stimulus-responsive block copolymer hyrogels. *Soft Matter* **2012**, *8*, 592–605. [CrossRef]

5.   Koetting, M.C.; Peters, J.P.; Steichen, S.D.; Peppas, N.A. Stimulus-responsive hydrogels: Theory, modern advances, and applications. *Mater. Sci. Eng. R Rep.* **2015**, *93*, 1–49. [CrossRef] [PubMed]

6.   Winnik, M.A.; Yekta, A. Associative Polymers in aqueous solution. *Curr. Opin. Colloid Interface Sci.* **1997**, *2*, 424–436. [CrossRef]

7.   Berret, J.-F.; Calvet, D.; Collet, A.; Viguier, M. Fluorocarbon associative polymers. *Curr. Opin. Colloid Interface Sci.* **2003**, *8*, 296–306. [CrossRef]

8.   Xu, C.; Kopeček, J. Self-assembling hydrogels. *Polym. Bull.* **2007**, *58*, 53–63. [CrossRef]

9.   Chassenieux, C.; Nicolai, T.; Benyahia, L. Rheology of associative polymer solutions. *Curr. Opin. Colloid Polym. Sci.* **2011**, *16*, 18–26. [CrossRef]

10. Kopeček, J.; Yang, Y. Hydrogels as smart biomaterials. *Polym. Int.* **2007**, *56*, 1078–1098. [CrossRef]

11. Ahn, S.-K.; Kasi, R.M.; Kim, S.-C.; Sharma, N.; Zhou, Y. Stimuli-responsive polymer gels. *Soft Matter* **2008**, *4*, 1151–1157. [CrossRef]

12. Tsitsilianis, C. Responsive reversible hydrogels from associative "smart" macromolecules. *Soft Matter* **2010**, *6*, 2372–2388. [CrossRef]

13. Pasparakis, G.; Vamvakaki, M. Multiresponsive polymers: Nano-sized assemblies, stimuli-sensitive gels and smart surfaces. *Polym. Chem.* **2011**, *2*, 1234–1248. [CrossRef]

14. Laschewsky, A.; Müller-Buschbaum, P.; Papadakis, C.M. Thermoresponsive amphiphilic di- and triblock copolymers based on poly(*N*-isopropylacrylamide) and poly(methoxy diethylene glycol acrylate): Aggregation and hydrogel formation in bulk solution and in thin films. *Progr. Colloid Polym. Sci.* **2013**, *140*, 15–34.

15. Yu, L.; Ding, J. Injectable hydrogels as unique biomedical materials. *Chem. Soc. Rev.* **2008**, *37*, 1473–1481. [CrossRef] [PubMed]

16. Nguyen, M.K.; Lee, D.S. Injectable biodegradable hydrogels. *Macromol. Biosci.* **2010**, *10*, 563–579. [CrossRef] [PubMed]

17. He, C.; Kim, S.W.; Lee, D.S. In-situ gelling stimuli-sensitive block copolymer hydrogels for drug delivery. *J. Control. Release* **2008**, *127*, 189–207. [CrossRef] [PubMed]

18. Drury, J.L.; Mooney, D.J. Hydrogels for tissue engineering: Scaffold design variables and applications. *Biomaterials* **2003**, *24*, 4337–4351. [CrossRef]

19. Jeong, B.; Kim, S.W.; Bae, Y.H. Thermosensitive sol–gel reversible hydrogels. *Adv. Drug Deliv. Rev.* **2002**, *54*, 37–51. [CrossRef]

20. Chassenieux, C.; Tsitsilianis, C. Recent trends in pH/thermoresponsive self-assembling hydrogels: From polyions to peptide-based polymeric gelators. *Soft Matter* **2016**, *12*, 1344–1359. [CrossRef] [PubMed]

21. Brassinne, J.; Fustin, C.-A.; Gohy, J.-F. Polymer gels constructed trough metal-ligand coordination. *J. Inorg. Organomet. Polym.* **2013**, *23*, 24–40. [CrossRef]

22. Brassinne, J.; Zhuge, F.; Fustin, C.-A.; Gohy, J.-F. Precise Control over the Rheological Behavior of Associating Stimuli-Responsive Block Copolymer Gels. *Gels* **2015**, *1*, 235–255. [CrossRef]

23. Hales, K.; Pochan, D.J. Using polyelectrolyte block copolymers to tune nanostructure self-assembly. *Curr. Opin. Colloid Interface Sci.* **2006**, *11*, 330–336. [CrossRef]

24. Bucur, C.B.; Sui, Z.; Schlenoff, J.B. Ideal mixing in polyelectrolyte complexes and multilayers: Entropy driven sssembly. *J. Am. Chem. Soc.* **2006**, *128*, 13690–13691. [CrossRef] [PubMed]

25. Cohen Stuart, M.A.; Hofs, B.; Voets, I.K.; de Keizer, A. Assembly of polyelectrolyte-containing block copolymers in aqueous media. *Curr. Opin. Colloid Interface Sci.* **2005**, *10*, 30–36. [CrossRef]

26. Voets, I.K.; de Keizer, A.; Cohen Stuart, M.A. Complex coacervate core micelles. *Adv. Colloid Interface Sci.* **2009**, *147*, 300–318. [CrossRef] [PubMed]

27. Van der Gucht, J.; Spruijt, E.; Lemmers, M.; Cohen Stuart, M.A. Polyelectrolyte complexes: Bulk phases and colloidal systems. *J. Colloid Interface Sci.* **2011**, *361*, 407–422. [CrossRef] [PubMed]

28. Priftis, D.; Laugel, N.; Tirrell, M. Thermodynamic characterization of polypeptide complex coacervation. *Langmuir* **2012**, *28*, 15947–15957. [CrossRef] [PubMed]

29. Fu, J.; Schlenoff, J.B. Driving forces for oppositely charged polyion association in aqueous solutions: Enthalpic, entropic, but not electrostatic. *J. Am. Chem. Soc.* **2016**, *138*, 980–990. [CrossRef] [PubMed]

30. Cohen Stuart, M.A.; Besseling, N.A.M.; Fokkink, R.G. Formation of micelles with complex coacervate cores. *Langmuir* **1998**, *14*, 6846–6849. [CrossRef]

31. Wang, W.; Schlenoff, J.B. The polyelectrolyte complex/coacervate continuum. *Macromolecules* **2014**, *47*, 3108–3116. [CrossRef]

32. Sfika, V.; Tsitsilianis, C. Association phenomena of poly(acrylic acid)-*b*-poly(2-vinylpyridine)-*b*-poly(acrylic acid) triblock polyampholyte in aqueous solutions: From transient network to compact micelles. *Macromolecules* **2003**, *36*, 4983–4988. [CrossRef]

33. Bossard, F.; Sfika, V.; Tsitsilianis, C. Rheological properties of physical gel formed by triblock polyampholyte in salt-fee aqueous solutions. *Macromolecules* **2004**, *37*, 3899–3904. [CrossRef]

34. Dyakonova, M.A.; Stavrouli, N.; Popescu, M.T.; Kyriakos, K.; Grillo, I.; Philipp, M.; Jaksch, S.; Tsitsilianis, C.; Papadakis, C.M. Physical hydrogels via charge driven self-organization of a triblock polyampholyte—Rheological and structural investigations. *Macromolecules* **2014**, *47*, 7561–7572. [CrossRef]

35. Dyakonova, M.A.; Berezkin, A.V.; Kyriakos, K.; Gkermpoura, S.; Popescu, M.T.; Filippov, S.K.; Štěpánek, P.; Di, Z.; Tsitsilianis, C.; Papadakis, C.M. Salt-induced changes in triblock polyampholyte hydrogels—Computer simulations, rheological, structural and dynamic characterization. *Macromolecules* **2015**, *48*, 8177–8189. [CrossRef]

36. Perry, S.L.; Li, Y.; Priftis, D.; Leon, L.; Tirrell, M. The effect of salt on the complex coacervation of vinyl polyelectrolytes. *Polymers* **2014**, *6*, 1756–1772. [CrossRef]

37. Lemmers, M.; Sprakel, J.; Voets, I.K.; van der Gucht, J.; Cohen Stuart, M.A. Multiresponsive reversible gels based on charge-driven assembly. *Angew. Chem. Int. Ed.* **2010**, *49*, 708–711. [CrossRef] [PubMed]

38. Lemmers, M.; Voets, I.K.; Cohen Stuart, M.A.; van der Gucht, J. Transient network topology of interconnected polyelectrolyte complex micelles. *Soft Matter* **2011**, *7*, 1378–1389. [CrossRef]

39. Lemmers, M.; Spruijt, E.; Beun, L.; Fokkink, R.; Leermakers, F.; Portale, G.; Cohen Stuart, M.A.; van der Gucht, J. The influence of charge ratio on transient networks of polyelectrolyte complex micelles. *Soft Matter* **2012**, *8*, 104–117. [CrossRef]

40. Ishii, S.; Kaneko, J.; Nagasaki, Y. Dual stimuli-responsive redox-active injectable gel by polyion complex based flower micelles for biomedical applications. *Macromolecules* **2015**, *48*, 3088–3094. [CrossRef]

41. Hunt, J.N.; Feldman, K.E.; Lynd, N.A.; Deek, J.; Campos, L.M.; Spruell, J.M.; Hernandez, B.M.; Kramer, E.J.; Hawker, C.J. Tunable, high-modulus hydrogels driven by ionic coacervation. *Adv. Mater.* **2011**, *23*, 2327–2331. [CrossRef] [PubMed]

42. Krogstad, D.V.; Lynd, N.A.; Choi, S.-H.; Spruell, J.M.; Hawker, C.J.; Kramer, E.J.; Tirrell, M.V. Effects of polymer and salt concentration on the structure and properties of triblock coacervate hydrogels. *Macromolecules* **2013**, *46*, 1512–1518. [CrossRef]

43. Audus, D.J.; Gopez, J.D.; Krogstad, D.V.; Lnyd, N.A.; Kramer, E.J.; Hawker, C.J.; Fredrickson, G.H. Phase behavior of electrostatically complexed polyelectrolyte gels using an embedded fluctuation model. *Soft Matter* **2015**, *11*, 1214–1225. [CrossRef] [PubMed]

44. Cui, H.; Zhuang, X.; He, C.; Wei, Y.; Chen, X. High performance and reversible ionic polypeptide hydrogel based on charge-driven assembly for biomedical applications. *Acta Biomater.* **2015**, *11*, 183–190. [CrossRef] [PubMed]

gels

MDPI

Article

Swelling Dynamics of a DNA-Polymer Hybrid Hydrogel Prepared Using Polyethylene Glycol as a Porogen

Ming Gao, Kamila Gawel and Bjørn Torger Stokke *

Biophysics and Medical Technology, Department of Physics, The Norwegian University of Science and Technology, Trondheim NO-7491, Norway; ming.gao@kemi.uu.se (M.G.); kamila.gawel@gmail.com (K.G.)
* Author to whom correspondence should be addressed; bjorn.stokke@ntnu.no;
  Tel.: +47-73-59-3434; Fax: +47-73-59-7710.

Academic Editor: Dirk Kuckling
Received: 7 September 2015; Accepted: 12 November 2015; Published: 18 November 2015

Abstract: DNA-polyacrylamide hybrid hydrogels designed with covalent and double-stranded (dsDNA) crosslinks respond to specific single-stranded DNA (ssDNA) probes by adapting new equilibrium swelling volume. The ssDNA probes need to be designed with a base pair sequence that is complementary to one of the strands in a dsDNA supported network junction. This work focuses on tuning the hydrogel swelling kinetics by introducing polyethylene glycol (PEG) as a pore-forming agent. Adding PEG during the preparation of hydrogels, followed by removal after polymerization, has been shown to improve the swelling dynamics of DNA hybrid hydrogels upon specific ssDNA probe recognition. The presence of porogen did not influence the kinetics of osmotic pressure-driven (2-acrylamido-2-methylpropane sulfonic acid)-*co*-acrylamide (AMPSA-*co*-AAm) hydrogels' swelling, which is in contrast to the DNA-sensitive hydrogels. The difference in the effect of using PEG as a porogen in these two cases is discussed in view of processes leading to the swelling of the gels.

Keywords: DNA competitive displacement; hydrogel swelling; interferometric readout; nanometer resolution; PEG porogen

---

1. Introduction

Hydrogels, crosslinked polymer networks imbibed with water, adapt an equilibrium swelling state depending on the molecular parameters of the polymer network and the solution conditions. They adopt swelling states depending on changes in the molecular properties of the network and also altered environmental conditions, such as pH [1–4], temperature [5–7], ionic strength [8,9], electric field [10,11], light [12] and surfactants [13]. Such responsive soft materials tuned to display various physicochemical properties, including swelling, mechanics, permeability and optical properties, have been targeted for various biosensing applications [14–19].

Changes in the equilibrium swelling volume and dynamics of hydrogels response can be determined using a high resolution interferometric technique. This has previously been applied for the characterization of various types of hydrogels and their swelling responses, including the ionic strength dependence of ionic hydrogels [20], carbohydrates and, in particular, glucose-induced swelling of phenylboronic acid-functionalized hydrogels [21–23], nucleotide-sensitive hydrogels [24–26], as well as the swelling response associated with the deposition of polymers [27] and surfactants [28]. The swelling of a responsive hydrogel deposited at the tip of an optical fiber is monitored by changes in the optical length of the hydrogel with 2-nm resolution and a sampling frequency at about 1 Hz. In addition, the swelling response of the material itself is rather quick (about 2 s [29]) due to the relatively small size of the hydrogel. The swelling rates, limited by network relaxation, are inversely proportional to the square of the hydrogel's size [30,31]. Thus, the hydrogel specimen being of only about ~60 μm in radius in the fiber optic-based readout platform yields a rapid swelling/deswelling kinetics that supports a much more rapid readout than methods exploiting larger sizes of the hydrogel specimen. This feature implies that the swelling kinetics of the polymer-DNA hybrid hydrogels in the presence of single-stranded DNA (ssDNA), having a time constant much longer than the 2 s [24,26], is not limited by the swelling response rate of the polymer network, but by other molecular aspects, e.g., associated with the transport and recognition process of the ssDNA probe. Such types of hydrogel are designed to respond through competitive displacement hybridization with the specific ssDNA oligonucleotide integrated in the network (Figure 1).

Two partially-hybridized oligonucleotides, referred to as sensing and blocking oligonucleotides, copolymerized within a hydrogel network, act as physical crosslinks in addition to the covalent ones. The physical crosslinks stabilized by Watson–Crick hydrogen bonds dissociate in the presence of the ssDNA probe with a longer complementary region to the sensing oligonucleotide than the blocking strand. This competitive displacement reaction is a complex process with a range of dynamics depending on the molecular parameters [32–34]. The completion of the competitive displacement results in the reduction of the hydrogel crosslink density and concomitant readjustment of the hydrogel swelling volume. The swelling kinetics of these hybrid DNA-polymer hydrogels during such competitive displacement is much slower than that of ionic hydrogels induced by changes in ionic strength [26,29]. Diffusion of ssDNA strands inside polymer-DNA hybrid hydrogel has been shown to be retarded by repeated association and dissociation of the probe with immobile oligonucleotide molecules [35]. Such retarded diffusion may affect the kinetics of swelling towards the equilibrium swelling state of the hydrogel. The repeated association and dissociation processes result in slower equilibration of the hybridization process [35]. The kinetics of polymer-DNA hydrogel biosensor response may nevertheless be tuned through the design of the sensing hydrogel material.

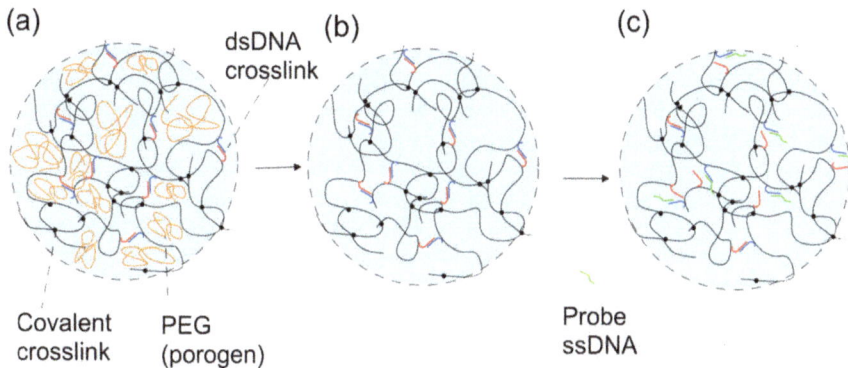

Figure 1. Schematic illustration of the effect of using polyethylene glycol (PEG) as a porogen in the synthesis of DNA hybrid hydrogel and single-stranded DNA competitive displacement of double-stranded, supported crosslinks (**a**–**c**). Hydrogels are prepared in the presence of polyethylene glycol, subsequently removed by washing in buffer, to yield hydrogels with an altered structure (**a**,**b**). DNA hybrid hydrogels prepared in the presence of polyethylene glycol are prepared where the physical crosslinks in the network are made up of double-stranded oligonucleotides with partial complementary bp sequences denoted as sensing (blue) and blocking (red). Swelling response is induced by the oligonucleotide complementary displacement reaction in the presence of an oligonucleotide probe (green), destabilizing the junctions (**b**,**c**), which results in swelling of the hydrogel, determined as the change in the optical length, Δl_{opt} using an interferometer.

It has been shown that the response rate of polymer-DNA hybrid hydrogel depends on such parameters as: concentration of the physical (double-stranded DNA (dsDNA)) [26] and covalent crosslinks [24], the length of the blocking region within DNA crosslinks and the length of the "toehold" on the ssDNA probe [24] and the presence of a mismatched base on a probe sequence [26]; and does not occur for non-complementary probes [26]. The swelling rate of such DNA-*co*-acrylamide (co-AAm) hybrid hydrogels depends also on the ratio and total concentration of covalent and dsDNA-supported crosslinks. Reduction of the number of base pairs and increasing the length of "toehold" sequences also increase the swelling rate on exposure to a certain concentration of specific ssDNA. The rate of swelling may be even further improved by application of the approaches reported for the preparation of fast-responsive hydrogels that comprise: the synthesis of hydrogels based on comb-type polymers having hydrophobic pendant chains [36] and/or the synthesis of hydrogels with an altered pore size distribution and enhanced porosity [37]. For example, temperature-responsive hydrogels with hydrophobic chains have been shown to shrink more quickly compared to conventional ones when exposed to a temperature stimulus; however, the increase in the collapse rate has not been reported to be associated with an increase in the swelling rate [36]. The approach of enhancing hydrogel porosity to improve the dynamics of both swelling and deswelling has been applied for various stimuli-responsive (pH, temperature, ionic strength) hydrogels. The swelling/deswelling dynamics enhancement is shown to result from a larger mesh size that allows solvent and solute to diffuse more rapidly through the hydrogel. The variety of techniques used to control hydrogel microstructure and porosity has been reviewed [38]. The application of polymers as porogen-forming agents appears to represent a strategy applied by various groups and will be explored here.

The aim of the present work is to explore the possible increase in the swelling kinetics of ssDNA-sensitive hydrogel and to offer more detailed and diversified information about the swelling behavior of porous hydrogels based on the high resolution interferometric technique. Toward this aim, we present how an altered hydrogel pore size distribution induced by adding a porogen during the synthesis facilitates changes in the dynamics of polymer-DNA hybrid hydrogel swelling. The hydrogel pore structure is affected at the stage of hydrogel synthesis by adding inert polyethylene glycol (PEG) during the co-polymerization followed by its removal after the network formation (Figure 1). The approach thus extends the use of PEG as a porogen also to polymer-DNA hybrid hydrogels. The experimental approach includes PEGs with two different molecular weights and introduced at various fractions in the hydrogel preparation procedure and that is subsequently removed by washing. For comparison, the swelling kinetics associated with step changes in ionic strength of the immersing solution for porous AMPSA (2-acrylamido-2-methylpropane sulfonic acid)-*co*-AAm anionic hydrogels is shown to be independent of the variations in the porogen during the synthesis. As swelling of polymer-DNA hybrid hydrogel is affected by the retarded diffusion of the ssDNA probe, this is compared to the kinetics of AMPSA-*co*-AAm hydrogel swelling, which do not possess the DNA-supported crosslink.

2. Results and Discussion

2.1. AMPSA-co-AAm Hydrogels Synthesized with Various Fractions of Porogens

AAm-*co*-AMPSA hydrogels with 3 mol % AMPSA and 3 mol % *N,N*'-methylenebisacrylamide (Bis) relative to AAm (10 wt %) were synthesized in the presence of two different molecular weight PEGs (M_n = 200, 950~1050 g/mol, denoted as: PEG200 and PEG950, respectively) as pore-forming agents representing 3 wt % of the final pre-gel solution. The synthesis of these follows the general scheme outlined in Figure 1 with the adoption of omitting the dsDNA as crosslinks and including a blend of the AMPSA monomer in the AAm pre-gel solution to obtain an anionic hydrogel. The charge and crosslink parameters were chosen to yield a readily-detectable swelling response, in line with the data previously reported for a polycationic hydrogel [29]. The swelling response of synthesized hydrogels to varied ionic strengths was measured using an interferometric technique for hemispherical hydrogels with a radius of about 60 μm attached to an optical fiber. Figure 2 displays the absolute and relative optical length *versus* the ionic strength in the immersion aqueous solution for the hydrogels synthesized in the presence of 3 wt % of PEG200 and PEG950. The relative changes in the optical lengths were obtained by normalizing with respect to the optical length of the cavity measured in 0.15 M NaCl. The swelling curves for ionic strength from 5×10^{-4} to 0.15 M showed good reproducibility within each molecular weight of PEG added during the synthesis of the hydrogel. Additionally, the swelling responses for the hydrogels prepared using the two molecular weights of PEG at 3 wt % are identical. The total change of the magnitude of optical length induced by such a difference in concentration of NaCl was about 25 μm. This represents about a 40% change in the optical length of the hydrogel attached to the optical fiber.

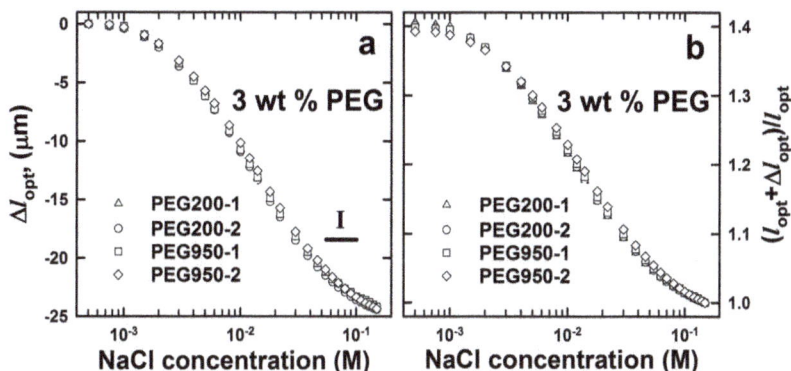

Figure 2. Changes in optical length (**a**) and deswelling ratio (**b**) of (2-acrylamido-2-methylpropane sulfonic acid)-*co*-acrylamide (AMPSA-*co*-AAm) hydrogels synthesized in the presence of pore-forming agents. AMPSA porous hydrogels consisted of 10 wt % AAm, 3 mol % Bis (relative to AAm) and 3 mol % AMPSA (relative to AAm) with 3 wt % PEG (M_n = 200 g/mol, PEG200, l_{opt} = 61.7 μm) and PEG (M_n = 950~1050 g/mol, PEG950, l_{opt} = 62.1 μm) removed after synthesis. All measurements were carried out in aqueous salt solution with the NaCl concentration in the range from 5×10^{-4} to 0.15 M and at least two times (labeled as -1 and -2 for each experimental series). Region I depicts the NaCl concentration range selected for displaying the time dependence of Δl_{opt} (Figure 3).

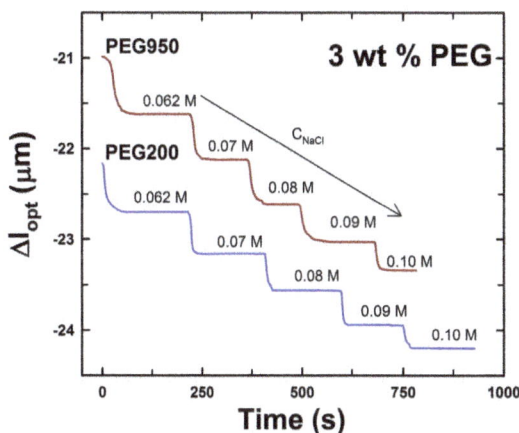

Figure 3. Deswelling kinetics of AMPSA-*co*-AAm hydrogels consisting of 10 wt % AAm, 3 mol % Bis (relative to AAm) and 3 mol % AMPSA (relative to AAm) with 3.33 wt % PEG200 (l_{opt} = 61.7 μm) and PEG950 PEG (l_{opt} = 62.1 μm) polymers removed right after synthesis. The data are selected for the NaCl concentration range (c_{NaCl}) from 0.05 to 0.1 M, as indicated (Region I in Figure 2).

Here, we report on the swelling changes of the hemispherical hydrogels attached to the optical fiber as changes in the optical length Δl_{opt} between a reference and the actual state, as deduced from the phase changes of the interference wave. The parameter Δl_{opt} is the primary observable and can be converted to standard measures of changes in hydrogel swelling parameters by applying the strategy as outlined in the following. The primary observable Δl_{opt} can be decomposed into changes due to the altered optical properties of the hydrogels and the change in the physical length,

$$\Delta l_{opt} = \langle n_2 \rangle l_2 - \langle n_1 \rangle l_1 \approx \langle n_1 \rangle \Delta l + l_1 \Delta n \tag{1}$$

where Indices 1 and 2 represent the two states to be compared, e.g., at two different NaCl concentrations in the case of the AAm-*co*-AMPSA hydrogels, and:

$$\langle n_i \rangle = l_i^{-1} \int_0^{l_i} n_i(l) dl \, , \ i = 1,2 \tag{2}$$

are the refractive indices averaged along the optical pathway of the two states, indicated by the indices. For most cases, the change in the physical length is the larger of these contributions, e.g., for cyclodextrin-induced swelling changes of cholesterol-modified pullulan hydrogels, we estimated that the relative changes due to altered optical properties were $l_1 \Delta n / l_{opt}$ up to 4×10^{-4} at a cyclodextrin concentration of 3 mM [39]. The finding that this was much less than $\Delta l_{opt}/l_{opt}$ led to the conclusion that changes in the physical length of the hydrogel along the optical axis are by far the dominating contribution (Equation (1)). The optical length within the hydrogel, l_{opt}, is determined from the frequency difference between two neighboring frequency peaks in the interference spectrum [29]. The experimentally-determined $\Delta l_{opt}/l_{opt}$ can be converted to relative volume changes, V/V_0, using the relation $V/V_0 \sim [(\Delta l_{opt} + l_{opt})/l_{opt}]^{2.6}$. The power law coefficient of 2.6, different from 3, arises due to the constraining of the hydrogel at the end of the optical fiber and is deduced from finite element analysis of such a hydrogel [40]. Data converted following this outline provide information similar to that by analyzing trends in the changes of the optical length directly.

Figure 3 presents deswelling kinetics for the two hydrogels PEG200 and PEG950 for the step changes in NaCl concentration from 5.4×10^{-2} to 0.1 M (concentration Range I depicted in Figure 2). The data indicate that deswelling kinetics did not depend on the molecular weight of the PEG added during the synthesis of the AMPSA hydrogels. The apparent equilibration time constants $\tau_{1/2}$ were found to be in the range of 10 to 15 s. These time constant values correspond to kinetic constants estimated for ionic strength-induced *N*-(3-dimethylaminopropyl)-*co*-AAm hydrogel deswelling in the ionic strength range from 0.03 to 0.074 M [29].

Whereas the molecular weight of PEG polymer did not influence either the swelling capacity or the kinetics of AMPSA-*co*-AAm hydrogels, the weight fraction of the pore-forming agent was shown to affect the swelling capacity (Figure 4).

Hydrogels were synthesized with four different fractions of PEG200 in the pre-gel solution: 0, 3, 10 and 17 wt %. These data (Figure 4) show that the ionic strength-induced swelling capacity of the AMPSA-*co*-AAm hydrogels was found to increase with the increasing fraction of the PEG200 included during the synthesis. This is consistent with the results reported by the others [41]. The presence of inert polymer, such as PEG, in the pre-gel solution during polymerization was shown to influence hydrogel network arrangement and result in changes in the hydrogel pore size distribution [42].

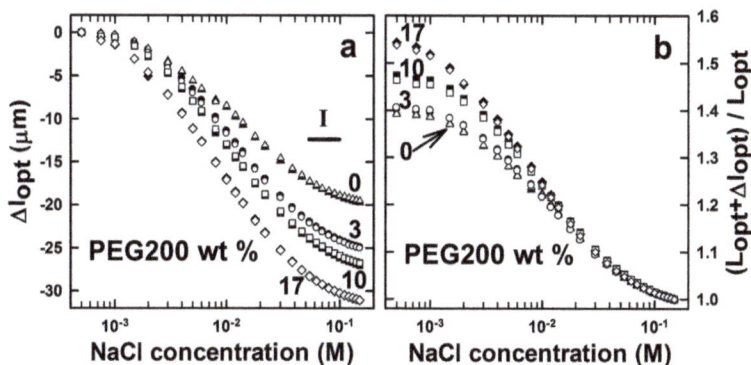

Figure 4. Changes in the optical length (**a**) and deswelling ratio (**b**) of AMPSA-*co*-AAm hydrogels consisting of 10 wt % AAm, 3 mol % Bis (relative to AAm) and 3 mol % AMPSA (relative to AAm) and with 0 (AMPSAGel0, l_{opt} = 50.3 μm), 3 wt % (AMPSAGel3, l_{opt} = 61.7 μm), 10 wt % (AMPSAGel10, l_{opt} = 57.6 μm) and 17 wt % (AMPSAGel17, l_{opt} = 57.2 μm) of PEG200 porogen removed right after synthesis. All measurements were carried out in NaCl concentrations ranging from 5×10^{-4} to 0.15 M and at least two times (open and filled symbols for each experimental series). Region I depicts the concentration range of NaCl selected for displaying the time dependence of Δl_{opt} (Figure 5).

The experimental procedure applied for the synthesis of AMPSA-*co*-AAm and DNA-*g*-AAm hydrogels assures a constant fraction of polymer for the hydrogels within the same compositional type (AMPSA- or DNA-type hydrogel series). This indicates that the total pore volume of the hydrogels is unchanged when including PEG in the mixed solvent during synthesis of the hydrogel, and the observed effects are suggested to originate from changes in the pore size distribution due to the various PEG contents present during the synthesis. The altered pore size distribution manifests itself in an increased hydrogel swelling ratio and swelling deswelling kinetics compared to hydrogels synthesized without PEG in the solvent [43]. The faster swelling/deswelling kinetics is believed to occur due to facilitated solvent and solute transport inside the hydrogel with an altered pore size distribution [30,43]. However, the kinetics of the ionic strength-induced deswelling of AMPSA-*co*-AAm hydrogels was comparable for all gels and independent of PEG content (Figure 5). The apparent deswelling time constants were at the order of a few seconds. For the hydrogel sizes used here, a hemispherical geometry with a radius of about 60 μm, it is reported that the kinetics of swelling with time constants of a few seconds corresponds to the network diffusion limit [29]. This indicates that variations in hydrogel structure induced by including PEG as the porogen do not influence the diffusivity of the hydrogel network.

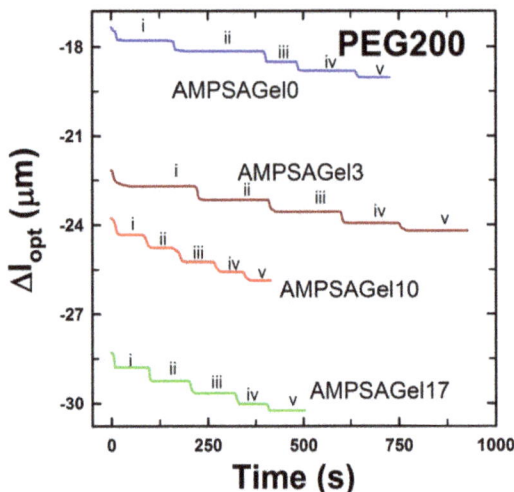

Figure 5. Deswelling kinetics of AMPSA-*co*-AAm hydrogels. The AMPSA-*co*-AAm hydrogels consisted of 10 wt % AAM, 3 mol % Bis (relative to AAm) and 3 mol % AMPSA (relative to AAm) and 0 (AMPSAGel0, l_{opt} = 50.3 µm), 3 wt % (AMPSAGel3, l_{opt} = 61.7 µm), 10 wt % (AMPSAGel10, l_{opt} = 57.6 µm) and 17 wt % (AMPSAGel17, l_{opt} = 57.2 µm) of PEG200 porogen removed after synthesis, respectively. The data are selected from Region I in Figure 4, and the actual NaCl concentrations depicted are 0.062 M (i), 0.07 M (ii), 0.08 M (iii), 0.09 M (iv) and 0.1 M (v).

2.2. DNA-g-AAm Hydrogels with Altered Pore Size Distribution

DNA-*g*-AAm hybrid hydrogels crosslinked with 0.6 mol % of covalent (Bis) and 0.4 mol % of physical DNA (sensing-blocking (S-B) oligonucleotide complex) crosslinks were synthesized in the presence of different fractions (0 (DNAGel0), 10 (DNAGel10), 17 (DNAGel17) wt %) of PEG200 inert polymer in the pre-gel solution. The base pair sequence of the S oligonucleotide was chosen at random and the B oligonucleotide with a 12-base pair complementarity to the S oligonucleotide (Table 1). The probe oligonucleotide was chosen to have a six base pair longer complementary sequence than the blocking oligonucleotide (six base pairs in the so-called "toe-hold" region beyond the 12 base pairs in the complementary region) (Table 1). Similar to that for the AMPSA-*co*-AAm hydrogels, the presence of PEG200 during the polymerization and subsequent removal was assumed to affect the experimentally-determined swelling dynamics by altering the structure and pore size characteristics of the DNA-*g*-AAm hybrid hydrogels. The swelling response of the prepared DNA-*g*-AAm hydrogels was explored in the presence of probe oligonucleotide P comprising 18 base pairs complementary to the sensing strand (S). The competitive displacement reaction between the S-B hybridized pair and probe P dissociates the S-B-supported crosslinks and leaves an S-P complex immobilized to the AAm network through the grafted S oligonucleotide. The reduction of the crosslink density associated with this competitive displacement reaction affects the swelling of the hydrogel. A schematic illustration of the response of a DNA hybrid hydrogel with an altered pore size distribution as synthesized at the end of the optical fiber is given in Figure 1.

Table 1. Sequence of oligonucleotides used for double-stranded (dsDNA) supported junctions in the DNA-*g*-AAm hydrogels with an altered pore size distribution. The oligonucleotide base pair (bp) sequence is depicted for the sensing (S, blue in Figure 1) and blocking (B, red in Figure 1) oligonucleotides of the dsDNA-supported junctions and probe oligonucleotide (P, green in Figure 1) used for assessing the hydrogel swelling response. The underlined parts of the bp sequences denote the complementary region.

Oligonucleotide	Base Pair Sequence
Sensing (blue)	5′-CTG ATC TA<u>A GTA ACT ACT AG</u>-3′
Blocking (red)	5′-CTC AGT CA<u>C TAG TAG TTA CT</u>-3′
Probe (green)	5′-<u>CTA GTA GTT ACT AG</u>ATC-3′

Figure 6 presents the swelling kinetics of DNA-*g*-AAm hydrogels when probe nucleotide P is added to the aqueous buffer (160 mM ionic strength, pH 7), yielding the probe concentration of 20 μM. The swelling process was observed to be faster for the DNA-*g*-AAm hydrogels for increasing PEG content in the pre-gel solution. Figure 6d compares the swelling equilibration time constants for samples DNAGel0, DNAGel10 and DNAGel17. The hydrogel DNAGel0 synthesized without the addition of pore-forming agent showed the longest equilibration time constant $\tau_{1/2}$ = 2360 s. The constant decreased to 1530 and 1430 s for the DNAGel10 and DNAGel17 samples, respectively. Equilibration time constants measured for DNA-*g*-AAm hydrogels were about two orders of magnitude higher than the time constants for ionic strength-induced AMPSA-*co*-AAm deswelling, found to be in the range of 10 to 15 s. This remarkable difference in kinetics results from the physico-chemical nature of the two processes. Diffusion of probe oligonucleotides inside the hydrogel network with grafted DNA strands is reported to be interrupted by subsequent association and dissociation processes of the probe with complementary base pairs with immobile DNA strands [35]. As a result, the diffusion of the probe is much slower than in hydrogels without complementary oligonucleotides grafted to the network. Changes in the DNA-*g*-AAm hydrogel porous structure induced by PEG may affect oligonucleotide transport by various mechanisms: by increasing hydrogel permeability and by decreasing the probability of probe-immobile strand interaction. It is also conceivable that an altered porous structure can affect the rate of the competitive displacement hybridization reaction. Steric constraints imposed by the proximity of the acrylamide backbone that are a possible obstacle for the displacement process may be affected by the change of the hydrogel pore distribution. Nevertheless, further experimental evidence would be required to assess the relative importance of the two described mechanisms on DNA-*g*-AAm hydrogels' swelling dynamics.

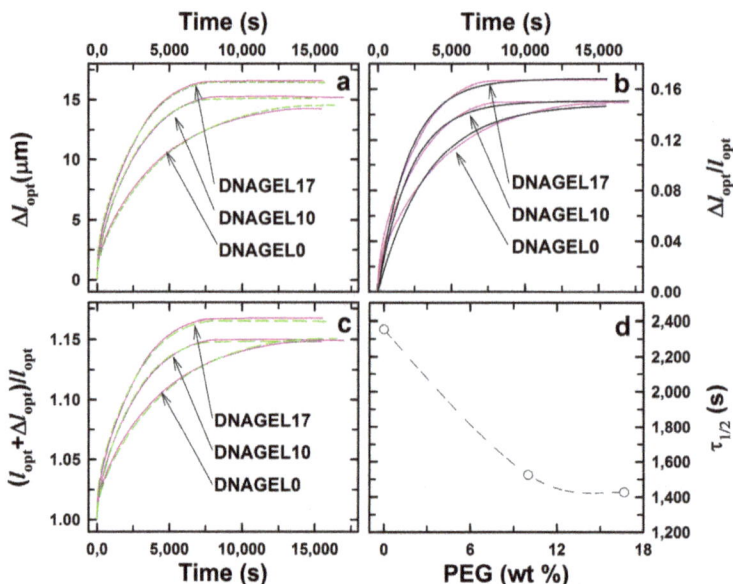

Figure 6. Optical length changes (Δl_{opt}) (**a**) and optical length changes relative to the optical length at reference state (($l_{opt} + \Delta l_{opt}$)/l_{opt} (**c**) of DNA hybrid hydrogels with an altered pore size distribution upon ssDNA probe recognition. Fits of first order swelling kinetics (black lines) to relative optical length changes ($\Delta l_{opt}/l_{opt}$, pink lines) (**b**) were the basis for the estimation of apparent equilibration time constants ($\tau_{1/2} = \ln 2/k$) (**d**). DNA hybrid porous hydrogels consisted of 10 wt % AAm, 0.6 mol % Bis (relative to AAm) and 0.4 mol % S-B double-stranded oligonucleotides (relative to AAm) synthesized in the presence of 0 (DNAGel0, l_{opt} = 96.2 µm), 10 (DNAGel10, l_{opt} = 101.8 µm) and 17 wt % (DNAGel17, l_{opt} = 99.1 µm) of PEG200. All measurements were carried out with a probe nucleotide concentration of 20 µM in buffer solution at room temperature and at least two times (red and green lines for each experimental series).

3. Conclusions

In this work, we present how the kinetics of the swelling response of DNA-*g*-AAm hydrogel to an ssDNA probe may be facilitated by introducing polyethylene glycol as a means to affect the DNA-*g*-AAm hybrid hydrogel porous structure. Swelling of DNA-*g*-AAm hydrogel in the presence of a complementary nucleotide probe occurs due to the destabilization of physical DNA crosslinks [26]. Swelling kinetics is governed by the kinetics of the competitive displacement reaction between the probe and DNA physical crosslinks, and the rate of this process is dependent on the diffusion rate of the probe within the hydrogel matrix. The ssDNA diffusion within the polymer-DNA hybrid hydrogel is slowed down by repeated association and dissociation of the probe with immobile DNA strands, which is known as "retarded diffusion" [35]. Quantitative modelling of the swelling kinetics, including also additional experimental approaches for the determination of relevant parameters, including these various identified molecular processes and network relaxation, is a topic for further exploration. Altering the hydrogel pore size distribution resulted in enhanced permeability of the material and a decreased probability of multiple probe-immobile DNA interactions, which led to faster hydrogel equilibration. The equilibration was faster the higher the weight fraction of PEG in the pre-gel solution was. This suggests that the hydrogel pore size distribution was showing an increased fraction of larger pores for higher PEG content. In contrast to the "retarded diffusion" limited swelling of DNA-*g*-AAm hydrogel, osmotic pressure-driven AMPSA-*co*-AAm hydrogel deswelling has not been affected by the weight fraction of porogen. The finding suggests that this hydrogel network diffusivity, which is a hydrogel deswelling limiting factor, is not strongly affected by changes in the pore size distribution of the network.

4. Experimental Section

DNA and AMPSA-*co*-acrylamide-bis-acrylamide hydrogels were synthesized at the end of optical fibers functionalized with 3-(trimethoxysilyl) propyl methacrylate (>98%, Aldrich, St Louis, MO, USA) according to the previously-described procedure [26]. Briefly, the DNA hybrid hydrogels (DNA-*g*-AAm) were synthesized using acrylamide (AAm, 99%, Sigma, St Louis, MO, USA), (*N*,*N*-methylene-bisacrylamide, (Bis, 99%+, Acros organics, Geel, Belgium) as the covalent crosslinker, hybridized oligonucleotides functionalized at the 5′ end (Acrydite, Integrated DNA Technologies, Coralville, IA, USA) acting as reversible physical crosslinks and in the presence of pore-forming agent polyethylene glycol (PEG, M_n = 200 g/mol, Aldrich), denoted as PEG200. The functionalized oligonucleotides contain an acrylic phosphoramidite as a 5′-modification of the oligonucleotide and can be incorporated into polyacrylamide gels upon polymerization. The incorporated oligonucleotides are connected with a 6-carbon spacer to the AAm network (Integrated DNA Technologies) [44]. AMPSA-*co*-AAm hydrogels were synthesized using 2-acrylamido-2-methyl-1-propane-sulfonic acid (AMPSA, 99%, Aldrich) as a co-monomer, 3 mol % Bis relative to AAm and in the presence of two PEG polymers with different molecular weights, M_n = 200 (PEG200) and 950~1050 g/mol (PEG950) (Aldrich), and different contents (0, 3, 10, 17 wt %). The hydrogels prepared with 0, 3, 10 and 17 wt % of PEG200 during the synthesis are denoted as AMPSAGel0, AMPSAGel3, AMPSAGel10 and AMPSAGel17, respectively.

The DNA-*g*-AAm hybrid hydrogels were synthesized using sensing oligonucleotide S with the nucleotide sequence chosen at random. The blocking oligonucleotide B was designed with 12 base pairs complementary to S. The probe oligonucleotide was chosen to have a 6 base pair longer complementary sequence than the blocking oligonucleotide (6 base pairs in the so-called "toe-hold" region beyond the 12 base pairs in the complementary region). The sequences of the oligonucleotides used were as follows: sensing S: 5′-CTGATCTAAGTAACTACTAG-3′, blocking B: 5′-CTCAGTCACTAGTAGTTACT-3′), probing P: 5′-CTAGTAGTTACTTAGATC-3′. Acrydite oligonucleotides were dissolved in AAm (10 wt %), 0.6 mol % Bis relative to AAm and PEG M_n = 200 g/mol (0, 10, 17 wt %) solution prepared in an aqueous buffer (pH 7.0, 10 mM Tris (99.8% Sigma Aldrich, St Louis, MO, USA), 1 mM EDTA (Sigma Aldrich, 99%), 150 mM NaCl (Sds, 99%), yielding a total ionic strength of about 160 mM). Hydrogels prepared using 0, 10 and 17 wt % of PEG200 are denoted as DNAGel0, DNAGel10 and DNAGel17, respectively. Appropriate volumes of the solutions to yield a 0.4 mol % nucleotide concentration of S and B were mixed together at least 3 h (room temperature) before the synthesis to allow S-B hybridization prior to the cross-linking co-polymerization.

Pre-gel solutions were deposited at the end of a functionalized optical fiber. UV-initiated polymerization was carried out for 6 min using Dymax Bluewave as the UV light source and a photo-initiator (hydroxycyclohexyl phenyl ketone, 99%, Aldrich). The pre-gel deposition at the end of the fiber and polymerization were carried out in a squalane oil (99%, Aldrich) to prevent water evaporation from the hydrogels [26]. Following the photo-polymerization, the fibers with attached hydrogels were protected with glass tubes with an inner diameter of about 1 mm. Hydrogels attached to the optical fiber were immersed in ultrapure water (AMSPSA hydrogels) or PBS buffer (DNA hydrogels) for enough time (24 h), changing the solutions from time to time. No further changes in the interferometer phase signal were employed as a guideline for reaching the equilibrium state of the hydrogel before proceeding with the characterization of the swelling response. It was confirmed by FT-IR (Figure 7) that such a washing and equilibration procedure yielded hydrogels without detectable signatures of PEG. The DNA-grafted hydrogels were pre-equilibrated at 23 °C in buffer solution prior to exposure to probe P buffered solution. The experiments were carried out as follows: the fibers/tubes were taken out of the buffer solution and immersed in 500 μL of the 20 μM oligonucleotide probe P in the buffered solutions thermally equilibrated at 23 °C.

Gels **2015**, *1*, 219–234

The AMPSA-*co*-AAm hydrogels with PEG content of 0, 3, 10, 17 wt % were synthesized and washed similar to the DNA-*g*-AAm hydrogels. The AMPSAGels were characterized by Fourier transform infrared spectra (FT-IR, PerkinElmer spectrum One FT-IR spectrometer) by depositing a droplet of pre-gel solution on the glass substrate, followed by UV polymerization for 3 min, and washed and stored in ultrapure water for 3 days with changing of the water from time to time. The samples were freeze-dried before FT-IR. The FT-IR spectra indicate a hydrogel structure comprising the monomers in the pre-gel solution (Figure 7). No specific curve of PEG (2870 cm^{-1}) observed by FT-IR proved the complete removal of porogen in AMPSA hydrogels.

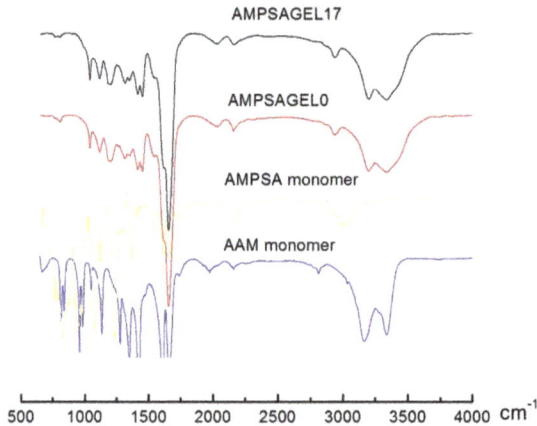

Figure 7. FT-IR of AAm monomer, AMPSA monomer, freezing-dried AMPSAGel0 and AMPSAGel17.

A typical FT-IR curve of AMPSA-*co*-AAm gels and their monomers is shown in Figure 7. The stretching vibration peaks at 3432 and 1647 cm^{-1} were respectively for amino and amidic carbonyl groups of AAm monomer and AMPSA gels. The peaks around 1039 and 1183 cm^{-1} were the characteristic peaks of the asymmetric and symmetric bands of sulfonate groups in the AMPSA unit, which also appeared for AMPSAGEL0 and AMPSAGEL17. The disappearance of the peak around 1600 cm^{-1} (vinyl groups) further confirmed the formation of AMPSA-*co*-AAm gels. The FT-IR spectra indicated a hydrogel structure comprising the monomers in the pre-gel solution.

The AMPSA-*co*-AAm hydrogels were immersed and equilibrated in 40 mL of 0.5 mM NaCl aq solution prior to the determination of the swelling properties. Hydrogel swelling of the AMPSA-*co*-AAm hydrogels was determined by stepwise adjusting the NaCl concentration in the immersing bath by pipetting aliquots of 2 M NaCl into the immersing solution under constant mixing.

Changes in the optical length of the hydrogels (Δl_{opt}) were determined using the high resolution interferometric readout technique described in detail previously [29]. The experimentally-determined phase change of the interference wave was used as the basis for the determination of Δl_{opt} due to its superior resolution compared to data extracted from the amplitude of the interference wave. The measurement was carried out until the hydrogel reached equilibrium, as manifested by the constant phase on the interferometric readout. Changes in the optical length of the hydrogel, Δl_{opt}, relative to the overall optical length of hydrogels l_{opt}, $\Delta l_{opt}/l_{opt}$ were employed as a measure of the hydrogel swelling behavior. Parameter Δl_{opt} is determined relative to l_{opt} selected as the reference state for each experimental series. The changes of the swelling ratio of the hydrogels following exposure to altered salt concentration or ssDNA probe (one concentration) were followed as a function of time. The $\Delta l_{opt}/l_{opt}$ was observed to change exponentially with time. An exponential decay was observed in the case of deswelling hydrogels exponentially increased to a new plateau maximum for the swelling gels. The kinetics of swelling was analyzed based on an apparent first order rate process of $\Delta l_{opt}/l_{opt}$.

The swelling/deswelling kinetics were quantitatively represented by an apparent equilibration time constant ($\tau_{1/2} = \ln2/k$) by fitting $\Delta l_{opt}(t)/l_{opt} = e^{-kt}$ to the experimental data.

Acknowledgments: This work was supported by the Norwegian Research Council, Contract Number 191818/V30.

Author Contributions: Ming Gao and Bjørn Torger Stokke designed the study. Ming Gao conducted the experiments. Ming Gao, Kamila Gawel and Bjørn Torger Stokke contributed to writing the manuscript. All authors approved the manuscript.

Conflicts of Interest: The authors declare no conflict of interest.

References

1. Ricka, J.; Tanaka, T. Swelling of ionic gels: Quantitative performance of the donnan theory. *Macromolecules* **1984**, *17*, 2916–2921. [CrossRef]
2. Brannon-Peppas, L.; Peppas, N.A. Dynamic and equilibrium swelling behaviour of pH-sensitive hydrogels containing 2-hydroxyethyl methacrylate. *Biomaterials* **1990**, *11*, 635–644. [CrossRef]
3. Brannon-Peppas, L.; Peppas, N.A. Time-dependent response of ionic polymer networks to pH and ionic strength changes. *Int. J. Pharm.* **1991**, *70*, 53–57. [CrossRef]
4. Beebe, D.J.; Moore, J.S.; Bauer, J.M.; Yu, Q.; Liu, R.H.; Devadoss, C.; Jo, B.-H. Functional hydrogel structures for autonomous flow control inside microfluidic channels. *Nature* **2000**, *404*, 588–590. [CrossRef] [PubMed]
5. Miyata, T.; Nakamae, K.; Hoffman, A.S.; Kanzaki, Y. Stimuli-sensitivities of hydrogels containing phosphate groups. *Macromol. Chem. Phys.* **1994**, *195*, 1111–1120. [CrossRef]
6. Yoshida, R.; Uchida, K.; Kaneko, Y.; Sakai, K.; Kikuchi, A.; Sakurai, Y.; Okano, T. Comb-type grafted hydrogels with rapid deswelling response to temperature changes. *Nature* **1995**, *374*, 240–242. [CrossRef]
7. Lee, W.F.; Yuan, W.Y. Thermoreversible hydrogels X: Synthesis and swelling behavior of the (N-isopropylacrylamide-co-sodium 2-acrylamido-2-methylpropyl sulfonate) copolymeric hydrogels. *J. Appl. Polym. Sci.* **2000**, *77*, 1760–1768. [CrossRef]
8. Schröder, U.P.; Oppermann, W. Properties of polyelectrolyte gels. In *Physical Properties of Polymeric Gels*; Wiley & Sons: Chichester, UK, 1996; pp. 19–38.
9. Park, T.G.; Hoffman, A.S. Sodium chloride-induced phase transition in nonionic poly(N-isopropylacrylamide) gel. *Macromolecules* **1993**, *26*, 5045–5048. [CrossRef]
10. Grimshaw, P.E.; Nussbaum, J.H.; Grodzinsky, A.J.; Yarmush, M.L. Kinetics of electrically and chemically induced swelling in polyelectrolyte gels. *J. Chem. Phys.* **1990**, *93*, 4462–4472. [CrossRef]
11. Sun, S.; Mak, A.F.T. The dynamical response of a hydrogel fiber to electrochemical stimulation. *J. Polym. Sci. B Polym. Phys.* **2001**, *39*, 236–246. [CrossRef]
12. Mamada, A.; Tanaka, T.; Kungwatchakun, D.; Irie, M. Photoinduced phase transition of gels. *Macromolecules* **1990**, *23*, 1517–1519. [CrossRef]
13. Eeckman, F.; Moes, A.J.; Amighi, K. Surfactant induced drug delivery based on the use of thermosensitive polymers. *J. Controll. Release* **2003**, *88*, 105–116. [CrossRef]
14. Holtz, J.H.; Asher, S.A. Polymerized colloidal crystal hydrogel films as intelligent chemical sensing materials. *Nature* **1997**, *389*, 829–832. [CrossRef]
15. Miyata, T.; Asami, N.; Uragami, T. A reversibly antigen-responsive hydrogel. *Nature* **1999**, *399*, 766–769. [CrossRef] [PubMed]
16. Alexeev, V.L.; Sharma, A.C.; Goponenko, A.V.; Das, S.; Lednev, I.K.; Wilcox, C.S.; Finegold, D.N.; Asher, S.A. High ionic strength glucose-sensing photonic crystal. *Anal. Chem.* **2003**, *75*, 2316–2323. [CrossRef] [PubMed]
17. Yang, Z.M.; Gu, H.W.; Fu, D.G.; Gao, P.; Lam, J.K.; Xu, B. Enzymatic formation of supramolecular hydrogels. *Adv. Mater.* **2004**, *16*, 1440–1444. [CrossRef]
18. Kim, H.; Cohen, R.E.; Hammond, P.T.; Irvine, D.J. Live lymphocyte arrays for biosensing. *Adv. Funct. Mater.* **2006**, *16*, 1313–1323. [CrossRef]
19. Yang, Z.; Ho, P.-L.; Liang, G.; Chow, K.H.; Wang, Q.; Cao, Y.; Guo, Z.; Xu, B. Using β-lactamase to trigger supramolecular hydrogelation. *J. Am. Chem. Soc.* **2007**, *129*, 266–267. [CrossRef] [PubMed]

20. Tierney, S.; Sletmoen, M.; Skjak-Braek, G.; Stokke, B.T. Interferometric characterization of swelling of covalently crosslinked alginate gel and changes associated with polymer impregnation. *Carbohydr. Polym.* **2010**, *80*, 828–832. [CrossRef]

21. Tierney, S.; Falch, B.M.H.; Hjelme, D.R.; Stokke, B.T. Determination of glucose levels using a functionalized hydrogel-optical fiber biosensor: Toward continuous monitoring of blood glucose *in vivo*. *Anal. Chem.* **2009**, *81*, 3630–3636. [CrossRef] [PubMed]

22. Tierney, S.; Volden, S.; Stokke, B.T. Glucose sensors based on a responsive gel incorporated as a fabry-perot cavity on a fiber-optic readout platform. *Biosens. Bioelectron.* **2009**, *24*, 2034–2039. [CrossRef] [PubMed]

23. Skjaervold, N.K.; Solligard, E.; Hjelme, D.R.; Aadahl, P. Continuous measurement of blood glucose validation of a new intravascular sensor. *Anesthesiology* **2011**, *114*, 120–125. [CrossRef] [PubMed]

24. Gao, M.; Gawel, K.; Stokke, B.T. Toehold of dsdna exchange affects the hydrogel swelling kinetics of a polymer-dsDNA hybrid hydrogel. *Soft Matter* **2011**, *7*, 1741–1746. [CrossRef]

25. Gawel, K.; Stokke, B.T. Logic swelling response of DNA-polymer hybrid hydrogel. *Soft Matter* **2011**, *7*, 4615–4618. [CrossRef]

26. Tierney, S.; Stokke, B.T. Development of an oligonucleotide functionalized hydrogel integrated on a high resolution interferometric readout platform as a label-free macromolecule sensing device. *Biomacromolecules* **2009**, *10*, 1619–1626. [CrossRef] [PubMed]

27. Gawel, K.; Gao, M.; Stokke, B.T. Impregnation of weakly charged anionic microhydrogels with cationic polyelectrolytes and their swelling properties monitored by a high resolution interferometric technique. Transformation from a polyelectrolyte to polyampholyte hydrogel. *Eur. Polym. J.* **2012**, *48*, 1949–1959. [CrossRef]

28. Gao, M.; Gawel, K.; Stokke, B.T. High resolution interferometry as a tool for characterization of swelling of weakly charged hydrogels subjected to amphiphile and cyclodextrin exposure. *J. Colloid Interface Sci.* **2013**, *390*, 282–290. [CrossRef] [PubMed]

29. Tierney, S.; Hjelme, D.R.; Stokke, B.T. Determination of swelling of responsive gels with nanometer resolution. Fiber-optic based platform for hydrogels as signal transducers. *Anal. Chem.* **2008**, *80*, 5086–5093. [CrossRef] [PubMed]

30. Zhao, B.; Moore, J.S. Fast pH- and ionic strength-responsive hydrogels in microchannels. *Langmuir* **2001**, *17*, 4758–4763. [CrossRef]

31. Bouklas, N.; Huang, R. Swelling kinetics of polymer gels: Comparison of linear and nonlinear theories. *Soft Matter* **2012**, *8*, 8194–8203. [CrossRef]

32. Reynaldo, L.P.; Vologodskii, A.V.; Neri, B.P.; Lyamichev, V.I. The kinetics of oligonucleotide replacements. *J. Mol. Biol.* **2000**, *297*, 511–520. [CrossRef] [PubMed]

33. Li, Q.Q.; Luan, G.Y.; Guo, Q.P.; Liang, J.X. A new class of homogeneous nucleic acid probes based on specific displacement hybridization. *Nucleic Acids Res.* **2002**, *30*, e5. [CrossRef] [PubMed]

34. Srinivas, N.; Ouldridge, T.E.; Sulc, P.; Schaeffer, J.M.; Yurke, B.; Louis, A.A.; Doye, J.P.K.; Winfree, E. On the biophysics and kinetics of toehold-mediated DNA strand displacement. *Nucleic Acids Res.* **2013**, *41*, 10641–10658. [CrossRef] [PubMed]

35. Livshits, M.A.; Mirzabekov, A.D. Theoretical analysis of the kinetics of DNA hybridization with gel-immobilized oligonucleotides. *Biophys. J.* **1996**, *71*, 2795–2801. [CrossRef]

36. Kaneko, Y.; Sakai, K.; Kikuchi, A.; Yoshida, R.; Sakurai, Y.; Okano, T. Influence of freely mobile grafted chain length on dynamic properties of comb-type grafted poly(*N*-isopropylacrylamide) hydrogels. *Macromolecules* **1995**, *28*, 7717–7723. [CrossRef]

37. Bin Imran, A.; Seki, T.; Takeoka, Y. Recent advances in hydrogels in terms of fast stimuli responsiveness and superior mechanical performance. *Polym. J.* **2010**, *42*, 839–851. [CrossRef]

38. Annabi, N.; Nichol, J.W.; Zhong, X.; Ji, C.D.; Koshy, S.; Khademhosseini, A.; Dehghani, F. Controlling the porosity and microarchitecture of hydrogels for tissue engineering. *Tissue Eng. Part B Rev.* **2010**, *16*, 371–383. [CrossRef] [PubMed]

39. Gao, M.; Toita, S.; Sawada, S.; Akiyoshi, K.; Stokke, B.T. Cyclodextrin triggered dimensional changes of polysaccharide nanogel integrated hydrogels at nanometer resolution. *Soft Matter* **2013**, *9*, 5178–5185. [CrossRef]

40. Prot, V.; Sveinsson, H.M.; Gawel, K.; Gao, M.; Skallerud, B.; Stokke, B.T. Swelling of a hemi-ellipsoidal ionic hydrogel for determination of material properties of deposited thin polymer films: An inverse finite element approach. *Soft Matter* **2013**, *9*, 5815–5827. [CrossRef]

41. Caykara, T.; Bulut, M.; Dilsiz, N.; Akyuz, Y. Macroporous poly(acrylamide) hydrogels: Swelling and shrinking behaviors. *J. Macromol. Sci. Pure Appl. Chem.* **2006**, *43*, 889–897. [CrossRef]

42. Caykara, T.; Bulut, M.; Demirci, S. Preparation of macroporous poly(acrylamide) hydrogels by radiation induced polymerization technique. *Nucl. Instrum. Methods Phys. Res. Sect. B* **2007**, *265*, 366–369. [CrossRef]

43. Gemeinhart, R.A.; Chen, J.; Park, H.; Park, K. pH-sensitivity of fast responsive superporous hydrogels. *J. Biomater. Sci. Polym. Ed.* **2000**, *11*, 1371–1380. [CrossRef] [PubMed]

44. Lin, D.C.; Yurke, B.; Langrana, N.A. Mechanical properties of a reversible, DNA-crosslinked polyacrylamide hydrogel. *J. Biomech. Eng. Trans. Asme* **2004**, *126*, 104–110.

Article

Precise Control over the Rheological Behavior of Associating Stimuli-Responsive Block Copolymer Gels

Jérémy Brassinne, Flanco Zhuge, Charles-André Fustin * and Jean-François Gohy *

Institute of Condensed Matter and Nanosciences (IMCN), Bio- and Soft Matter (BSMA) Division, Université catholique de Louvain (UCL), Place L. Pasteur 1, Louvain-la-Neuve 1348, Belgium; jeremy.brassinne@uclouvain.be (J.B.); flanco.zhuge@uclouvain.be (F.Z.)

* Authors to whom correspondence should be addressed; jean-francois.gohy@uclouvain.be (J.-F.G.); charles-andre.fustin@uclouvain.be (C.-A.F.);
Tel.: +32-10-479-269 (J.-F.G.); +32-10-479-345 (C.-A.F.); Fax: +32-10-479-178 (J.-F.G.).

Academic Editor: Dirk Kuckling
Received: 13 October 2015; Accepted: 30 November 2015; Published: 7 December 2015

Abstract: "Smart" materials have considerably evolved over the last few years for specific applications. They rely on intelligent macromolecules or (supra-)molecular motifs to adapt their structure and properties in response to external triggers. Here, a supramolecular stimuli-responsive polymer gel is constructed from heterotelechelic double hydrophilic block copolymers that incorporate thermo-responsive sequences. These macromolecular building units are synthesized via a three-step controlled radical copolymerization and then hierarchically assembled to yield coordination micellar hydrogels. The dynamic mechanical properties of this particular class of materials are studied in shear flow and finely tuned via temperature changes. Notably, rheological experiments show that structurally reinforcing the micellar network nodes leads to precise tuning of the viscoelastic response and yield behavior of the material. Hence, they constitute promising candidates for specific applications, such as mechano-sensors.

Keywords: supramolecular; stimuli-responsive; micellar gel; associating polymer; terpyridine; rheology

---

1. Introduction

"Smart" materials, *i.e.*, whose properties can be significantly changed in a controlled fashion by external triggers, are experiencing an unprecedented development over the last few years [1]. As a mirror of modern society, the practical demand for "smart" devices has overwhelmed all other forms of computing and communications in a very short time. They consist of systems that can operate to some extent interactively and autonomously to fulfill individual requirements. Analogously, "smart" materials have been designed to support a range of properties pertaining to use in various environments. In this regard, the most relevant focus having attracted much attention concerns "smart" polymer gels [2–4]. Due to their large, often macroscopic responses, they may indeed find numerous technological applications as mechano-sensors or soft actuators [5].

At the basis of "smart" materials lie intelligent polymer sequences or (supra-)molecular motifs having the ability to adapt their conformations and properties in response to external triggers, such as temperature [6–9], light [10–12] or pH [13–15]. Those variables can notably reverse the solvophilicity of synthetic macromolecules incorporating responsive groups, which pave the way for stimuli-sensitive materials. Another strategy to impart responsiveness to materials involves the use of secondary interactions, like hydrogen bonding [16,17], ionic interactions [18,19], π-stacking [20] or metal-ligand coordination [21]. They are indeed weaker, but dynamically more labile than primary covalent bonds [22,23], which impart adaptive properties to non-covalent assemblies [8,24].

In the field of environmentally-adaptive polymers, the thermo-sensitive poly(N-isopropylacrylamide) (PNIPAAm) constitutes an intensively-used polymer whose aqueous solutions show a lower critical solution temperature (LCST) around 32 °C [25]. Besides, many other thermo-responsive polymers have also demonstrated their applicability in the preparation of smart materials [6]. Among them, poly(2-(dimethylamino)ethyl methacrylate) (PDMAEMA) is particularly attractive, since it further shows pH responsiveness. By raising the hydrophilic nature of PDMAEMA through protonation of tertiary amino-groups [26], the overall hydrogen bonding ability of the macromolecules, as well as their electrostatic repellency are indeed increased, which leads to higher transition temperatures [26–29].

Although synthetically challenging, stimuli-responsive multi-block copolymers produce well-defined materials through the triggered or autonomous non-covalent associations of one or more blocks. In turn, they provide appealing flexibility for controlling the material micro-structure and physical properties through the application of proper stimuli [30–32]. In this context, our research has been focused on the synthesis of stimuli-responsive sequenced copolymers [33–35] and their self-assembly into "smart" materials with controlled structure and properties [36–43]. Our design strategy relies on a combination of metal-ligand and hydrophobic interactions, whose strength, density and dynamics can be adjusted depending on the material [36–38] and environmental [39–41] variables.

In the present paper, the purpose is to demonstrate the possibility of finely controlling the rheological behavior of this particular class of stimuli-responsive polymer gels. To this aim, a heterotelechelic double hydrophilic block copolymer is designed by means of controlled radical copolymerization techniques. This macromolecular building block is then hierarchically organized in aqueous media to yield a metallo-supramolecular micellar network. At its core, this work aims at studying in detail the rheological response of the self-assembled material in response to temperature changes.

2. Results and Discussion

In the last decade, the interest in polymers incorporating supramolecular motifs has dramatically risen, since they can easily find advanced applications as smart materials [44]. In the following, we describe the synthesis of a terpyridine end-capped polystyrene-*block*-poly(N-isopropylacrylamide)-*block*-poly(2-(dimethylamino)ethyl methacrylate) triblock terpolymer (PS-*b*-PNIPAAm-*b*-PDMAEMA-*b*-tpy) and its hierarchical assembly into a coordination micellar network. Rotational rheometry is then used as a characterization tool to probe the thermo-mechanical properties of the supramolecular hydrogel.

2.1. Synthesis of Functional Building Block

Among coordination motifs, the 2,2';6',2''-terpyridine ligand is particularly attractive, since it forms stable *bis*-complexes in combination with various transition metal ions [45,46]. Being the most prominent representative of its family [47,48], this N-heteroaromatic ligand can be easily introduced into macromolecular architectures by post-modification [49–51], or via the use of modified comonomers [52–55] or chain initiators [56,57]. More recently, the use of terpyridine-modified chain transfer agents (CTA) has been developed as a straightforward approach toward functional block copolymers [33,34]. In this study, a dithiobenzoate is selected due to its compatibility with various functional monomers [58], affording the possibility to sequentially copolymerize methacrylate, acrylamide and styrene monomers [59,60]. This control agent is derived from commercially available 4-(4-cyanopentanoic acid) dithiobenzoate (CPAD) according to a procedure reported elsewhere [34].

The synthesis of the PS-*b*-PNIPAAm-*b*-PDMAEMA-tpy triblock copolymer is sequentially achieved by reversible addition-fragmentation chain transfer (RAFT) polymerization, as depicted in Figure 1. This copolymer is designed to produce self-assembled hydrogels that would combine the association strength of polystyrene stickers and metal-terpyridine complexes, with the stimuli-responsiveness of both PNIPAAm and PDMAEMA blocks. Following a procedure reported elsewhere [34], a terpyridine end-capped double hydrophilic PNIPAAm$_{73}$-b-PDMAEMA$_{103}$ copolymer is first synthesized, the number in subscript referring to the average degree of polymerization of each block. In a third step, styrene is polymerized in the presence of the PNIPAAm-b-PDMAEMA-tpy copolymer as macro-CTA and 2,2'-azobis(isobutyronitrile) (AIBN) as a source of primary radicals (Figure 1). In order to ensure control over the polymerization process, the initial ratio between styrene, macro-CTA and AIBN is set to 3000:7.5:1. In practice, the reaction is conducted at 80 °C in dry 1,4-dioxane and stopped at a predetermined time interval to afford a polystyrene block of a few units. In our strategy, the length of the PS segment is deliberately kept short to allow direct dissolution of the triblock copolymer in aqueous media.

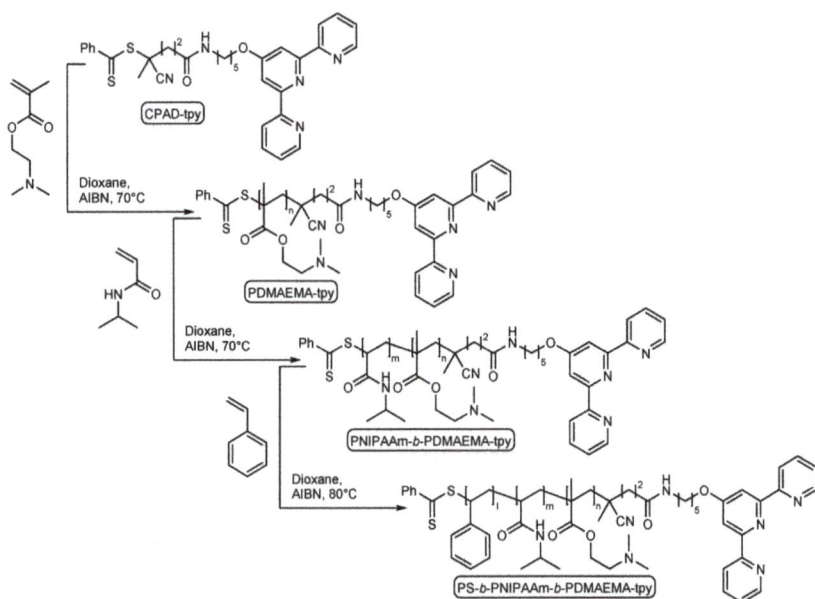

Figure 1. Sequential reversible addition-fragmentation chain transfer (RAFT) copolymerization leading to polystyrene-*block*-poly(N-isopropylacrylamide)-*block*-poly(2-(dimethylamino)ethyl methacrylate) triblock terpolymer (PS-b-PNIPAAm-b-PDMAEMA-b-tpy). CPAD, 4-(4-cyanopentanoic acid) dithiobenzoate; AIBN, 2,2'-azobis(isobutyronitrile).

After the polymerization step, the synthesized triblock copolymer is purified by precipitation of the crude reaction mixture, followed by isolation of the precipitate and subsequent drying. The composition of the copolymer is determined to be PS$_{12}$-b-PNIPAAm$_{73}$-b-PDMAEMA$_{103}$-tpy by means of proton nuclear magnetic resonance (^{1}H-NMR) spectroscopy. As shown in Figure 2, the ^{1}H-NMR spectrum analysis reveals characteristic broad signals of polystyrene and further attests to the presence of terpyridine ligand in the chain architecture. Practically, the length of the polystyrene block is estimated from the ratio between the peak area of PS aromatic protons and isopropyl/aliphatic ester protons of the PNIPAAm and PDMAEMA block around 4 ppm.

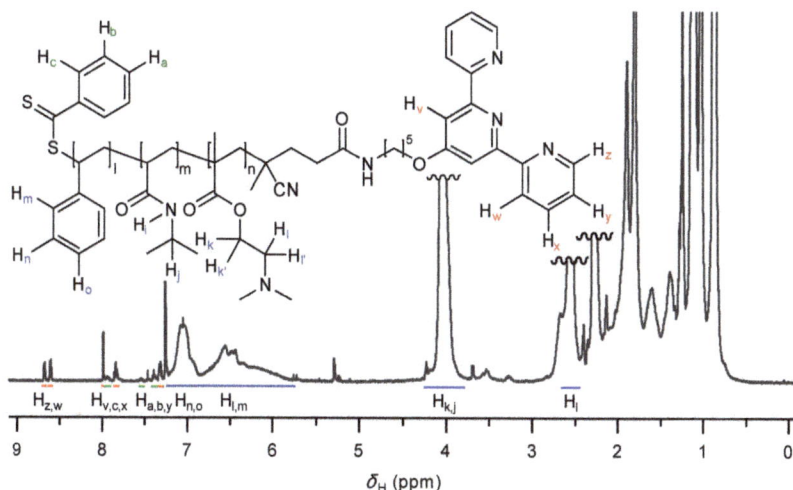

Figure 2. ^{1}H-NMR spectrum of pure PS-*b*-PNIPAAm-*b*-PDMAEMA-tpy triblock terpolymer in deuterated chloroform as the solvent.

As shown in Figure 3, the chain extension of PNIPAAm-*b*-PDMAEMA-tpy macro-CTA into PS-*b*-PNIPAAm-*b*-PDMAEMA-tpy triblock terpolymers is further evidenced by size exclusion chromatography (SEC). This analysis indeed shows a significant shift of the SEC trace to lower elution times after polymerization of the third block, indicating an increase in molar masses. Last, but not least, the SEC analysis reveals still narrow molar mass dispersity ($Đ$ = 1.29), attesting to the control over the polymerization process.

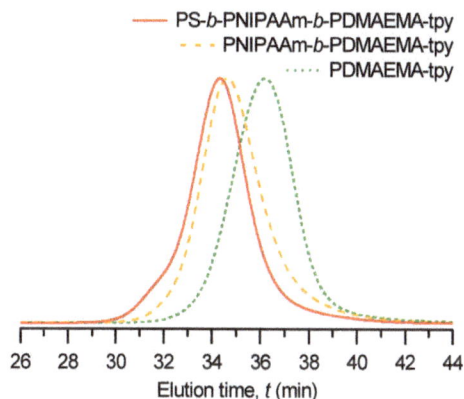

Figure 3. Size exclusion chromatography (SEC) elugrams of PDMAEMA-tpy homopolymer, derived PNIPAAm-*b*-PDMAEMA-tpy diblock copolymer and PS-*b*-PNIPAAm-*b*-PDMAEMA-tpy triblock terpolymer.

2.2. Self-Assembly into Metallo-Supramolecular Hydrogel

The synthesized heterotelechelic associating copolymer is then used as precursor of a metallo-supramolecular hydrogel. As schematized in Figure 4, the first level of assembly is achieved upon direct dissolution of the triblock copolymer in ultra-pure water, which presumably leads to the formation of micelles due to the high incompatibility of polystyrene with aqueous media. Thanks to the shortness of this hydrophobic block, the copolymer is found to be easily dispersible in water at low to room temperature and a concentration ranging in the semi-dilute regime. The pH of the solution is adjusted around neutral to ensure a good solubility of the PDMAEMA block, which further avoids its co-precipitation above the LCST of the PNIPAAm block [34]. At this point, a clear free-flowing concentrated solution is obtained, and no gelation occurred, which can be explained by the lack of entanglement between coronal chains of neighboring micelles.

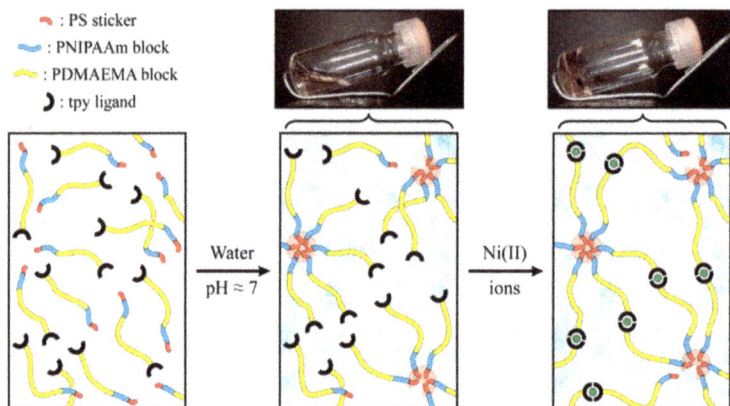

Figure 4. Schematic representations and illustrations of the hierarchical assembly of PS-*b*-PNIPAAm-*b*-PDMAEMA-tpy associating copolymers into a coordination micellar gel.

In practice, the formation of triblock copolymer micelles is tested by dynamic light scattering (DLS) measurements, which provide information about the size distributions of samples by the analysis of scattered light intensity. As shown in Figure 5a, the distribution of hydrodynamic radii (R_h) essentially reveals the presence of micellar nanostructures in the triblock terpolymer solution, with an apparent radius size around 18 nm. Small proportions of isolated chains (unimers), as well as aggregates are also evidenced in the investigated concentration range, respectively around a few and a few hundred nanometers in size. To provide further insight into the micellization process, the solution behavior of the PS-*b*-PNIPAAm-*b*-PDMAEMA-tpy associating terpolymer is compared to the one of the parent diblock copolymer. As shown in Figure 5b, DLS measurement on the PNIPAAm-*b*-PDMAEMA-tpy diblock copolymer solution mainly reveals the presence of an isolated chain, accompanied by larger aggregates that could in fact arise from slow diffusion modes of the polyelectrolyte. However, the micellar nano-objects are no longer detected, which also explains the very low intensity of light scattered by the diblock copolymer solution. On the whole, these experiments indicate that the formation of micelles is indeed driven by the aggregation of the PS segment in aqueous triblock terpolymer solution. Furthermore, the weak dependence of the DLS signal as a function of the scattered angle strongly suggests the formation of spherical micelles, as expected, given the balance between the hydrophobic and hydrophilic blocks.

As a second level of assembly, the stoichiometric amount of half an equivalent of transition metal ions dissolved in water is added to the concentrated micellar solutions to reach a final weight-to-volume fraction (C) of 5% *w/v*. In this concentration regime, the presence of metal ions in the media would result in inter-micellar complexation between the segregated PS nano-domains formed in the aqueous environment (Figure 4). Of course, the choice of the metal ion used for the terpyridine complexation constitutes a powerful means to control the mechanical properties of terpyridine-based gels [61,62]. This consideration was addressed in our previous study, where the effect of different transition metal ions was tested on the rheological properties of metallo-supramolecular micellar hydrogels [36].

In this study, nickel(II) ions are selected, since they afford one of the more stable *bis*-complexes in combination with the terpyridine ligand [63], thereby providing efficient bridges towards the formation of a supramolecular network. In practice, the establishment of a percolated network structure is evidenced by the tube inversion test at room temperature. As shown in Figure 4, the initially free-flowing concentrated solution of PS-*b*-PNIPAAm-*b*-PDMAEMA-tpy copolymers indeed turns into a free-standing supramolecular gel upon the addition of metal ions, which occurs within minutes.

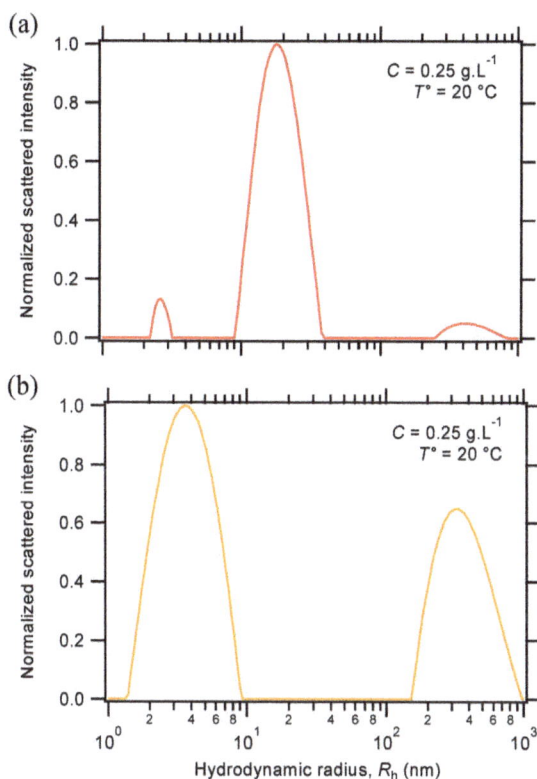

Figure 5. Distribution of hydrodynamic radii of (**a**) the PS-*b*-PNIPAAm-*b*-PDMAEMA-tpy triblock terpolymer and (**b**) the parent PNIPAAm-*b*-PDMAEMA-tpy diblock copolymer in aqueous solution.

Structurally speaking, a free-standing supramolecular hydrogel is obtained thanks to the aggregation of hydrophobic polystyrene segments that provide branching points within the supramolecular network [36,40]. Upon heating, the structure of the triblock micellar network is expected to be further reinforced by the collapse of PNIPAAm segments onto micellar nodes, as illustrated in Figure 6. However, no phase separation is visually observed when heating the PS-*b*-PNIPAAm-*b*-PDMAEMA-tpy hydrogel, even well above the LCST of the PNIPAAm block. Actually, the partial collapse of coronal chains would result in the formation of three-layer micelles with a hydrophobic PS core, a collapsed PNIPAAm shell and a remaining PDMAEMA corona (Figure 6). In this picture, the soluble PDMAEMA segment in the network architecture ensures the swelling of the gel phase, in a temperature range that exceeds the upper solubility limit of the PNIPAAm block. Given the characteristics of the triblock copolymer, a core–shell-corona structure is realistically expected for the micelles at elevated temperature. Nevertheless, the actual morphology that the microphase-separated structures may adopt in the gel phase remains an open question that would require deeper analyses.

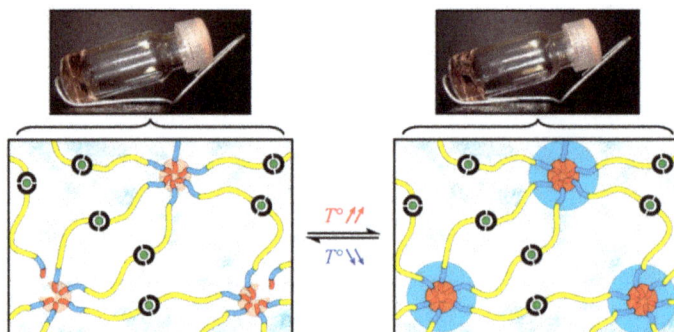

Figure 6. Schematic representations and illustrations of the thermo-response of the PS-*b*-PNIPAAm-*b*-PDMAEMA-tpy triblock copolymer gel.

2.3. Characterization of the Rheological Response

As the main characterization tool, rotational shear rheometry is conducted on the PS-*b*-PNIPAAm-*b*-PDMAEMA-tpy hydrogels to elucidate the hypothetical effect of temperature ($T°$) on their rheological behavior. This technique allows determining the fraction of shear energy that is stored in elastic distortions of the polymer network, as measured by the storage modulus (G'). Complementary to this, the loss modulus (G'') measures the fraction of deformation energy that is dissipated due to relaxations that occur on the timescale of the experiment. Being intimately interrelated, the magnitude of those moduli is thus affected by each relaxation process of the polymer network, whose rates are determined by the molecular mechanism involved. Information about the structure and dynamics of the material can be thus accessed by following the evolution of dynamic moduli while varying the frequency and amplitude of the imposed stress.

In parallel, additional information about the structures of the assembly can be obtained in steady shear flow measurements. This complementary approach allows monitoring the equilibrium flow by maintaining a constant stress (σ) or shear rate ($\dot{\gamma}$) for a sufficient time in one direction to allow dynamic equilibrium to be achieved in the fluid. In particular, monitoring the material viscosity (η) as a function of the shear rate allows evaluating the resistance of the material to being deformed, which is crucial for processing.

In the following, the evolution of dynamic storage and loss moduli upon thermal variations is first followed in the form of temperature sweeps, at a given oscillatory frequency (ω) and low stress (σ_0) or strain (γ_0) amplitude. Then, frequency sweeps are performed under small amplitude oscillatory shear, over a temperature range that covers the phase transition of the PNIPAAm block. After that, the non-linear viscoelastic response of the gel is investigated by amplitude strain sweeps operating at different temperatures. Finally, steady shear flow measurements are conducted on the supramolecular gel, below and above the phase transition temperature.

2.3.1. Oscillatory Temperature Sweep

Due to an enhanced relaxation of the transient cross-links, supramolecular gels often suffer a loss in their mechanical properties when tested at elevated temperatures [8,64]. In sharp contrast, the PS-*b*-PNIPAAm-*b*-PDMAEMA-tpy hydrogel shows a strengthened viscoelastic response to shear as the temperature increases, as attested by dynamic temperature sweeps (Figure 7). In these experiments, the evolution of dynamic storage and loss modulus against temperature is monitored as an indication of the gel strengthening. During measurements, the temperature is varied between 20 and 60 °C (first run), followed by cooling at the same rate immediately after heating (second run). Each temperature ramp is performed at a low stress amplitude of 10 Pa, a fixed frequency of 1 rad/s, with a heating/cooling rate of 2 °C/min.

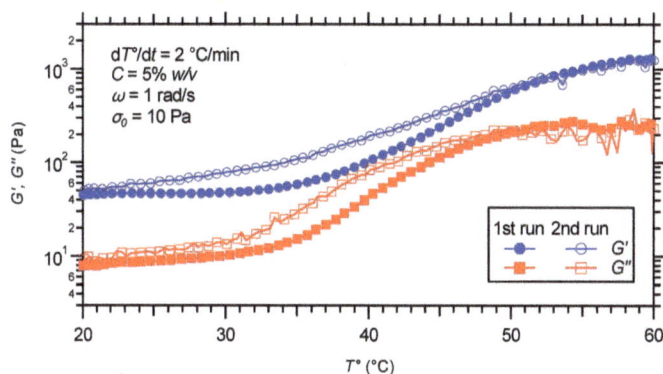

Figure 7. Temperature dependence of dynamic moduli for a hydrogel prepared from the PS-*b*-PNIPAAm-*b*-PDMAEMA-tpy copolymer and Ni(II) ions: first run upon heating; second run upon cooling.

Under ambient conditions, the gel strength results from the aggregation of PS segments into hydrophobic domains that are further bridged by metal-ligand associations, thereby allowing elastic stretching. Above a certain temperature, core-shell-corona micelles are formed with the PS blocks forming micellar cores and the PNIPAAm segments constituting the shell layer. This temperature-induced transition, centered at 40 °C, stems from the solution behavior of the thermo-sensitive middle block of the triblock copolymer. When the temperature is raised above the LCST of PNIPAAm segments, they undergo a hydration-to-dehydration transition that reinforces the network structure, as illustrated in Figure 6. For both G' and G'', the thermal transition forms a continuum extending over almost the entire investigated temperature range, with a net increase in moduli approaching one and a half orders of magnitude. Then, decreasing temperature dissolves the PNIPAAm blocks and weakens the network, resulting in the reverse transition with almost no hysteresis (Figure 7).

2.3.2. Oscillatory Frequency Sweep

To get further insight into the thermo-responsive behavior of PS-*b*-PNIPAAm-*b*-PDMAEMA-tpy triblock copolymer gels, frequency sweeps are conducted using a low stress amplitude of 10 Pa, at selected temperatures covering the strengthening transition range (Figure 8). Under ambient conditions, a rubber-like behavior is observed for the triblock copolymer gel, as indicated by the weak frequency-dependent plateaus in both moduli over the entire investigated frequency range. This signature evidences that neither the micellar cores nor the metal-ligand bridges between them relax on the experiment timescale. The absence of core relaxation can be attributed to the glassy nature of the polystyrene association. Indeed, block copolymer micelles having glassy cores are known to be kinetically frozen and are thus expected to have very slow unimer exchange kinetics [65]. As reported by several authors [66,67], the high energy barrier for unimer exchange between micelles having cores made of even short polystyrene segments leads to negligible exchange of chains at ambient temperature. In addition, the large incompatibility between hydrophobic segments and polyelectrolyte corona creates high interfacial tensions that are sufficient to freeze micellar aggregates, even above the glass transition temperature [68]. In parallel, the apparent inertness of coordination bridges is in good agreement with the exchange rates reported for metal–terpyridine bonds [45,46,69]. In this regard, Ni(II) ions are expected to form long-lifetime complexes in combination with the terpyridine ligand. This particular stability has been used in the strategic formation of metallo-supramolecular polymers with high structural integrity [34,63,70,71] or metallo-supramolecular networks with a delayed terminal relaxation [36,40].

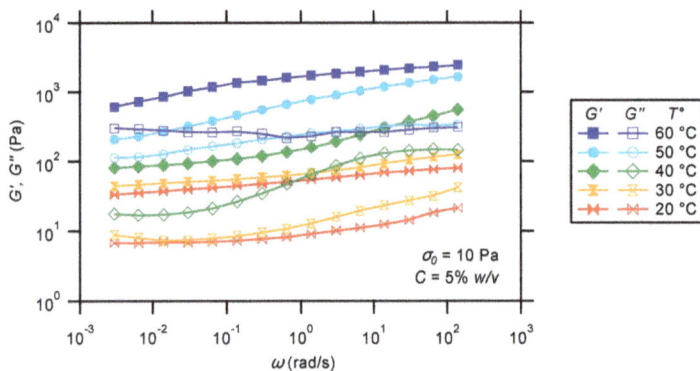

Figure 8. Frequency dependence of dynamic moduli for a hydrogel prepared from the PS-*b*-PNIPAAm-*b*-PDMAEMA-tpy copolymer and Ni(II) ions, at different temperatures.

In agreement with the temperature sweep, the plateaus in moduli further increase by around one and a half orders of magnitude when the temperature is increased to 60 °C. During the transition, the increase in moduli is however more pronounced in the high-frequency regime than in the low-frequency region, especially around 40 °C. Although a maximum in G'' is not observed, the presence of an apparent fast relaxation process is postulated to account for the decrease in G' in this temperature range. Experimentally, the presence of multiple relaxation modes in thermo-rheologically-complex materials indeed leads to frequency dependence that does not always show a maximum in G'', especially when the relaxation processes are not well-defined or resolved. At higher temperatures, both G' and G'' finally tend to show again much weaker frequency dependencies. Although unconventional, attempts are made to construct a master curve using a negative temperature shift factor, but result in a poor superposition of the curves compared to what is expected for thermo-rheologically-simple materials. Intuitively, the construction of a master curve is based on the assumption that the material behavior is thermo-rheologically simple; otherwise, the time-temperature superposition principle is not applicable. Here, this assumption is not verified, since the viscoelastic material functions are dramatically altered upon heating. Indeed, an increasing temperature not only affects the relaxation spectra of the material, but mainly causes the precipitation of the PNIPAAm blocks.

To explain the thermal transition in moduli, it is assumed that the relaxation being observed at intermediate temperatures corresponds to local and diffusional motions of PNIPAAm segments. At a low temperature, relaxation of the chain segments in solution is extremely prompt due to micro-Brownian motions or detachment from the weakly-associated micellar nodes, which do not contribute to the modulus. Being part of the fast relaxation spectrum of the material, these motions are not directly accessible by experiment, since they involve extremely short time scales and low activation energy barriers. Upon heating, the collapse of PNIPAAm blocks onto preformed micellar cores progressively restricts the motions of chain segments in solution. In turn, the decreasing degree of motional freedom and reinforcement of hydrophobic nodes, caused by intra- and inter-chain association of PNIPAAm in the collapsed state, results in a gradual shift of this relaxation mode to lower frequencies. At elevated temperatures, the advanced dehydration of the PNIPAAm block leads to a dramatic increase in the conformational rigidity of the network, therefore disabling the motions of PNIPAAm segments, as schematized in Figure 6.

Of course, the collapse of the PNIPAAm blocks also inevitably results in a non-trivial change in the structure of the material: at low temperature, the hydrogel presents the characteristic structural features of telechelic associating networks, each building elements bearing discrete associating units at each chain end; at an elevated temperature, the structure of the material evolves to the one of micellar gels, consisting of segregated nano-objects that are inter-connected by metal-ligand bridges. This transition can be plausibly accompanied by a modification in the rheology of the material structural elements or a variation in the number of elastically-active chains between them, occurring at a specific or over multiple assembly scales. In this regard, it is known that the rheological behavior of micelles is significantly influenced by their shape [41,72–74], which is in turn dictated by the hydrophobic-hydrophilic balance in the block copolymers and the nature of the solvating media. However, this situation is fundamentally encompassed by considering the overall change in the dynamic and number of network active chains, or network elements, in the multiscale assembled material.

2.3.3. Oscillatory Strain Sweep

To further look into the gel characteristics, amplitude strain sweeps are conducted on the triblock copolymer gel at various temperatures and a given frequency of 1 rad/s. As illustrated in Figure 9, the studied material displays a long linear viscoelastic response over the whole covered temperature range. Under small amplitude oscillatory shear, moduli indeed remain nearly invariant with respect to strain amplitude, indicating no breakup of the network structure. At larger deformations, both moduli drop dramatically with strain on the sample, which is highly characteristic of a non-linear response. Interestingly, the yield of the materials is observed around comparable strain amplitudes in nearly the entire temperature range. In practice, a significant deviation to lower yield strain is only observed at a temperature of 60 °C.

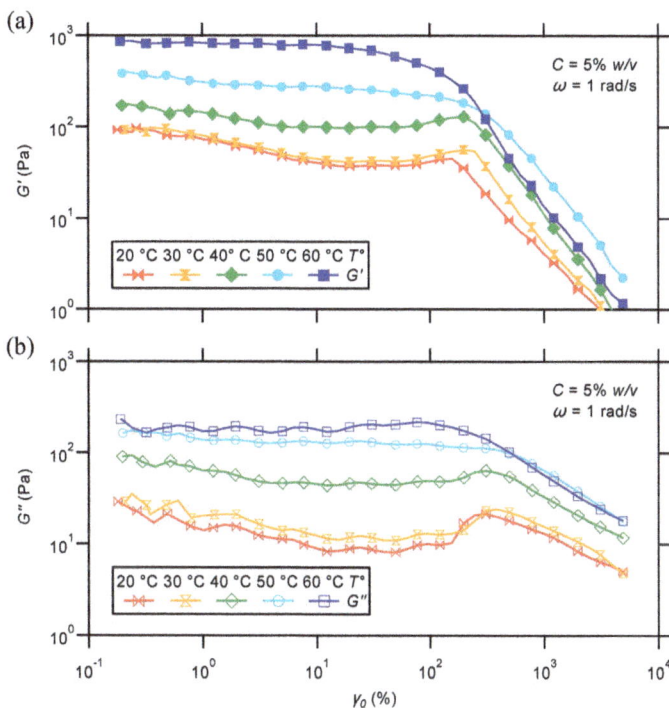

Figure 9. Strain dependence of (**a**) storage and (**b**) loss moduli for a hydrogel prepared from the PS-*b*-PNIPAAm-*b*-PDMAEMA-tpy copolymer and Ni(II) ions, at different temperatures.

The dynamic strain sweep study also reveals significant thermal variations in the behavior of the gels at the limit of the linear range (Figure 9). At low temperatures, a hardening response to strain is observed prior to yielding that is particularly marked in the loss modulus. As suggested by Tam and coworkers [75], such a behavior may result from the incorporation of finite-sized aggregates into the percolated hydrogel, which is possible due to reorganization of the weakly-associated network structure. As the temperature is increased, the collapse of PNIPAAm segments onto micellar nodes is thought to reduce the tolerance and flexibility of the same structure against high shear deformation. As a consequence, strain softening is only observed above 40 °C, as shown in Figure 9.

2.3.4. Steady Shear Viscometry

Last, but not least, steady shear flow experiments are performed on the triblock terpolymer gel, below and above the phase transition temperature measured in oscillatory shear. In practice, the apparent viscosity of the material is monitored as a function of the rate at which it is sheared, thereby providing important information about processing. Indeed, the viscosity is the measure of a fluid's internal flow resistance, which may sensibly vary when the material is sheared. As reported in Figure 10, the steady shear analysis has taken the data between 1 and 1000 s^{-1} shear rates, at 20 and 60 °C. Fundamentally, the results reveal that the supramolecular polymer gel is highly shear thinning, *i.e.*, its viscosity decreases with shear rate, at both temperatures. At a 1-s^{-1} shear rate, the material has an apparent viscosity order of magnitude higher than water. As the shear rate is increased, the apparent viscosity however decreases dramatically, which indicates the excellent processability of the gels. Since the data show linearity in the double-logarithmic plot, the viscosity dependence can be described by a power law: $\eta = K. \dot{\gamma}^{n-1}$, with shear-thinning indexes, n, close to zero.

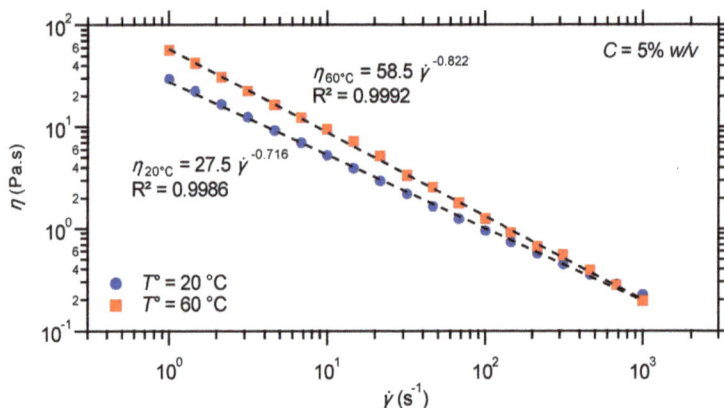

Figure 10. Shear rate dependence of steady shear viscosity for a hydrogel prepared from the PS-*b*-PNIPAAm-*b*-PDMAEMA-tpy copolymer and Ni(II) ions, at different temperatures.

At 60 °C, the material only shows a slight increase of apparent viscosity in the whole investigated shear rate range, which might be in fact due to non-negligible evaporation at this temperature. However, the shear-thinning behavior is more pronounced than under ambient conditions, as indicated by the slope in the viscosity-shear rate graph (Figure 10). Fundamentally, this behavior can be rationalized by the decreasing volume fraction of polymer chains in solution at elevated temperature due to the collapse of the PNIPAAm segments [76]. In addition, this observation strongly supports the formation of well-dispersed spherical core-shell network nodes upon heating, as hypothetically depicted in Figure 6. Indeed, a transition in micelle morphology to, e.g., rods or continuous phases, would have resulted in a large increase in viscosity and a hindered flow due to additional contact between the extended hydrophobic domains [77], which is not observed here.

3. Conclusions

In conclusion, we described here the synthesis and self-assembly of a heterotelechelic double hydrophilic block copolymer, one extremity being ended by a short associating sticker, the other bearing a chelating ligand. Precisely, a terpyridine end-functionalized polystyrene-*block*-poly(*N*-isopropylacrylamide)-*block*-poly(2-(dimethylamino)ethyl methacrylate) triblock terpolymer was synthesized via sequential controlled radical copolymerization, using a modified chain transfer agent. Through the hydrophobic and coordination terminal moieties, the associating copolymer was hierarchically assembled into a supramolecular network with a precisely-controlled architecture. The gelation was induced by the formation of metal-ligand bridges between micellar nodes resulting from the aggregation of polystyrene segments in aqueous media. Due to the tailored network structure, these transient nodes can be shielded by the thermo-induced collapse of the PNIPAAm blocks, which opened a way to finely tune the mechanical properties of the gel.

The shear viscosity, frequency- and strain-dependences of the viscoelastic properties of the gel were investigated as a function of the temperature. Notably, results suggested that the high degree of freedom of the PNIPAAm segment in the hydrated state allowed partial relaxation of mechanical stress. This wasted contribution to the elastic response was however modulated by the solubility of the middle block at higher temperatures. By ultimately suppressing the segmental relaxation, the collapse of the PNIPAAm block also led to stronger hydrophobic associations that were less easily disrupted by mechanical forces. Finally, the response of the gels under large oscillatory shear was sensibly varied across the investigated temperature range. Precisely, an overshoot in both moduli was observed below the collapse transition temperature of the PNIPAAm segment, while the same materials exhibited strain thinning only at higher temperatures.

Combining valuable mechanical properties and the ability to modulate them via temperature changes, the investigated material constitutes a promising candidate for specific applications, such as mechano-sensors or actuators. In this continuity, the influence of several parameters, like, e.g., the length of the different blocks and the pH of the media, will be investigated on the dynamic mechanical properties of this particular class of stimuli-responsive gels. In parallel, future works will focus on the structural characterization of the triblock terpolymer hydrogel. In this regard, the small-angle neutron scattering technique can be identified as a powerful characterization technique for monitoring the change in material structure along the thermal transition.

4. Experimental Section

4.1. Materials

All chemicals were purchased from Acros (Acros Organics Belgium, Geel, Belgium) or Aldrich (Sigma-Aldrich, Belgium, Diegem, Belgium) and were of the highest purity grade. All chemicals were used as received unless otherwise specified. 2,2'-Azobis(isobutyronitrile) was recrystallized from methanol. Styrene and 2-(dimethylamino)ethyl methacrylate were dried and vacuum-distilled over calcium hydride. *N*-isopropylacrylamide was recrystallized from *n*-hexane and dried overnight in a vacuum oven at 35 °C prior to use. Dichloromethane and 1,4-dioxane were distilled over calcium hydride. $NiCl_2$ transition metal salt was dried before use.

4.2. Instrumentation

All ^1H nuclear magnetic resonance spectra were recorded on a Bruker 300 MHz Avance II spectrometer (Bruker Belgium, Brussels, Belgium) in deuterated solvents containing tetramethylsilane as an internal standard. Chemical shifts (δ) were reported in parts per million downfield from the internal standard. Size exclusion chromatography was performed in *N,N*-dimethylformamide containing 2.5 mM NH_4PF_6 to determine molecular weight distributions with respect to polystyrene standards (Polymer Standard Service (PSS), Mainz, Germany). The measurements were carried out on a system composed of two PSS Gram columns (100 and 1000 Å) connected to a Waters 410 differential refractive index detector operating at 0.5-mL/min flow rate and a temperature of 35 °C. Dynamic light scattering experiments were performed on a Malvern CGS-3 apparatus equipped with a He–Ne laser with a wavelength of 632.8 nm and a thermostat. The size distribution histograms of the self-assembled nanostructures were obtained using the constrained regularization method for inverting data (CONTIN), which is based on an inverse-Laplace transformation of the data. Shear rheological experiments were performed on a Kinexus Ultra (Malvern Instruments, Hoeilaart, Belgium) rheometer equipped with a heat exchanger and modified with a solvent trap. Measurements were carried out using a 20-mm plate-plate geometry, in a water-saturated atmosphere, in order to minimize evaporation of the solvent. The gap was adjusted between, so that the geometry was completely filled. Normal forces were checked to be relaxed prior to any measurement.

4.3. Synthesis of PDMAEMA$_{103}$-tpy

CPAD-tpy (50.3 mg, 84 μmol), DMAEMA (4.27 mL, 25.3 mmol) and AIBN (2.8 mg, 17 μmol) were dissolved in 1,4-dioxane (9.1 mL). The solution was degassed three times by freeze-pump-thaw, filled with argon and stirred in a preheated paraffin oil bath at 70 °C. After 6 h, the polymerization was stopped by placing the Schlenk tube into liquid nitrogen. The monomer conversion was evaluated around 40% from ^{1}H-NMR integration. The homopolymer solution was precipitated twice into a 20-times excess of cold *n*-hexane. The precipitate was isolated by centrifugation at 2000 rpm and dried under vacuum at room temperature to afford a red semi-solid.

^{1}H-NMR (300 MHz, CDCl$_3$) δ_H: 8.69 (d, 2H), 8.61 (d, 2H), 8.01 (s, 2H), 7.91 (t, 2H), 7.85 (td, 2H), 7.56 (t, 1H), 7.38 (t, 2H), 7.33 (dd, 2H), 5.70 (br, 1H), 4.24 (t, 2H), 4.07 (br, 206H), 3.31 (q, 2H), 2.54 (t, 2H), 2.53 (br, 206H), 2.52 (m, 2H), 2.26 (br, 618H), 1.88 (q, 2H), 1.88–1.80 (br, 206H), 1.03–0.87 (br, 309H), 1.78 (q, 2H), 1.70 (q, 2H), 0.88 (s, 3H).

M_n (SEC) = 20,000 g·mol^{-1}, M_w (SEC) = 23,950 g·mol^{-1}, M_w/M_n (SEC) = 1.20; M_n (NMR) = 16,750 g·mol^{-1}.

4.4. Synthesis of PNIPAAm$_{73}$-b-PDMAEMA$_{103}$-tpy

PDMAEMA-tpy (750 mg, 45 μmol), NIPAAm (1.81 g, 16 mmol) and AIBN (1.3 mg, 8 μmol) were dissolved in 1,4-dioxane (7.44 mL). The solution was degassed three times by freeze-pump-thaw, filled with argon and stirred in a preheated paraffin oil bath at 70 °C. After 2 h, the polymerization was stopped by placing the Schlenk tube into liquid nitrogen. The monomer conversion was evaluated around 20% from ^{1}H-NMR integration. The copolymer solution was precipitated five times into a 10-times excess of *n*-hexane. The precipitate was isolated by centrifugation at 2000 rpm and dried under vacuum at room temperature to afford a pale red solid.

^{1}H-NMR (300 MHz, CDCl$_3$) δ_H: 8.69 (d, 2H), 8.61 (d, 2H), 8.01 (s, 2H), 7.91 (d, 2H), 7.85 (td, 2H), 7.56 (tt, 1H), 7.38 (t, 2H), 7.33 (dd, 2H), 6.30 (br, 73H), 5.70 (br, 1H), 4.24 (t, 2H), 4.07 (br, 206H), 3.99 (br, 73H), 3.31 (q, 2H), 2.54 (t, 2H), 2.53 (br, 206H), 2.52 (m, 2H), 2.26 (br, 618H), 2.08 (br, 73H), 1.88 (q, 2H), 1.88–1.80 (br, 206H), 1.82–1.62–1.33 (br, 146H), 1.12 (br, 438H), 1.03-0.87 (br, 309H), 1.78 (q, 2H), 1.70 (q, 2H), 0.88 (s, 3H).

M_n (SEC) = 29,650 g·mol^{-1}, M_w (SEC) = 36,450 g·mol^{-1}, M_w/M_n (SEC) = 1.23; M_n (NMR) = 25,000 g·mol^{-1}.

4.5. Synthesis of PS$_{12}$-b-PNIPAAm$_{73}$-b-PDMAEMA$_{103}$-tpy

PNIPAAm-*b*-PDMAEMA-tpy (200.5 mg, 8 μmol), styrene (368 μL, 3.2 mmol) and AIBN (0.175 mg, 1.05 μmol) were dissolved in 1,4-dioxane (1.549 mL). The solution was degassed three times by freeze-pump-thaw, filled with argon and stirred in a preheated paraffin oil bath at 80 °C. After 1 h, the polymerization was stopped by placing the Schlenk tube into liquid nitrogen. The copolymer solution was precipitated three times into a 10-times excess of *n*-hexane. The precipitate was isolated by centrifugation at 2000 rpm and dried under vacuum at room temperature to afford a pale red solid.

^{1}H-NMR (300 MHz, CDCl$_3$) δ_H: 8.69 (d, 2H), 8.61 (d, 2H), 8.01 (s, 2H), 7.91 (d, 2H), 7.85 (td, 2H), 7.56 (tt, 1H), 7.38 (t, 2H), 7.33 (dd, 2H), 7.05 (br, 24H), 6.95 (br, 12H), 6.50 (br, 24H), 6.30 (br, 73H), 5.70 (br, 1H), 4.24 (t, 2H), 4.07 (br, 206H), 3.99 (br, 73H), 3.31 (q, 2H), 2.54 (t, 2H), 2.53 (br, 206H), 2.52 (m, 2H), 2.30–1.70 (br, 12H), 2.26 (br, 618H), 2.08 (br, 73H), 1.88 (q, 2H), 1.88–1.80 (br, 206H), 1.82–1.62–1.33 (br, 146H), 1.12 (br, 438H), 1.03-0.87 (br, 309H), 1.78 (q, 2H), 1.70 (q, 2H), 1.90–1.30 (br, 24H), 0.88 (s, 3H).

M_n (SEC) = 31,300 g·mol^{-1}, M_w (SEC) = 40,400 g·mol^{-1}, M_w/M_n (SEC) = 1.29; M_n (NMR) = 26,250 g·mol^{-1}.

4.6. Sample Preparation

The hydrogels were prepared by dissolving given amounts of block copolymer in ultrapure water. The sealed reaction vessels was placed in a fridge and shaken periodically to form a homogeneous concentrated solution after a few days. The pH of the solution was adjusted via the addition of hydrochloric acid aqueous solution of a given molarity. The gel was then readily obtained by adding the stoichiometric amount of half an equivalent of transition metal ions (with respect to the terpyridine content) dissolved in a defined amount of ultra-pure water to the concentrated solution. Lastly, the reaction vessel was placed again in the fridge over three days to ensure homogenization and stabilization of the material. The final concentration of copolymer in the sample was 5% *w/v*.

4.7. Loading and Testing Protocol

For each test, around 50 μL of the material were loaded onto the stationary bottom plate of the rheometer preheated at 20 °C. By stepwise lowering the gap between the two plates, the samples were compressed and forced to spread over the geometry, so that the gap was completely filled. Equilibration of the samples was followed by monitoring the evolution of normal force, storage and loss moduli with time, under small amplitude oscillatory shear. Rheological tests were started when both moduli reached constant values and the normal force had relaxed to <0.05 N.

Acknowledgments: The authors are grateful to Marie Curie Innovative Training Network (ITN) 2013 "Supolen" 607937 and to the Precision Polymer Materials (P2M) Programme from the European Science Foundation (ESF). Charles-André Fustin is a research associate of the Fonds de la Recherche Scientifique (FRS-FNRS). Jérémy Brassinne thanks Fonds pour la Formation à la Recherche dans l'Industrie et dans l'Agriculture (FRIA) for a PhD thesis grant. Christian Bailly and Evelyne van Ruymbeke are acknowledged for providing access to rheological facilities.

Author Contributions: Jérémy Brassinne performed the copolymer synthesis, the sample preparation, the rheological characterization and wrote the paper. Flanco Zhuge contributed to the manuscript revision through rheological characterization. Jean-François Gohy and Charles-André Fustin supervised the research and contributed to reagents and analytical tools.

Conflicts of Interest: The authors declare no conflict of interest.

References

1. Dai, L. *Intelligent Macromolecules for Smart Devices*; Springer: New York, NY, USA, 2003; p. 496.
2. Ahn, S.K.; Kasi, R.M.; Kim, S.C.; Sharma, N.; Zhou, Y.X. Stimuli-responsive polymer gels. *Soft Matter* **2008**, *4*, 1151–1157. [CrossRef]
3. Pasparakis, G.; Vamvakaki, M. Multiresponsive polymers: Nano-sized assemblies, stimuli-sensitive gels and smart surfaces. *Polym. Chem.* **2011**, *2*, 1234–1248. [CrossRef]
4. Kopeček, J.; Yang, J.Y. Hydrogels as smart biomaterials. *Polym. Int.* **2007**, *56*, 1078–1098. [CrossRef]
5. Osada, Y.; Gong, J. Stimuli-responsive polymer gels and their application to chemomechanical systems. *Prog. Polym. Sci.* **1993**, *18*, 187–226. [CrossRef]
6. Roy, D.; Brooks, W.L.A.; Sumerlin, B.S. New directions in thermoresponsive polymers. *Chem. Soc. Rev.* **2013**, *42*, 7214–7243. [CrossRef] [PubMed]
7. Liu, R.X.; Fraylich, M.; Saunders, B.R. Thermoresponsive copolymers: From fundamental studies to applications. *Colloid Polym. Sci.* **2009**, *287*, 627–643. [CrossRef]

8. Beck, J.B.; Rowan, S.J. Multistimuli, multiresponsive metallo-supramolecular polymers. *J. Am. Chem. Soc.* **2003**, *125*, 13922–13923. [CrossRef] [PubMed]

9. Noro, A.; Matsushita, Y.; Lodge, T.P. Thermoreversible supramacromolecular ion gels via hydrogen bonding. *Macromolecules* **2008**, *41*, 5839–5844. [CrossRef]

10. Jochum, F.D.; Theato, P. Temperature- and light-responsive smart polymer materials. *Chem. Soc. Rev.* **2013**, *42*, 7468–7483. [CrossRef] [PubMed]

11. Burnworth, M.; Tang, L.M.; Kumpfer, J.R.; Duncan, A.J.; Beyer, F.L.; Fiore, G.L.; Rowan, S.J.; Weder, C. Optically healable supramolecular polymers. *Nature* **2011**, *472*, 334–230. [CrossRef] [PubMed]

12. Zhang, Q.; Qu, D.-H.; Wu, J.; Ma, X.; Wang, Q.; Tian, H. A dual-modality photoswitchable supramolecular polymer. *Langmuir* **2013**, *29*, 5345–5350. [CrossRef] [PubMed]

13. Dai, S.; Ravi, P.; Tam, K.C. pH-Responsive polymers: Synthesis, properties and applications. *Soft Matter* **2008**, *4*, 435–449. [CrossRef]

14. Charbonneau, C.L.; Chassenieux, C.; Colombani, O.; Nicolai, T. Controlling the dynamics of self-assembled triblock copolymer networks via the pH. *Macromolecules* **2011**, *44*, 4487–4495. [CrossRef]

15. Yao, X.; Chen, L.; Chen, X.; He, C.; Zhang, J.; Chen, X. Metallo-supramolecular nanogels for intracellular pH-responsive drug release. *Macromol. Rapid. Commun.* **2014**, *35*, 1697–1705. [CrossRef] [PubMed]

16. Ten Brinke, G.; Ruokolainen, J.; Ikkala, O. Supramolecular materials based on hydrogen-bonded polymers. *Adv. Polym. Sci.* **2007**, *207*, 113–177.

17. Cordier, P.; Tournilhac, F.; Soulie-Ziakovic, C.; Leibler, L. Self-healing and thermoreversible rubber from supramolecular assembly. *Nature* **2008**, *451*, 977–980. [CrossRef] [PubMed]

18. Faul, C.F.; Antonietti, M. Ionic self-assembly: Facile synthesis of supramolecular materials. *Adv. Mater.* **2003**, *15*, 673–683. [CrossRef]

19. Faul, C.F.J. Ionic self-assembly for functional hierarchical nanostructured materials. *Acc. Chem. Res.* **2014**, *47*, 3428–3438. [CrossRef] [PubMed]

20. Hoeben, F.J.M.; Jonkheijm, P.; Meijer, E.W.; Schenning, A.P.H.J. About supramolecular assemblies of π-conjugated systems. *Chem. Rev.* **2005**, *105*, 1491–1546. [CrossRef] [PubMed]

21. Brassinne, J.; Fustin, C.-A.; Gohy, J.-F. Polymer gels constructed through metal-ligand coordination. *J. Inorg. Organomet. Polym. Mater.* **2013**, *23*, 24–40. [CrossRef]

22. Goshe, A.J.; Crowley, J.D.; Bosnich, B. Supramolecular recognition: Use of cofacially disposed bis-terpyridyl square-planar complexes in self-assembly and molecular recognition. *Helv. Chim. Acta* **2001**, *84*, 2971–2985. [CrossRef]

23. Goshe, A.J.; Steele, I.M.; Ceccarelli, C.; Rheingold, A.L.; Bosnich, B. Supramolecular recognition: On the kinetic lability of thermodynamically stable host-guest association complexes. *Proc. Natl. Acad. Sci. USA* **2002**, *99*, 4823–4829. [CrossRef] [PubMed]

24. Paulusse, J.M.J.; Huijbers, J.P.J.; Sijbesma, R.P. Quantification of ultrasound-induced chain scission in Pd^{II}–phosphine coordination polymers. *Chem. Eur. J.* **2006**, *12*, 4928–4934. [CrossRef] [PubMed]

25. Heskins, M.; Guillet, J.E. Solution properties of poly(*N*-isopropylacrylamide). *J. Macromol. Sci. Chem.* **1968**, *2*, 1441–1455. [CrossRef]

26. Plamper, F.A.; Ruppel, M.; Schmalz, A.; Borisov, O.; Ballauff, M.; Muller, A.H.E. Tuning the thermoresponsive properties of weak polyelectrolytes: Aqueous solutions of star-shaped and linear poly(*N*,*N*-dimethylaminoethyl methacrylate). *Macromolecules* **2007**, *40*, 8361–8366. [CrossRef]

27. Li, F.-M.; Chen, S.-J.; Du, F.-S.; Wu, Z.-Q.; Li, Z.-C. Stimuli-responsive behavior of *N*,*N*-dimethylaminoethyl methacrylate polymers and their hydrogels. *ACS Symp. Ser.* **1999**, *726*, 266–276.

28. Liu, Q.; Yu, Z.; Ni, P. Micellization and applications of narrow-distribution poly[2-(dimethylamino)ethyl methacrylate]. *Colloid Polym. Sci.* **2004**, *282*, 387–393. [CrossRef]

29. Butun, V.; Armes, S.; Billingham, N. Synthesis and aqueous solution properties of near-monodisperse tertiary amine methacrylate homopolymers and diblock copolymers. *Polymer* **2001**, *42*, 5993–6008. [CrossRef]

30. He, Y.; Lodge, T.P. Thermoreversible ion gels with tunable melting temperatures from triblock and pentablock copolymers. *Macromolecules* **2008**, *41*, 167–174. [CrossRef]

31. Sugihara, S.; Kanaoka, S.; Aoshima, S. Stimuli-responsive ABC triblock copolymers by sequential living cationic copolymerization: Multistage self-assemblies through micellization to open association. *J. Polym. Sci. Part A Polym. Chem.* **2004**, *42*, 2601–2611. [CrossRef]

32. Dyakonova, M.A.; Stavrouli, N.; Popescu, M.T.; Kyriakos, K.; Grillo, I.; Philipp, M.; Jaksch, S.; Tsitsilianis, C.; Papadakis, C.M. Physical hydrogels via charge driven self-organization of a triblock polyampholyte—Rheological and structural investigations. *Macromolecules* **2014**, *47*, 7561–7572. [CrossRef]

33. Piogé, S.; Fustin, C.-A.; Gohy, J.-F. Temperature-responsive aqueous micelles from terpyridine end-capped poly(*N*-isopropylacrylamide)-*block*-polystyrene diblock copolymers. *Macromol. Rapid. Commun.* **2012**, *33*, 534–539. [CrossRef] [PubMed]

34. Brassinne, J.; Poggi, E.; Fustin, C.-A.; Gohy, J.-F. Synthesis and self-assembly of terpyridine end-capped poly(*N*-isopropylacrylamide)-*block*-poly(2-(dimethylamino)ethyl methacrylate) diblock copolymers. *Macromol. Rapid Commun.* **2015**, *36*, 610–615. [CrossRef] [PubMed]

35. Mugemana, C.; Guillet, P.; Fustin, C.-A.; Gohy, J.-F. Metallo-supramolecular block copolymer micelles: Recent achievements. *Soft Matter* **2011**, *7*, 3673–3678. [CrossRef]

36. Brassinne, J.; Stevens, A.M.; van Ruymbeke, E.; Gohy, J.-F.; Fustin, C.-A. Hydrogels with dual relaxation and two-step gel–sol transition from heterotelechelic polymers. *Macromolecules* **2013**, *46*, 9134–9143. [CrossRef]

37. Jochum, F.D.; Brassinne, J.; Fustin, C.-A.; Gohy, J.-F. Metallo-supramolecular hydrogels based on copolymers bearing terpyridine side-chain ligands. *Soft Matter* **2013**, *9*, 2314–2320. [CrossRef]

38. Mugemana, C.; Joset, A.; Guillet, P.; Appavou, M.-S.; de Souza, N.; Fustin, C.-A.; Leyh, B.; Gohy, J.-F. Structure of metallo-supramolecular micellar gels. *Macromol. Chem. Phys.* **2013**, *214*, 1699–1709. [CrossRef]

39. Guillet, P.; Mugemana, C.; Stadler, F.J.; Schubert, U.S.; Fustin, C.-A.; Bailly, C.; Gohy, J.-F. Connecting micelles by metallo-supramolecular interactions: Towards stimuli responsive hierarchical materials. *Soft Matter* **2009**, *5*, 3409–3411. [CrossRef]

40. Brassinne, J.; Gohy, J.-F.; Fustin, C.-A. Controlling the cross-linking density of supramolecular hydrogels formed by heterotelechelic associating copolymers. *Macromolecules* **2014**, *47*, 4514–4524. [CrossRef]

41. Brassinne, J.; Mugemana, C.; Guillet, P.; Bertrand, O.; Auhl, D.; Bailly, C.; Fustin, C.-A.; Gohy, J.-F. Tuning micellar morphology and rheological behaviour of metallo-supramolecular micellar gels. *Soft Matter* **2012**, *8*, 4499–4506. [CrossRef]

42. Brassinne, J.; Fustin, C.-A.; Gohy, J.-F. Thermo-responsive metallo-supramolecular gels based on terpyridine end-functionalized amphiphilic diblock copolymers. *Mater. Res. Soc. Symp. Proc.* **2013**, *1499*. [CrossRef]

43. Brassinne, J.; Bourgeois, J.-P.; Fustin, C.-A.; Gohy, J.-F. Thermo-responsive properties of metallo-supramolecular block copolymer micellar hydrogels. *Soft Matter* **2014**, *10*, 3086–3092. [CrossRef] [PubMed]

44. Yan, X.; Wang, F.; Zheng, B.; Huang, F. Stimuli-responsive supramolecular polymeric materials. *Chem. Soc. Rev.* **2012**, *41*, 6042–6065. [CrossRef] [PubMed]

45. Holyer, R.H.; Hubbard, C.D.; Kettle, S.F.A.; Wilkins, R.G. The kinetics of replacement reactions of complexes of the transition metals with 2,2′,2″-terpyridine. *Inorg. Chem.* **1966**, *5*, 622–625. [CrossRef]

46. Hogg, R.; Wilkins, R. Exchange studies of certain chelate compounds of the transitional metals. Part VIII. 2,2′,2″-terpyridine complexes. *J. Chem. Soc.* **1962**, 341–350. [CrossRef]

47. Schubert, U.; Hofmeier, H.; Newkome, G.R. *Modern Terpyridine Chemistry*; Wiley-VCH: Weinheim, Germany, 2006; p. 229.

48. Schubert, U.; Winter, A.; Newkome, G.R. *Terpyridine-Based Materials: For Catalytic, Optoelectronic and Life Science Applications*; Wiley-VCH: Weinheim, Germany, 2011; p. 522.

49. Aamer, K.A.; Tew, G.N. Supramolecular polymers containing terpyridine-metal complexes in the side chain. *Macromolecules* **2007**, *40*, 2737–2744. [CrossRef]

50. Ott, C.; Ulbricht, C.; Hoogenboom, R.; Schubert, U.S. Metallo-supramolecular materials based on amine-grafting onto polypentafluorostyrene. *Macromol. Rapid. Commun.* **2012**, *33*, 556–561. [CrossRef] [PubMed]

51. Hackelbusch, S.; Rossow, T.; van Assenbergh, P.; Seiffert, S. Chain dynamics in supramolecular polymer networks. *Macromolecules* **2013**, *46*, 6273–6286. [CrossRef]

52. Calzia, K.J.; Tew, G.N. Methacrylate polymers containing metal binding ligands for use in supramolecular materials: Random copolymers containing terpyridines. *Macromolecules* **2002**, *35*, 6090–6093. [CrossRef]

53. El-ghayoury, A.; Hofmeier, H.; de Ruiter, B.; Schubert, U.S. Combining covalent and noncovalent cross-linking: A novel terpolymer for two-step curing applications. *Macromolecules* **2003**, *36*, 3955–3959. [CrossRef]

54. Hofmeier, H.; Schubert, U.S. Supramolecular branching and crosslinking of terpyridine-modified copolymers: Complexation and decomplexation studies in diluted solution. *Macromol. Chem. Phys.* **2003**, *204*, 1391–1397. [CrossRef]

55. Brassinne, J.; Jochum, F.; Fustin, C.-A.; Gohy, J.-F. Revealing the supramolecular nature of side-chain terpyridine-functionalized polymer networks. *Int. J. Mol. Sci.* **2015**, *16*, 990–1007. [CrossRef] [PubMed]

56. Ott, C.; Lohmeijer, B.G.G.; Wouters, D.; Schubert, U.S. Terpyridine-terminated homo and diblock copolymer LEGO units by nitroxide-mediated radical polymerization. *Macromol. Chem. Phys.* **2006**, *207*, 1439–1449. [CrossRef]

57. Lohmeijer, B.G.G.; Schubert, U.S. The LEGO toolbox: Supramolecular building blocks by nitroxide-mediated controlled radical polymerization. *J. Polym. Sci. Part A Polym. Chem.* **2005**, *43*, 6331–6344. [CrossRef]

58. Keddie, D.J. A guide to the synthesis of block copolymers using reversible-addition fragmentation chain transfer (RAFT) polymerization. *Chem. Soc. Rev.* **2014**, *43*, 496–505. [CrossRef] [PubMed]

59. Liu, L.; Wu, C.; Zhang, J.; Zhang, M.; Liu, Y.; Wang, X.; Fu, G. Controlled polymerization of 2-(diethylamino)ethyl methacrylate and its block copolymer with *N*-isopropylacrylamide by RAFT polymerization. *J. Polym. Sci. Part A Polym. Chem.* **2008**, *46*, 3294–3305. [CrossRef]

60. Nuopponen, M.; Ojala, J.; Tenhu, H. Aggregation behaviour of well defined amphiphilic diblock copolymers with poly(*N*-isopropylacrylamide) and hydrophobic blocks. *Polymer* **2004**, *45*, 3643–3650. [CrossRef]

61. Rossow, T.; Habicht, A.; Seiffert, S. Relaxation and dynamics in transient polymer model networks. *Macromolecules* **2014**, *47*, 6473–6482. [CrossRef]

62. Rossow, T.; Seiffert, S. Supramolecular polymer gels with potential model-network structure. *Polym. Chem.* **2014**, *5*, 3018–3029. [CrossRef]

63. Chiper, M.; Meier, M.A.R.; Kranenburg, J.M.; Schubert, U.S. New insights into nickel(II), iron(II), and cobalt(II) bis-complex-based metallo-supramolecular polymers. *Macromol. Chem. Phys.* **2007**, *208*, 679–689. [CrossRef]

64. Vermonden, T.; van Steenbergen, M.J.; Besseling, N.A.M.; Marcelis, A.T.M.; Hennink, W.E.; Sudhölter, E.J.R.; Stuart, M.A.C. Linear rheology of water-soluble reversible neodymium(III) coordination polymers. *J. Am. Chem. Soc.* **2004**, *126*, 15802–15808. [CrossRef] [PubMed]

65. Nicolai, T.; Colombani, O.; Chassenieux, C. Dynamic polymeric micelles *versus* frozen nanoparticles formed by block copolymers. *Soft Matter* **2010**, *6*, 3111–3118. [CrossRef]

66. Van Stam, J.; Creutz, S.; de Schryver, F.C.; Jérôme, R. Tuning of the exchange dynamics of unimers between block copolymer micelles with temperature, cosolvents, and cosurfactants. *Macromolecules* **2000**, *33*, 6388–6395. [CrossRef]

67. Wang, Y.; Balaji, R.; Quirk, R.P.; Mattice, W.L. Detection of the rate of exchange of chains between micelles formed by diblock copolymers in aqueous solution. *Polym. Bull.* **1992**, *28*, 333–338. [CrossRef]

68. Jacquin, M.; Muller, P.; Talingting-Pabalan, R.; Cottet, H.; Berret, J.; Futterer, T.; Theodoly, O. Chemical analysis and aqueous solution properties of charged amphiphilic block copolymers PBA-*b*-PAA synthesized by MADIX®. *J. Colloid Interface Sci.* **2007**, *316*, 897–911. [CrossRef] [PubMed]

69. Farina, R.D.; Hogg, R.; Wilkins, R.G. Rate-pH profile for the dissociation of iron(II)-and cobalt(II)-2,2′,2″-terpyridine complexes. *Inorg. Chem.* **1968**, *7*, 170–172. [CrossRef]

70. Mugemana, C.; Guillet, P.; Hoeppener, S.; Schubert, U.S.; Fustin, C.-A.; Gohy, J.-F. Metallo-supramolecular diblock copolymers based on heteroleptic cobalt(III) and nickel(II) bis-terpyridine complexes. *Chem. Commun.* **2010**, *46*, 1296–1298. [CrossRef] [PubMed]

71. Mugemana, C.; Gohy, J.-F.; Fustin, C.-A. Functionalized nanoporous thin films from metallo-supramolecular diblock copolymers. *Langmuir* **2012**, *28*, 3018–3023. [CrossRef] [PubMed]

72. Fielding, L.A.; Lane, J.A.; Derry, M.J.; Mykhaylyk, O.O.; Armes, S.P. Thermo-responsive diblock copolymer worm gels in non-polar solvents. *J. Am. Chem. Soc.* **2014**, *136*, 5790–5798. [CrossRef] [PubMed]

73. Dreiss, C.A. Wormlike micelles: Where do we stand? Recent developments, linear rheology and scattering techniques. *Soft Matter* **2007**, *3*, 956–970. [CrossRef]

74. Clausen, T.; Vinson, P.; Minter, J.; Davis, H.; Talmon, Y.; Miller, W. Viscoelastic micellar solutions: Microscopy and rheology. *J. Phys. Chem.* **1992**, *96*, 474–484. [CrossRef]

75. Tam, K.; Jenkins, R.; Winnik, M.; Bassett, D. A structural model of hydrophobically modified urethane-ethoxylate (HEUR) associative polymers in shear flows. *Macromolecules* **1998**, *31*, 4149–4159. [CrossRef]

76. Howe, A.J.; Howe, A.M.; Routh, A.F. The viscosity of dilute poly(*N*-isopropylacrylamide) dispersions. *J. Colloid Interface Sci.* **2011**, *357*, 300–307. [CrossRef] [PubMed]

77. Porter, M.R. *Handbook of Surfactants*; Springer: New York, NY, USA, 1991; p. 227.

Article

Peptide Drug Release Behavior from Biodegradable Temperature-Responsive Injectable Hydrogels Exhibiting Irreversible Gelation

Kazuyuki Takata [1], Hiroki Takai [1], Yuta Yoshizaki [2], Takuya Nagata [1], Keisuke Kawahara [1], Yasuyuki Yoshida [1,3], Akinori Kuzuya [1,2] and Yuichi Ohya [1,2,*

[1] Department of Chemistry and Materials Engineering, Faculty of Chemistry, Materials and Bioengineering, 3-3-35 Yamate, Suita, Osaka 564-8680, Japan; kazuyuki.takata@shionogi.co.jp (K.T.); k266872@kansai-u.ac.jp (H.T.); k161109@kansai-u.ac.jp (T.N.); kawaharakeisuke43@gmail.com (K.K.); k831414@gmail.com (Y.Y.); kuzuya@kansai-u.ac.jp (A.K.)

[2] Organization for Research and Development of Innovative Science and Technology (ORDIST), Kansai University, Suita, Osaka 564-8680, Japan; y-yoshi@kansai-u.ac.jp

[3] Research Fellow of Japan Society for the promotion of Science, Kojimachi, Chiyoda-ku, Tokyo 102-0083, Japan

*   Correspondence: yohya@kansai-u.ac.jp; Tel.: +81-6-6368-1121

Received: 4 October 2017; Accepted: 13 October 2017; Published: 15 October 2017

Abstract: We investigated the release behavior of glucagon-like peptide-1 (GLP-1) from a biodegradable injectable polymer (IP) hydrogel. This hydrogel shows temperature-responsive irreversible gelation due to the covalent bond formation through a thiol-ene reaction. In vitro sustained release of GLP-1 from an irreversible IP formulation (**F(P1/D+PA$_{40}$)**) was observed compared with a reversible (physical gelation) IP formulation (**F(P1)**). Moreover, pharmaceutically active levels of GLP-1 were maintained in blood after subcutaneous injection of the irreversible IP formulation into rats. This system should be useful for the minimally invasive sustained drug release of peptide drugs and other water-soluble bioactive reagents.

Keywords: injectable polymers; sustained release; sol-to-gel transition; hydrogel; peptide drug delivery

---

1. Introduction

Various temperature-responsive water-soluble polymers have been investigated for application in drug delivery systems (DDSs) [1–9]. Of these, several polymers in aqueous solution exhibit a temperature-responsive sol-to-gel transition between room temperature (r.t.) and body temperature and can thus be used as injectable polymer (IP) systems. Such polymer solutions can be mixed with water-soluble bioactive reagents such as proteins, peptides, or living cells before injection, and form a hydrogel entrapping these reagents at the injection site in the body. If the polymer can be hydrolyzed to low-molecular-weight compounds that can be metabolized or excreted from the body, such biodegradable IP systems could potentially act as implantable minimally invasive sustained drug release systems [10–12].

There have been many reports on drug release from hydrogels. The release rates of drugs from hydrogels are influenced by several factors, such as the hydrophilicity/hydrophobicity (solubility) and molecular weight of the drug, the degradation rates of the hydrogel, the mesh size of the network, and the diffusion constant of the drug in the hydrogel [13]. In general, temperature-responsive biodegradable IP hydrogels, once formed in the body, are likely to quickly revert to the sol state (typically in less than 24 h) at the injection site, where there is a large amount of body fluid [14]. This is because the gelation of such an IP system is caused by non-covalent (hydrophobic) interactions, and gel formation is an equilibrium process affected by local conditions such as concentration, pH, and temperature. This phenomenon may cause rapid disappearance of the hydrogel and more rapid release of bioactive agents than intended.

To address this problem, we recently reported the generation of biodegradable temperature-triggered covalent gelation systems exhibiting longer and controllable durations of the gel state by using a "mixing strategy" [14–16]. We synthesized a tri-block copolymer of poly(caprolactone-*co*-glycolic acid) (PCGA), poly(ethylene glycol) (PEG), PCGA-*b*-PEG-*b*-PCGA (tri-PCG), and tri-PCG with acryloyl groups attached at both termini (tri-PCG-Acryl) (Figure 1). A mixture of tri-PCG-Acryl micelle solution and tri-PCG micelle solution containing dipentaerythritolhexakis(3-mercaptopropionate) (DPMP) (Figure 1) as a hydrophobic hexa-functional polythiol exhibited an irreversible sol-to-gel transition by covalent cross-linking using a bio-orthogonal Michael-addition-type thiol-ene reaction in response to a temperature increase [15,16]. The mixed micelle solution remained in the sol state just after mixing at r.t., but underwent gelation in response to a temperature increase. Once formed, the hydrogel stayed in the gel state even after cooling. In this system, the acryloyl groups at the copolymer termini and the thiol groups of DPMP existed separately in different micelles just after mixing. Covalent bond formation between the acryloyl and thiol groups occurred only upon sol-to-gel transition induced by a temperature increase, since this temperature rise induced inter-micellar aggregation due to hydrophobic interactions, resulting in a physically cross-linked hydrogel. During the aggregation process, the micelle cores fused and subsequently the thiol groups of DPMP in the tri-PCG micelle core covalently cross-linked with the acryloyl groups via the Michael-addition-type thiol-ene reaction. This system existed for a longer and controllable duration time in the gel state. The duration time of the gel state after subcutaneous injection in vivo could be altered easily from 1 day to more than 60 days simply by changing the mixing ratio of DPMP/tri-PCG and tri-PCG-Acryl [15].

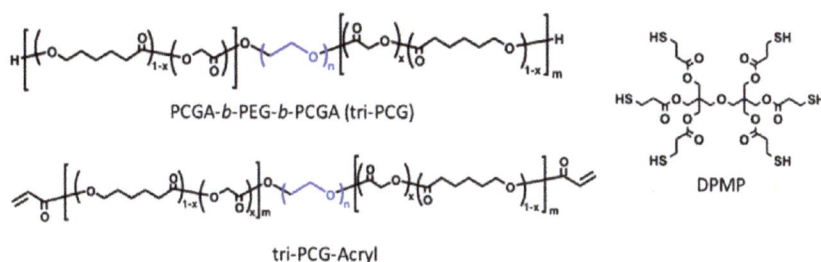

PCGA-*b*-PEG-*b*-PCGA (tri-PCG)

DPMP

tri-PCG-Acryl

Figure 1. Structures of the polymers and polythiol used in this study.

Peptides are becoming increasingly important as drugs due to their activity, target specificity, tolerability, and availability. However, peptides must be administered via the parenteral route because their half-life in the body is extremely short and their oral absorption is poor. Consequently, when continuous exposure to the peptide drug is needed, continuous infusion or multiple injections are required to achieve therapeutic efficacy, which is inconvenient and distressing for patients. Thus, there is need to develop sustained release parenteral formulations, where a single injection allows the drug to be released over a period of weeks, months, or even years [17,18].

As described above, we previously developed a biodegradable temperature-triggered IP system exhibiting covalent gelation using a bio-orthogonal reaction and controllable duration times of the gel state. This system should have potential utility as a drug deposition method allowing the sustained release of peptide drugs. As far as we were aware, there has been no report on the drug release behavior using temperature-triggered covalent gelation system. In this study, we evaluated this IP system as a sustained peptide drug release device by studying the release behavior of a peptide from hydrogels prepared using our IP system. We chose glucagon-like peptide-1 (7-36 amide) (GLP-1) as the peptide model drug. GLP-1 holds promise for the treatment of type 2 diabetes but must be continuously infused or administered by multiple injections because of its extremely short half-life [19–21]. This is the first report on the drug release behavior using biodegradable temperature-responsive covalent gelation system exhibiting long-term release of peptide drugs.

2. Results and Discussion

Tri-PCGs (tri-PCG-1 and tri-PCG-2) and tri-PCG-Acryl were successfully synthesized according to the methods reported previously [15] (Schemes S1 and S2). Several characteristics of the polymers are shown in Supplementary Materials (Tables S1 and S2). The IP formulations were typically prepared by mixing DPMP-loaded tri-PCG micelles in phosphate buffered saline (PBS) (Solution A) and tri-PCG-Acryl micelles in PBS (Solution B) at a mixing ratio A/B = 3/2, with the weight content of tri-PCG-Acryl in the total polymer being 40%. This IP formulation is denoted as F(P1/D+PA$_{40}$), where P1, /D, and +PA$_{40}$ denote the presence of tri-PCG-1 added, the presence of DPMP added, and the amount tri-PCG-Acryl added was 40 wt % in total polymers, respectively. The control formulation, prepared using only tri-PCG-1 or tri-PCG-Acryl, is denoted as F(P1) or F(PA). The phase diagrams for tri-PCG-1 and tri-PCG-Acryl are shown in Supplementary Materials (Figure S1). Both these polymer solutions (25 wt %) and their mixtures adopted a gel state at 37 °C.

The sol-to-gel transition behavior of the F(P1/D+PA$_{40}$) and F(P1) formulations containing GLP-1 are shown in Figure 2. Both formulations showed a sol-to-gel transition in response to a temperature increase from 25 to 37 °C. The presence of GLP-1 had almost no effects on the sol-to-gel transitions of these formulations (Figure S2), and the transition temperatures were between 25 and 37 °C. After cooling to 4 °C, F(P1) containing GLP-1 adopted the sol state, showing that the sol-to-gel transition was reversible. In contrast, F(P1/D+PA$_{40}$) containing GLP-1 remained in the gel state after cooling, showing that the sol-to-gel transition was irreversible. These behaviors are in accordance with our earlier observations in the absence of GLP-1 [15]. Consequently, the presence of GLP-1 had no effect on the phase transition irreversibility of the formulation.

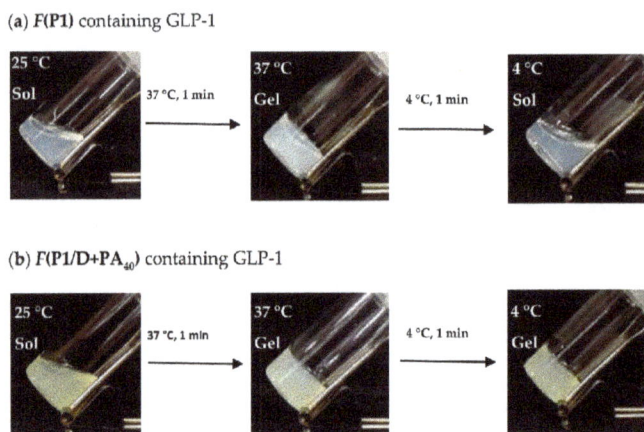

(a) F(P1) containing GLP-1

(b) F(P1/D+PA$_{40}$) containing GLP-1

Figure 2. Photographs of (a) F(P1) hydrogel containing GLP-1 and (b) F(P1/D+PA$_{40}$) hydrogel containing GLP-1 after heating at 37 °C for 1 min and subsequent cooling at 4 °C for 1 min.

We next investigated the in vitro release behavior of GLP-1 from the formulations. Figure 3 shows the cumulative amount of GLP-1 released from F(P1/D+PA$_{40}$) and F(P1) hydrogels at 37 °C in vitro. Both hydrogels showed rapid release of GLP-1 during the first two days, and the release rate from F(P1) was higher than that from F(P1/D+PA$_{40}$). On Day 2, the cumulative amount of GLP-1 released from F(P1) and F(P1/D+PA$_{40}$) was about 59% and 38%, respectively. We previously reported that irreversible F(P1/D+PA$_{x}$) IP hydrogels with higher DPMP and tri-PCG-Acryl content showed a lower swelling ratio compared with hydrogel without cross-linking (**F(P1)**) and compared with hydrogels with lower DPMP and tri-PCG-Acryl content. This lower swelling was due to the higher cross-linking densities of these gels [15]. The mesh size of a hydrogel increases due to swelling and cross-linking density correlates with the swelling ratio [22]. The main reason for the more rapid release of GLP-1 from **F(P1)** compared to **F(P1/D+PA$_{40}$)** hydrogel is likely due to differences in the swelling ratios and diffusion coefficients of the two gels. Figure 4 shows photographs of the hydrogels taken during the release tests. The **F(P1)** hydrogel adopts a highly swollen state within 2 h and the formulation is essentially a sol after Day 1. By Day 2, the release of GLP-1 from **F(P1)** becomes gradual. As shown in Figure 4, although **F(P1)** reverts to the sol state completely by Day 30, 100% cumulative release is not attained. The GLP-1 molecule can adsorb onto the polymer chains by hydrophobic interactions [23], and GLP-1 may be unstable due to spontaneous hydrolysis; consequently, 100% of the GLP-1 molecules could not be detected by reversed-phase high-performance liquid chromatography (RP-HPLC) analysis. On the other hand, the release of GLP-1 from the **F(P1/D+PA$_{40}$)** hydrogel became very slow after Day 2. As shown in Figure 4, **F(P1/D+PA$_{40}$)** remained in the gel state even after 30 days. The storage modulus (G′) (4202 Pa) of **F(P1/D+PA$_{40}$)** was larger than the loss modulus (G″) (326 Pa) and was similar to that of the sample on Day 0 (4696 Pa) (Table 1). As described above for **F(P1)** above, not all the GLP-1 molecules could be detected by RP-HPLC analysis. Regardless, some GLP-1 remained inside the hydrogel and was released very slowly.

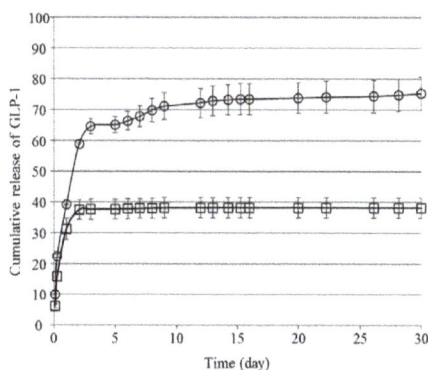

Figure 3. Cumulative release of GLP-1 (%) in vitro from **F(P1)** hydrogel containing GLP-1 (○) and **F(P1/D+PA$_{40}$)** hydrogel containing GLP-1 (□). The data are shown as the mean ± SD (*n* = 3).

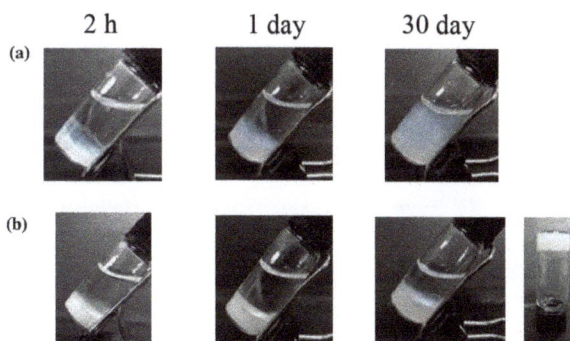

Figure 4. Photographs of (a) **F(P1)** hydrogel containing GLP-1 and (b) **F(P1/D+PA$_{40}$)** hydrogel containing GLP-1 during the in vitro release test.

Table 1. Physical properties of the hydrogels during release tests and in vivo experiments.

Formulation	In Vitro or Vivo	Day	G' (Pa)	G'' (Pa)
F(P1)	in vitro	0	353	99
		30	N.D. [1]	N.D. [1]
	in vivo	25	N.D. [2]	N.D. [2]
F(P1/D+PA$_{40}$)	in vitro	0	4694	569
		30	4202	326
	in vivo	25	1147	364

[1] not detected because of being dissolved, [2] not detected because of disappearance.

To evaluate the cytocompatibilities of the IP formulations, we investigated the cytotoxicity of each formulation and their components towards L929 mouse fibroblast cells. The experiments were under dilute conditions (<1 wt %), below the critical gelation concentration of each polymer. The results are shown in Figure 5. **F(P1)** (tri-PCG solution) exhibited no cytotoxicity over the concentration range tested. In contrast, **F(P1/D)** (tri-PCG micelle entrapping DPMP: solution (A) and **F(PA)** (tri-PCG-Acryl micelle solution: Solution (B) showed weak cytotoxicity at concentrations above 0.01 wt % and 0.1 wt %, respectively. Liu et al. reported that oligo-thiols such as dithiothreitol (DTT) are cytotoxic towards NIH/3T3 fibroblasts and rat bone marrow mesenchymal stem cells (BMSCs) [24]. Klouda et al. reported that polymers with multivalent acryloyl groups showed cytotoxicity [25]. Our results are therefore in agreement with these earlier reports. The thiol groups of DPMP in **F(P1/D)** and the acryloyl groups in **F(PA)** exhibited cytotoxicity, probably because of their interaction with the cell membrane. In contrast, the cytotoxicity of **F(P1/D+PA$_{50}$)** was negligible, similar to that of **F(P1)**. These results suggest that the thiol groups of DPMP and the acryloyl groups of tri-PCG-Acryl reacted with each other, decreasing the number of both functional groups, and that the reaction products exhibited no cytotoxicity. Consequently, **F(P1/D+PA$_{50}$)** showed no significant cytotoxicity.

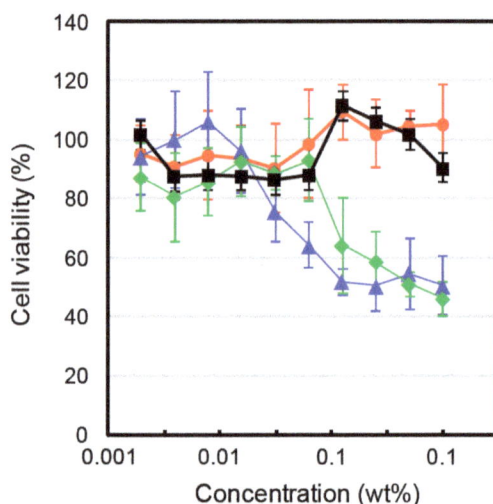

Figure 5. Cell viability of L929 fibroblast cells incubated in the presence of **F(P1)** (●), **F(P1/D)** (◆), **F(PA)** (▲) and **F(P1/D+PA50)** (■) in E-MEM containing 10% Fetal calf serum (FCS) at 37 °C for 21 h.

We investigated the in vivo plasma concentration of GLP-1 after subcutaneous injection of the formulations into rats (Figure 6). We used an ELISA system to detect only active GLP-1; inactivated

GLP-1 is not detectable by this system. GPL-1 disappeared from the plasma very quickly following the injection of the GLP-1 solution: some GLP-1 was detected in the plasma after 2 h, but essentially none was detected after 1 day. This is as expected because the half-life of GLP-1 after subcutaneous injection is very short ($t_{1/2}$ = 9 min) [21]. The injection of **F(P1)** formulations resulted in much higher plasma GLP-1 levels compared to the GLP-1 solution at 2 h, suggesting that the **F(P1)** hydrogel provided a delayed effect compared to the solution, allowing GLP-1 to be absorbed from the injection site and transferred to the blood circulation. The plasma GLP-1 level subsequently decreased rapidly, to 100 ng/L after 19 days, which is slightly higher than the level found in the control rats. The plasma GLP-1 level decreased to the control level 25 days after the injection of **F(P1)** hydrogel. These results are in relatively good agreement with the in vitro release tests (Figure 3): rapid, early release of GLP-1 from **F(P1)** hydrogel, then continuous sustained release from the sol state polymer chains acting as adsorbents in the subcutaneous space. **F(P1/D+PA$_{40}$)** hydrogel also provided a relatively high plasma GLP-1 level 2 h after injection, similar to that of the GLP-1 solution and much lower than that of the **F(P1)** hydrogel. This result suggests that the initial burst release of GLP-1 was suppressed by the lower swelling of the gel resulting from covalent cross-linking, in good agreement with the in vitro results. Interestingly, although **F(P1/D+PA$_{40}$)** hydrogel exhibited no detectable release of GLP-1 after three days in vitro, rats in the **F(P1/D+PA$_{40}$)** group showed higher plasma GLP-1 levels (about 300 ng/L) between Day 7 and Day 25 compared with the **F(P1)** and control rats. These results suggest that the hydrogel network in **F(P1/D+PA$_{40}$)** might be partially hydrolyzed by body fluids or by autocatalytic effects [26], resulting in the slow release of the GLP-1 entrapped in the hydrogel matrix in vivo. This speculation is supported by the results of a physical strength study (Table 1). The G' value of subcutaneously implanted **F(P1/D+PA$_{40}$)** hydrogel was 1147 Pa after 25 days, which is much lower than the G' value obtained in vitro after 30 days' incubation (4202 Pa), suggesting partial degradation. Kim et al. reported that the blood glucose level decreased significantly at a plasma GLP-1 level of around 200 ng/L [27]. Therefore, the plasma GLP-1 level (about 300 ng/L) obtained following the injection of **F(P1/D+PA$_{40}$)** hydrogel containing GLP-1 should be sufficient to be pharmacologically active over a period of 25 days. Figure 7 shows photographs of the sites where the formulations were injected subcutaneously in rats 25 days earlier. All **F(P1)** hydrogel injected disappeared within 25 days, whereas all **F(P1/D+PA$_{40}$)** hydrogel remained at the injection site after 25 days and was in the gel state (G' (1147 Pa) was larger than G'' (364 Pa); (Table 1)). We can, therefore, expect the continuous release of GLP-1 from **F(P1/D+PA$_{40}$)** hydrogel even after 25 days.

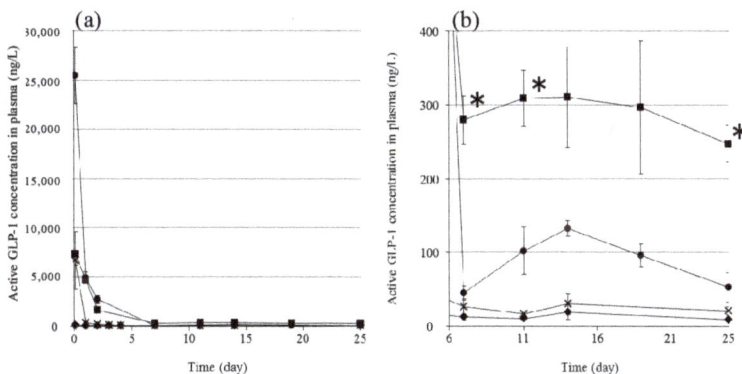

Figure 6. (**a**) Active GLP-1 concentration in plasma after subcutaneous injection for 25 days. (**b**) Magnification of the area between Days 6 and 25. GLP-1 solution (×), **F(P1)** containing GLP-1 (●), **F(P1/D+PA40)** containing GLP-1 (■), and **F(P1/D+PA40)** without GLP-1 (◆). The data are shown as the mean ± SD (*n* = 3–6). * $p < 0.05$ vs. **F(P1)** containing GLP-1.

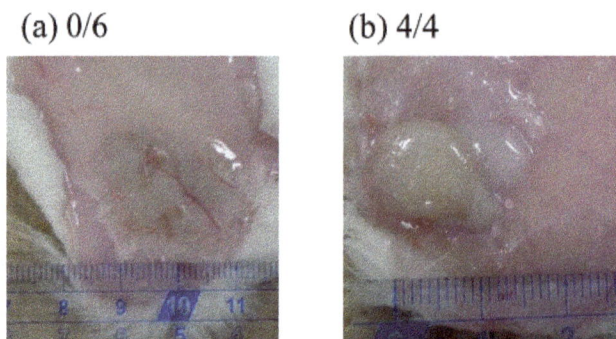

Figure 7. Photographs of hydrogels (**a**) **F(P1)**, (**b**) **F(P1/D+PA$_{40}$)** 25 days after subcutaneous injection in rats. The values indicate the number of rats harboring hydrogel/all rats.

3. Conclusions

We achieved long-term maintenance levels of plasma GLP-1 in vivo by using a biodegradable IP system that shows irreversible gelation due to covalent bond formation. Compared with a conventional physical gelation system, this irreversible hydrogel system exhibits a lower swelling ratio early in the drug release process, a suppressed initial burst, and sustained release of the encapsulated drug. Moreover, the system remained in the gel state for a longer time and gradually degraded after subcutaneous injection, which could help maintain a therapeutic blood drug level for over 1 month. This IP formulation did not exhibit severe cytotoxicity. Therefore, this irreversible IP hydrogel system holds promise for use in a minimally invasive sustained drug release system for hydrophilic compounds such as peptides and proteins. Especially, the IP system with GLP-1 can be a new effective therapeutic system for type 2 diabetes providing good quality of life of patients without frequent injections.

4. Materials and Methods

4.1. Materials

Tri-PCG-1 and tri-PCG-2 were synthesized by the ring-opening polymerization of ε-caprolactone (CL) glycolide (GL) in the presence of PEG (molecular weight (MW) = 1500 Da) according to the methods reported previously (Scheme S1) [15]. The Mn of PCGA segments, the total Mn, and the molar ratio of glycolic acid (GA) units to CL units in the tri-PCG-1 copolymer (CL/GA) were 1950, 5400 Da, and 3.4, respectively, and for tri-PCG-2 the values were 1250, 4000 Da, and 3.9, respectively (Table S1). Tri-PCG-Acryl was synthesized from tri-PCG-2 by the method described previously [15]. The total Mn and the degree of substitution by acryloyl groups were 4200 Da and 91%, respectively (Table S2). GLP-1 (7–36 amide) was purchased from Aviva Systems Biology, Corp. (San Diego, CA, USA). DPMP was a gift from SC Organic Chemical Co., Ltd. (Osaka, Japan). Fetal calf serum (FCS) was obtained from Thermo Fisher Scientific (Waltham, MA, USA). Eagle's minimum essential medium (E-MEM) was purchased from Nissui Pharmaceutical Co. (Tokyo, Japan). Mouse fibroblast NCTC clone 929 (L929) cells were obtained from the Health Science Research Resources Bank (HSRRB, Osaka, Japan). Spague-Dawley (SD) rats (7 weeks old, female, 180 g average body weight) were purchased from Japan SLC, Inc. (Hamamatsu, Japan). Water was purified using a Milli-Q (Merck Millipore, Billerica, MA, USA) system. All other reagents and organic solvents were of commercial grade and were used without further purification.

4.2. Preparation of the IP Formulations

The IP formulations were prepared as reported previously [15]. DPMP-loaded tri-PCG micelle solution (Solution A) was prepared as follows. Tri-PCG and DPMP were placed in a glass vial and dissolved with a small amount of acetone at r.t. The solution was dropped into pure water in a flask stirred at r.t. for 10 min, and then evaporated to remove the acetone. The aqueous solution was lyophilized to obtain powdery DPMP-loaded tri-PCG micelles. The powder was placed in a glass vial and a predetermined amount of PBS was added. After mixing with a vortex mixer for 1 min at r.t., the obtained suspension was heated to 65 °C and kept at 65 °C for 1 min, then mixed using a vortex mixer for 1 min at r.t. The glass vial was immersed in ice-cold water for 2 min and mixed with a vortex mixer for 1 min at r.t. These procedures were repeated until no insoluble particles were observed to give DPMP-loaded tri-PCG micelle solution (Solution A).

Tri-PCG-Acryl micelle solution (Solution B) was prepared as follows. Tri-PCG-Acryl was placed in a glass vial and PBS was added. After mixing with a vortex mixer for 1 min at r.t., the obtained suspension was heated to 65 °C and kept at 65 °C for 10 s, then further mixed using a vortex mixer for 1 min at r.t. The glass vial was immersed in ice-cold water for 2 min and mixed with a vortex mixer for 1 min at r.t. These procedures were repeated until no insoluble particles were observed to give tri-PCG-Acryl micelle solution (Solution B).

Finally, Solution A and Solution B were mixed at desired ratios to give IP formulations. The IP formulations are expressed as $F(P1/D+PA_x)$, where P1, /D, and $+PA_x$ denote the presence of tri-PCG-1 added, the presence of DPMP added, and the amount of tri-PCG-Acryl added was x wt % in total polymers.

4.3. In Vitro Release Test of GLP-1

IP formulations $F(P1/D+PA_{40})$ containing GLP-1 were prepared as follows. A predetermined amount of GLP-1 was dissolved in Solution A (total polymer concentration = 26 wt %) by mixing and sonication, and the solution was then mixed with Solution B (total polymer concentration = 26 wt %). The pH was adjusted to 7.4 with 1N NaOH aqueous solution or HCl aqueous solution, and the total polymer concentration was adjusted to 25 wt % by the addition of PBS. The mixing ratio of solution A/solution B was 3/2, with the content of tri-PCG-Acryl in the total polymer being 40 wt %, $F(P1/D+PA_{40})$. As a control, IP formulation containing only tri-PCG-1 (without DPMP or tri-PCG-Acryl) and GLP-1 $F(P1)$ was prepared by a similar method using only Solution A. The GLP-1 concentration in each IP formulation was 7.5 mg/mL.

Each formulation (200 μL) was placed in a glass vial and incubated at 37 °C for 30 min to obtain a hydrogel, then 1 mL of PBS as a release medium was gently added. At each sampling time, 0.6 mL of supernatant was removed and measured, and 0.6 mL fresh PBS was added to the vial, and the sample was then further incubated at 37 °C. The amount of GLP-1 in the sample solution was determined using a reversed-phase high-performance liquid chromatography (RP-HPLC) system (Waters, Milford, MA, USA) (column: Vydac 218TP54 (4.6 mm × 250 mm), eluent: acetonitrile containing 0.1% trifluoroacetic acid (TFA)/water containing 0.1% TFA, 1/4 to 4/1 linear gradient for 25 min; flow rate: 0.8 mL/min; detector: UV at 210 nm).

4.4. Cytotoxicity

The viability of L929 mouse fibroblast cells after incubation with each sample for 24 h was investigated using a WST-8 assay (Dojindo, Tokyo, Japan). L929 mouse fibroblast cells (100 μL, 2.5×10^3 cells/well) in E-MEM containing 10% fetal bovine serum (FBS) were seeded in a 96-well microplate and cultured in a humidified atmosphere containing 5% CO_2 at 37 °C. After preincubation for 24 h, all the medium supernatant was removed and added to 90 μL of fresh medium. Then, 10 μL of medium containing an IP formulation was added to each well and further incubated for 21 h. The medium supernatant was removed again from the wells, and the cells in the wells were washed with PBS twice. Thereafter, 90 µL of fresh medium and WST-8 reagent (10 µL) were added to the wells, and incubation was continued for a further 3 h. The microplates were read at 450 nm using a microplate reader. The average background absorbance from the control wells was subtracted from the sample data. The values for each sample were in the linear region of the standard curve for the WST-8 assay. Data are expressed as the means and SD ($n = 6$). Cell viability was calculated using the following equation:

$$\text{Cell viability (\%)} = Nt/Nc \times 100$$

where Nt and Nc are the number of cells with or without IP formulation after 21 h of incubation, respectively.

4.5. In Vivo Experiments

The **F(P1/D+PA$_{40}$)** formulation (500 µL) containing GLP-1 was administrated by syringe with a 25 G needle subcutaneously in the back neck of a rat after anesthetizing with isoflurane. **F(P1)** containing GLP-1, **F(P1/D+PA$_{40}$)** without GLP-1, and GLP-1 in PBS (pH 7.4) were used as controls. The volume of all samples was 500 µL, and the concentration of GLP-1 was 7.5 mg/mL.

At each sampling time (2 h–25 days), 250 µL blood samples were obtained from the tail vein using a blood collection tube (BD Microtainer with K2EDTA, Becton, Dickinson and Company, Franklin Lakes, NJ, USA). The blood samples were treated with 5 µL of dipeptidyl peptidase IV inhibitor (Merck Millipore, Billerica, MA, USA) and centrifuged (9100 g, 10 min, 4 °C) to obtain the plasma. The amount of active GLP-1 in the plasma was determined using an ELISA kit (GLP-1 active form assay kit, Immuno-Biological Laboratories Co., Ltd., Shizuoka, Japan). The results were expressed as mean ± SE ($n = 3$–6). Statistical comparisons were made using a Student's *t*-test. A value of $p < 0.05$ was considered significant. Photographs of the rats injected with **F(P1)** without GLP-1 (top) and **F(P1/D+PA 40)** without GLP-1 just after injection, and after 1 day were shown in Figure S3 for references. These experiments followed the guidelines for animal experiments at Kansai University. The experiment was approved by the Ethical Committee for Animal Experiments of Faculty of Chemistry, Materials and Bioengineering, Kansai University (17 April, 2017, Identification number 1709).

4.6. Rheological Measurements

The physical properties of the formulations after soaking in PBS (release test) or after subcutaneous injection into rats were investigated at 37 °C by rheological measurements using a dynamic rheometer (Thermo HAAKE RS600, Thermo Fisher Scientific, Waltham, MA, USA). A solvent trap was used to prevent solvent vaporization. The diameter of the parallel plate was 35 mm, and the gap was 0.2 mm. The controlled stress and frequency were 0.4 Pa and 1.0 rad/s, respectively.

Supplementary Materials: The following are available online at www.mdpi.com/2310-2861/3/4/38/s1. Table S1: Characterization of PCGA-*b*-PEG-*b*-PCGA triblock copolymers (tri-PCGs); Table S2: Characterization of tri-PCG-Acryl; Scheme S1: Synthesis of PCGA-b-PEG-b-PCGA triblock copolymer (tri-PCG); Scheme S2: Synthesis of tri-PCG-Acryl; Figure S1: Phase diagrams of (**a**) tri-PCG and (**b**) tri-PCG-Acryl. ●: sol; ●: gel; ●: sol (syneresis). The gelation temperature (T_{gel}) of each concentration is indicated; Figure S2: Comparison of the gelation temperature in the presence or absence of GLP-1 for (**a**) tri-PCG and (**b**) tri-PCG-Acryl. ●: sol; ●: gel; ●: sol (syneresis). The polymer concentration = 25 wt%. The gelation temperature (T_{gel}) of each sample is indicated; Figure S3: Photographs of the rats injected with **F(P1)** without GLP-1 (top) and **F(P1/D+PA 40)** without GLP-1 just after injection, and after 1 day.

Acknowledgments: This work was financially supported in part by Private University Research Branding Project: Matching Fund Subsidy from Ministry of Education, Culture, Sports, Science and Technology (MEXT) Japan (2016–2020), a Grant-in-Aid for Scientific Research (16H01854) from the Japan Society for the Promotion of Science (JSPS), and by the Kansai University Outlay Support for Establishing Research Centers, 2016. The authors thank SC Organic Chemical Co. Ltd. for providing DPMP.

Author Contributions: Yuichi Ohya conceived and designed the experiments; Kazuyuki Takata, Hiroki Takai, Keisuke Kawahara, Yuta Yoshizaki, Takuya Nagata and Yasuyuki Yoshida performed the experiments; Yuta Yoshizaki and Akinori Kuzuya analyzed the data; Kazuyuki Takata and Yuichi Ohya wrote the paper.

Conflicts of Interest: The authors declare no conflict of interest.

References

1. Zentner, G.M.; Rathi, R.; Shih, C.; McRea, J.C.; Seo, M.; Oh, H.; Rhee, B.G.; Mestecky, J.; Moldoveanu, Z.; Morgan, M.; et al. Biodegradable block copolymers for delivery of proteins and water-insoluble drugs. *J. Control. Release* **2001**, *72*, 203–215. [CrossRef]

2. Ruel-Gariepy, E.; Leroux, J.C. In situ-forming hydrogels-review of temperature-sensitive systems. *Eur. J. Pharm. Sci.* **2004**, *58*, 409–426. [CrossRef] [PubMed]

3. Packhaeuser, C.B.; Schnieders, J.; Oster, C.G.; Kissel, T. In situ forming parenteral drug delivery systems: An overview. *Eur. J. Pharm. Sci.* **2004**, *58*, 445–455. [CrossRef] [PubMed]

4. Van Tomme, S.R.; Storm, G.; Hennink, W.E. In situ gelling hydrogels for pharmaceutical and biomedical applications. *Int. J. Pharm.* **2008**, *355*, 1–18. [CrossRef] [PubMed]

5. Yu, L.; Ding, J. Injectable hydrogels as unique biomedical materials. *Chem. Soc. Rev.* **2008**, *37*, 1473–1481. [CrossRef] [PubMed]

6. Nguyen, M.K.; Lee, D.S. Injectable biodegradable hydrogels. *Macromol. Biosci.* **2010**, *10*, 563–579. [CrossRef] [PubMed]

7. Overstreet, D.J.; Dutta, D.; Stabenfeldt, S.E.; Vernon, B.L. Injectable hydrogels. *J. Polym. Sci. Pol. Phys.* **2012**, *50*, 881–903. [CrossRef]

8. Kempe, S.; Mader, K. In situ forming implants—An attractive formulation principle for parenteral depot formulations. *J. Control. Release* **2012**, *161*, 668–679. [CrossRef] [PubMed]

9. Gong, C.; Qi, T.; Wei, X.; Qu, Y.; Wu, Q.; Luo, F.; Qian, Z. Thermosensitive polymeric hydrogels as drug delivery system. *Curr. Med. Chem.* **2013**, *20*, 79–94. [CrossRef] [PubMed]

10. Choi, S.; Kim, S.W. Controlled release of insulin from injectable biodegradable triblock copolymer depot in ZDF rats. *Pharm. Res.* **2003**, *20*, 2008–2010. [CrossRef] [PubMed]

11. Huynh, D.P.; Im, G.J.; Chae, S.Y.; Lee, K.C.; Lee, D.S. Controlled release of insulin from pH/temperature-sensitive injectable pentablock copolymer hydrogel. *J. Control. Release* **2009**, *137*, 20–24. [CrossRef] [PubMed]

12. Yu, L.; Li, K.; Liu, X.; Chen, C.; Bao, Y.; Ci, T.; Chen, Q.; Ding, J. In vitro and in vivo evaluation of a once-weekly formulation of an antidiabetic peptide drug exenatide in an injectable thermogel. *J. Pharm. Sci.* **2013**, *102*, 4140–4149. [CrossRef] [PubMed]

13. Huynh, D.P.; Shim, W.S.; Kim, J.H.; Lee, D.S. pH/temperature sensitive poly(ethylene glycol)-based biodegradable polyester block copolymer hydrogels. *Polymer* **2006**, *47*, 7918–7926. [CrossRef]

14. Yoshida, Y.; Kawahara, K.; Inamoto, K.; Mitsumune, S.; Ichikawa, S.; Kuzuya, A.; Ohya, Y. Biodegradable injectable polymer systems exhibiting temperature-responsive irreversible sol-to-gel transition by covalent bond formation. *ACS Biomater. Sci. Eng.* **2017**, *3*, 56–67. [CrossRef]

15. Yoshida, Y.; Takai, H.; Kawahara, K.; Mitsumune, S.; Takata, K.; Kuzuya, A.; Ohya, Y. Biodegradable Injectable Polymer Systems Exhibiting Longer and Controllable Duration Time of the Gel State. *Biomat. Sci.* **2017**, *5*, 1304–1314. [CrossRef] [PubMed]

16. Yoshida, Y.; Takata, K.; Takai, H.; Kawahara, K.; Kuzuya, A.; Ohya, Y. Extemporaneously Preparative Biodegradable Injectable Polymer Systems Exhibiting Temperature-Responsive Irreversible Gelation. *J. Biomat. Sci. Polym. Ed.* **2017**, *28*, 1427–1443. [CrossRef] [PubMed]

17. Cleland, J.L.; Daugherty, A.; Mrsny, R. Emerging protein delivery methods. *Curr. Opin. Biotechnol.* **2001**, *12*, 212–219. [CrossRef]

18. Antosova, Z.; Mackova, M.; Kral, V.; Macek, T. Therapeutic application of peptides and proteins: Parenteral forever? *Trends Biotechnol.* **2009**, *27*, 628–635. [CrossRef] [PubMed]

19. Huotari, A.; Xu, W.; Monkare, J.; Kovalainen, M.; Herzig, K.H.; Lehto, V.P.; Jarvinen, K. Effect of surface chemistry of porous silicon microparticles on glucagon-like peptide-1 (GLP-1) loading, release and biological activity. *Int. J. Pharm.* **2013**, *454*, 67–73. [CrossRef] [PubMed]

20. Hanato, J.; Kuriyama, K.; Mizumoto, T.; Debari, K.; Hatanaka, J.; Onoue, S.; Yamada, S. Liposomal formulations of glucagon-like peptide-1: Improved bioavailability and anti-diabetic effect. *Int. J. Pharm.* **2009**, *382*, 111–116. [CrossRef] [PubMed]

21. Lee, S.H.; Lee, S.; Youn, Y.S.; Na, D.H.; Chae, S.Y.; Byun, Y.; Lee, K.C. Synthesis, characterizaiton, and pharmacokinetic studies of PEGylated glucagon-like peptide-1. *Bioconj. Chem.* **2005**, *16*, 377–382. [CrossRef] [PubMed]
22. Canal, T.; Peppas, N.A. Correlation between mesh size and equilibrium degree of swelling of polymeric networks. *J. Biomed. Mater. Res.* **1989**, *23*, 1183–1193. [CrossRef] [PubMed]
23. Castillo, G.M.; Reichstetter, S.; Bolotin, E.M. Extending residence time and stability of peptides by protected graft copolymer (PGC) excipient: GLP-1 example. *Pharm. Res.* **2012**, *29*, 306–318. [CrossRef] [PubMed]
24. Liu, Z.Q.; Wei, Z.; Zhu, X.L.; Huang, G.Y.; Xu, F.; Yang, J.H.; Osada, Y.; Zrinyi, M.; Li, J.H.; Yong, M.C. Dextran-based hydrogel formed by thiol-michael addition reaction for 3D cell encapsulation. *Colloids Surf. B* **2015**, *128*, 140–148. [CrossRef] [PubMed]
25. Klouda, L.; Hacker, M.C.; Kretlow, J.D.; Mikos, A.G. Cytocompatibility evaluation of amphiphilic, thermally responsive and chemically crosslinkable macromers for in situ forming hydrogels. *Biomaterials* **2009**, *30*, 4558–4566. [CrossRef] [PubMed]
26. Lu, L.; Peter, S.J.; Lyman, M.D.; Lai, H.L.; Leite, S.M.; Tamada, J.A.; Uyama, S.; Vacanti, J.P.; Langer, R.; Mikos, A.G. In vitro and in vivo degradation of porous poly(DL-lactic-*co*-glycolic acid) foams. *Biomaterials* **2000**, *21*, 1837–1845. [CrossRef]
27. Choi, S.; Baudys, M.; Kim, S.W. Control of blood glucose by novel GLP-1 delivery using biodegradable triblock copolymer of PLGA-PEG-PLGA in type 2 diabetic rats. *Pharm. Res.* **2004**, *21*, 827–831. [CrossRef] [PubMed]

Review

Functional Stimuli-Responsive Gels: Hydrogels and Microgels

Coro Echeverria [1,*], Susete N. Fernandes [2], Maria H. Godinho [2], João Paulo Borges [2] and Paula I. P. Soares [2,*]

1   Instituto de Ciencia y Tecnología de Polímeros, ICTP-CSIC, Calle Juan de la Cierva 3, Madrid 28006, Spain
2   I3N/CENIMAT, Department of Materials Science, Faculty of Science and Technology, Universidade NOVA de Lisboa, Campus de Caparica, Caparica 2829-516, Portugal; sm.fernandes@fct.unl.pt (S.N.F.); mhg@fct.unl.pt (M.H.G.); jpb@fct.unl.pt (J.P.B.)
*   Correspondence: cecheverria@ictp.csic.es (C.E.); pi.soares@fct.unl.pt (P.I.P.S.); Tel.: +351-212948564 (P.I.P.S.)

Received: 4 May 2018; Accepted: 8 June 2018; Published: 12 June 2018

Abstract: One strategy that has gained much attention in the last decades is the understanding and further mimicking of structures and behaviours found in nature, as inspiration to develop materials with additional functionalities. This review presents recent advances in stimuli-responsive gels with emphasis on functional hydrogels and microgels. The first part of the review highlights the high impact of stimuli-responsive hydrogels in materials science. From macro to micro scale, the review also collects the most recent studies on the preparation of hybrid polymeric microgels composed of a nanoparticle (able to respond to external stimuli), encapsulated or grown into a stimuli-responsive matrix (microgel). This combination gave rise to interesting multi-responsive functional microgels and paved a new path for the preparation of multi-stimuli "smart" systems. Finally, special attention is focused on a new generation of functional stimuli-responsive polymer hydrogels able to self-shape (shape-memory) and/or self-repair. This last functionality could be considered as the closing loop for smart polymeric gels.

Keywords: functional gels; hydrogels; microgels; hybrid microgels; stimuli-responsive; shape memory hydrogels; self-healing gels

---

1. Introduction

Gels have pervaded our everyday life in a variety of forms. The wet soft solids that we encounter in the form of commercial products such as soap, shampoo, toothpaste, hair gel and other cosmetics, as well as contact lenses and gel pens, etc., are all gels derived from polymeric compounds. Polymer gels have been known for centuries and used for application in fields as diverse as food, medicine, materials science, cosmetics, pharmacology, and sanitation, among others. In general, gels are viscoelastic solid-like materials comprised of an elastic cross-linked network and a solvent, which is the major component. The solid-like appearance of a gel is a result of the entrapment and adhesion of the liquid in the large surface area solid three-dimensional (3D) matrix [1].

Pierre-Gilles de Gennes during his Nobel lecture in 1991 recognized the place of gels among the broad category of "soft matter". Focusing on the phenomenological characteristics, polymer gels are 3D networks swollen by a large amount of solvent. Solid-like gels are characterized by: (i) the absence of an equilibrium modulus; (ii) the presence of a storage modulus, $G'(u)$, which exhibits a pronounced plateau extending to times at least of the order of seconds; and (iii) the presence of a loss modulus, $G''(u)$, which is considerably smaller than the storage modulus in the plateau region. The presence of liquid-like behavior on molecular length scales combined with solid-like macroscopic properties makes them very unique systems. Polymer gels are classified as chemical and physical gels, depending on the nature of crosslinks (Figure 1) [2].

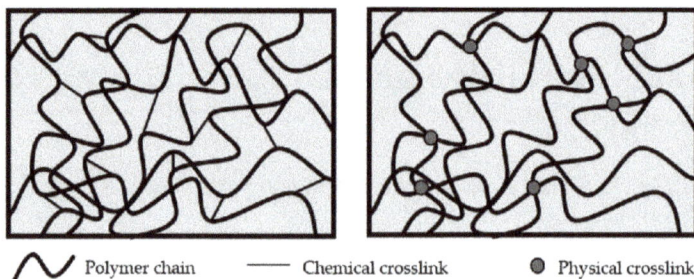

Figure 1. Schematic representation of polymer gels formed by chemical crosslinks (**left** image) or physical crosslinks (**right** image) [2].

Considering the current needs in the field of materials science, the strategy that has gained attention in the last decades is the understanding and further mimicking of structures and behaviours found in nature in order to develop materials with additional functionalities. These materials will have the ability to respond to several stimuli and could be used for biomedical purposes or in everyday life applications. In this context, the so-called "smart" or responsive polymers are becoming increasingly important. This kind of polymer is able to transform external stimuli into shape change or movement by taking advantage of the surrounding environment. The interest in the stimuli-responsive polymers has continued over many decades and it is still a hot topic in materials science. Indeed, researchers' interest in the ability that this kind of gels present when subjected to external stimuli can be seen in the literature since the early 1990's (Figure 2a) and is transversal to several scientific subjects (Figure 2b). In this regard, polymer gels have a special place in the field of stimuli responsive systems, and have attracted much attention due to their composition, versatility, and wide variety of potential applications, in particular in the biomedical field. However, we can also find its use within fields as catalysis, mechanical, chemical or optical sensors, or actuators, among others.

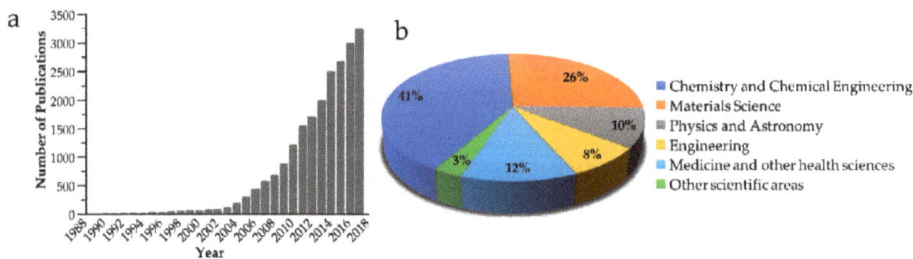

Figure 2. (**a**) Number of publications per year and (**b**) publications by subject featuring stimuli-responsive gel. Databased used for the bibliographic analysis: Scopus® Elsevier, Amsterdam, The Netherlands consulted on the 2nd of May 2018.

The nature of the stimulus can be classified as physical, chemical or biological [3]. Physical stimuli influence the systems molecular interactions, while the last two types of stimuli directly affect the interactions between the polymer chains (intramolecular) or with other constituents of the system. However, in both cases, a phase transition is observed. Gels can be divided into macrogels, microgels and nanogels and can be assembled to give rise to thin-films, brushes, membranes, colloidal particles, foams, among others, as is well explained by Stuart et al. [4].

In this review, we focus attention on recent advances found within the field of stimuli responsive functional polymeric gels. In the first section of the review, we will devote our attention to earlier (last year) developments on stimuli-responsive polymer hydrogels. Some examples are presented where stimuli as temperature, pH, light and redox, or analyte-responsive hydrogels and 3D printing applications are highlighted. The second section of the review will focus on the most recent studies regarding multifunctional stimuli-responsive microgels, in particular hybrid microgels systems. Special attention will be paid to (multi)responsiveness induced by inorganic nanoparticles incorporated within the microgel matrix, such as, metallic, magnetic, or carbonaceus particles. Finally, the last part of this review aims to highlight the new generation of functional stimuli-responsive polymer hydrogels—self-shape (shape memory) or self-healing hydrogels, as a step-forward in the field of "smart" materials.

2. Stimuli-Responsive Functional Hydrogels

Hydrogel (water containing gels) can be defined as a polymeric 3D network insoluble in water. This polymeric network is composed of a polar and hydrophilic natural and/or synthetic polymer that is physically or chemically crosslinked and can uptake a large amount of water. If within the polymer hydrogel network, one element can act as a receptor unit or is spectroscopically active, this gel is called a stimuli-responsive or smart material [1,2]. The effect that an external stimulus induces in the hydrogel network can be observed/measured often on a macroscopic scale, in different forms as change in shape and size, changes in its optical, wettability, electric and mechanical properties [4].

Several reviews regarding the production and application of hydrogels can be found in literature in recent years. These documents can be generic [5–7] or divided by the type of polymer (as for instance natural polymers [8], cellulose nanocrystals [9], DNA [10]), mechanical properties [11], applications (drug delivery [12,13], drug delivery and tissue engineering [14]), fabrication process and for instance a combination of type of polymer and application ((poly(N-isopropylacrylamide) (PNIPAAm) and biomedical applications [15], biomolecules in medicine [16], sensing devices and drug delivery [17,18], sensing molecular targets [19], PNIPAAm as photonic sensor [20]). It is important to note that due to the high number of publications on a specific sub-topic (for instance the use of PNIPAAm within the hydrogels), only a small selection of the most recent articles will be presented. With this selection, the authors intend to show the reader a broader view of this type of systems, its properties, and capabilities.

2.1. Thermoresponsive Hydrogels

Temperature is one of the most employed stimuli within these 3D systems since it is easier to regulate and can be used for in vitro and in vivo testing. The temperature point at which a response is observed is called the critical temperature (Tcr) or volume phase transition temperature (TvPT). At this point, a phase change occurs between the polymer and solvent within the system (depending on the system composition) and the phase is enhanced in one of the constituents. If this phase separation occurs above the critical temperature, the polymer exhibits a lower critical solution temperature (LCST) behavior; if this transition is observed below Tcr, an upper critical solution temperature (UCST) is shown by the polymer. Systems with LCST behavior are the most studied.

Polymer gels containing PNIPAAm are amongst the most investigated since its lower LCST in water (~32 °C) favors its potential use in such an important area as biomedicine. Micro/nanogels and hydrogels of PNIPAAm contract upon heating and swell with water upon cooling, and this transition is fully reversible [21,22]. The study of this polymer is so important that a constant development in the experimental reaction conditions can be found in the literature. Keeping this in mind, Tang and co-workers [23] recently synthesized hydrophobic crosslink, 4,4'-dihydroxybiphenyl diacrylate that enables the production of PNIPAAm hydrogels through free radical polymerization. With this system, the authors were able to decrease hydrogels' LCST (up to 4.3 °C when compared with the poly(ethylene glycol) (PEG) 400 dimethacrylate-PNIPAAm hydrogel control), while maintaining a reversible thermal-responsive.

Nevertheless, the interest in PNIPAAm-based hydrogel has attracted scientists in other areas as shown recently by Kim et al. [24]. The authors produced a nature-inspired multi-functional hydrogel based in PNIPAAm and poly(pyrrole) (PPY), which combines the thermoresponsive nature of the first polymer with the conductive nature of the second one. To make this system, the authors prepare a photo-crosslinked thermoresponsive PNIPAAm membrane with specific patterns, in accordance with its envisioned use. This membrane was then deswelled and subsequently swelled with PPY monomer solution. The conductive layer was finally formed through in situ polymerization by addition of ammonium persulfate and phytic acid solution. This hydrogel system presents a controllable thermoresponsive deformability and electric conductivity. Moreover, upon heating, the hydrogel acts as a filtration membrane for the separation of proteins from water, mimicking the properties observed on functional features of plants such as bending of the pulvinus in *M. pudica* (genus Mimosa), the gating of stomata in leaf, and the selective filtering of the cellular membrane.

More recently, D'Eramo et al. [25] using a photo-patterning technique simultaneously grafted, patterned and crosslinked PNIPAAm onto plane substrate. With this biotechnologic approach that consists mainly of three steps (photo-patterning of hydrogel films, thiol-modification of substrates and closure of the device with an optical adhesive or silicone polymer), the authors were able to produce a microfluidic valve. This microvalve is controlled with temperature, i.e., above the LCST the valve is open. The authors were able to generate a platform of 7800 cages that were efficient in capturing and releasing solutes on-demand with 100-millisecond-scale response times. With these systems, the authors also showed that this hydrogel technology could be used in single-cell trapping/release and DNA amplification. For example, it was shown as an effective amplification of the human gene synaptojanin-1, linked with Parkinson disease.

Exploration of new methods to produce PNIPAAm-based hydrogels is constant as shown by the recent work of Han et al. [26]. In this essay, the authors were able to obtain 3D PNIPAAm micro-structures, using a high-resolution projection micro-stereolithography (PμSL), as schematic presented in Figure 3a. Materials with different shapes (Figure 3b,c) were prepared and could present deformation upon pre-programmed thermal-stimulus. With temperature, hydrogel's swelling was tunable by changes on the crosslinker to monomer ratio and shrinking controlled by monomer concentration. The authors, by controlling the printing parameter, were also able to produce materials with an anisotropic swelling behavior. Using these features, the authors created a structure with layers that present different swelling ratios above the hydrogel's temperature transition (Figure 3c). The different swelling ratios lead to a strain in the structure and a bending deformation can be observed as the temperature is cycled.

Grafting of PNIPAAm in a natural polymer backbone, as in the chitosan (CS), has also attracted attention from the scientific community due to their biocompatibility, low toxicity and their high content of functional groups. Luckanagul et al. [27] were able to prepare nanogels with control efficiency drug delivery. It was also possible to control the drug loading capacity of the gels by changing the degree of substitution of PNIPAAm in CS.

Another interesting example of thermo-responsive hydrogel is the approach proposed by Fang and co-workers [28]. In this essay, PNIPAAm hydrogel was coated with a layer of cell-adherent arginine-glycine-aspartate (RGD) peptides that successfully rupture cancer cells attached to the hydrogel surface. The authors explored the physical force of the expanding stimuli-responsive hydrogel to obtain such an interesting result with this simple approach.

One can also find another type of thermoresponsive hydrogel polymers as the biocompatible, fully synthetic polymer polyisocyanopeptide [29], which exhibits a LCST behavior. Zimoch and co-workers [30] were able to show that the synthetic GLY-ARG-GLY-ASP-SER peptide conjugated to the polymer backbone as a hydrogel was able to support complex cell culture, such as adipogenic differentiation and formation of blood capillaries. The thermoresponsive behavior of this system was able to induce a quick and easy retraction of viable cells from the hydrogel structure, eliminating the need for enzymes or other time-consuming procedures.

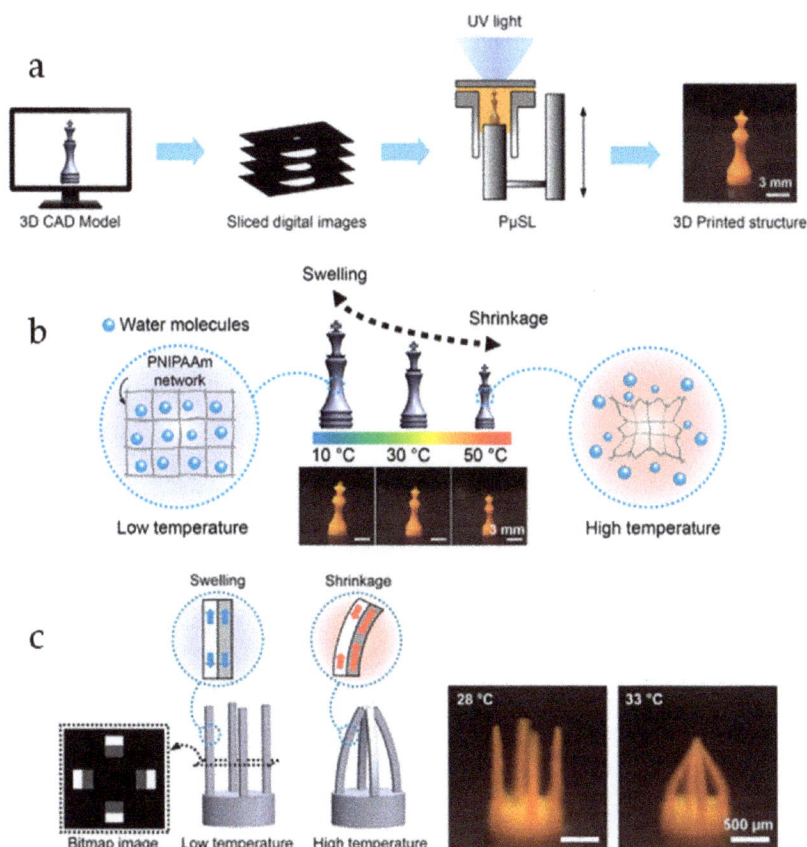

Figure 3. 3D printing of PNIPAAm thermo-responsive hydrogel: (**a**) schematic illustrating of the printing process that goes from the drawing of the 3D model with computer-aided-design (CAD) software, that is then digitally sliced into a series of cross-sectional images. The 3D structure is obtained with PμSL by curing the liquid resin with a series of masks (derived from each slice) optically patterned with ultra-violet light; (**b**) schematic drawing of the temperature response of the 3D printed hydrogel with a phase change when the temperature decreases or increases, observed by swelling and shrinkage, respectively. In the middle (photographs), one can see this effect on the printed 3D structure (at T of 10, 20 and 50 °C); (**c**) in this sequence of sketches and pictures one can see, from left to right, a slice of the gripper with two arms and two different levels of gray to be able to produce a 4 arm 3D structure. The arms of this hydrogel bend when the temperature is higher than the temperature transition, depicted in the central region, and photos of the bending responsive observed when the 3D gel is placed at 33 °C (Figure adapted under the terms of the CC BY 4.0 License) [26].

2.2. pH-Responsive Hydrogels

If a in a polymer backbone ionizable side chains (groups that can donate and accept ions) are grafted, a system that responds to external pH changes is developed (normally observed when pH is raised above the pK_a of the gel). These systems can form polyelectrolytes with water and are classified as anionic, cationic or amphiphilic. For instance, poly(acrylic acid) (PAAc) and poly(4-vinylpyridine) are a weak polyacid and a weak polybase, respectively. The first polymer accepts protons at low pH and releases protons at neutral and high pH. However, in the case of the weak polybase, the polymer will be protonated at high pH or positively ionized at low pH values. Controlling the pH around the hydrogel, and considering the hydrodynamic changes that occur in the polymer chain due to this pH variation, a change on the state of the material from a collapsed to an expanded state can be obtained [31]. The design of the polymer's pendant groups is mostly related to the desirable application.

Development of hydrogels at the nanoscale has driven the group of N. Peppas [32] to recently foster systems with pH-responsive moieties for oral delivery of hydrophobic therapeutics. In this work, a set of nanoscale hydrogels, obtained from the photo-emulsion co-polymerization (ultraviolet (UV) radiation) of the methacrylic acid (MAAc) and poly(ethylene glycol monoethyl ether methacrylate) (PEGMMAA) (1:1 ratio) with the hydrophobic groups *tert*-butyl methacrylate, *n*-butyl methacrylate, *n*-butyl acrylate, and *n*-methyl methacrylate were prepared. All the hydrogels could present a transition in size induced by pH variations (4 to 7) and good physicochemical properties assessed by in vitro testing, where no considerable cytotoxicity was observed. Moreover, the same group was able to show that these nanoscale hydrogels improve the solubility and permeability of hydrophobic therapeutic doxorubicin by in vitro testing [33].

If one combines the thermal responsiveness of PNIPAAm and a pH-sensitive component as the acrylates, a pH- and thermal-responsive hydrogel is obtained. This is the basic idea behind the work of Jin et al. [34], where the authors produced smart hydrogels, based on the co-polymerization of NIPAAm, itaconic acid (IA) and methacrylate lignosulfonate. The hydrogels showed temperature-sensitive behavior, with the TvPT around the body temperature, and pH sensitivity in the range of 3.0 to 9.1. Since a wide sensitive-pH range was achieved, the authors suggest a multitude of possible applications ranging from agriculture to medicine.

2.3. Light-Responsive Hydrogels

The development of light-responsive materials has many advantages. For starters, light is non-invasive, and it does not require contact with the material. Moreover, light is highly accurate and has a low thermal effect [35]. Light-responsive hydrogels can be produced using different strategies, usually composed of a polymeric network and a photoreactive moiety. The photochromic molecule captures the optical signal and converts it to a chemical signal through a photoreaction (isomerization, cleavage, or dimerization) [36,37].

Azobenzenes are amongst the mostly used photo-switchable molecules. These molecules undergo *trans*-to-*cis* or *cis*-to-*trans* isomerization when irradiated with UV- and blue-light, respectively. Heat can also induce isomerization in azobenzenes due to the thermodynamic stability of the *trans* isomer [38]. Several authors have used these photoswitchable molecules to develop light-responsive hydrogels. A recent review highlights different types of light switchable azobenzene containing-macromolecules and their potential applications [39], where some examples of hydrogels were underlined.

Recently several interesting examples can be found in the literature where the authors prepared hydrogels containing azobenzene moieties. Su et al. [40] prepared starch-based hydrogel with an azo group incorporated. This hydrogel has dual-stimuli-response: photo-response due to the presence of azo groups; and pH-response due to acrylic acid and poly(vinyl alcohol) (PVA)-based macromonomers. Rastogi et al. [41] functionalized four-arm, amine-terminated PEG with azobenzoic acid. The authors used this light-responsive hydrogel to enhance the release of Alexa Fluor® 750, a near infrared fluorescent dye, triggered by UV light irradiation. In both cases, the UV light irradiation induces photoisomerization of the crosslinkers from *trans*-to-*cis* isomer, which is reversible upon irradiation with visible light or dark conditions. Another research group produced a light-responsive hydrogel based on crosslinked polymers of 2-hydroxyethyl methacrylate functionalized with azobenzene groups. Upon light irradiation, the photoisomerization of azobenzene induces changes in the hydrogel polarity. In this manner, the authors were able to control the degree and rate of swelling by light irradiation [35].

Zhao et al. [42] functionalized a PEG hydrogel with *ortho*-fluoroazobenzenes as crosslinkers. Upon irradiation with blue and green light, this hydrogel exhibits reversible photo-modulation of its elasticity. This system avoids the use of UV light irradiation, which not only affects the photostability of the materials, but can also be a problem for its use in biological applications.

More recently, a new class of hydrogels have emerged. In these systems, supramolecular hydrogels are formed by self-aggregation of low molecular mass compounds to form entangled self-assembled fibrillary networks through a combination of non-covalent interactions. Since these networks are formed by weak interactions, they can be readily transformed in a fluid (sol) upon exposure to heat [1]. Mandl et al. [38] produced a supramolecular hydrogel based on PAAc-azobenzene copolymer using deoxycholate-β-cyclodextrin as a crosslinker. The authors incorporated $LiYF_4:Tm^{3+}/Yb^{3+}$ upconverting nanoparticles in the supramolecular hydrogel, which can emit UV light upon NIR irradiation. Among the possible non-covalent interactions to form supramolecular hydrogels, the host-guest interaction is an important one. The most commonly used host molecules are cyclodextrins due to their low toxicity and better-inducing abilities [43,44]. The incorporation of an azobenzene group in its *trans* isomer into a cyclodextrin is stable in water. Upon UV light irradiation, the azobenzene group converts to *cis* isomer and repels cyclodextrin. Wang et al. [45] incorporated azobenzene into an anionic surfactant producing 6-(4-dimethylaminoazobenzene-4'-oxy) hexanoate sodium (DAH). The authors included DAH in its *trans* form in α-cyclodextrin, producing a supramolecular hydrogel by promoting hydrogen bonds between α-cyclodextrin and carboxylate in DAH and water. The presence of the dimethylamino group shifts the responsive wavelength to the region of the visible light (red), thus avoiding the use of UV light. Upon irradiation with visible light, this hydrogel showed gel-sol reversible phase transition.

Light-responsive hydrogels can also be produced through the incorporation of chlorophyllin as chromophore unit in the polymeric matrix. The incorporation of this molecule in PNIPAAm was first reported in 1990 [46]. A few years later, the same authors produced a covalently crosslinked network of PNIPAAm, sodium acrylate and chlorophyllin that can be activated using visible light, and deactivated using pH, temperature and light [47]. More recently, Xu et al. [48] produced an interpenetrated network using PAAm and PAAc with chlorophyllin incorporated. These hydrogels are not only able to respond to visible light, but also have enhanced mechanical properties caused by the incorporation of the chromophore unit.

Light-responsive hydrogels also find application in microfluidic devices. Particularly, PNIPAAm hydrogel is widely used to produce microvalves and micromixers. Using this type of material not only reduces the production cost of such devices, but also decreases the energy consumption of the device. For example, spirobenzopyran is often incorporated in PNIPAAm hydrogel to act as a molecular photo-switch triggered by light. The reader can find more useful information on this topic in a recent review [49].

2.4. Redox-Responsive Hydrogels

Redox-responsive hydrogels are materials able to respond to reduction and oxidation of their constituent molecular components. The activation of such systems can be chemical or electrochemical. The electrochemical activation starts ion mobility between two electrodes, thus producing a signal. More commonly, conducting polymers such as polyaniline and PPY are used to produce redox-responsive hydrogels. In these polymers, oxidation of a portion of their subunits within the polymer backbone occurs. This leads to an influx of counterions to balance the newly formed charges and a swelling response of the material [50,51].

Other polymers such as chitosan are also used to produce redox-responsive systems. Liu et al. [52] produced a synbio system able to generate two intermediates: the first redox-active chemical intermediate is an output; and a second one, a redox-capacitor that transduces the biological output response into an amplified electrical signal. The first intermediate is *p*-aminophenol and the second one is based on cathecol and chitosan. To produce the redox-capacitor, two steps are necessary: first a cathodic electrodeposition of the pH-responsive self-assembly chitosan; then an anodic electrochemical grafting of cathecol onto the deposited chitosan. The resulting system is non-conductive, but redox-active. Fu et al. [53] produced a supramolecular hydrogel composed of chitosan and zinc ions. They reported for the first time the use of this system as a disposable electrochemical sensing platform to determine hydroxyl radicals and hydrogen peroxide. In the presence of these molecules, the chitosan hydrogel disassembles, thus releasing the zinc ions. This release causes changes in the current, which is used as a signal to determine the analyte concentration.

Wojciechowski et al. [54] also produced a supramolecular redox-responsive hydrogel but with a different composition. The authors used N,N'-dibenzoyl-L-cystine as a redox-active supramolecular generator. This molecule self-assembles when triggered by pH in the presence of a reducing agent, which later enables the system disassembly. Consequently, a transient hydrogel is produced that has autonomous sol-gel-sol transitions based on the kinetic control of competing chemical reactions. The lifetimes of the transient hydrogen can be tuned either by pH or by the concentration of the reducing agent.

PNIPAAm also finds application in this type of stimuli-responsive systems. A PNIPAAm matrix incorporating covalently-bound phenylboronic acids was produced to be used as a saccharide sensing unit and a redox-active $[Ru(bpy)_3]^{2+}$ luminophore. The redox activity of this system can be reversed by changing the sequential stimuli (fructose and temperature), thus controlling the swelling and deswelling of the hydrogel [55].

In a different study, Zhang and coworkers [56] produced a mechanically redox-tunable hydrogel that acts by the valent transformation between ferric and ferrous ions. The hydrogel is composed of PAAc and hexadecyl methacrylate, and is reinforced by the hydrophobic association of these two molecules and by the metal ion coordination. The latter occurs between the carboxylate groups of PAAc and iron (III). The mechanical strength of the hydrogel is tuned by changing the complexation degree between PAAc and iron (III). When exposed to UV light, iron (III) is reduced to iron (II) in the presence of UV-sensitive citric acid molecules. This reaction can be reverted in the presence of oxygen, thus modulating the mechanical strength of the produced hydrogel.

2.5. Analyte-Responsive Hydrogels

Analyte-responsive systems are a scientific research interest since they mimic nature's ability of molecular recognition. To design these 3D polymer networks, biomolecules capable of recognizing target molecules are incorporated in the hydrogel. These molecules can be enzymes, peptides and nuclei acids. Target analyte can be used to induce hydrogel's assembly, disassembly, syneresis, swell and also to prompt molecular reorientation and displacement [18]. With the vast knowledge in the hydrogel field, a wide range of sensing applications have been suggested—immobilizing scaffolds for biomolecules, optical or photonic sensors, electrical or magnetic transducers or pressure-responsive, just to state a few [17,18,57].

Glucose-responsive hydrogels have been widely studied for its use in diabetes' treatment. Moreover, the design of enzyme-responsive hydrogels is very attractive since this can act directly upon the target molecule. More recently, the production of hydrogels using a template molecule created molecularly imprinted polymers. The use of the templated molecule produces binding sites in the polymer, leading to its recognition by the desired target. Peppas' group has recently published a review about these three classes of analyte-responsive hydrogels [7]. The reader can also find useful information about the design and production of analyte-responsive hydrogels in another recent review from the same group [18].

Pujol-Vila et al. [58] developed a very interesting biomaterial for fast antibiotic-susceptibility evaluation. This biomaterial is based on alginate complexed with iron (III) ions. Basically, the bacteria under evaluation are entrapped and pre-concentrated in the hydrogel matrix. This occurs through oxidation of iron (II) to iron (III), leading to in situ formation of the alginate hydrogel in less than two minutes in soft experimental conditions. Afterwards, the hydrogel is incubated with the antibiotic and ferricyanide is added to the biomaterial. If the bacteria are resistant to the antibiotic, they will remain alive. In this case, bacteria will reduce ferricyanide to ferrocyanide, which reacts with iron (III) ions present in the hydrogel. This reaction originates Prussian blue molecules, leading to a blue color development that is visible to the naked eye (Figure 4).

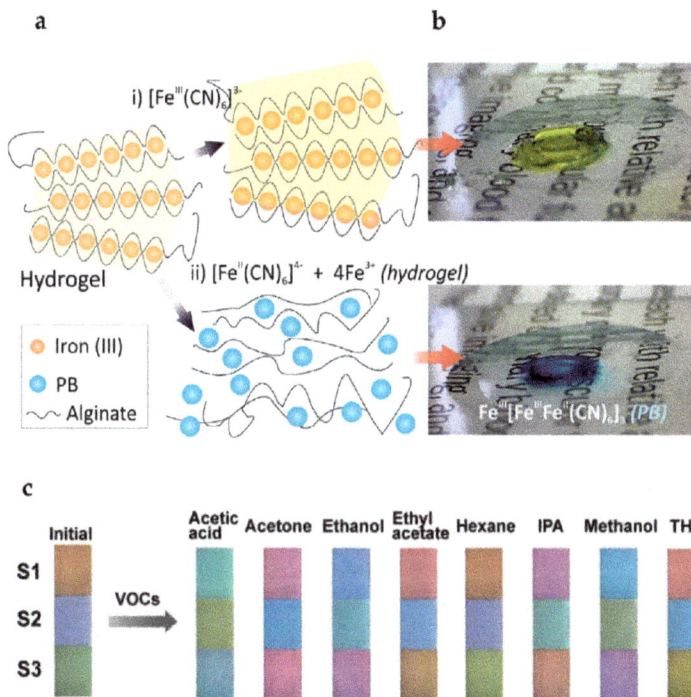

Figure 4. Analyte-responsive hydrogels: (**a**) schematic representation and (**b**) photographs of a bacterial metabolic activity-response hydrogel, based on an electrochromic iron(III)-complexed alginate system with antibiotic-susceptibility. In this biomaterial, the reduction of ferricyanide to ferrocyanide by the bacteria will lead to the formation of Prussian blue molecules, macroscopically observed by the strong chromatic change of the hydrogel from yellow to blue [58] (Reprinted from ref. [58]. Copyright 2018 Elsevier); (**c**) sensor array composed of poly(2-hydroxyethyl methacrylate-*co*-AAc) deposit on top of a reflective substrate with tree different thickness (234, 290 nm and 362 nm for S1, S2 and S3, respectively). When the system is expose to different volatile organic compounds one can observe a change in coloration due to the diverse swelling behavior observed for each VOC. For each VOC the system response gives rise to a specific pattern, even for similar analytes such as alcohols (methanol, ethanol, and IPA) (Reproduced from ref. [59]. Copyright 2017 John Wiley and Sons).

Keeping in mind the importance of nitric oxide (NO) as a signaling molecule in neurons and in the immune system, Park and co-workers [60] recently developed a therapeutic gas-responsive hydrogel. The NO-responsive hydrogel is based on the polymerization of the monomer acrylamide (AAm) as a monomer and a new synthesized NO-cleavable crosslinker the *N,N′*-(2-amino-1,4-phenylene) diacrylamide. This crosslinker act as a NO scavenger and when expose to NO gas the gels react with the diffuse-in NO molecules to form a benzotriazole group moiety that is then hydrolyzed, resulting in swelling of the hydrogel. The authors also showed that this system can incorporate an enzyme-triggering NO donor.

Cellulose nanocrystals (CNC) obtained from the acid hydrolysis of cellulose fibers have attracted a lot of attention from the scientific community as they are renewable, biocompatible, have low density, with high specific surface area and present interesting liquid crystalline and mechanical properties, among other [61]. Oechsle et al. [62] recently described the production of a CO_2-responsive hydrogel based on CNC, where to a suspension of positively charge nanoparticles, an imidazole solution is added. The presence of the analyte CO_2 triggers the gelation of the hydrogel in a pH-responsive fashion, since CO_2 reacts with water to form carbonic acid decreasing the pH of the solution and imidazolium salt. The authors showed that if the gel was exposed to N_2, the 3D network collapses. The combination of CNC characteristics and its response in the form of the responsive hydrogel allowed the authors to envisage a wide range of applications, starting from switchable absorbents, and flocculants to tissue engineering.

Nature has several interesting examples where animals adapt their coloration in response to the changes in its environment. For instance, Cphalopods *Euprymna scolopes* can alter the thickness of its iridophores by stretching it, leading to a change in its iridescent coloration as a form of camouflage. In addition, structural coloration variation can be observed on the longhorn beetle *Tmesisternus isabellae*, since they change their coloration from green to red when the ambient relative humidity increases. Inspired by such examples, Qin et al. [59] developed an interferometer with adaptive coloration to be used as a vapor chemical sensor. The color adaptive sensor was produced with a simple design that consists on a single layer of hydrogel film of poly(2-hydroxyethyl methacrylate-*co*-AAc) deposit on top of a reflective substrate. The resulting colors arise from the interference of light waves reflected from the air–hydrogel interface and the hydrogel–substrate interface. The authors were able to demonstrate a color change in the presence of volatile organic compounds (VOC) as acetic acid, acetone, ethanol, ethyl acetate, hexane among others, as shown in Figure 4c. Moreover, the authors were also able to demonstrate with an on-demand patterning that these systems can be used to produce information encryption devices, triggered by humidity.

Analyte-responsive hydrogels can be key versatile diagnostic tools, as they can also create systems that are easy to use. As an example, one can see the use of a photonic biosensor array based on a interpenetrated cholesteric liquid crystal/hydrogel polymer network [63]. This system is prepared in a multi-step procedure that first give rise to a solid cholesteric liquid crystal film in which a hydrogel interpenetrated array is then built using a UV photomask template. In this set, the enzyme is then immobilized, giving rise to the biosensor array. The complexity of the preparation set-up is antagonistic to the oversimplified nature of its use since the biosensor uses a change in color system to detect the target analyte, as the authors demonstrated for the analyte urea.

2.6. Hydrogels for 3D Printing

If the interesting characteristics of stimuli-responsive hydrogels presented above are considered, one can envisage its use in 3D printing systems or bioprinting or biofabrication, in which one can accurately develop structures, as scaffolds for tissue engineering, or printing cells or even organs [64,65]. The most commonly used hydrogels inks comprise calcium of alginate, or include colloidal suspensions and polyelectrolyte gels [66]. One should note that a hydrogel can be printable if it demonstrates shear-thinning behavior under pressure, this is, if it acts as a non-Newtonian fluid (its viscosity is dependent on the applied shear) allowing, for instance, normal thick liquids or gels to flow freely. However, the knowledge of printing parameters is of extremely importance in order to be able to precisely control the hydrogel depositions, especially if one considered the printing of 3D structures that mimic human organs. Keeping this in mind, He et al. [65] were able to establish relationships between printing parameters, such as air pressure, feed-rate, or even printing distance and printing quality of the expected structures.

Tan et al. [64] recently presented an innovative set-up to produce 3D printing structures that mimic the mechanical properties of the softest tissues found in the human body (for instance brain and lung tissues). In this work, a liquid ink of PVA:Phytagel (agar substitute gelling agent) is cryogenic printed, with an extrusion-based method, on an isopropanol bath with dry ice and solid CO_2. In this set-up, the authors use the liquid-to-solid phase change to obtain 3D hydrogel structures, by rapidly cooling the ink below its freezing point. When coated with collagen type I, the hydrogels show good cell attachment and viability. The use of cryogenics to obtained printed super soft hydrogels led the authors to postulate several possible applications, which range from soft tissue for surgical training and simulations to mechanobiology and tissue engineering.

Multi-stimuli-responsive hydrogels inks for direct-write 3D printing were developed by Karis and co-workers [66]. In this essay, the authors presented to the reader cleverly designed triblock copolymer hydrogels based on living anionic ring-opening polymerization of glycidyl ethers, as allyl glycidyl ether (AGE), with isopropyl glycidyl ether (iPGE) from a PEG macroinitiator. In doing so, the authors could combine the thermo-responsiveness of PiPGE-*b*-PEG-*b*-PiPGE with the cross-linkable functionality of PAGE-*b*-PEG-*b*-PAGE and form a series of PiPGE-*stat*-PAGE-*b*-PEG-*b*-PiPGE-*stat*-PAGE macromolecules. These polymers can self-assemble in water and give rise to hydrogels that respond to temperature, pressure, and UV light. These features help produce a robust free-standing 3D printed object.

Yang et al. [67] recently presented a functional hybrid ink for 3D printing. The authors, using PNIPAAm hydrogel within a water-rich silica-alumina (SiO_2/Al_2O_3)-based gel matrix, were able to obtain a hydrogel that can take up to 70 wt. % in water, respond to temperature and electric stimuli, and be printed using a commercial 3D printer. The hybrid hydrogel is transparent below PNIPAAm's LCST and opaque above, due to the gels' dehydration. Using this feature, the authors printed this hybrid system into a flexible electro-thermochromic device. This device allows the authors to display the electric/temperature tunable transmittance of the systems, with lower response times when compared to other reported systems, as well as its bending properties (the system sustain bending angles of 180° without feature failure).

3. Multi-Stimuli-Responsive Hybrid Microgels

Polymeric microgels are in an intermediate state between the branched polymers and macroscopic networks, with a molecular weight (Mw) comparable to that of a linear high Mw polymer, however with a structure resembling that of a macroscopic network. This indicates that in an appropriate solvent, they swell instead of dissolving, forming a colloidal dispersion [68]. Depending on the composition, microgels can be sensitive to external stimuli, a feature that has generated much interest because of its versatility for potential application. Pelton and Chibante [69] synthesize for the first time thermoresponsive PNIPAAm microgel particles that shrinked upon heating due to the intrinsic LCST that NIPAAm monomer possesses. This kind of thermoresponsive microgels is defined as negative thermorsensitive. However, there are also positive thermosensitive microgels, albeit less studied. An example of the latter is the system formed by the poly(acrylamide-*co*-acrylic acid) (PAAm-*co*-AAc) copolymer, which swells upon heating due to the breakage of hydrogen bond interactions (the so-called zipper effect) [70].

PNIPAAm is undoubtedly the most studied thermo-responsive microgel [71–77]. The NIPAAm monomer is often used to produce microgels using precipitation polymerization. In this technique, the initially produced polymers are insoluble in the used solvent (usually water) at the reaction temperature (above LCST). The initial nucleation period determines the number of particles, and is followed by the growth period, in which the freshly polymerized polymer adds onto the preformed nuclei. In the presence of a suitable crosslinker, the polymer chains remain in the network and only when the temperature decreases (below LCST), the PNIPAAm microgels swell in water [78]. More information related to microgel synthesis can be found in a recently published book [79].

Although plain PNIPAAm microgels are often used in several applications, the incorporation of additional functional monomers gives raise to more complex and interesting microgels. The addition of co-monomers usually affects the swelling behavior of the microgel. PNIPAAm, in particular, is often functionalized with organic acids (e.g., AAc [80,81], MAAc [82,83]), which introduce changes in the polymer network, thus influencing its swelling behavior [76]. The use of acrylamide-based monomers is also advantageous due to the presence of at least one vinyl group that participates in the polymerization reaction and/or crosslinking process. Furthermore, non-acrylamide based monomers such as *N*-vinylcaprolactam originate pH-responsive microgels [84].

Core-shell microgels of PNIPAAm have also been explored due to the possibility of combining the different characteristics of both the core and the shell. For example, core-shell microgels composed of a polystyrene core and a PNIPAAm shell [85] have been studied for catalysis applications [86] and to produce composite membranes with tunable permeability [87]. In a different approach, Brugnoni et al. [88] produced core-shell-shell microgels using silica nanoparticles as the core, poly(*N*-isopropylmethacylamide) (PNIPMAAm) as the inner shell, and PNIPAAm as the outer shell. In addition, by dissolution of the silica core, the authors produced hollow microgels, which enable the production of a series of microgels with diverse thermal response from the initial composition. Zha et al. [89] produced thermoresponsive microcontainers for delivery applications, also using silica nanoparticles as templates with a PNIPAAm shell.

Using these few examples, we tried to give the reader a glance of the multiple possiblities of homo- and co-polymeric microgels constructs. It is not our intention to cover every type of microgel and its application. Several reviews have been published recently covering different aspects of microgel design and synthesis, and their applications [75,76,78,84,85,90–94]. As indicated in the introduction, much effort has been devoted to the obtention of advanced materials able to responde to several external stimuli. In this regard, the strategy followed consisted of the formation of hybrid systems. By definition, a hybrid material results from a combination at the nanometric and molecular level of an inorganic and an organic component, allowing the preparation of new multifunctional materials [95]. The properties of hybrid materials emerge not only from the individual properties of each component, but also from synergetic effects from the interaction of the components. In the case of hybrid polymeric gels, the inorganic component is generally based on nanoparticles not only due to their quantum size effects but also to take advantage of their high surface to volume ratio [96]. Therefore, in this section, we intend to describe the most recent stimuli-responsive hybrid polymer microgels, focusing on the type of nanoparticles embedded in the polymeric matrix.

3.1. Metallic Nanoparticles

3.1.1. Gold Nanoparticles

Gold nanoparticles (AuNPs) have been used since 1857 when Faraday discovered the different properties of colloidal gold and bulk gold [97]. Nowadays, AuNPs are used in analytical chemistry [98], electronics [99], as well as biology and nanomedicine due to their characteristic properties such as tunable optical properties, biocompatibility, high surface area, and can undergo surface modification [100].

The most common organic component of a hybrid polymer gel with gold nanoparticles is PNIPAAm. Several research groups have recently used AuNPs incorporated into PNIPAAm microgels for catalytic applications. Shi et al. [101] used AuNPs with 3.8 nm in diameter homogeneously incorporated into thiol-functionalized PNIPAAm microgels by in situ reduction of $HAuCl_4$ aqueous solution based in Au-thiol chemistry. By using this methodology, the authors were able to obtain control over the distribution of AuNPs within the microgel. In another case [102], the authors used AuNPs incorporated into a PNIPAAm-based copolymer (poly(NIPAAm-*co*-allylamine)), which responds to both temperature and pH stimuli. With this hybrid system, the authors were not only able to modulate the catalytic activity of AuNPs, but also allow label-free in situ localized surface plasmon resonance (LSPR) monitoring of the catalyzed chemical reaction. The modulation is performed by varying the solution temperature, while the monitoring is made through stimuli-responsive volume phase transitions of microgels, thus changing the immediate physicochemical microenvironment of AuNPs, resulting in an alternation of the LSPR. Similarly, Rehman and coworkers [103] crosslinked NIPAAm and *N,N'*–dimethylaminoethylmetacrylate (DMAEMA) in water to obtain microgels responsive to temperature and pH. The poly(NIPAAm-*co*-DMAEMA) microgels were used as nanoreactor for the synthesis of AuNPs. These microgels demonstrated catalytic properties for the reduction of toxic 4-nitrophenol (4-NP). Moreover, the catalystic activity can be tuned using the microgels' thermobehaviour. Chen and his team used photoluminescent gold nanodots with 1.8 nm in diameter incorporated into PNIPAAm microgels for the detection of mercury ions (Hg^{2+}). This detection is based in the formation of Au-Hg amalgam and hybrid microgel aggregates in the presence of Hg^{2+}, leading to decreased photoluminescence of the hybrid microgels. By using these hybrid microgels, the authors were able to sensitively and selectively detect Hg^{2+} ions in environmental and biological samples [104]. Tang et al., used the same principle to produce hybrid microgels for the detection of Hg^{2+} by using poly(NIPAAm-*co*-2-(dimethylamino)ethylmethacrylate) microgels [105]. Mackiewicz and coworkers [106] produced a new multifunctional microcomposite by combining three components: PNIPAAm microgels, polyaniline (PANI) fibers and AuNPs (Figure 5a). PANI is one of the most common and interesting conducting polymers. The combination of PANI nanofibers with AuNPs led to a substantial increase in conductivity and electroactivity of the microcomposite. Thus, the microcomposite showed much stronger electrocatalystic properties for ethanol oxidation reaction in alkaline medium compared to bare gold electrodes.

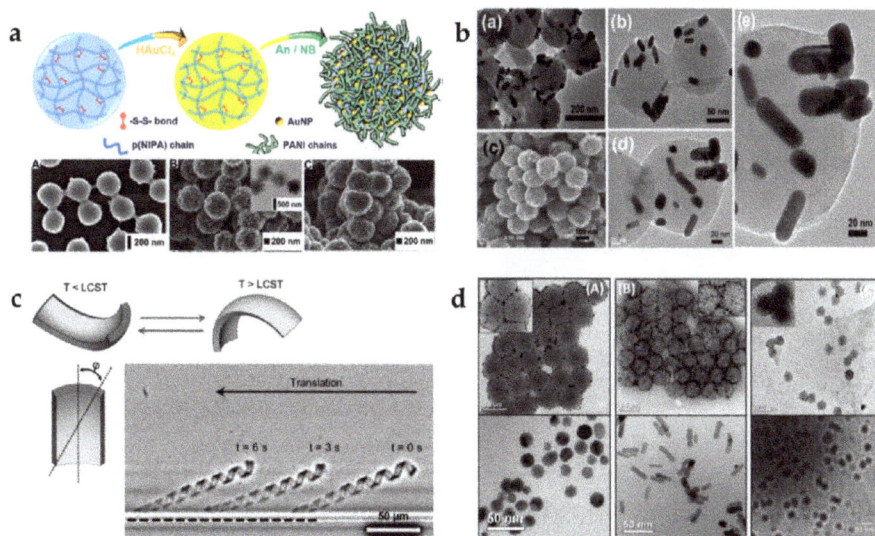

Figure 5. (**a**) Scheme for the preparation of PNIPAAm microgels, polyaniline (PANI) fibers and AuNPs microcomposite and the respective scanning electron microscopy images (Reproduced from ref. [106]. Copyright 2016 The Royal Society of Chemistry); (**b**) Transmission electron microscopy images of hybrid microgel particles composed of PNIPAAm microgels and AuNRs. (a,b,d,e) The figure show spherical microgels with a rough surface with Au NRs attached; (c) Field emission scanning electron microscopy image of hybrid microgels (Reproduced from ref. [107]. Copyright 2013 John Wiley and Sons); (**c**) Schematic illustration of the helix direction controlled by the principle curvatures 1/R1, 1/R2 and the mismatch of φ toward the length axis of the ribbon. Directing the rotational motion to a linear translocation when the oscillating helix is confined close to a flat wall that impedes the rotation around the axis normal to the helix direction. The helix contour length is 160 μm, the dashed line indicates the wall position (Reproduced from ref. [108] Copyright 2015 John Wiley and Sons); (**d**) Transmission electron microscopy images of (A) PNIPAAm-Au nanospheres (B) PNIPAAm-AuNR and (C) PNIPAAm-Ag nanospheres [109] (Reprinted from ref. [109]. Copyright 2017 Elsevier).

Other authors used gold nanorods (AuNR) located at the surface of the microgel instead of spherical nanoparticles incorporated into PNIPAAm microgels (Figure 5b). In both cases, the stimuli-responsiveness of the microgel (temperature and pH), which causes a significant change in the aspect ratio (length/width) of the hybrid microgel, created an important shift in the UV-VIS absorption intensity and a shift in the longitudinal surface plasmon bands of the gold nanorods [107,110].

Mourran et al. [108] used the particle replication in non-wetting templates (PRINT) technique to produce PNIPAAm microgels with gold nanorods to act as soft microbots. Temperature-responsive thin PNIPAAm microgel bodies can undergo bending and torsional motions upon sweeling and unsweeling (Figure 5c). The addition of AuNRs promote very fast temperature jumps of more than 20 °C within less than miliseconds by their photothermal heating. The authors demonstrated that for the proposelfully shaped microgels, the kinematics of a cyclic body shape variation can be tuned to differentiate between the forward and the backward motion.

A common strategy for the incorporation of gold NPs into polymeric microgels is through the in situ reduction method. Liz-Marzán and coworkers [111–113] have developed a method to encapsulate AuNPs within PNIPAAm microgels. The authors found that thermo-responsiveness behavior and crosslinking density of the microgel allow efficient regulation of the catalytic activity of the AuNP. Moreover, this composite can be reusable for several cycles with its catalytic property almost perfectly reproducible [114]. Shi et al., used P(NIPAAm-*co*-MAAc) to obtain thermo-, pH-, and light-responsive hybrid microgels. The authors produced the polymeric microgels, then thiol-functionalized them and obtained the hybrid microgels by in situ reduction of the gold precursor in the presence of the microgels based on Au-thiol chemistry, achieving a fluff-like structure due to the numerous AuNPs distributed in the interior of the microgel [115]. In a different study, Agrawal and her team used *N*-vinylcaprolactam/acetoacetoxyethyl methacrylate/acrylic acid-based microgels (P(VCL-AAEM-AAc)) as a polymer template for in situ reduction of the gold precursor. The colloidal gold nanostructure is selectively formed in the core of the microgel and the hybrid structure is used as a noble metal catalyst [116]. PVCL has a similar thermosensitivity to PNIPAAm, i.e., it exhibits LCST in aqueous media and goes through coil-to-globule transitions upon heating. However, PVCL is more biocompatible than PNIPAAm since upon hydrolysis, the resultant monomers are less cytotoxic. Consequently, some authors have used PVCL for the construction of hybrid microgels incorporating AuNPs for biomedical purposes [117,118].

In a different study, thermoresponsive PNIPAAm-metal hybrids were produced using PNIPAAm microgels as a microreactor. Metal NPs of different shapes were embbebed in the microgel: Au nanorods, Au nanospheres, and silver (Ag) nanospheres (Figure 5d). These core-shell microgels were used to catalytically convert 4-NP to 4-aminophenol (4-AP). The authors conclude PNIPAAm-AuNPs and AuNRs have better catalytic activity, as well as recyclability compared to that of PNIPAAm-AgNPs [109]. Similarly, Shah et al. [119] produced multiresponsive tercopolymer microgels (poly(NIPAAm-*co*-MAA-*co*-2-hydroxymethylmethacrylate) with homogeneously dispersed Ag and AuNPs. These microgels exhibit sensitivity towards temperature and pH. In addition, terpolymer microgels are efficient catalystics for the reduction of 4-NP, congo red and methylene blue in waste waters.

3.1.2. Silver Nanoparticles

Silver nanoparticles (AgNPs) being one of the noble metals, have unique optical, electrical and surface properties. These nanoparticles are commonly used as sensor, optical switches, antimicrobial agents, and catalysts [120,121].

Similarly to AuNPs, silver-based hybrid microgels are generally composed of thermosensitive polymers, most commonly PNIPAAm. In addition, a copolymer is often used to obtain a dual-stimuli-responsive microgel, for example, a temperature- and pH-sensitive microgel. Liu et al. [122] prepared PNIPAAm microgels with AgNPs by in situ reduction of Ag^+ coordinated into monodisperse PNIPAAm microgels, in which the catalytic activity can be tuned with four stages of change versus temperature. With the same purpose, Tang et al. [123] used P(NIPAAm-*co*-DMA) for in situ reduction of Ag^+ ions pre-dispersed into the microgels. These hybrid microgels possess thermo- and pH-responsiveness, as well as catalytic activity. In a different study, Khan et al. [120] used poly(NIPAAm-*co*-allylacetic acid) (P(NIPAAm-*co*-AAAc)) with AgNPs incorporated into the microgel by in situ reduction to be used as catalysts for reduction of nitrobenzene into aniline, exploring various conditions of temperature, catalyst dose and concentration of the reagents. The authors found that the catalytic activity of the hybrid microgels can be tuned by changing the temperature of the medium.

As previously stated, AgNPs can also be used in sensor devices. For example for glucose sensing, Khan et al. [124] performed in situ reduction of silver nitrate in the presence of glucose using P(NIPAAm-*co*-AAc) microgels as a template, while Zhou et al. [125] used AgNPs immobilized in a glucose-imprinted boronate derivated gel network of poly(NIPAAm-*co*-acrylamide-*co*-4-vinylphenylboronic acid) (P(NIPAAm-*co*-AAm-*co*-VPBA)) as a glucose sensing element. Another research group used AgNPs incorporated into poly(3-acrylamidophenylboronic acid-*co*-2-(dimethylamino)ethyl acrylate) microgels as glucose sensors. The glucose-sensing performance of the hybrid system was improved by tailoring the phase behavior from monotonous shrinking upon adding glucose, as well as the following change in the signaling manner from "turn-off" to "turn-on" [126].

PNIPAAm-based hybrid microgels are also used as sensors for other molecules, for example H_2O_2 colorimetric sensing. Han and coworkers developed P(NIPAAm-*co*-AAc) microgels with AgNPs incorporated that went through autocatalytic oxidization in the presence of H_2O_2. The authors found that the absorbance at 400 nm was linearly dependent of H_2O_2 concentration and that was selective [127].

Kumacheva's group produced AgNPs in pH-responsive microgels of P(NIPAAm-AAc-2-hydroxyethyl acrylate) and found that the AgNPs produced in situ, i.e., through photoreduction of Ag^+ ions were fluorescent, while the AgNPs produced from conventional method do not exhibit this property [128]. Similarly, other research groups have used this property of in situ produced AgNPs to obtain hybrid microgels for fluorescent detection [129,130].

Finally, AgNPs-based hybrid microgels are also used for plasmonic sensing due to the coupling between the oscillating electromagnetic fields on the AgNPs surface originating from plasmon resonance, which provides significant enhancement of the local electromagnetic fields. As such, these systems can be used as substrates for surface-enhancement raman scattering sensing. For this application, the plasmonic properties and surface electromagnetic field enhacement effects of the AgNPs present in the hybrid microgels are tuned by temperature and pH, for example, thus resulting as mobile plasmonic or SERS microsensors to detect environmental temperature or pH variation [121,131].

3.2. Magnetic Nanoparticles

Magnetic nanoparticles such as superparamagnetic iron oxide nanoparticles (Fe_3O_4, SPIONs) are widely used for biomedical applications due to their biocompatibility and interesting magnetic properties. The main biomedical applications of this type of nanoparticles are magnetic hyperthermia agents or contrast agents for magnetic resonance image [132–137]. The combination of polymeric microgels with Fe_3O_4 NPs allows the construction of a composite system with the properties of the polymer with magnetic-responsive properties. For example, several authors incorporated Fe_3O_4 NPs into PNIPAAm microgels, obtaining a thermo- and magnetic-responsive system. In some cases, the iron oxide NPs act as a crosslinker for PNIPAAm microgels, assuring a higher dispersibility of the NPs in the polymeric network [138,139]. In addition, the heating ability of the Fe_3O_4 NPs may provide an in vitro burst release of chemotherapeutic drugs [140,141].

Boularas et al. [142,143] were able to encapsulate a high content of magnetic NPs (up to 33 wt. %) into poly(di(ethylene glycol)) methyl ether methacrylate-*co*-oligo(ethylene glycol) methyl ether methacrylate-*co*-methacrylic acid) microgels, obtaining thermo-, pH-, and magnetic-responsive microgels. For the same purpose, other authors used different polymers for the production of the microgels, such as P(VLC-*co*-IA) based microgels [144], poly(2-(2-methoxyethoxy) ethyl methacrylate-*co*-oligo(ethylene glycol) methacrylate-*co*-AAc) incorporated with attapulgite/Fe_3O_4 NPs [145,146], among others.

Zhou et al. [147] developed a facile one-pot synthesis method of Fe_3O_4-PVA microbeads for drug delivery applications. The beads are prepared by dropwise addition of mixed aqueous solution of iron salts and PVA solution into alkaline solution. To add a thermo-responsive functionality to the beads, the authors added PNIPAAm to the mixture, creating magnetic gel beads with high drug loading capacity that have magnetic and temperature-responsiveness.

Other authors produced poly(vinyl phosphonic acid) nanogels with silica-coated Fe_3O_4 NPs. They also produced porous nanogels by using triethoxyvinylsilane as co-monomer during polymerization, followed by removing the silica by hydrofluoric acid treatment after polymerization. These nanogels were used for two main applications: (i) drug delivery systems of zuclopenthixol and phenazopyridine hydrocholride as models drugs; and (ii) absorbents for removal of organic contaminants such as 4-NP, 1,1'-dimethyl-4,4'-bipyridinum dichloride (Paraquat), methylene blue, and rhodamine 6 G from aqueous media [148].

The combination of polymer gels with Fe_3O_4 NPs can also be used for catalytic applications. Liu et al. [149] prepared Fe_3O_4-P(MBAAm-*co*-MAAc) microspheres that contained carboxyl groups for application as magnetic catalyst support to load a series of metallic nanoparticles such as Ag, Pt, and Au. The Fe_3O_4 NPs loaded into the microgels were able to maintain their superparamagnetic behavior, together with their high saturation magnetization, providing a convenient way for separating metallic nanoparticles from a solution. In a different study, Nabid et al. [150] incorporated Fe_3O_4 NPs into P2VP microgels where the magnetic NPs acted as crosslinkers. These hybrid microgels are able to trap metal ions such as Pd^{2+} via complex formation of P2VP with metal ions, and the reduction of the Palladium ions by sodium borohydride led to the formation of PdNPs@P2VP-Fe_3O_4 hybrid microgels. These systems may be employed as nanocatalysts toward oxidation reaction of alcohols, which can be modulated by the volume phase transitions of the microgel.

3.3. Carbon-Based Materials

Carbon nanotubes (CNTs) are often used as reinforcement agents to create porous hydrogels with tunable mechanical properties. The incorporation of CNTs into hydrogels increases not only later elastic modulus, but also the thermal stability and electric conductivity of the resulting composites [151,152].

Shin et al. [152] used reinforced CNT-gelatin methacrylate hybrid as a biocompatible, cell-responsive hydrogel platform to create cell-laden three-dimensional constructs (Figure 6). These hybrids are photopatternable, which allow easy fabrication of microscale structures without harsh processes. In a different study, Spizzirri and coworkers produced gelatin/CNT hybrid hydrogels as effective diclofenac sodium salt carriers. The drug is released in response to an electric stimulation [151]. Cui and coworkers [153] produced conductive gel composites using only colloidal particles as building blocks. These composites are composed of mixed dispersions of vinyl-functionalized pH-responsive microgel particles and multi-walled CNTs. The microgel particles act as dispersant and macrocrosslinkers, while CNTs provide electrical conductivity and modulus enhancement. These gel composites showed promising properties to be used as injectable gels for application in soft tissue repair and electronic skin.

Zhou et al. [154] used the technique previously described for AgNPs [125], but instead of silver, the authors went further and added carbon dots as the inorganic part to the polymeric matrix (P(NIPAAm-AAm-VPBA)) to produce hybrid microgels for optical sensing of glucose. This new sensor is able to continuously monitor glucose level change in physiological conditions.

Among carbon-based materials, graphene oxide (GO) obtained from chemical modification of graphite is considered a molecular two-dimensional (2D) platform with a high aspect-ratio, also being water-dispersible. This material has several advantages such as high mechanical strength, pH sensitivity, photosensitivity, and low toxicity. In fact, GO shows higher photothermal sensitivity than CNTs, and is capable of efficiently converting optical energy into thermal energy [155,156].

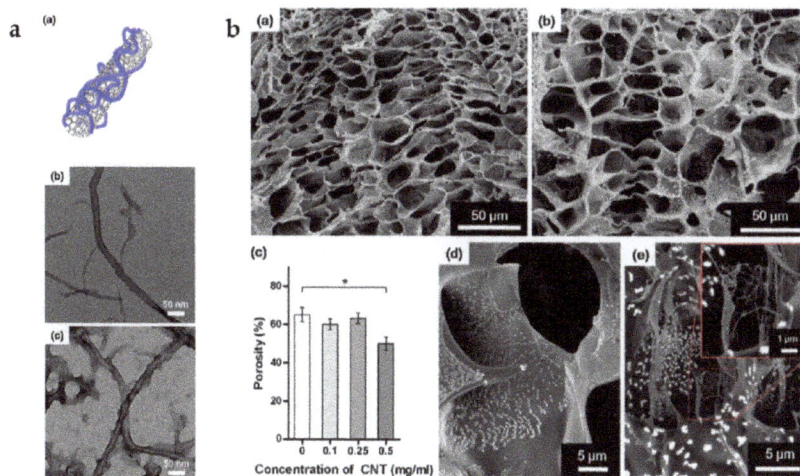

Figure 6. (**a**) CNT-gelatin methacrylate (GelMA) hybrids: (a) Schematic diagram of a GelMA-coated CNT and the respective high resolution transmission electron microscopy images of bare CNTs (b) and GelMA coated CNTs (c); (**b**) Scanning electron images of cross sections of GelMA (a) and CNT GelMA (b) hydrogels with 0.5 mg/mL CNTs; (c) CNT-GelMA hydrogels porosity with different concentrations of CNTs (* $p < 0.05$); Magnified images show parts of a cross section of (d) GelMA and (e) CNT GelMA hydrogel with nanofiber junctions inside a porous structure. (Reprinted with permission from [152]. Copyright 2012 American Chemical Society).

Lu et al. [157] incorporated chemically-reduced GO into PNIPAAm microgel for the preparation of photo- and thermoresponsive composite microgels for drug delivery applications. Wang et al. [156] used the same idea but the chemically reduced GO was incorporated into a thermosensitive nanogel composed of acrylated chitosan, NIPAAm, and PEG diacrylate incorporated with doxorubicin. The composite nanogel exhibited a near infrared (NIR)-induced thermal effect with a high doxorubicin loading capacity. Moreover, doxorubicin release appears to be faster when the composite nanogel is irradiated with NIR light. Graphene oxide composite microgels also found applications as selective glucose-responsive systems for photoluminescent detection of glucose in blood serum [158] or as biocompatible injectable composite gels to repair load supporting soft tissue [155], among others.

Finally, Girard et al. [159] synthetized thermoresponsive nanodiamonds hybrid microgels using PNIPAAm as the organic component. The results showed that the microgels kept the thermosensitivity of PNIPAAm, while achieving a 2D organized array of nanodiamonds, allowing the cost-limited nanoscaled nanodiamonds organization on a substrate with great flexibility for technological or optical applications.

3.4. Other Inorganic Materials

Among the different inorganic materials used to produce hybrid polymeric microgels, silica nanoparticles are of great interest for the production of smart self-catalyzing systems. Agrawal et al. [160] produced composite microgel particles containing silica NPs by simultaneously converting PEG-poly(ethoxysiloxane) and deposite silica in the microgels. In this case, the incorporation of silica NPs increased the rigidity of the microgels and reduced their thermosensitivity. Similarly to other inorganic materials, PNIPAAm is the most commonly used polymer for the production of silica-based hybrid microgel systems [77,161], for example for drug delivery applications [162]. Dechézelles et al. [163] synthesized thermosensitive raspberry-like hybrid microgels consisting of a PNIPAAm core decorated with in situ formed silica particles. These hybrid microgels provide a convenient basis for additional surface modifications through silane coupling agents.

Nanoclays such as Laponite have been introduced into microgel systems for the development of responsive scavenger systems due to the cation-exchange capabilities of clays, thus allowing charging of the microgel with (cationic) metal precursors. For this purpose, Contin et al. [164] developed a ternary system composed of Laponite NPs incorporated into poly(VCL-*co*-acetoacetoxyethyl methacrylate) PVCL/AAEM microgels loaded with different metal NPs: Pd^{2+}, Au$^+$, Pt$^+$. The results showed that some of the synthesized microgels are active catalysts for the Suzuki reaction of aryl iodides in a water-rich medium. In a different application, Du and coworkers [165] produced PAAm/hectorite (Laponite® XLS) double-network hydrogels containing neutral PAAm microgels, obtaining a tough and highly stretchable microgel-reinforced hydrogel with superb mechanical strength. In this case, Laponite acts as a physical crosslinker of the hydrogels.

Another class of interesting inorganic materials are quantum dots (QDs), a semiconductor nanostructure that can be used for bioimaging and labeling probes. Their combination with polymer microgels extends the range of application for optical sensing, imaging diagnostics, and controlled drug release [166]. For example, Wu et al. [166] produced hybrid nanogels based on in situ immobilization of CdSe QDs in CS-PMAAc networks. These hybrid nanogels have excellent colloidal and structural stability and are pH-responsive. Moreover, the results suggest that covalently crosslinked nanogels are more suitable for optical pH-sensing and bioimaging with low-cytotoxicity, compared to physically associated hybrid nanogels. In a different study, Gui and coworkers [167] produced CdTe/ZnS QDs incorporated into P(NIPAAm-*co*-AAc) microgels by facile electrostatic attractions at room temperature. These hybrid microgels are thermosensitive, with almost fully reversible photoluminescence between 25 °C and 45 °C, and bright fluorescence imaging. Similarly, Cai et al. [168] incorporate cysteamine-capped CdTe QDs in P(NIPAAm-*co*-AAc) microgels. The photoluminescence intensity and color of the hybrid microgels can be tuned by adjusting the QDs content. In a different approach, Lai et al. [169] used microfluidic electrospray technology to produce alginate-based multicompartment microgel beads with CdTe QDs incorporated (Figure 7). These microgel beads can effectively separate incompatible drugs during co-delivery and significantly prolong the time of fluorescence emission from QDs.

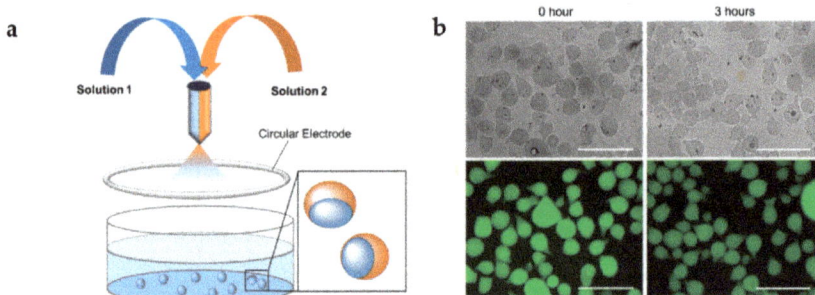

Figure 7. Fabrication of the alginate-based multi-compartment microgel beads with CdTe QDs incorporated: (**a**) Schematic diagram showing fabrication of multi-compartment microgel beads using the microfluidic electrospray technology; (**b**) Representative phase-contrast and fluorescence images of the QD-loaded microgel beads captured at different time points. The scale bar is 500 μm (Reprinted with permission from [169]. Copyright 2016 American Chemical Society.)

Cui et al. [170] produced hybrid polymeric microgels using a different type of polymer. The authors used a imidazolium-based poly(ionic liquid) to prepare monodisperse microgels by means of microfluidics. They found that the imidazolium units in the microgel network may be exploited as reactive sites for functionalization by a simple counteranion-exchange or conversion reaction. As a proof-of-concept, the authors produced three types of isotropic functional particles, including metal-polymer hybrid particles, conductive composite particles, and catalytic particles, from poly(ionic liquid) microgels, demonstrating the versatility of these systems.

3.5. Multifunctional Hybrid Fibrillary Gels

In a different approach, multi-stimuli-responsive systems can be produced using stimuli-responsive microgels as active sites incorporated into polymeric fibers. Using colloidal electrospinning, it is possible to design multifunctional, highly porous, and biocompatible membranes suitable for several applications.

Electrospinning is a simple, fast and easy to scale-up technique to produce functional materials based on polymeric fibers with diameters ranging from 2 nm to several micrometers. Electrospun micro/nanofibers have a high surface-to-volume ratio, tunable porosity, and can be produced from natural or synthetic polymers [171–174]. This technique involves the application of an electrostatic force to generate a polymer jet towards a collector electrode. Despite the simplistic setup, the theory behind this technique is more complex and occurs in three stages: initiation of the jet, elongation, and fiber formation [172,175].

A variation of traditional electrospinning technique is colloidal electrospinning, where the polymeric solution is replaced by a colloidal dispersion. The presence of particles in the spinning solution allows the formation of core-shell fibers from a single nozzle, offering a simpler setup than that used in coaxial electrospinning, which needs two or more needle-tips [176,177]. The presence of small amounts of a fiber forming polymer, acting as a template, is usually required to produce composite fibers from colloidal dispersions containing either inorganic or organic (polymeric) particles [176].

Colloidal electrospinning allows the formation of materials with hierarchical levels of nanostructures, given by the particles and fiber morphology. Moreover, the immobilization of particles in nanofibers allows easy handling and separation from a reaction medium, in contrast to particle suspensions. More commonly, inorganic particles are used in colloidal electrospinning. These particles have higher electronic density than the polymer templates, and can be easily localized in the fibers by electron microscopy [178]. However, the confinement of stimulus-sensitive microgels in fibers by means of colloidal electrospinning could be an interesting approach towards the production of multifunctional fibers with fast thermoresponsive behavior and super-hydrophobic tunable surfaces. This may be used in drug delivery systems, bio-sensing, chemical separation, catalysis, and optics [179]. Few studies reported the confinement of crosslinked PNIPAAm microgels inside nanofibers. For instance, Nieves et al., produced composite electrospun fibers of PNIPAAm microgels (up to 40% of microgels per-centage mass) using poly(vinyl pirrolidone) (PVP) (which is a hydrogel itself) as fiber template with a mean fiber diameter of 0.9 μm [179]. Tunable surfaces of electrospun non-woven mats with PNIPAAm microgels/poly(L-lactic acid) fibers, in which the production of fibers with a mean fiber diameter of 284 nm connected to bead sizes of 3.4 μm with a spindle-like structure, was reported by Gu et al. [180]. Recently, PNIPAAm based microgels were confined in electrospun fibers using colloidal electrospinning technique. In the first case, PNIPAAm and PNIPAAm-CS crosslinked microgel dispersions were prepared by means of surfactant-free emulsion polymerization using N,N'-methylene bisacrylamide as a crosslinking agent. These thermosensitive microgels were successfully confined in poly(ethylene oxide) (PEO) fibers via colloidal electrospinning, resulting in beads randomly distributed over the fibers with a typical "bead-on-a-string" morphology. Furthermore, by performing a statistical analysis, the relationship of the processing variable over the fiber diameter was evaluated using the response surface methodology. Using the set of optimized parameters, PEO fibers with an average diameter of 63 nm containing PNIPAAm microgels, were produced (Figure 8) [71].

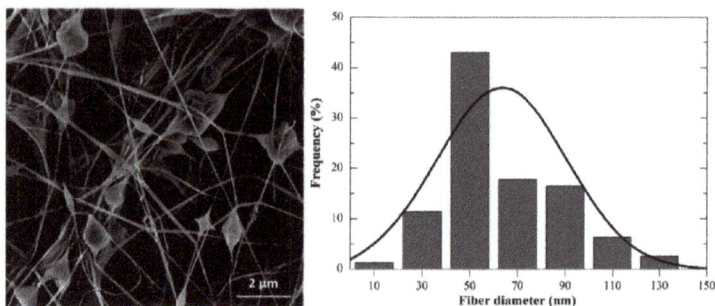

Figure 8. Scanning electron microscopy image optimized electrospun fibers containing PNIPAAm microgels (on the **right**) and fiber diameter distribution with a mean fiber diameter of 63 ± 25 nm (on the **left**) (Reprinted from ref. [71]).

In a later study, PVP was used as a fiber template to produce multifunctional fibrillary gels using PNIPAAm-based microgels as active sites. PVP has the advantages of being biocompatible and water-soluble, thus avoiding the use of organic solvents. In addition, this polymer can be easily crosslinked using UV radiation, therefore avoiding its immediate dissolution under physiological conditions. Composite fibers produced by colloidal electrospinning and crosslinked for 30 min were able to swell about 30 times their weight after 1 h in aqueous medium [73].

These works are a step forward in the design and development of new and improved smart systems that can be tailored for the desired application —biomedical or industrial. One example is to use different types of microgels to create a multiresponsive membrane. Another approach will be to use hybrid microgels instead, thus taking advantage of the dual-responsivess of these type of systems, combined with the fiber's properties.

4. New Generation of Functional Stimuli-Responsive Polymer Hydrogels: Self-Shape (Shape-Memory) and Self-Healable Systems

4.1. Stimuli-Responsive Shape Memory Hydrogels

Materials that show shape memory effect are able to deform on a new temporary shape and store in this state with a further controlled recovery of its original shape when an external stimuli (typically temperature) is applied above a critical value [181,182]. For instance, Shape Memory Polymer (SMP) systems have facilitated the development of devices for biomedical applications such as stents, dialysis needles, drug carriers, drug delivery systems, etc. [183–185]. However, the problem of SMP systems is the difficulty to be applied for a biological media in a living system, mainly due to the high temperatures that are needed to deform the polymers. Such temperatures could damage the surrounding tissues and even the drugs' activity that are carried out and further delivered. In order to solve such problems, in recent years, shape memory hydrogels that are predominantly water, have postulated as alternative candidates as potential systems for biomedical applications due to the use of water as the driving force for the desired shape memory effect [186–188].

The development of shape memory hydrogels faces some difficulties in terms of glass transition temperature, which is usually the phase transition used for the shape stabilization and shape recuperation in SMP. Basically, the presence of a large content of water molecules carries a loss of efficiency in hydrogels. The main strategy to overcome this problem is to promote the formation of new crosslinking in the different deformation states of the hydrogel. Such new links allow the storage of shape. Breaking the new crosslinking points will end up in recovery of the initial shape.

In 1995, Osada et al. [189] reported for the first time the phenomenon of "shape memory" effect in polymer hydrogels. They showed how the temperature-dependent poly(AAc-*co-n*-stearyl acrylate) swells with temperature. At temperatures below 25 °C, the gel behaves as a hard system, whereas above 50 °C it, becomes soft and stretchable (up to 1.5 times its initial length). Authors suggested that the shape memory effect came from the temperature-induced order-disorder transition associated with alkyls side chains and stearyl acrylate unit interactions. In addition, authors highlighted that the shape "remembered" by the gel was dependent on the container where it was first formed.

Despite this pioneer study, hydrogels that can undergo shape transitions (temporary-to-permanent) while containing large amount of water solvent are still scarce. Here, we report some of the most recent and relevant achievements (last five years) in the subject.

Among the polymer matrices constituting shape memory hydrogels, three are more commonly used: PAAm, PVA and PEG. For instance, Zhang et al. [190] performed a study regarding the water content dependent shape memory behavior of P(AAm-*co*-AAc) commercial hydrogel. Authors differentiate two aspects: (i) Shape Change Effect (SCE), which is usually observed in electroactive polymer or piezoelectric materials and (ii) Shape Memory Effect (SME). Both properties were directly related to hydrogel water content. In fact, they found that at low water content hydrogels possessed shape memory effect, whereas at higher water content PAAm-*co*-AAc commercial hydrogel showed both rubber like shape memory effect and water induced shape change. With these findings, they also understand that programming of the system could be performed through the deformation of the hydrogel above Tg or by dehydration after previous deformation of the rubbery-like hydrogel. Such results indicated that the commercial hydrogel PAAm-*co*-AAc could be used as an elastic matrix in a hybrid system to induce the shape memory effect [190]. Following this work, Zhang et al. [191] performed a similar investigation in a double network or semi-interpenetrated network nanocomposite hydrogel composed of PAAm and poly(2-acrylamide-2-methylpropanesulfonic acid). Such hydrogel did not show a remarkable swelling ratio. After the systematic study, authors found that this interesting hydrogel possessed mechano-responsive SCE and water-induced SME properties that are not found in brittle hydrogels with higher swelling ratios [191].

Fan et al. developed a novel dual-responsive shape memory hydrogel of P(AAm-*co*-AAc) with a characteristic thermoplasticity. More concretely, authors prepared hydrogels by simple ternary copolymerization of AAm and AAc with low amounts of a cationic surfactant monomer, in the absence of organic crosslinkers. The resultant hydrogels could exhibit shape memory effects via two different strategies: (i) From an ionic/complex binding between carboxyl groups and ions, which provided additional physical crosslinking points, necessary to lock a temporary shape; and (ii) By salt-strengthened hydrophobic association that enables the formation of a trapped shape. Such physical hydrogel exhibited novel characteristic thermoplasticity, which allowed the change of its permanent shape upon heating and the storage of the shape after cooling, which is a strong step forward compared to the conventional chemically crosslinked shape memory hydrogels [192].

Hu et al. [193,194], also using PAAm as a matrix, fabricated and further optimized [194] a multi-triggered shape memory hydrogel that owns its properties to the presence of two different crosslinkers and stabilizing agents with two different functions: (i) One transforms the shaped hydrogel into an randomly shaped quasi-liquid state; (ii) The other crosslinker, which is present in the quasi-liquid, provides an internal memory that returns the original shaped hydrogel due to the regeneration of the second crosslinker when stimulated. By following the above strategy, authors were able to fabricate two pH sensitive shape memory hydrogels composed of acrylamide chains, crosslinked by duplex DNA and pH-sensitive triplex DNA crosslinking units. The final shape memory hydrogels were indeed stable hydrogels at neutral pH, but dissociated to quasi-liquid states either at acidic or basic pH values. Authors took a step forward and prepared a hybrid system with two different domains, each being selectively triggered to undergo hydrogel/quasi-liquid shape memory dictated transitions [193].

Li et al. [195] were able to develop a chemically crosslinked polymer gel that showed shape memory and self-healing ability. Here, the approach used by the author consists of the development of an interpenetrated network composed of a chemically crosslinked PEG and physically crosslinked

PVA hydrogel deriving in a chemical/physical double network (Figure 9a). The PEG/PVA system was subjected to freeze-thawing cycles to obtain the physical entanglement of PVA within the PEG matrix (Figure 9b). By playing with deformations and further freeze-thawing cycles, authors were able to fix a shape after hydrogel deformation (Figure 9c). On removal of the external force, they evaluated the recovery of the hydrogel upon heating [195] (Figure 9e).

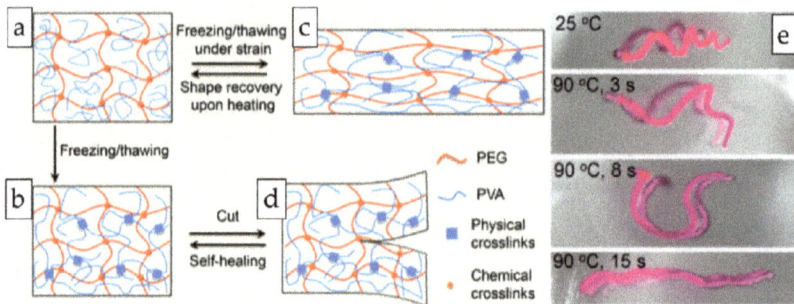

Figure 9. Schematic description of PVA/PEG double-network hydrogels showing both self-shape and self-healing ability (**a**) covalently crosslinked network of PEG mixed with PVA, the "precursor" of the second physical network; (**b**) Using the freezing/thawing treatment, the self-healable double-network hydrogel is produced; (**c,d**) the resultant hydrogel gains shape memory effect when the hydrogel is mechanically strong enough to be processed to a deformed state, followed by the formation of the physical network in the hydrogel under deformation; (**e**) photos showing the fast thermally activated shape recovery of the PVA/PEG hydrogel from a temporary twisted shape to the initial straight state. (Adapted with permission from ref. [195]. Copyright 2015 American Chemical Society).

By following a similar strategy, Li et al. also developed a PVA-based shape memory hydrogel. In this case, authors fabricated a melamine-enhanced PVA physical hydrogel [196]. They found that the incorporation of a small amount of melamine allow the formation of two type of physical crosslinks in two separates states. In addition to improved mechanical properties and enhanced biocompatibility, such advantage allows the deformation of hydrogel (65% water) and further fixation of the achieved temporary shape. In this case, authors found that heating induced by therapeutic ultrasound was able to trigger the shape memory effect of the hydrogel.

With the aim of obtaining tougher and stronger hydrogel films, Zhang et al. [197] prepared a nanocomposite of CS and GO. Authors applied a water evaporation self-assembly method and a further crosslinking of the solution to obtain a layered CS/GO hydrogel film with bio-inspired nacre-like brick- and mortar-structure, composed of organic and inorganic layers. Besides, the obtained hydrogel showed pH-driven shape memory ability. This interesting feature make this hydrogel film a promising candidate for biomedical applications, such as actuators, biosensors, or even wound dressings. More recently, Ma et al. [198] also inspired by nature, developed a macroscopic anisotropic composite polymeric hydrogel, by the combination of a thermoresponsive GO reinforced PNIPAAm-based hydrogel sheet with a pH-sensitive polyethyleneimine-based fluorescent hydrogel sheet. This combination allowed the fabrication of a bilayer anisotropic hydrogel actuator with on–off switchable and color-tunable fluorescence behaviors. The shape was controllable via temperature changes and the fluorescence color was tunable through pH.

Shape memory hydrogels can also be triggered upon irradiation as recently reported by Huang et al. [199]. In this work, authors fabricated a hybrid (organic-inorganic) shape memory hydrogel (containing more than 76% of water), which comprises of an interpenetrated double network of chemically crosslinked PAAm network and a physically crosslinked gelatin network containing GO. The presence of GO allowed the hydrogel to be triggered upon NIR irradiation, due to its fast and efficient photo-thermal transformation. The thermal-reversible gelatin network was responsible for fixing the temporary shape of the hydrogels, whereas the presence of GO in the chemically crosslinked network permitted the recovery of the permanent shape by irradiation with NIR laser. Authors performed a deep analysis of the effect of the composition in the shape memory effect as well as in the mechanical properties to obtain an optimized composition that could show a stable temporary shape with fast recovery to the permanent shape.

4.2. Self-Healing Hydrogels

Since the last decade, self-healable systems have emerged as a new important class of materials whose origin is the ability to heal or repair itself [200,201]. The interest and vitality of this area of research is highlighted by a nearly exponential increase in the number of publications on the subject over the past 20 years (Figure 10).

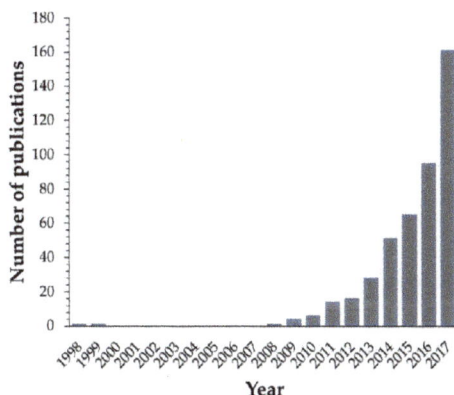

Figure 10. Evolution of publications over the past 20 years on the subject "self-healing hydrogels" (database used for the bibliographic analysis: Scopus® (Elsevier, Amsterdam, The Netherlands consulted on the 3th of May 2018).

This is a complete change in focus from materials that resist environmental damage to responsive, dynamic and smart systems. The development of self-healing polymeric materials has been largely based on mimicking biological healing [202], which is inspired from its innate ability to sense and repair damage in order to restore functionality. Applications of self-healing materials are expected almost entirely in all industries in the future since they effectively enhance the lifetime use of the product and has desirable economic and human safety attributes. The few applications being developed to date are mainly in the automotive, aerospace, and building industries. For instance, self-healing systems are also being adapted for use in paints and coatings, both through the incorporation of micro- and nano-encapsulation of healing agents.

Focusing on polymer hydrogels, the interesting and tunable properties of this functional materials give rise to excellent performances and a wide range of potential applications, as mentioned above. However, such outstanding properties can be lost when the 3D network structure of the gel is damaged (i.e., cracked), resulting mainly in a loss of mechanical properties and eventually in the failure of structure for the desired application. To avoid the consequences of this event, huge efforts have been made in the past years to develop polymer gels with the ability to self-repair or self-heal and thus restore their initial structure and functionalities. The approaches used for the development of self-healable hydrogels are mainly based on the constitutional dynamic chemistry (CDC) concept, key factors of which are dynamicity and reversibility, involving covalent (chemical self-healing gels) and non-covalent chemistry (physical self-healing gels) [203,204]. Self-healing gels based in this CDC concept can recover their original shape and restore their initial properties, in multiple cycles, due to the reversibility of the process. In both cases, the presence of functional groups able to perform such physical or chemical dynamic interactions, is necessary for self-healing purposes [102,200,205]. The advantage that gels present over polymers for self-healing is noteworthy, since the presence of the solvent enables the formation of a mobile structure around the damage area, which could eventually facilitate the healing.

Phadke et al. [206] reported for the first time a rapid self-healable hydrogel, in which the repair occurred within seconds of the damage or breakage into two hydrogel pieces (glue of two parts) and in an aqueous environment. The prepared hydrogel presented reversible healing, externally controllable by pH changes, with multiple cycles of healing-separation without affecting mechanical properties and repair kinetics. To induce such ability, they decorated the polymer hydrogel (chemical gel) prepared from acryloyl-6-aminocaproic acid precursors with dangling hydrocarbon side chains containing polar functional groups that would induce hydrogen bonds responsible for the polymer healing. At low pH, carboxyl groups are mostly protonated, which allows them to form hydrogen bonds with other carboxyl groups or amide groups, resulting in the repair of the hydrogel. However, at pH > pKa of the precursor monomer, carboxyl groups of the polymer are protonated, inhibiting the hydrogen bond formation and thus, the healing of the material. In addition, authors found that the side chain length was affecting the healing ability of the system. Short side chains do not favor hydrogen bond interactions among functional groups due to the limited availability of the carboxyl groups. In contrast, when the side chains exceed in length, they begin to front a larger steric hindrance to the interactions between the carboxyl and amide groups. This result highlighted the importance that the balance of hydrophilic-hydrophobic interaction has in hydrogels to present a self-healing property.

Along these lines, Tuncaboylu et al. [207–209] developed a self-healing hydrogel by the micellar free-radical polymerization of hydrophilic AAm in the presence of long alkyl chains of hydrophobic monomers. Interaction between hydrophobic blocks avoids dissolution in water and the dynamic nature of the junction points between the network chains (non-associated hydrophobic blocks) provides a self-healing efficiency to elongation at break of about 100% together with a high degree of toughness [207,208]. Following this trend, Gulyuz et al. [210] prepared self-healing physical PAAc hydrogels by micellar copolymerization of the AAc with an alkyl monomer (stearyl methacrylate, C18) in a surfactant micellar solution (sodium dodecyl sulfate). Authors found that by replacing AAm with AAc monomer, the obtained hydrogel results in stronger self-healing due to the cooperative hydrogen bonds formed between the carboxyl groups stabilizing the hydrophobic domains. In addition to the self-healing property, the obtained hydrogels exhibit high modulus (6–53 kPa), high fracture stress (41–173 kPa) and high elongation ratios at break (1800–5000%).

Zhang et al. [211] worked with the well-known PVA physical hydrogel (prepared using the freezing/thawing method) and found that it can autonomously self-heal at room temperature without the need for any stimulus or healing agent. The obtained self-healable hydrogel was mechanically strong, showing high fracture stress. The autonomous self-healing ability of the PVA hydrogel was caused by the formation of hydrogen bonding between the PVA chains. In fact, the authors indicated that the key factor to obtaining fast and efficient self-healing of PVA hydrogel was the balance between a sufficient amount of free hydroxyl groups of PVA on cut surfaces required to form interchain H-bonds and enough chain mobility ensuring chain diffusion across the interface.

As indicated in previous sections of this review, thermoresponsive polymer PNIPAAm played an important role in the field of polymer gels due to the numerous stimuli-responsive hydrogels and microgels developed involving this polymer. However, approaches to develop PNIPAAm and PNIPAAm-based hydrogels with self-healing abilities are still scarce due to the side groups of PNIPAAm that hinder secondary interactions. Along these lines, Gulyuz et al. [212] once again took advantage of the micellar radical copolymerization and presented for the first time the preparation of PNIPAAm hydrogels with self-healing property. In fact, the hydrogels have autonomous self-healing ability with an efficiency of up to 100%, as demonstrated by mechanical measurements. However, the obtained hydrogel loses its self-healing ability upon swelling due to micellar structure disintegration that increases hydrophobic interactions. Authors overcomed this problem by the incorporation of AAc (replacing some NIPAAm monomers) that binds with the surfactant so that it remains limited in the polymer. The addition of AAc also contributes to the improvement of mechanical properties.

The evolution on hydrogel research has brought about the development of materials that exhibit improved mechanical properties. However, most of the reported hydrogels that exhibit self-healing properties are in general mechanically weak. To overcome this problem, several authors hybridized self-healable hydrogels with GO, similarly to what was also observed in the previous section regarding shape memory hydrogels [213–215]. In this respect, Liu et al. [216] proposed the preparation of chemical self-healing PAAm hydrogel synthesis in the presence of GO. Authors suggested that the formation of hydrogen bonds between the PAAm polymer chains and the GO sheets were a determinant for self-healable hydrogels to present strong mechanical properties [217]. Such strong interactions between PAAm and GO, being difficult to be broken, gave stability to the hydrogel during swelling and deformation (application of a load). GO sheets have been used as fillers in polymer composites, improving the mechanical properties of polymer substrates. However, the use of GO sheets assembled into hydrogels can act not only as a crosslinking agent but also contribute to the self-healing ability through the dissipation of crack energy [213].

At the beginning of this section, we have indicated that the ability of self-repair is desired property in fieldls like aeronautics, coatings, paintings, etc. However, to confer the self-healing ability into smart hydrogels can also be useful in biomedical applications. For instance, when considering smart hydrogels to be used as implants in the body, it should be noted that these implants are under constant movement and thus at risk of mechanical damage. As a consequence, this fact could directly affect the functionality of the implant or even be the cause of an infection. In this regard, as stated by Li et al., developing a hydrogel able to self-heal could contribute to extending the application of the hydrogel and the lifetime of the material. These authors developed a mussel-inspired self-healing injectable hydrogel with anti-biofouling and anti-microbial properties [218–220]. For the development of such self-healing hydrogels, the author took into consideration the self-repair ability of mussel's byssal threads, which is mainly attributed to the reversible metal-catechol coordination (catechol group of an amino acid). In the work of Li et al., authors based the hydrogel preparation on the self-assembly of a ABA tri-block copolymer. As block A, authors used catechol-functionalized PNIPAAm as the thermosensitive block and as hydrophilic block B of the copolymer PEO. The temperature sensitivity provided by PNIPAAm-based block, A, confers control over the liquid or solid-like state of the copolymer. For instance, at temperatures below a critical value, the hydrogel presents a liquid-like behavior, but when heating above certain temperatures, the gelification and hence, hydrogel formation occurs. Rheological studies to determine the self-healing ability of the material as well as visual evidences demonstrated that the hydrogel can heal after being damaged within seconds. Authors attributed the self-healing mechanism to catechol-mediated hydrogen bonding.

The mussel-inspired strategy and chemistry was also followed by other authors. For instance, Hou et al. [221] developed supramolecular stimuli-responsive fast self-healing hydrogels, with the particularity of being highly stretchable depending on composition and deformation conditions. Compared to covalent bonds, supramolecular interactions are reversible and can respond to different external stimuli as pH, temperature, and external force, a property that imparts dynamic features and hence, self-healing capacity to this kind of hydrogels. In this case, the obtained hydrogel based in a monomer bearing a catechol group, can be cast and photo-crosslinked into a desire shape in just one step. For the same, authors prepared supramolecular hydrogels using photopolymerizable precoordinated catechol$-Fe^{3+}$ complexes as the multifunctional cross-linkers and AAm as the monomer. The self-healing ability of the obtained supramolecular hydrogel follows a dynamic process due to the re-association capacity of precoordinated metal$-$catechol interactions.

4.3. Stimuli-Responsive Self-Healing Microgels

In this field of self-healing polymer hydrogels, it is of importance to highlight the growing interest regarding the use of polymeric microgel and nanogel particles in the development of autonomous self-healing systems [222,223]. The ease of synthesizing microgels to display a wide range of stimuli-responsiveness and mechanical properties at the particle level and in assemblies provide microgel-based materials with real potential for self-healing applications.

Latnikova et al. [222] proposed a new approach towards "active" self-healing coatings with the use of microgels as reservoirs of "Green" corrosion inhibitors. In this case, microgels act as reinforcement while retaining the healing agent to be delivered when coating is damaged. Lyon et al. went a step ahead and used microgels as building blocks for self-healing materials, not as carriers or capsules. The authors were able to fabricate a self-healing hydrogel composed of microgel of P((NIPAAm-AAc) by means of a layer-by-layer deposition technique over a polycation. They built a 3D network (film) upon the self-assembly of microgel layers sustained by the different surfaces charges among microgels and polycation surfaces [224–227]. The peculiarity of this system is based on the dynamics that microgels have within the layers. When a damage or crack is generated in the hydrogel film, there is a lack of equilibrium in terms of electrical charges. When the system is solvated, this allows the reconfiguration of microgels toward the less energetic state, which entails the healing of the damage.

5. Conclusions

In the field of materials science and engineering, the design, research and development of new and improved smart structures and systems is currently a hot topic. In this context, functional stimuli-responsive gels are of great interest due to their versatility and possible application in fields as diverse as biomedicine, catalysis, or even biosensors.

The relevance of the subject dealt in this review is clearly reflected in the considerable number of relevant publications. In this review, we intended to cover the most recent and high-impact publications using functional stimuli-responsive gels. The review started from macro to micro, i.e., from hydrogels to microgels, ending with a new generation of functional gels. Stimuli-responsive functional hydrogels and microgels are used for a broad range of applications, and several publications can be found in literature. PNIPAAm is the most used thermoresponsive polymer. In many cases, PNIPAAm is combined with another polymer to add a new functionality to the hydrogel. In microgels, PNIPAAm is most commonly combined with an inorganic component, ranging from metallic nanoparticles to carbonaceous particles. In the last part of the review, the new generation of functional stimuli-responsive gels, able to self-repair or/and possess shape memory property, are described. Among others, PAAm, PVA and PEG are the commonly used polymers in this new generation of functional gels. The development of Shape Memory Polymers suitable for biological media in a living system is a still a challenge today. In this regard, shape memory hydrogels that are predominantly water stand out as alternative candidates for biomedical applications due to the use of water as a driving force for the desired shape memory effect. The advantages and attributes of functional gels (macro and micro) have been highlighted throughout the entire review, but the most promising is the ability to self-heal, expected in almost all industries in the near future. Self-healable hydrogels go through improvements in their mechanical properties without affecting their intrinsic properties, and importantly, their responsiveness to external stimuli.

In summary, this review demonstrates the prominent role of smart functional gels in materials science; its versatility and tunable properties allows usage in a wide range of applications.

Author Contributions: All authors contributed the same for the review.

Acknowledgments: This work is funded by National Funds through FCT—Portuguese Foundation for Science and Technology, Reference UID/CTM/50025/2013 and FEDER funds through the COMPETE 2020 Program under the project number POCI-01-0145-FEDER-007688. C. Echeverria acknowledges the Juan de la Cierva contract (IJCI-2015-26432) from MINECO—Spanish Ministry of Economy, Industry and Competitiveness.

Conflicts of Interest: The authors declare no conflict of interest.

References

1. Sangeetha, N.M.; Maitra, U. Supramolecular gels: Functions and uses. *Chem. Soc. Rev.* **2005**, *34*, 821–836. [CrossRef] [PubMed]

2. Ahn, S.K.; Kasi, R.M.; Kim, S.C.; Sharma, N.; Zhou, Y.X. Stimuli-responsive polymer gels. *Soft Matter* **2008**, *4*, 1151–1157. [CrossRef]

3. Hoffman, A.S.; Stayton, P.S.; Bulmus, V.; Chen, G.H.; Chen, J.P.; Cheung, C.; Chilkoti, A.; Ding, Z.L.; Dong, L.C.; Fong, R.; et al. Really smart bioconjugates of smart polymers and receptor proteins. *J. Biomed. Mater. Res.* **2000**, *52*, 577–586. [CrossRef]

4. Stuart, M.A.; Huck, W.T.; Genzer, J.; Muller, M.; Ober, C.; Stamm, M.; Sukhorukov, G.B.; Szleifer, I.; Tsukruk, V.V.; Urban, M.; et al. Emerging applications of stimuli-responsive polymer materials. *Nat. Mater.* **2010**, *9*, 101–113. [CrossRef] [PubMed]

5. Buwalda, S.J.; Boere, K.W.; Dijkstra, P.J.; Feijen, J.; Vermonden, T.; Hennink, W.E. Hydrogels in a historical perspective: From simple networks to smart materials. *J. Control Release* **2014**, *190*, 254–273. [CrossRef] [PubMed]

6. Bauri, K.; Nandi, M.; De, P. Amino acid-derived stimuli-responsive polymers and their applications. *Polym. Chem.* **2018**, *9*, 1257–1287. [CrossRef]

7. Koetting, M.C.; Peters, J.T.; Steichen, S.D.; Peppas, N.A. Stimulus-responsive hydrogels: Theory, modern advances, and applications. *Mater. Sci. Eng. R Rep.* **2015**, *93*, 1–49. [CrossRef] [PubMed]

8. Del Valle, L.; Díaz, A.; Puiggalí, J. Hydrogels for biomedical applications: Cellulose, chitosan, and protein/peptide derivatives. *Gels* **2017**, *3*, 27. [CrossRef]

9. De France, K.J.; Hoare, T.; Cranston, E.D. Review of hydrogels and aerogels containing nanocellulose. *Chem. Mater.* **2017**, *29*, 4609–4631. [CrossRef]

10. Kahn, J.S.; Hu, Y.; Willner, I. Stimuli-responsive DNA-based hydrogels: From basic principles to applications. *Acc. Chem. Res.* **2017**, *50*, 680–690. [CrossRef] [PubMed]

11. Haq, M.A.; Su, Y.; Wang, D. Mechanical properties of PNIPAM based hydrogels: A review. *Mater. Sci. Eng. C Mater. Biol. Appl.* **2017**, *70*, 842–855. [CrossRef] [PubMed]

12. Sosnik, A.; Seremeta, K. Polymeric hydrogels as technology platform for drug delivery applications. *Gels* **2017**, *3*, 25. [CrossRef]

13. Li, D.; van Nostrum, C.F.; Mastrobattista, E.; Vermonden, T.; Hennink, W.E. Nanogels for intracellular delivery of biotherapeutics. *J. Control Release* **2017**, *259*, 16–28. [CrossRef] [PubMed]

14. Knipe, J.M.; Peppas, N.A. Multi-responsive hydrogels for drug delivery and tissue engineering applications. *Regen. Biomater.* **2014**, *1*, 57–65. [CrossRef] [PubMed]

15. Lanzalaco, S.; Armelin, E. Poly(*N*-isopropylacrylamide) and copolymers: A review on recent progresses in biomedical applications. *Gels* **2017**, *3*, 36. [CrossRef]

16. Sharifzadeh, G.; Hosseinkhani, H. Biomolecule-responsive hydrogels in medicine. *Adv. Healthc. Mater.* **2017**, *6*. [CrossRef] [PubMed]

17. Peppas, N.A.; Van Blarcom, D.S. Hydrogel-based biosensors and sensing devices for drug delivery. *J. Control Release* **2016**, *240*, 142–150. [CrossRef] [PubMed]

18. Culver, H.R.; Clegg, J.R.; Peppas, N.A. Analyte-responsive hydrogels: Intelligent materials for biosensing and drug delivery. *Acc. Chem. Res.* **2017**, *50*, 170–178. [CrossRef] [PubMed]

19. Battista, E.; Causa, F.; Netti, P. Bioengineering microgels and hydrogel microparticles for sensing biomolecular targets. *Gels* **2017**, *3*, 20. [CrossRef]

20. Li, X.; Gao, Y.; Serpe, M. Stimuli-responsive assemblies for sensing applications. *Gels* **2016**, *2*, 8. [CrossRef]

21. Dhanya, S.; Bahadur, D.; Kundu, G.C.; Srivastava, R. Maleic acid incorporated poly-(*N*-isopropylacrylamide) polymer nanogels for dual-responsive delivery of doxorubicin hydrochloride. *Eur. Polym. J.* **2013**, *49*, 22–32. [CrossRef]

22. Patenaude, M.; Hoare, T. Injectable, degradable thermoresponsive poly(*N*-isopropylacrylamide) hydrogels. *ACS Macro Lett.* **2012**, *1*, 409–413. [CrossRef]

23. Tang, S.; Floy, M.; Bhandari, R.; Sunkara, M.; Morris, A.J.; Dziubla, T.D.; Hilt, J.Z. Synthesis and characterization of thermoresponsive hydrogels based on *N*-isopropylacrylamide crosslinked with 4,4′-dihydroxybiphenyl diacrylate. *ACS Omega* **2017**, *2*, 8723–8729. [CrossRef] [PubMed]

24. Kim, H.; Kim, K.; Lee, S.J. Nature-inspired thermo-responsive multifunctional membrane adaptively hybridized with PNIPAM and PPY. *NPG Asia Mater.* **2017**, *9*, e445. [CrossRef]

25. D'Eramo, L.; Chollet, B.; Leman, M.; Martwong, E.; Li, M.X.; Geisler, H.; Dupire, J.; Kerdraon, M.; Vergne, C.; Monti, F.; et al. Microfluidic actuators based on temperature-responsive hydrogels. *Microsyst. Nanoeng.* **2018**, *4*, 17069. [CrossRef]

26. Han, D.; Lu, Z.; Chester, S.A.; Lee, H. Micro 3D printing of a temperature-responsive hydrogel using projection micro-stereolithography. *Sci. Rep.* **2018**, *8*, 1963. [CrossRef] [PubMed]

27. Luckanagul, J.A.; Pitakchatwong, C.; Ratnatilaka Na Bhuket, P.; Muangnoi, C.; Rojsitthisak, P.; Chirachanchai, S.; Wang, Q.; Rojsitthisak, P. Chitosan-based polymer hybrids for thermo-responsive nanogel delivery of curcumin. *Carbohydr. Polym.* **2018**, *181*, 1119–1127. [CrossRef] [PubMed]

28. Fang, Y.; Tan, J.J.; Lim, S.; Soh, S.L. Rupturing cancer cells by the expansion of functionalized stimuli-responsive hydrogels. *NPG Asia Mater.* **2018**, *10*, e465. [CrossRef]

29. Kouwer, P.H.; Koepf, M.; Le Sage, V.A.; Jaspers, M.; van Buul, A.M.; Eksteen-Akeroyd, Z.H.; Woltinge, T.; Schwartz, E.; Kitto, H.J.; Hoogenboom, R.; et al. Responsive biomimetic networks from polyisocyanopeptide hydrogels. *Nature* **2013**, *493*, 651–655. [CrossRef] [PubMed]

30. Zimoch, J.; Padial, J.S.; Klar, A.S.; Vallmajo-Martin, Q.; Meuli, M.; Biedermann, T.; Wilson, C.J.; Rowan, A.; Reichmann, E. Polyisocyanopeptide hydrogels: A novel thermo-responsive hydrogel supporting pre-vascularization and the development of organotypic structures. *Acta Biomater.* **2018**, *70*, 129–139. [CrossRef] [PubMed]

31. Gil, E.S.; Hudson, S.M. Stimuli-reponsive polymers and their bioconjugates. *Prog. Polym. Sci.* **2004**, *29*, 1173–1222. [CrossRef]

32. Puranik, A.S.; Pao, L.P.; White, V.M.; Peppas, N.A. Synthesis and characterization of pH-responsive nanoscale hydrogels for oral delivery of hydrophobic therapeutics. *Eur. J. Pharm. Biopharm.* **2016**, *108*, 196–213. [CrossRef] [PubMed]

33. Puranik, A.S.; Pao, L.P.; White, V.M.; Peppas, N.A. In vitro evaluation of pH-responsive nanoscale hydrogels for the oral delivery of hydrophobic therapeutics. *Ind. Eng. Chem. Res.* **2016**, *55*, 10576–10590. [CrossRef]

34. Jin, C.; Song, W.J.; Liu, T.; Xin, J.N.; Hiscox, W.C.; Zhang, J.W.; Liu, G.F.; Kong, Z.W. Temperature and pH responsive hydrogels using methacrylated lignosulfonate cross-linker: Synthesis, characterization, and properties. *ACS Sustain. Chem. Eng.* **2018**, *6*, 1763–1771. [CrossRef]

35. Unger, K.; Salzmann, P.; Masciullo, C.; Cecchini, M.; Koller, G.; Coclite, A.M. Novel light-responsive biocompatible hydrogels produced by initiated chemical vapor deposition. *ACS Appl. Mater. Interfaces* **2017**, *9*, 17408–17416. [CrossRef] [PubMed]

36. Fomina, N.; Sankaranarayanan, J.; Almutairi, A. Photochemical mechanisms of light-triggered release from nanocarriers. *Adv. Drug Deliv. Rev.* **2012**, *64*, 1005–1020. [CrossRef] [PubMed]

37. Tomatsu, I.; Peng, K.; Kros, A. Photoresponsive hydrogels for biomedical applications. *Adv. Drug Deliv. Rev.* **2011**, *63*, 1257–1266. [CrossRef] [PubMed]

38. Mandl, G.A.; Rojas-Gutierrez, P.A.; Capobianco, J.A. A NIR-responsive azobenzene-based supramolecular hydrogel using upconverting nanoparticles. *Chem. Commun.* **2018**, *54*, 5847–5850. [CrossRef] [PubMed]

39. Weis, P.; Wu, S. Light-switchable azobenzene-containing macromolecules: From UV to near infrared. *Macromol. Rapid Commun.* **2018**, *39*, 1700220. [CrossRef] [PubMed]

40. Su, X.; Xiao, C.; Hu, C. Facile preparation and dual responsive behaviors of starch-based hydrogel containing azo and carboxylic groups. *Int. J. Biol. Macromol.* **2018**, *115*, 1189–1193. [CrossRef] [PubMed]

41. Rastogi, S.K.; Anderson, H.E.; Lamas, J.; Barret, S.; Cantu, T.; Zauscher, S.; Brittain, W.J.; Betancourt, T. Enhanced release of molecules upon ultraviolet (UV) light irradiation from photoresponsive hydrogels prepared from bifunctional azobenzene and four-arm poly(ethylene glycol). *ACS Appl. Mater. Interfaces* **2017**. [CrossRef] [PubMed]

42. Zhao, F.; Bonasera, A.; Nochel, U.; Behl, M.; Bleger, D. Reversible modulation of elasticity in fluoroazobenzene-containing hydrogels using green and blue light. *Macromol. Rapid Commun.* **2018**, *39*, 1700527. [CrossRef] [PubMed]

43. Harada, A.; Takashima, Y.; Yamaguchi, H. Cyclodextrin-based supramolecular polymers. *Chem. Soc. Rev.* **2009**, *38*, 875–882. [CrossRef] [PubMed]

44. Tamesue, S.; Takashima, Y.; Yamaguchi, H.; Shinkai, S.; Harada, A. Photoswitchable supramolecular hydrogels formed by cyclodextrins and azobenzene polymers. *Angew. Chem.* **2010**, *122*, 7623–7626. [CrossRef]

45. Wang, J.; Li, Q.; Yi, S.; Chen, X. Visible-light/temperature dual-responsive hydrogel constructed by α-cyclodextrin and an azobenzene linked surfactant. *Soft Matter* **2017**, *13*, 6490–6498. [CrossRef] [PubMed]

46. Suzuki, A.; Tanaka, T. Phase-transition in polymer gels induced by visible-light. *Nature* **1990**, *346*, 345–347. [CrossRef]

47. Suzuki, A.; Ishii, T.; Maruyama, Y. Optical switching in polymer gels. *J. Appl. Phys.* **1996**, *80*, 131–136. [CrossRef]

48. Xu, Y.; Ghag, O.; Reimann, M.; Sitterle, P.; Chatterjee, P.; Nofen, E.; Yu, H.; Jiang, H.; Dai, L.L. Development of visible-light responsive and mechanically enhanced "smart" UCST interpenetrating network hydrogels. *Soft Matter* **2017**, *14*, 151–160. [CrossRef] [PubMed]

49. Ter Schiphorst, J.; Saez, J.; Diamond, D.; Benito-Lopez, F.; Schenning, A. Light-responsive polymers for microfluidic applications. *Lab Chip* **2018**, *18*, 699–709. [CrossRef] [PubMed]

50. Greene, A.F.; Danielson, M.K.; Delawder, A.O.; Liles, K.P.; Li, X.S.; Natraj, A.; Wellen, A.; Barnes, J.C. Redox-responsive artificial molecular muscles: Reversible radical-based self-assembly for actuating hydrogels. *Chem. Mater.* **2017**, *29*, 9498–9508. [CrossRef]

51. Stejskal, J. Conducting polymer hydrogels. *Chem. Pap.* **2017**, *71*, 269–291. [CrossRef]

52. Liu, Y.; Tsao, C.Y.; Kim, E.; Tschirhart, T.; Terrell, J.L.; Bentley, W.E.; Payne, G.F. Using a redox modality to connect synthetic biology to electronics: Hydrogel-based chemo-electro signal transduction for molecular communication. *Adv. Healthc. Mater.* **2017**, *6*, 1600908. [CrossRef] [PubMed]

53. Fu, L.; Wang, A.W.; Lyu, F.C.; Lai, G.S.; Yu, J.H.; Lin, C.T.; Liu, Z.; Yu, A.M.; Su, W.T. A solid-state electrochemical sensing platform based on a supramolecular hydrogel. *Sens. Actuators B Chem.* **2018**, *262*, 326–333. [CrossRef]

54. Wojciechowski, J.P.; Martin, A.D.; Thordarson, P. Kinetically controlled lifetimes in redox-responsive transient supramolecular hydrogels. *J. Am. Chem. Soc.* **2018**, *140*, 2869–2874. [CrossRef] [PubMed]

55. Li, H.; Voci, S.; Ravaine, V.; Sojic, N. Tuning electrochemiluminescence in multistimuli responsive hydrogel films. *J. Phys. Chem. Lett.* **2018**, *9*, 340–345. [CrossRef] [PubMed]

56. Zhang, Y.; Li, Y.F.; Gao, Z.J.; Gao, G.H.; Duan, L.J. Mechanically redox-tunable hydrogels reinforced by hydrophobic association and metal ion coordination. *Mater. Chem. Phys.* **2018**, *207*, 175–180. [CrossRef]

57. Tavakoli, J.; Tang, Y.H. Hydrogel based sensors for biomedical applications: An updated review. *Polymers* **2017**, *9*, 364. [CrossRef]

58. Pujol-Vila, F.; Dietvorst, J.; Gall-Mas, L.; Diaz-Gonzalez, M.; Vigues, N.; Mas, J.; Munoz-Berbel, X. Bioelectrochromic hydrogel for fast antibiotic-susceptibility testing. *J. Colloid Interface Sci.* **2018**, *511*, 251–258. [CrossRef] [PubMed]

59. Qin, M.; Sun, M.; Bai, R.; Mao, Y.; Qian, X.; Sikka, D.; Zhao, Y.; Qi, H.J.; Suo, Z.; He, X. Bioinspired hydrogel interferometer for adaptive coloration and chemical sensing. *Adv. Mater.* **2018**, *30*, e1800468. [CrossRef] [PubMed]

60. Park, J.; Pramanick, S.; Park, D.; Yeo, J.; Lee, J.; Lee, H.; Kim, W.J. Therapeutic-gas-responsive hydrogel. *Adv. Mater.* **2017**, *29*. [CrossRef] [PubMed]

61. Almeida, A.P.C.; Canejo, J.P.; Fernandes, S.N.; Echeverria, C.; Almeida, P.L.; Godinho, M.H. Cellulose-based biomimetics and their applications. *Adv. Mater.* **2018**, *30*, e1703655. [CrossRef] [PubMed]

62. Oechsle, A.L.; Lewis, L.; Hamad, W.Y.; Hatzikiriakos, S.G.; MacLachlan, M.J. CO_2-switchable cellulose nanocrystal hydrogels. *Chem. Mater.* **2018**, *30*, 376–385. [CrossRef]

63. Noh, K.-G.; Park, S.-Y. Biosensor array of interpenetrating polymer network with photonic film templated from reactive cholesteric liquid crystal and enzyme-immobilized hydrogel polymer. *Adv. Funct. Mater.* **2018**, *28*, 1707562. [CrossRef]

64. Tan, Z.; Parisi, C.; Di Silvio, L.; Dini, D.; Forte, A.E. Cryogenic 3D printing of super soft hydrogels. *Sci. Rep.* **2017**, *7*, 16293. [CrossRef] [PubMed]

65. He, Y.; Yang, F.; Zhao, H.; Gao, Q.; Xia, B.; Fu, J. Research on the printability of hydrogels in 3D bioprinting. *Sci. Rep.* **2016**, *6*, 29977. [CrossRef] [PubMed]

66. Karis, D.G.; Ono, R.J.; Zhang, M.S.; Vora, A.; Storti, D.; Ganter, M.A.; Nelson, A. Cross-linkable multi-stimuli responsive hydrogel inks for direct-write 3D printing. *Polym. Chem.* **2017**, *8*, 4199–4206. [CrossRef]

67. Zhou, Y.; Layani, M.; Wang, S.C.; Hu, P.; Ke, Y.J.; Magdassi, S.; Long, Y. Fully printed flexible smart hybrid hydrogels. *Adv. Funct. Mater.* **2018**, *28*, 1705365. [CrossRef]
68. Murray, M.J.; Snowden, M.J. The preparation, characterization and applications of colloidal microgels. *Adv. Colloid Interface Sci.* **1995**, *54*, 73–91. [CrossRef]
69. Pelton, R.H.; Chibante, P. Preparation of aqueous latices with *N*-isopropylacrylamide. *Colloids Surf.* **1986**, *20*, 247–256. [CrossRef]
70. Echeverria, C.; López, D.; Mijangos, C. UCST responsive microgels of poly(acrylamide−acrylic acid) copolymers: Structure and viscoelastic properties. *Macromolecules* **2009**, *42*, 9118–9123. [CrossRef]
71. Marques, S.C.S.; Soares, P.I.P.; Echeverria, C.; Godinho, M.H.; Borges, J.P. Confinement of thermoresponsive microgels into fibres via colloidal electrospinning: Experimental and statistical analysis. *RSC Adv.* **2016**, *6*, 76370–76380. [CrossRef]
72. Echeverria, C.; Soares, P.; Robalo, A.; Pereira, L.; Novo, C.M.M.; Ferreira, I.; Borges, J.P. One-pot synthesis of dual-stimuli responsive hybrid PNIPAAm-chitosan microgels. *Mater. Des.* **2015**, *86*, 745–751. [CrossRef]
73. Faria, J.; Echeverria, C.; Borges, J.P.; Godinho, M.H.; Soares, P.I.P. Towards the development of multifunctional hybrid fibrillary gels: Production and optimization by colloidal electrospinning. *RSC Adv.* **2017**, *7*, 48972–48979. [CrossRef]
74. Saunders, B.R. On the structure of poly(*N*-isopropylacrylamide) microgel particles. *Langmuir* **2004**, *20*, 3925–3932. [CrossRef] [PubMed]
75. Saunders, B.R.; Laajam, N.; Daly, E.; Teow, S.; Hu, X.; Stepto, R. Microgels: From responsive polymer colloids to biomaterials. *Adv. Colloid Interface Sci.* **2009**, *147–148*, 251–262. [CrossRef] [PubMed]
76. Karg, M.; Hellweg, T. New "smart" poly(Nipam) microgels and nanoparticle microgel hybrids: Properties and advances in characterisation. *Curr. Opin. Colloid Interface Sci.* **2009**, *14*, 438–450. [CrossRef]
77. Karg, M.; Wellert, S.; Prevost, S.; Schweins, R.; Dewhurst, C.; Liz-Marzan, L.M.; Hellweg, T. Well defined hybrid PNIPAM core-shell microgels: Size variation of the silica nanoparticle core. *Colloid Polym. Sci.* **2011**, *289*, 699–709. [CrossRef]
78. Plamper, F.A.; Richtering, W. Functional microgels and microgel systems. *Acc. Chem. Res.* **2017**, *50*, 131–140. [CrossRef] [PubMed]
79. Pich, A.; Richtering, W. *Chemical Design of Responsive Microgels*; Springer: Berlin, Germany, 2011; pp. 4–23.
80. Kratz, K.; Hellweg, T.; Eimer, W. Influence of charge density on the swelling of colloidal poly(*N*-isopropylacrylamide-*co*-acrylic acid) microgels. *Colloid Surf. A* **2000**, *170*, 137–149. [CrossRef]
81. Debord, J.D.; Lyon, L.A. Synthesis and characterization of pH-responsive copolymer microgels with tunable volume phase transition temperatures. *Langmuir* **2003**, *19*, 7662–7664. [CrossRef]
82. Culver, H.R.; Sharma, I.; Wechsler, M.E.; Anslyn, E.V.; Peppas, N.A. Charged poly(*N*-isopropylacrylamide) nanogels for use as differential protein receptors in a turbidimetric sensor array. *Analyst* **2017**, *142*, 3183–3193. [CrossRef] [PubMed]
83. Zhou, S.Q.; Chu, B. Synthesis and volume phase transition of poly(methacrylic acid-*co*-N-isopropylacrylamide) microgel particles in water. *J. Phys. Chem. B* **1998**, *102*, 1364–1371. [CrossRef]
84. Thorne, J.B.; Vine, G.J.; Snowden, M.J. Microgel applications and commercial considerations. *Colloid Polym. Sci.* **2011**, *289*, 625–646. [CrossRef]
85. Naseem, K.; Begum, R.; Wu, W.; Irfan, A.; Farooqi, Z.H. Advancement in multi-functional poly(styrene)-poly(*N*-isopropylacrylamide) based core–shell microgels and their applications. *Polym. Rev.* **2018**, *58*, 288–325. [CrossRef]
86. Ballauff, M.; Lu, Y. "Smart" nanoparticles: Preparation, characterization and applications. *Polymer* **2007**, *48*, 1815–1823. [CrossRef]
87. Barth, M.; Wiese, M.; Ogieglo, W.; Go, D.; Kuehne, A.J.C.; Wessling, M. Monolayer microgel composite membranes with tunable permeability. *J. Membr. Sci.* **2018**, *555*, 473–482. [CrossRef]
88. Brugnoni, M.; Scotti, A.; Rudov, A.A.; Gelissen, A.P.H.; Caumanns, T.; Radulescu, A.; Eckert, T.; Pich, A.; Potemkin, I.I.; Richtering, W. Swelling of a responsive network within different constraints in multi-thermosensitive microgels. *Macromolecules* **2018**, *51*, 2662–2671. [CrossRef]
89. Zha, L.S.; Zhang, Y.; Yang, W.L.; Fu, S.K. Monodisperse temperature-sensitive microcontainers. *Adv. Mater.* **2002**, *14*, 1090. [CrossRef]
90. Gandhi, A.; Paul, A.; Sen, S.O.; Sen, K.K. Studies on thermoresponsive polymers: Phase behaviour, drug delivery and biomedical applications. *Asian J. Pharm. Sci.* **2015**, *10*, 99–107. [CrossRef]

91. Pelton, R. Temperature-sensitive aqueous microgels. *Adv. Colloid Interface Sci.* **2000**, *85*, 1–33. [CrossRef]
92. Gracia, L.; Snowden, M.J. Preparation, properties and applications of colloidal microgels. In *Handbook of Industrial Water Soluble Polymers*; Williams, P.A., Ed.; John Wiley & Sons: Hoboken, NJ, USA, 2007.
93. Suzuki, D.; Horigome, K.; Kureha, T.; Matsui, S.; Watanabe, T. Polymeric hydrogel microspheres: Design, synthesis, characterization, assembly and applications. *Polym. J.* **2017**, *49*, 695–702. [CrossRef]
94. Agrawal, G.; Agrawal, R. Stimuli-responsive microgels and microgel-based systems: Advances in the exploitation of microgel colloidal properties and their interfacial activity. *Polymers* **2018**, *10*, 418. [CrossRef]
95. Soares, P.I.P.; Echeverria, C.; Baptista, A.C.; João, C.F.C.; Fernandes, S.N.; Almeida, A.P.C.; Silva, J.C.; Godinho, M.H.; Borges, J.P. 4-hybrid polysaccharide-based systems for biomedical applications. In *Hybrid Polymer Composite Materials*; Woodhead Publishing: Sawston, UK, 2017; pp. 107–149.
96. Karg, M. Multifunctional inorganic/organic hybrid microgels. *Colloid Polym. Sci.* **2012**, *290*, 673–688. [CrossRef]
97. Edwards, P.P.; Thomas, J.M. Gold in a metallic divided state—From faraday to present-day nanoscience. *Angew. Chem. Int. Ed.* **2007**, *46*, 5480–5486. [CrossRef] [PubMed]
98. Haes, A.J.; Stuart, D.A.; Nie, S.M.; Van Duyne, R.P. Using solution-phase nanoparticles, surface-confined nanoparticle arrays and single nanoparticles as biological sensing platforms. *J. Fluoresc.* **2004**, *14*, 355–367. [CrossRef] [PubMed]
99. Murray, R.W. Nanoelectrochemistry: Metal nanoparticles, nanoelectrodes, and nanopores. *Chem. Rev.* **2008**, *108*, 2688–2720. [CrossRef] [PubMed]
100. Vivero-Escoto, J.L.; Huang, Y.T. Inorganic-organic hybrid nanomaterials for therapeutic and diagnostic imaging applications. *Int. J. Mol. Sci.* **2011**, *12*, 3888–3927. [CrossRef] [PubMed]
101. Shi, S.; Zhang, L.; Wang, T.; Wang, Q.M.; Gao, Y.; Wang, N. Poly(N-isopropylacrylamide)-Au hybrid microgels: Synthesis, characterization, thermally tunable optical and catalytic properties. *Soft Matter* **2013**, *9*, 10966–10970. [CrossRef]
102. Xiao, C.F.; Wu, Q.S.; Chang, A.P.; Peng, Y.H.; Xu, W.T.; Wu, W.T. Responsive au@polymer hybrid microgels for the simultaneous modulation and monitoring of au-catalyzed chemical reaction. *J. Mater. Chem. A* **2014**, *2*, 9514–9523. [CrossRef]
103. Rehman, S.U.; Khan, A.R.; Shah, A.; Badshah, A.; Siddiq, M. Preparation and characterization of poly(N-isoproylacrylamide-co-dimethylaminoethyl methacrylate) microgels and their composites of gold nanoparticles. *Colloid Surf. A* **2017**, *520*, 826–833. [CrossRef]
104. Chen, L.Y.; Ou, C.M.; Chen, W.Y.; Huang, C.C.; Chang, H.T. Synthesis of photoluminescent Au ND-PNIPAM hybrid microgel for the detection of Hg^{2+}. *ACS Appl. Mater. Interfaces* **2013**, *5*, 4383–4388. [CrossRef] [PubMed]
105. Tang, Y.C.; Ding, Y.; Wu, T.; Lv, L.Y.; Yan, Z.C. A turn-on fluorescent probe for Hg^{2+} detection by using gold nanoparticle-based hybrid microgels. *Sens. Actuators B Chem.* **2016**, *228*, 767–773. [CrossRef]
106. Mackiewicz, M.; Karbarz, M.; Romanski, J.; Stojek, Z. An environmentally sensitive three-component hybrid microgel. *RSC Adv.* **2016**, *6*, 83493–83500. [CrossRef]
107. Khan, A.; Alhoshan, M. Preparation and characterization of pH-responsive and thermoresponsive hybrid microgel particles with gold nanorods. *J. Polym. Sci. A* **2013**, *51*, 39–46. [CrossRef]
108. Mourran, A.; Zhang, H.; Vinokur, R.; Moller, M. Soft microrobots employing nonequilibrium actuation via plasmonic heating. *Adv. Mater.* **2017**, *29*, 1604825. [CrossRef] [PubMed]
109. Satapathy, S.S.; Bhol, P.; Chakkarambath, A.; Mohanta, J.; Samantaray, K.; Bhat, S.K.; Panda, S.K.; Mohanty, P.S.; Si, S. Thermo-responsive PNIPAM-metal hybrids: An efficient nanocatalyst for the reduction of 4-nitrophenol. *Appl. Surf. Sci.* **2017**, *420*, 753–763. [CrossRef]
110. Fernandez-Lopez, C.; Polavarapu, L.; Solis, D.M.; Taboada, J.M.; Obelleiro, F.; Contreras-Caceres, R.; Pastoriza-Santos, I.; Perez-Juste, J. Gold nanorod-PNIPAM hybrids with reversible plasmon coupling: Synthesis, modeling, and sers properties. *ACS Appl. Mater. Interfaces* **2015**, *7*, 12530–12538. [CrossRef] [PubMed]
111. Contreras-Caceres, R.; Sanchez-Iglesias, A.; Karg, M.; Pastoriza-Santos, I.; Perez-Juste, J.; Pacifico, J.; Hellweg, T.; Fernandez-Barbero, A.; Liz-Marzan, L.M. Encapsulation and growth of gold nanoparticles in thermoresponsive microgels. *Adv. Mater.* **2008**, *20*, 1666–1670. [CrossRef]

112. Fernandez-Lopez, C.; Perez-Balado, C.; Perez-Juste, J.; Pastoriza-Santos, I.; de Lera, A.R.; Liz-Marzan, L.M. A general LbL strategy for the growth of PNIPAM microgels on Au nanoparticles with arbitrary shapes. *Soft Matter* **2012**, *8*, 4165–4170. [CrossRef]

113. Perez-Juste, J.; Pastoriza-Santos, I.; Liz-Marzan, L.M. Multifunctionality in metal@microgel colloidal nanocomposites. *J. Mater. Chem. A* **2013**, *1*, 20–26. [CrossRef]

114. Carregal-Romero, S.; Buurma, N.J.; Perez-Juste, J.; Liz-Marzan, L.M.; Herves, P. Catalysis by Au@pNIPAM nanocomposites: Effect of the cross-linking density. *Chem. Mater.* **2010**, *22*, 3051–3059. [CrossRef]

115. Shi, S.; Wang, Q.; Wang, T.; Ren, S.; Gao, Y.; Wang, N. Thermo-, pH-, and light-responsive poly(N-isopropylacrylamide-co-methacrylic acid)–Au hybrid microgels prepared by the in situ reduction method based on au-thiol chemistry. *J. Phys. Chem. B* **2014**, *118*, 7177–7186. [CrossRef] [PubMed]

116. Agrawal, G.; Schurings, M.P.; van Rijn, P.; Pich, A. Formation of catalytically active gold-polymer microgel hybrids via a controlled in situ reductive process. *J. Mater. Chem. A* **2013**, *1*, 13244–13251. [CrossRef]

117. Jia, H.; Schmitz, D.; Ott, A.; Pich, A.; Lu, Y. Cyclodextrin modified microgels as "nanoreactor" for the generation of Au nanoparticles with enhanced catalytic activity. *J. Mater. Chem. A* **2015**, *3*, 6187–6195. [CrossRef]

118. Hou, L.; Wu, P.Y. The effect of added gold nanoparticles on the volume phase transition behavior for PVCL-based microgels. *RSC Adv.* **2014**, *4*, 39231–39241. [CrossRef]

119. Shah, L.A.; Haleem, A.; Sayed, M.; Siddiq, M. Synthesis of sensitive hybrid polymer microgels for catalytic reduction of organic pollutants. *J. Environ. Chem. Eng.* **2016**, *4*, 3492–3497. [CrossRef]

120. Khan, S.R.; Farooqi, Z.H.; Waheed-uz-Zaman; Ali, A.; Begum, R.; Kanwal, F.; Siddiq, M. Kinetics and mechanism of reduction of nitrobenzene catalyzed by silver-poly(N-isopropylacryl amide-co-allylacetic acid) hybrid microgels. *Mater. Chem. Phys.* **2016**, *171*, 318–327. [CrossRef]

121. Liu, X.Y.; Wang, X.Q.; Zha, L.S.; Lin, D.L.; Yang, J.M.; Zhou, J.F.; Zhang, L. Temperature- and pH-tunable plasmonic properties and sers efficiency of the silver nanoparticles within the dual stimuli-responsive microgels. *J. Mater. Chem. C* **2014**, *2*, 7326–7335. [CrossRef]

122. Liu, Y.Y.; Liu, X.Y.; Yang, J.M.; Lin, D.L.; Chen, X.; Zha, L.S. Investigation of Ag nanoparticles loading temperature responsive hybrid microgels and their temperature controlled catalytic activity. *Colloid Surf. A* **2012**, *393*, 105–110. [CrossRef]

123. Tang, Y.C.; Wu, T.; Hu, B.T.; Yang, Q.; Liu, L.; Yu, B.; Ding, Y.; Ye, S.Y. Synthesis of thermo- and pH-responsive Ag nanoparticle-embedded hybrid microgels and their catalytic activity in methylene blue reduction. *Mater. Chem. Phys.* **2015**, *149*, 460–466. [CrossRef]

124. Khan, A.; El-Toni, A.M.; Alrokayan, S.; Alsalhi, M.; Alhoshan, M.; Aldwayyan, A.S. Microwave-assisted synthesis of silver nanoparticles using poly-N-isopropylacrylamide/acrylic acid microgel particles. *Colloid Surf. A* **2011**, *377*, 356–360. [CrossRef]

125. Wu, W.; Shen, J.; Li, Y.; Zhu, H.; Banerjee, P.; Zhou, S. Specific glucose-to-SPR signal transduction at physiological pH by molecularly imprinted responsive hybrid microgels. *Biomaterials* **2012**, *33*, 7115–7125. [CrossRef] [PubMed]

126. Ye, T.; Jiang, X.M.; Xu, W.T.; Zhou, M.M.; Hu, Y.M.; Wu, W.T. Tailoring the glucose-responsive volume phase transition behaviour of Ag@poly(phenylboronic acid) hybrid microgels: From monotonous swelling to monotonous shrinking upon adding glucose at physiological pH. *Polym. Chem.* **2014**, *5*, 2352–2362. [CrossRef]

127. Han, D.M.; Zhang, Q.M.; Serpe, M.J. Poly(N-isopropylacrylamide)-co-(acrylic acid) microgel/Ag nanoparticle hybrids for the colorimetric sensing of H₂O₂. *Nanoscale* **2015**, *7*, 2784–2789. [CrossRef] [PubMed]

128. Zhang, J.G.; Xu, S.Q.; Kumacheva, E. Photogeneration of fluorescent silver nanoclusters in polymer microgels. *Adv. Mater.* **2005**, *17*, 2336–2340. [CrossRef]

129. Naeem, H.; Farooqi, Z.H.; Shah, L.A.; Siddiq, M. Synthesis and characterization of p(NIPAM-AA-AAm) microgels for tuning of optical properties of silver nanoparticles. *J. Polym. Res.* **2012**, *19*, 1–10. [CrossRef]

130. Tang, F.; Ma, N.; Tong, L.; He, F.; Li, L. Control of metal-enhanced fluorescence with pH- and thermoresponsive hybrid microgels. *Langmuir* **2012**, *28*, 883–888. [CrossRef] [PubMed]

131. Liu, X.Y.; Zhang, C.; Yang, J.M.; Lin, D.L.; Zhang, L.; Chen, X.; Zha, L.S. Silver nanoparticles loading pH responsive hybrid microgels: pH tunable plasmonic coupling demonstrated by surface enhanced raman scattering. *RSC Adv.* **2013**, *3*, 3384–3390. [CrossRef]

132. Soares, P.I.P.; Ferreira, I.M.M.; Igreja, R.A.G.B.N.; Novo, C.M.M.; Borges, J.P.M.R. Application of hyperthermia for cancer treatment: Recent patents review. In *Recent Patents on Anti-Cancer Drug Discovery*; Bentham Science Publishers: Emirate of Sharjah, UAE, 2012; Volume 7, pp. 64–73.

133. Soares, P.I.; Alves, A.M.; Pereira, L.C.; Coutinho, J.T.; Ferreira, I.M.; Novo, C.M.; Borges, J.P. Effects of surfactants on the magnetic properties of iron oxide colloids. *J. Colloid Interface Sci.* **2014**, *419*, 46–51. [CrossRef] [PubMed]

134. Soares, P.I.P.; Laia, C.A.T.; Carvalho, A.; Pereira, L.C.J.; Coutinho, J.T.; Ferreira, I.M.M.; Novo, C.M.M.; Borges, J.P. Iron oxide nanoparticles stabilized with a bilayer of oleic acid for magnetic hyperthermia and mri applications. *Appl. Surf. Sci.* **2016**, *383*, 240–247. [CrossRef]

135. Soares, P.I.; Lochte, F.; Echeverria, C.; Pereira, L.C.; Coutinho, J.T.; Ferreira, I.M.; Novo, C.M.; Borges, J.P. Thermal and magnetic properties of iron oxide colloids: Influence of surfactants. *Nanotechnology* **2015**, *26*, 425704. [CrossRef] [PubMed]

136. Soares, P.I.; Machado, D.; Laia, C.; Pereira, L.C.; Coutinho, J.T.; Ferreira, I.M.; Novo, C.M.; Borges, J.P. Thermal and magnetic properties of chitosan-iron oxide nanoparticles. *Carbohydr. Polym.* **2016**, *149*, 382–390. [CrossRef] [PubMed]

137. Zamora-Mora, V.; Soares, P.; Echeverria, C.; Hernández, R.; Mijangos, C. Composite chitosan/agarose ferrogels for potential applications in magnetic hyperthermia. *Gels* **2015**, *1*, 69–80. [CrossRef]

138. Chen, T.; Cao, Z.; Guo, X.; Nie, J.; Xu, J.; Fan, Z.; Du, B. Preparation and characterization of thermosensitive organic–inorganic hybrid microgels with functional Fe_3O_4 nanoparticles as crosslinker. *Polymer* **2011**, *52*, 172–179. [CrossRef]

139. Laurenti, M.; Guardia, P.; Contreras-Caceres, R.; Perez-Juste, J.; Fernandez-Barbero, A.; Lopez-Cabarcos, E.; Rubio-Retama, J. Synthesis of thermosensitive microgels with a tunable magnetic core. *Langmuir* **2011**, *27*, 10484–10491. [CrossRef] [PubMed]

140. Regmi, R.; Bhattarai, S.R.; Sudakar, C.; Wani, A.S.; Cunningham, R.; Vaishnava, P.P.; Naik, R.; Oupicky, D.; Lawes, G. Hyperthermia controlled rapid drug release from thermosensitive magnetic microgels. *J. Mater. Chem.* **2010**, *20*, 6158–6163. [CrossRef]

141. Echeverria, C.; Mijangos, C. UCST-like hybrid PAAm-AA/Fe_3O_4 microgels. Effect of Fe_3O_4 nanoparticles on morphology, thermosensitivity and elasticity. *Langmuir* **2011**, *27*, 8027–8035. [CrossRef] [PubMed]

142. Boularas, M.; Gombart, E.; Tranchant, J.F.; Billon, L.; Save, M. Design of smart oligo(ethylene glycol)-based biocompatible hybrid microgels loaded with magnetic nanoparticles. *Macromol. Rapid Commun.* **2015**, *36*, 79–83. [CrossRef] [PubMed]

143. Boularas, M.; Deniau-Lejeune, E.; Alard, V.; Tranchant, J.F.; Billon, L.; Save, M. Dual stimuli-responsive oligo(ethylene glycol)-based microgels: Insight into the role of internal structure in volume phase transitions and loading of magnetic nanoparticles to design stable thermoresponsive hybrid microgels. *Polym. Chem.* **2016**, *7*, 350–363. [CrossRef]

144. Medeiros, S.F.; Filizzola, J.O.C.; Oliveira, P.F.M.; Silva, T.M.; Lara, B.R.; Lopes, M.V.; Rossi-Bergmann, B.; Elaissari, A.; Santos, A.M. Fabrication of biocompatible and stimuli-responsive hybrid microgels with magnetic properties via aqueous precipitation polymerization. *Mater. Lett.* **2016**, *175*, 296–299. [CrossRef]

145. Wang, Y.; Dong, A.J.; Yuan, Z.C.; Chen, D.J. Fabrication and characterization of temperature-, pH- and magnetic-field-sensitive organic/inorganic hybrid poly (ethylene glycol)-based hydrogels. *Colloid Surf. A* **2012**, *415*, 68–76. [CrossRef]

146. Yuan, Z.C.; Wang, Y.; Chen, D.J. Preparation and characterization of thermo-, pH-, and magnetic-field-responsive organic/inorganic hybrid microgels based on poly(ethylene glycol). *J. Mater. Sci.* **2014**, *49*, 3287–3296. [CrossRef]

147. Chengjun, Z.; Qinglin, W. Recent development in applications of cellulose nanocrystals for advanced polymer-based nanocomposites by novel fabrication strategies. In *Nanocrystals-Synthesis, Characterization and Applications*; Neralla, S., Ed.; Intech: Rijeka, Croatia, 2012; pp. 103–118.

148. Sengel, S.B.; Sahiner, N. Poly(vinyl phosphonic acid) nanogels with tailored properties and their use for biomedical and environmental applications. *Eur. Polym. J.* **2016**, *75*, 264–275. [CrossRef]

149. Liu, B.; Zhang, W.; Yang, F.K.; Feng, H.L.; Yang, X.L. Facile method for synthesis of Fe_3O_4@polymer microspheres and their application as magnetic support for loading metal nanoparticles. *J. Phys. Chem. C* **2011**, *115*, 15875–15884. [CrossRef]

150. Nabid, M.R.; Bide, Y.; Aghaghafari, E.; Rezaei, S.J.T. PdNPs@P2VP-Fe$_3$O$_4$ organic–inorganic hybrid microgels as a nanoreactor for selective aerobic oxidation of alcohols. *Catal. Lett.* **2013**, *144*, 355–363. [CrossRef]

151. Spizzirri, U.G.; Hampel, S.; Cirillo, G.; Nicoletta, F.P.; Hassan, A.; Vittorio, O.; Picci, N.; Iemma, F. Spherical gelatin/CNTs hybrid microgels as electro-responsive drug delivery systems. *Int. J. Pharm.* **2013**, *448*, 115–122. [CrossRef] [PubMed]

152. Shin, S.R.; Bae, H.; Cha, J.M.; Mun, J.Y.; Chen, Y.C.; Tekin, H.; Shin, H.; Farshchi, S.; Dokmeci, M.R.; Tang, S.; et al. Carbon nanotube reinforced hybrid microgels as scaffold materials for cell encapsulation. *ACS Nano* **2012**, *6*, 362–372. [CrossRef] [PubMed]

153. Cui, Z.; Zhou, M.; Greensmith, P.J.; Wang, W.; Hoyland, J.A.; Kinloch, I.A.; Freemont, T.; Saunders, B.R. A study of conductive hydrogel composites of pH-responsive microgels and carbon nanotubes. *Soft Matter* **2016**, *12*, 4142–4153. [CrossRef] [PubMed]

154. Wang, H.; Yi, J.; Velado, D.; Yu, Y.; Zhou, S. Immobilization of carbon dots in molecularly imprinted microgels for optical sensing of glucose at physiological ph. *ACS Appl. Mater. Interfaces* **2015**, *7*, 15735–15745. [CrossRef] [PubMed]

155. Cui, Z.; Milani, A.H.; Greensmith, P.J.; Yan, J.; Adlam, D.J.; Hoyland, J.A.; Kinloch, I.A.; Freemont, A.J.; Saunders, B.R. A study of physical and covalent hydrogels containing pH-responsive microgel particles and graphene oxide. *Langmuir* **2014**, *30*, 13384–13393. [CrossRef] [PubMed]

156. Wang, C.; Mallela, J.; Garapati, U.S.; Ravi, S.; Chinnasamy, V.; Girard, Y.; Howell, M.; Mohapatra, S. A chitosan-modified graphene nanogel for noninvasive controlled drug release. *Nanomedicine* **2013**, *9*, 903–911. [CrossRef] [PubMed]

157. Lu, N.Y.; Liu, J.J.; Li, J.L.; Zhang, Z.X.; Weng, Y.Y.; Yuan, B.; Yang, K.; Ma, Y.Q. Tunable dual-stimuli response of a microgel composite consisting of reduced graphene oxide nanoparticles and poly(*N*-isopropylacrylamide) hydrogel microspheres. *J. Mater. Chem. B* **2014**, *2*, 3791–3798. [CrossRef]

158. Zhou, M.M.; Xie, J.D.; Yan, S.T.; Jiang, X.M.; Ye, T.; Wu, W.T. Graphene@poly(phenylboronic acid)s microgels with selectively glucose-responsive volume phase transition behavior at a physiological pH. *Macromolecules* **2014**, *47*, 6055–6066. [CrossRef]

159. Girard, H.A.; Benayoun, P.; Blin, C.; Trouve, A.; Gesset, C.; Arnault, J.C.; Bergonzo, P. Encapsulated nanodiamonds in smart microgels toward self-assembled diamond nanoarrays. *Diam. Relat. Mater.* **2013**, *33*, 32–37. [CrossRef]

160. Agrawal, G.; Schurings, M.; Zhu, X.M.; Pich, A. Microgel/SiO$_2$ hybrid colloids prepared using a water soluble silica precursor. *Polymer* **2012**, *53*, 1189–1197. [CrossRef]

161. Li, Z.B.; Chen, T.Y.; Nie, J.J.; Xu, J.T.; Fan, Z.Q.; Du, B.Y. P(Nipam-*co*-Tmspma)/silica hybrid microgels: Structures, swelling properties and applications in fabricating macroporous silica. *Mater. Chem. Phys.* **2013**, *138*, 650–657. [CrossRef]

162. Chai, S.G.; Zhang, J.Z.; Yang, T.T.; Yuan, J.J.; Cheng, S.Y. Thermoresponsive microgel decorated with silica nanoparticles in shell: Biomimetic synthesis and drug release application. *Colloid Surf. A* **2010**, *356*, 32–39. [CrossRef]

163. Dechezelles, J.F.; Malik, V.; Crassous, J.J.; Schurtenberger, P. Hybrid raspberry microgels with tunable thermoresponsive behavior. *Soft Matter* **2013**, *9*, 2798–2802. [CrossRef]

164. Contin, A.; Biffis, A.; Sterchele, S.; Dormbach, K.; Schipmann, S.; Pich, A. Metal nanoparticles inside microgel/clay nanohybrids: Synthesis, characterization and catalytic efficiency in cross-coupling reactions. *J. Colloid Interface Sci.* **2014**, *414*, 41–45. [CrossRef] [PubMed]

165. Du, Z.S.; Hu, Y.; Gu, X.Y.; Hu, M.; Wang, C.Y. Poly(acrylamide) microgel-reinforced poly(acrylamide)/hectorite nanocomposite hydrogels. *Colloid Surf. A* **2016**, *489*, 1–8. [CrossRef]

166. Wu, W.; Shen, J.; Banerjee, P.; Zhou, S. Chitosan-based responsive hybrid nanogels for integration of optical pH-sensing, tumor cell imaging and controlled drug delivery. *Biomaterials* **2010**, *31*, 8371–8381. [CrossRef] [PubMed]

167. Gui, R.J.; An, X.Q.; Gong, J.; Chen, T. Thermosensitive, reversible luminescence properties and bright fluorescence imaging of water-soluble quantum dots/microgels nanocompounds. *Mater. Lett.* **2012**, *88*, 122–125. [CrossRef]

168. Cai, Y.T.; Du, G.L.; Gao, G.R.; Chen, J.; Fu, J. Colour-tunable quantum dots/poly(Nipam-*co*-aac) hybrid microgels based on electrostatic interactions. *RSC Adv.* **2016**, *6*, 98147–98152. [CrossRef]

169. Lai, W.F.; Susha, A.S.; Rogach, A.L. Multicompartment microgel beads for co-delivery of multiple drugs at individual release rates. *ACS Appl. Mater. Interfaces* **2016**, *8*, 871–880. [CrossRef] [PubMed]

170. Cui, J.C.; Gao, N.; Li, J.; Wang, C.; Wang, H.; Zhou, M.M.; Zhang, M.; Li, G.T. Poly(ionic liquid)-based monodisperse microgels as a unique platform for producing functional materials. *J. Mater. Chem. C* **2015**, *3*, 623–631. [CrossRef]

171. Baptista, A.C.; Martins, J.I.; Fortunato, E.; Martins, R.; Borges, J.P.; Ferreira, I. Thin and flexible bio-batteries made of electrospun cellulose-based membranes. *Biosens. Bioelectron.* **2011**, *26*, 2742–2745. [CrossRef] [PubMed]

172. Bhardwaj, N.; Kundu, S.C. Electrospinning: A fascinating fiber fabrication technique. *Biotechnol. Adv.* **2010**, *28*, 325–347. [CrossRef] [PubMed]

173. Baptista, A.; Soares, P.; Ferreira, I.; Borges, J.P. Nanofibers and nanoparticles in biomedical applications. In *Bioengineered Nanomaterials*; Tiwari, A., Ed.; CRC Press: New York, NY, USA, 2013; pp. 98–100.

174. Canejo, J.P.; Borges, J.P.; Godinho, M.H.; Brogueira, P.; Teixeira, P.I.C.; Terentjev, E.M. Helical twisting of electrospun liquid crystalline cellulose micro- and nanofibers. *Adv. Mater.* **2008**, *20*, 4821–4825. [CrossRef]

175. Agarwal, S.; Greiner, A.; Wendorff, J.H. Functional materials by electrospinning of polymers. *Prog. Polym. Sci.* **2013**, *38*, 963–991. [CrossRef]

176. Crespy, D.; Friedemann, K.; Popa, A.M. Colloid-electrospinning: Fabrication of multicompartment nanofibers by the electrospinning of organic or/and inorganic dispersions and emulsions. *Macromol. Rapid Commun.* **2012**, *33*, 1978–1995. [CrossRef] [PubMed]

177. Elahi, M.F.; Lu, W. Core-shell fibers for biomedical applications—A review. *J. Bioeng. Biomed. Sci.* **2013**, *3*, 1–14. [CrossRef]

178. Jiang, S.; He, W.; Landfester, K.; Crespy, D.; Mylon, S.E. The structure of fibers produced by colloid-electrospinning depends on the aggregation state of particles in the electrospinning feed. *Polymer* **2017**, *127*, 101–105. [CrossRef]

179. Diaz, J.E.; Barrero, A.; Marquez, M.; Fernandez-Nieves, A.; Loscertales, I.G. Absorption properties of microgel-pvp composite nanofibers made by electrospinning. *Macromol. Rapid Commun.* **2010**, *31*, 183–189. [PubMed]

180. Gu, S.Y.; Wang, Z.M.; Li, J.B.; Ren, J. Switchable wettability of thermo-responsive biocompatible nanofibrous films created by electrospinning. *Macromol. Mater. Eng.* **2010**, *295*, 32–36. [CrossRef]

181. Liu, C.; Qin, H.; Mather, P.T. Review of progress in shape-memory polymers. *J. Mater. Chem.* **2007**, *17*, 1543–1558. [CrossRef]

182. Hager, M.D.; Bode, S.; Weber, C.; Schubert, U.S. Shape memory polymers: Past, present and future developments. *Prog. Polym. Sci.* **2015**, *49–50*, 3–33. [CrossRef]

183. Wache, H.M.; Tartakowska, D.J.; Hentrich, A.; Wagner, M.H. Development of a polymer stent with shape memory effect as a drug delivery system. *J. Mater. Sci. Mater. M* **2003**, *14*, 109–112. [CrossRef]

184. Lendlein, A.; Behl, M.; Hiebl, B.; Wischke, C. Shape-memory polymers as a technology platform for biomedical applications. *Expert Rev. Med. Devices* **2010**, *7*, 357–379. [CrossRef] [PubMed]

185. Yakacki, C.M.; Gall, K. Shape-memory polymers for biomedical applications. In *Shape-Memory Polymers*; Lendlein, A., Ed.; Springer: Berlin, Germany, 2010; pp. 147–175.

186. Du, H.Y.; Zhang, J.H. Solvent induced shape recovery of shape memory polymer based on chemically cross-linked poly(vinyl alcohol). *Soft Matter* **2010**, *6*, 3370–3376. [CrossRef]

187. Cui, Y.; Tan, M.; Zhu, A.; Guo, M. Mechanically strong and stretchable peg-based supramolecular hydrogel with water-responsive shape-memory property. *J. Mater. Chem. B* **2014**, *2*, 2978–2982. [CrossRef]

188. Miyamae, K.; Nakahata, M.; Takashima, Y.; Harada, A. Self-healing, expansion-contraction, and shape-memory properties of a preorganized supramolecular hydrogel through host-guest interactions. *Angew. Chem. Int. Ed.* **2015**, *54*, 8984–8987. [CrossRef] [PubMed]

189. Osada, Y.; Matsuda, A. Shape memory in hydrogels. *Nature* **1995**, *376*, 219. [CrossRef] [PubMed]

190. Zhang, J.L.; Huang, W.M.; Lu, H.B.; Sun, L. Thermo-/chemo-responsive shape memory/change effect in a hydrogel and its composites. *Mater. Des.* **2014**, *53*, 1077–1088. [CrossRef]

191. Zhang, J.L.; Huang, W.M.; Gao, G.R.; Fu, J.; Zhou, Y.; Salvekar, A.V.; Venkatraman, S.S.; Wong, Y.S.; Tay, K.H.; Birch, W.R. Shape memory/change effect in a double network nanocomposite tough hydrogel. *Eur. Polym. J.* **2014**, *58*, 41–51. [CrossRef]

192. Fan, Y.; Zhou, W.; Yasin, A.; Li, H.; Yang, H. Dual-responsive shape memory hydrogels with novel thermoplasticity based on a hydrophobically modified polyampholyte. *Soft Matter* **2015**, *11*, 4218–4225. [CrossRef] [PubMed]

193. Hu, Y.W.; Lu, C.H.; Guo, W.W.; Aleman-Garcia, M.A.; Ren, J.T.; Willner, I. A shape memory acrylamide/DNA hydrogel exhibiting switchable dual pH-responsiveness. *Adv. Funct. Mater.* **2015**, *25*, 6867–6874. [CrossRef]

194. Hu, Y.; Guo, W.; Kahn, J.S.; Aleman-Garcia, M.A.; Willner, I. A shape-memory DNA-based hydrogel exhibiting two internal memories. *Angew. Chem. Int. Ed.* **2016**, *55*, 4210–4214. [CrossRef] [PubMed]

195. Li, G.; Zhang, H.; Fortin, D.; Xia, H.; Zhao, Y. Poly(vinyl alcohol)-poly(ethylene glycol) double-network hydrogel: A general approach to shape memory and self-healing functionalities. *Langmuir* **2015**, *31*, 11709–11716. [CrossRef] [PubMed]

196. Li, G.; Yan, Q.; Xia, H.; Zhao, Y. Therapeutic-ultrasound-triggered shape memory of a melamine-enhanced poly(vinyl alcohol) physical hydrogel. *ACS Appl. Mater. Interfaces* **2015**, *7*, 12067–12073. [CrossRef] [PubMed]

197. Zhang, Y.; Zhang, M.; Jiang, H.; Shi, J.; Li, F.; Xia, Y.; Zhang, G.; Li, H. Bio-inspired layered chitosan/graphene oxide nanocomposite hydrogels with high strength and pH-driven shape memory effect. *Carbohydr. Polym.* **2017**, *177*, 116–125. [CrossRef] [PubMed]

198. Ma, C.; Lu, W.; Yang, X.; He, J.; Le, X.; Wang, L.; Zhang, J.; Serpe, M.J.; Huang, Y.; Chen, T. Bioinspired anisotropic hydrogel actuators with on-off switchable and color-tunable fluorescence behaviors. *Adv. Funct. Mater.* **2018**, *28*, 1704568. [CrossRef]

199. Huang, J.; Zhao, L.; Wang, T.; Sun, W.; Tong, Z. Nir-triggered rapid shape memory pam-go-gelatin hydrogels with high mechanical strength. *ACS Appl. Mater. Interfaces* **2016**, *8*, 12384–12392. [CrossRef] [PubMed]

200. Wool, R.P. Self-healing materials: A review. *Soft Matter* **2008**, *4*, 400–418. [CrossRef]

201. Wu, D.Y.; Meure, S.; Solomon, D. Self-healing polymeric materials: A review of recent developments. *Prog. Polym. Sci.* **2008**, *33*, 479–522. [CrossRef]

202. Toohey, K.S.; Sottos, N.R.; Lewis, J.A.; Moore, J.S.; White, S.R. Self-healing materials with microvascular networks. *Nat. Mater.* **2007**, *6*, 581–585. [CrossRef] [PubMed]

203. Lehn, J.M. From supramolecular chemistry towards constitutional dynamic chemistry and adaptive chemistry. *Chem. Soc. Rev.* **2007**, *36*, 151–160. [CrossRef] [PubMed]

204. Caulder, D.L.; Raymond, K.N. Supermolecules by design. *Acc. Chem. Res.* **1999**, *32*, 975–982. [CrossRef]

205. Amaral, A.J.R.; Pasparakis, G. Stimuli responsive self-healing polymers: Gels, elastomers and membranes. *Polym. Chem.* **2017**, *8*, 6464–6484. [CrossRef]

206. Phadke, A.; Zhang, C.; Arman, B.; Hsu, C.C.; Mashelkar, R.A.; Lele, A.K.; Tauber, M.J.; Arya, G.; Varghese, S. Rapid self-healing hydrogels. *Proc. Natl. Acad. Sci. USA* **2012**, *109*, 4383–4388. [CrossRef] [PubMed]

207. Tuncaboylu, D.C.; Sari, M.; Oppermann, W.; Okay, O. Tough and self-healing hydrogels formed via hydrophobic interactions. *Macromolecules* **2011**, *44*, 4997–5005. [CrossRef]

208. Tuncaboylu, D.C.; Argun, A.; Sahin, M.; Sari, M.; Okay, O. Structure optimization of self-healing hydrogels formed via hydrophobic interactions. *Polymer* **2012**, *53*, 5513–5522. [CrossRef]

209. Tuncaboylu, D.C.; Argun, A.; Algi, M.P.; Okay, O. Autonomic self-healing in covalently crosslinked hydrogels containing hydrophobic domains. *Polymer* **2013**, *54*, 6381–6388. [CrossRef]

210. Gulyuz, U.; Okay, O. Self-healing polyacrylic acid hydrogels. *Soft Matter* **2013**, *9*, 10287–10293. [CrossRef]

211. Zhang, H.J.; Xia, H.S.; Zhao, Y. Poly(vinyl alcohol) hydrogel can autonomously self-heal. *ACS Macro Lett.* **2012**, *1*, 1233–1236. [CrossRef]

212. Gulyuz, U.; Okay, O. Self-healing poly(N-isopropylacrylamide) hydrogels. *Eur. Polym. J.* **2015**, *72*, 12–22. [CrossRef]

213. Cong, H.P.; Wang, P.; Yu, S.H. Highly elastic and superstretchable graphene oxide/polyacrylamide hydrogels. *Small* **2014**, *10*, 448–453. [CrossRef] [PubMed]

214. Cong, H.-P.; Wang, P.; Yu, S.-H. Stretchable and self-healing graphene oxide–polymer composite hydrogels: A dual-network design. *Chem. Mater.* **2013**, *25*, 3357–3362. [CrossRef]

215. Zhang, E.; Wang, T.; Zhao, L.; Sun, W.; Liu, X.; Tong, Z. Fast self-healing of graphene oxide-hectorite clay-poly(N,N-dimethylacrylamide) hybrid hydrogels realized by near-infrared irradiation. *ACS Appl. Mater. Interfaces* **2014**, *6*, 22855–22861. [CrossRef] [PubMed]

216. Liu, J.; Song, G.; He, C.; Wang, H. Self-healing in tough graphene oxide composite hydrogels. *Macromol. Rapid Commun.* **2013**, *34*, 1002–1007. [CrossRef] [PubMed]

217. Liu, J.; Chen, C.; He, C.; Zhao, J.; Yang, X.; Wang, H. Synthesis of graphene peroxide and its application in fabricating super extensible and highly resilient nanocomposite hydrogels. *ACS Nano* **2012**, *6*, 8194–8202. [CrossRef] [PubMed]

218. Li, L.; Smitthipong, W.; Zeng, H.B. Mussel-inspired hydrogels for biomedical and environmental applications. *Polym. Chem.* **2015**, *6*, 353–358. [CrossRef]

219. Li, L.; Yan, B.; Yang, J.; Chen, L.; Zeng, H. Novel mussel-inspired injectable self-healing hydrogel with anti-biofouling property. *Adv. Mater.* **2015**, *27*, 1294–1299. [CrossRef] [PubMed]

220. Li, L.; Yan, B.; Yang, J.; Huang, W.; Chen, L.; Zeng, H. Injectable self-healing hydrogel with antimicrobial and antifouling properties. *ACS Appl. Mater. Interfaces* **2017**, *9*, 9221–9225. [CrossRef] [PubMed]

221. Hou, S.; Ma, P.X. Stimuli-responsive supramolecular hydrogels with high extensibility and fast self-healing via precoordinated mussel-inspired chemistry. *Chem. Mater.* **2015**, *27*, 7627–7635. [CrossRef] [PubMed]

222. Latnikova, A.; Grigoriev, D.; Schenderlein, M.; Mohwald, H.; Shchukin, D. A new approach towards "active" self-healing coatings: Exploitation of microgels. *Soft Matter* **2012**, *8*, 10837–10844. [CrossRef]

223. Park, C.W.; South, A.B.; Hu, X.B.; Verdes, C.; Kim, J.D.; Lyon, L.A. Gold nanoparticles reinforce self-healing microgel multilayers. *Colloid Polym. Sci.* **2011**, *289*, 583–590. [CrossRef]

224. Serpe, M.J.; Jones, C.D.; Lyon, L.A. Layer-by-layer deposition of thermoresponsive microgel thin films. *Langmuir* **2003**, *19*, 8759–8764. [CrossRef]

225. South, A.B.; Lyon, L.A. Autonomic self-healing of hydrogel thin films. *Angew. Chem. Int. Ed.* **2010**, *49*, 767–771. [CrossRef] [PubMed]

226. Gaulding, J.C.; Spears, M.W.; Lyon, L.A. Plastic deformation, wrinkling, and recovery in microgel multilayers. *Polym. Chem.* **2013**, *4*, 4890–4896. [CrossRef] [PubMed]

227. Spears, M.W., Jr.; Herman, E.S.; Gaulding, J.C.; Lyon, L.A. Dynamic materials from microgel multilayers. *Langmuir* **2014**, *30*, 6314–6323. [CrossRef] [PubMed]

gels

MDPI

Review

Stimuli-Responsive Assemblies for Sensing Applications

Xue Li, Yongfeng Gao and Michael J. Serpe *

Department of Chemistry, University of Alberta, Edmonton, AB T6G 2G2, Canada;
xue13@ualberta.ca (X.L.); yg2@ualberta.ca (Y.G.)
* Correspondence: serpe@ualberta.ca; Tel.: +1-780-492-5778

Academic Editor: Dirk Kuckling
Received: 9 December 2015; Accepted: 1 February 2016; Published: 16 February 2016

Abstract: Poly (*N*-isopropylacrylamide) (pNIPAm)-based hydrogels and hydrogel particles (microgels) have been extensively studied since their discovery a number of decades ago. While their utility seems to have no limit, this feature article is focused on their development and application for sensing small molecules, macromolecules, and biomolecules. We highlight hydrogel/microgel-based photonic materials that have order in one, two, or three dimensions, which exhibit optical properties that depend on the presence and concentration of various analytes. A particular focus is put on one-dimensional materials developed in the Serpe Group.

Keywords: stimuli-responsive polymers; photonic materials; poly (*N*-isopropylacrylamide)-based microgels; etalons; optical sensing

1. Introduction

Polymer-based stimuli responsive materials have been of great interest over the years due to their ability to convert external chemical and/or physical stimuli into observable changes of the material itself [1–3]. Some of the most important of these materials are responsive hydrogels, which are hydrophilic crosslinked polymer networks capable of changing their solvation state in response to various stimuli [4–6]. Hydrogel particles (e.g., microgels) can also be synthesized, and typically have diameters of 100–2000 nm [7,8]. Over the years, many polymers, and polymer-based materials, have been identified that exhibit a specific response to a variety of stimuli. Some of those stimuli include temperature [9,10], light [11,12], electric [13], and magnetic fields [14]. These responsivities make hydrogels very useful for many applications, e.g., sensing [15,16], drug delivery [17,18], artificial muscles [19,20], tissue engineering [21,22], and self-healing materials [23,24]. Among these, thermoresponsive materials have been the most extensively studied, and poly (*N*-isopropylacrylamide) (pNIPAm)-based hydrogels and microgels are the most well-known and extensively studied thermoresponsive materials [25–27]. PNIPAm is fully soluble in water below ~32 °C, and transitions to an "insoluble" state when the temperature is above 32 °C. This transition is observed as a coil-to-globule transition, where the polymer transitions from an extended to collapsed state, respectively [28]. The conformational change is also accompanied by a water exchange process. That is, when pNIPAm undergoes the coil-to-globule transition, water is "expelled", while water is "absorbed" when pNIPAm undergoes a globule-to-coil transition. Similarly, crosslinked pNIPAm-based hydrogels and microgels contract upon heating, and swell with water upon cooling [29–31]. This swelling-deswelling transition is fully reversible over multiple heating/cooling cycles.

While there are many uses of pNIPAm hydrogels and microgels, a majority of this review focuses on their use for sensing applications. Specifically, this review focuses on the use of microgels and hydrogels as components of photonic material (PM) assemblies. A specific example of a PM is a photonic crystal (PC); PCs are composed of materials of varying refractive indices arranged in an ordered fashion in one, two, or three dimensions (1D, 2D, 3D). There are many examples of PCs in nature, most commonly associated with the vibrant colors of butterfly wings and the opal gemstone. These materials are unique because, unlike many other colored materials found in nature that exhibit color due to the absorbance of light by small molecule chromophores, PCs are colored due to their structure. Specifically, the opal gemstone is composed of a close-packed array of colloids (typically silica), which are capable of interacting with wavelengths of light in the visible region of the electromagnetic spectrum. These interactions lead to constructive and destructive interference of the light in the assembly, leading to specific wavelengths of light being reflected, which leads to the observed color [32]. A major goal of many research groups around the world is to generate synthetic colloidal crystals. This is typically done by "forcing" colloids of high refractive index into an ordered array in a matrix of relatively low refractive index (e.g., air, water, polymer). If the particle periodicity (*i.e.*, refractive index periodicity) is on the order of visible wavelengths of light, then the device will appear colored. This is a direct result of light refraction, reflection, and diffraction off the material's particles, which leads to light interference, and hence color [32–34]. PM and PCs are of great interest for various applications, including optics [35,36], actuators [37,38], sensors [39], controlled drug delivery [40] and for display devices [41,42].

In this submission, we will first discuss examples of PMs generated from inorganic components (such as silica particles) and block copolymers, and 1D, 2D, and 3D PCs fabricated from them. Then we will discuss their use for sensing and biosensing, with a particular focus on 1D PCs constructed by our group from pNIPAm-based microgels.

2. Photonic Materials

As discussed above, the opal gemstone is composed of particles packed into an ordered array—these structures are sometimes referred to as colloidal crystal arrays (CCAs). Both natural and synthetic CCAs exist, and can yield extremely colorful materials. The color the materials exhibit depends on the spacing between the array elements (and other parameters), according to Equation (1):

$$m\lambda = 2nd \cdot \sin\theta \tag{1}$$

where *m* is the order of diffraction, λ is the wavelength of incident light, *n* is the refractive index of the optical components, *d* is the interplanar spacing, and θ is the angle between the incident light and the diffracting crystal planes, which are oriented parallel to the crystal surface in the prepared CCA. Since the color the material exhibits depends directly on the array element spacing, expansion/contraction of responsive polymers coupled to CCAs can be used to tune the spacing, and hence the color of the materials. One of the most extensively used responsive polymer for this purpose is pNIPAm, and early examples from the Asher Group showed that the volume changes that pNIPAm undergoes as a function of temperature can be used to tune the visual color of colloidal crystals. The Asher Group [43] showed that the optical properties of these materials, referred to as polymerized crystalline colloidal arrays (PCCAs), could be tuned quite dramatically with temperature. Specifically, as shown in Figure 1, the Bragg peak could be tuned between 704 and 460 nm by variation in the temperature.

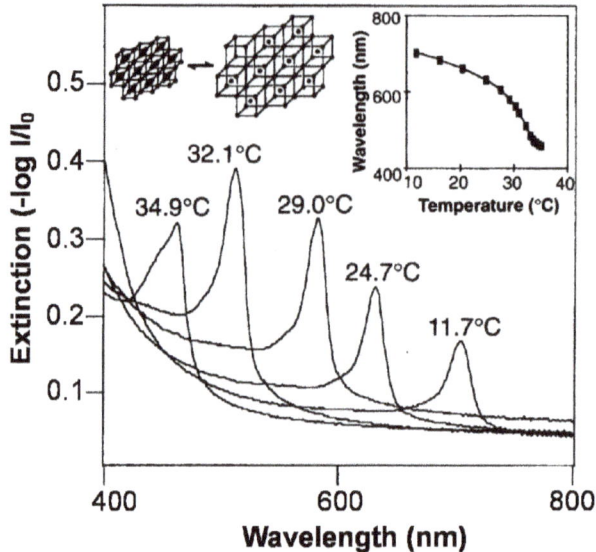

Figure 1. Temperature tuning of Bragg diffraction from a 125-μm-thick polymerized crystalline colloidal array (PCCA) film of 99-nm polystyrene (PS) spheres embedded in a poly (*N*-isopropylacrylamide) (pNIPAm) gel. The shift of the diffraction wavelength results from the temperature-induced volume change of the gel, which alters the lattice spacing. Spectra were recorded in a UV-visible-near IR spectrophotometer with the sample placed normal to the incident light beam. The inset shows the temperature dependence of the diffracted wavelength for this PCCA film when the incident light is normal to the (110) plane of the lattice. Reprinted with the permission from [43] Copyright © 1996, American Association for the Advancement of Science.

In many subsequent investigations, Asher and coworkers entrapped CCAs in various hydrogel materials, which changed volume (and hence optical properties) in the presence of various analytes, and upon exposure to a variety of stimuli [6,44,45]. In a more recent example, the Asher group developed a novel two-dimensional (2D) CCA for the visual detection of amphiphilic molecules in water [46]. These 2D photonic crystals were placed on a mirrored surface (liquid Hg), and exhibited intense diffraction that enabled them to be used for detection of analytes by observation of visual color changes. Figure 2 shows a schematic illustration of the 2D photonic crystal. A monolayer of 2D close-packed polystyrene (PS) particles was embedded in a pNIPAm-based hydrogel film. Binding of surfactant molecules increased the charge density in the hydrogel. This resulted in a swelling of the pNIPAm-based hydrogel, and a concomitant change in the distance between the array elements. The resulting increase in particle spacing red shifts the 2D diffracted light according to Equation (2):

$$m\lambda = 3^{1/2} \times d \sin \theta \tag{2}$$

where *m* is the diffraction order, λ is the wavelength of the diffracted light, *d* is the nearest neighboring particle spacing, and θ is the angle between the incident light and the normal to the 2D array. For a fixed angle of incidence (θ), the diffracted wavelength (λ) is proportional to the 2D particle spacing (*d*). As shown in Figure 3, normalized and smoothed diffraction spectra of 2D pNIPAm-based sensors were obtained at different concentrations of aqueous, sodium dodecyl sulfate (SDS) solutions and diffraction wavelengths *versus* different concentration of SDS.

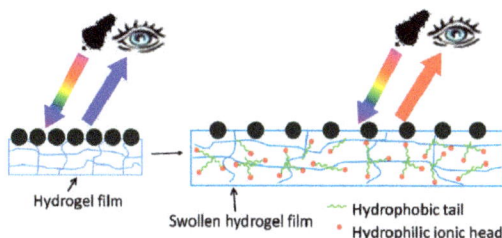

Figure 2. Schematic illustration of a 2D photonic crystal sensor formed by polymerization of a pNIPAm hydrogel network onto a 2D array of 490 nm PS particles, which was added to a liquid Hg surface. The pNIPAm hydrogel swells upon binding of surfactant molecules. The 2D particle spacing increases upon swelling of the pNIPAm hydrogel, red shifting the diffracted wavelength. Reprinted with the permission of [46] Copyright © 2012, American Chemical Society.

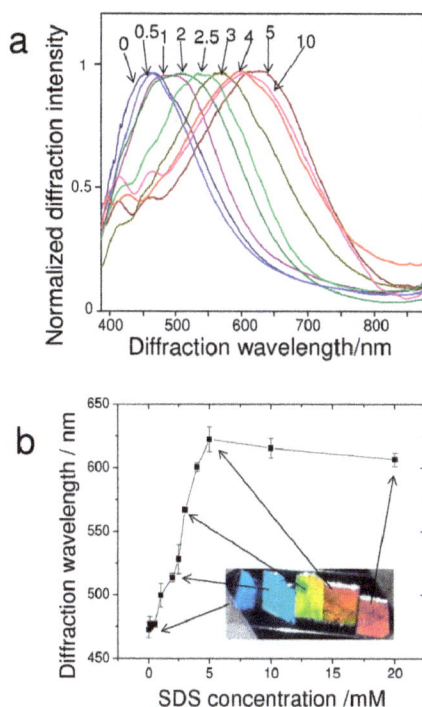

Figure 3. (**a**) Normalized and smoothed diffraction spectra of 2D pNIPAm-based sensors at different concentrations of aqueous sodium dodecyl sulfate (SDS) solutions. Each number in the graph corresponds to the different concentration in the unit of mM. The measurement angle between the probe and the normal to the 2D array was 28°; (**b**) Wavelength of diffracted peak *versus* SDS concentration. The inset shows photographs taken close to the Littrow configuration at an angle of 28° between the source and camera to the 2D array normal. Reprinted with the permission of [46] Copyright © 2012, American Chemical Society.

In another example, Fudouzi and coworkers developed an elastic poly(dimethylsiloxane) (PDMS) sheet with a thin layer of cubic close packed polystyrene particles embedded. This material exhibited structural color, which could be tuned as a function of the extent of stretching [47]. Specifically,

Figure 4A shows that the sheet can be stretched "horizontally", which decreases the size of the material in the vertical direction leading to the decrease in the lattice spacing of (111) planes of the array, resulting in a blue shift of the device's reflectance peaks and a concomitant visual color change, as shown in Figure 4B,C. The ability of these materials to change color as a function of PDMS elongation makes them well suited for quantifying mechanical strains on materials. Ultimately, this could be used for quantifying the fidelity of structures such as buildings and bridges.

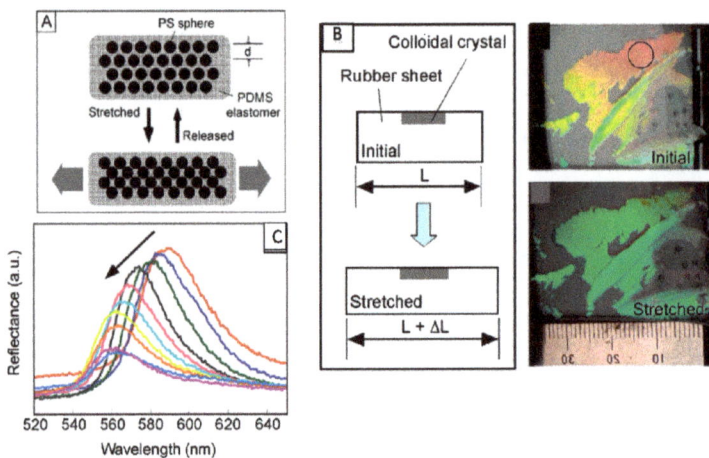

Figure 4. (**A**) Reversible tuning by stretching of the lattice distance of a PS colloidal crystal embedded in a poly(dimethylsiloxane) (PDMS) elastomer matrix; (**B**) Changes in the structural color of the colloidal crystal film covering the silicone rubber sheet; (**C**) Relationship between the reflectance peak position and elongation of the silicone rubber sheet by stretching. The peak position shifted from 590 to 560 nm, and the reflectance intensity decreased gradually as indicated by the arrow. Reprinted with the permission of [47] Copyright © 2005, American Chemical Society.

Recently, there has been a growing interest in using block copolymer-based photonic gels for sensing and biosensing applications. Block copolymers offer the flexibility of fabricating 1D, 2D, and 3D photonic materials through self-assembly, making them relatively easy to fabricate. Some photonic gels are extremely sensitive to a change in charge and/or charge density in the gel matrix, as well as the dielectric environment. Proteins are highly charged dielectric materials, and thus the electrostatic and dielectric environment of photonic gels can change abruptly upon protein binding. This property has been harnessed by modifying the PCs with molecules that can bind specific targets, e.g., proteins, and the change in the local environment upon binding of the target can change the spacing between the material's array elements. In one example, Kang and coworkers [48] have generated PCs using the ability of polystyrene-b-quaternized poly(2-vinyl pyridine) (PS-b-QP2VP) to self-assemble into a photonic gel, and modified the structure with biotin. This is shown schematically in Figure 5A,B for the "on-gel" and "in-gel" configurations they investigated. Upon binding streptavidin the spacing of the array elements changed, and a visual color change with streptavidin binding was observed for the in-gel photonic materials, as shown in Figure 6.

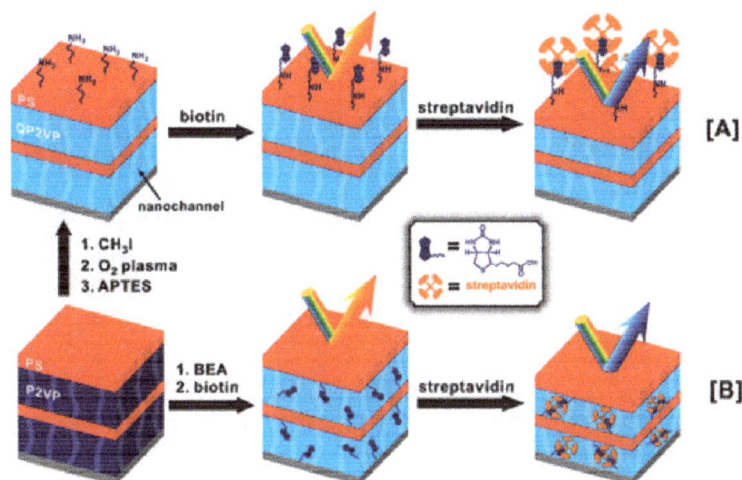

Figure 5. Preparation and use of biotinylated photonic gels. Procedure used for biotinylating the (**A**) surface of photonic gel films, and (**B**) the inside of photonic gel films. Reprinted with the permission of [48] Copyright © 2012, The Polymer Society of Korea and Springer Science + Business Media Dordrecht.

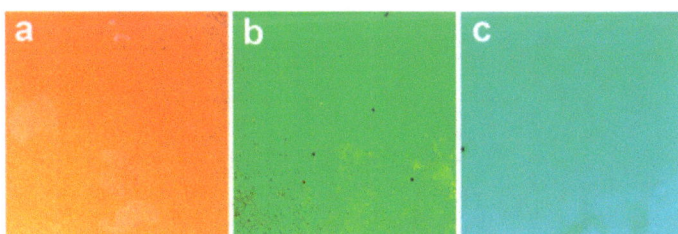

Figure 6. Color changes of the photonic gels with inside biotinylation in response to streptavidin; (**a**) [S] = 0 M; (**b**) [S] = 0.5 Mm; (**c**) [S] = 10 mM. Reprinted with the permission of [48] Copyright © 2012, The Polymer Society of Korea and Springer Science + Business Media Dordrecht.

3. PNIPAm Microgel-Based 1D PCs and Their Application for Sensing and Biosensing

Compared with other pNIPAm microgel-based photonic materials, which exhibit order in 2D or 3D [46,49,50], the Serpe Group discovered color tunable materials (etalons) that exhibit structure in 1D. This was accomplished by sandwiching a pNIPAm microgel-based layer between two thin Au layers, which act as mirrors. The devices exhibit visual color, and unique multipeak reflectance spectra—the position of the peaks in the reflectance spectra primarily depend the thickness of the microgel layer, according to Equation (3),

$$m\lambda = 2nd \cdot \cos\theta \qquad (3)$$

where m is the peak order, n is the refractive index of the dielectric material, d is the distance between Au layers, and θ is the angle of incidence [51,52]. The structure of the device and representative reflectance spectra are shown in Figure 7. Since the optical properties primarily depend on the thickness of the microgel layer (controls the mirror-mirror distance), the microgel's response to a stimulus results in changes in the optical properties [53–59]. This property is extremely important for sensing applications, since the solvation state of microgels can be made to depend on many different stimuli.

a

b

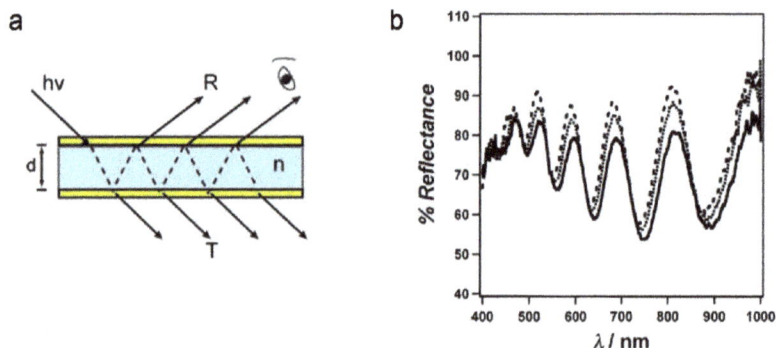

Figure 7. (a) Structure of the microgel-based etalon. Two reflective metal layers (in this case Au, top and bottom layers) sandwiching a dielectric (middle); (b) Representative reflectance spectra from a pNIPAm microgel-based etalon. Reprinted with the permission of [54] Copyright © 2012, Royal Society of Chemistry.

The most basic response of the devices is to temperature, due to the collapse of the microgels in water at $T > 32$ °C. This leads to a decrease in the thickness of the microgel layer, and hence Equation (3) predicts a blue shift in the device's reflectance peaks. As can be seen in Figure 8, the device's reflectance peaks show a blue shift (e.g., the star-labelled peaks) with increasing temperature in aqueous solution. This is due to microgel deswelling, resulting in the decrease of the distance between the device's Au layers [60].

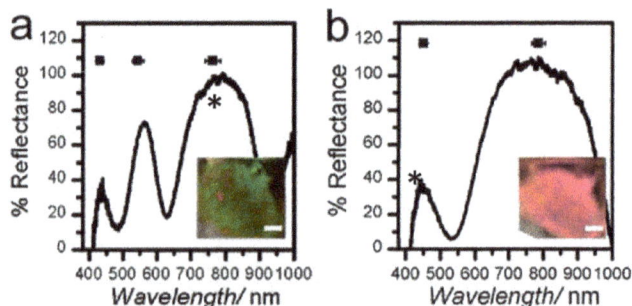

a

b

Figure 8. Representative spectra for an etalon in pH 3 solution (2 mM ionic strength (I.S.), I.S. adjusted with NaCl) at (a) 25 °C, (b) 37 °C. The star labelled peak is for the same reflection order from the device. The insets show the corresponding photographs of the devices. Reprinted with the permission of [60] Copyright © 2013, Royal Society of Chemistry.

By copolymerizing acrylic acid (AAc) (pK$_a$ ~ 4.25) with NIPAm, pH responsive pNIPAm-co-AAc microgels and microgel-based etalons could be generated. The pH responsivity is a result of Coulombic repulsion and osmotic swelling in the microgel layer at pH > pKa as a result of the charges on the deprotonated AAc groups. Furthermore, we sought to determine if spatially isolated regions of pNIPAm-co-AAc microgel-based etalons can be independently modulated [54]. Figure 9 shows an etalon with solutions of different pH spotted on spatially isolated regions. As can be seen, the spots appear different colors, which can be changed as a function of temperature and pH, independently. In another example, we have shown that pNIPAm-co-AAc microgel-based etalons could be attached to the surface of a quartz crystal microbalance, and used for the very sensitive detection of solution pH [55,61].

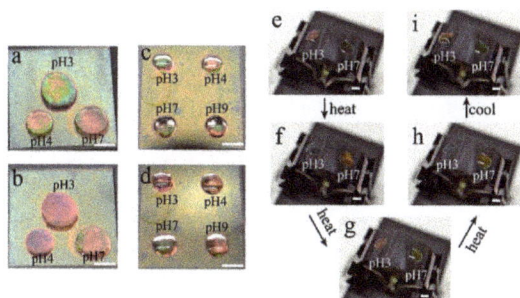

Figure 9. Photographs of an etalon with solutions of various pH spotted on a single surface (**a, c, e, i**) 25 °C and (**b, d, f, g, h**) 37 °C; (**f**) ~3 min after heating; (**g**) ~5 min after heating; (**h**) ~6 min after heating. In each panel, the scale bar is 5 mm. Reprinted with the permission of [54] Copyright © 2012, Royal Society of Chemistry.

Microgel-based etalons were also fabricated that could detect the concentration of glucose in solution. This was done by fabricating 3-aminophenylboronic acid (APBA) functionalized microgels [53]. As illustrated in Figure 10, glucose binding with the boron atom will promote more boron atoms to become negatively charged resulting in swelling of the microgel layer. Because λ is proportional to the distance between the two mirrors, d, the swelling of microgels gives rise to a red shift of the reflectance peaks.

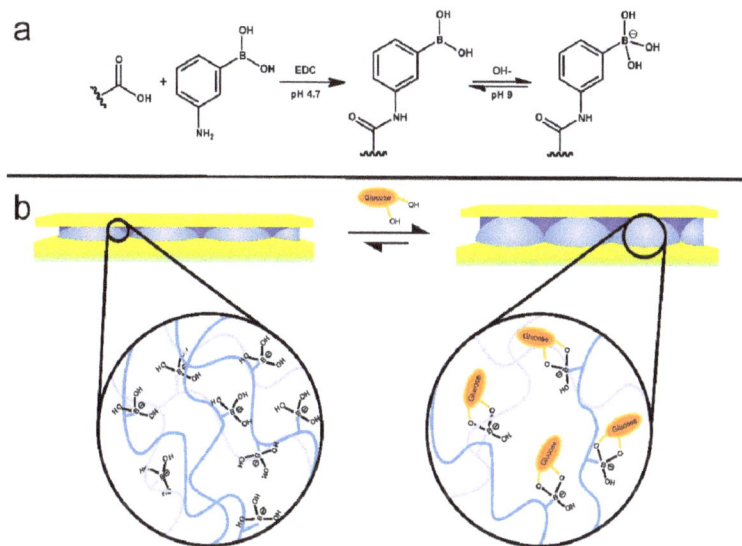

Figure 10. (**a**) Reaction scheme for the functionalization of the acrylic acid moieties on the microgel with 3-aminophenylboronic acid (APBA) followed by the activation of the boronic acid at high pH and (**b**) a cartoon depiction of the glucose responsivity of an APBA functionalized microgel etalon at pH 9. Reprinted with the permission of [53] Copyright © 2012, Springer Verlag.

Etalons composed of either pNIPAm-co-AAc or pNIPAm-co-N-(3-aminopropyl)methacrylamide hydrochloride (pNIPAm-co-APMAH) microgels were also constructed [62]. We investigated their response to the presence of linear polycations and/or polyanions. When the etalon was at a pH that renders the microgels multiply charged, the microgel layer of the etalon deswells in the presence of the oppositely charged linear polyelectrolyte; it is unresponsive to the presence of the like charged polyelectrolyte. Furthermore, the etalon's response depended on the thickness of the Au overlayer. For example, low molecular weight (MW) polyelectrolyte could penetrate all Au overlayer thicknesses, while high MW polyelectrolytes could only penetrate the etalons fabricated from thin Au overlayers, as shown in Figure 11a. We hypothesize that this is due to a decrease in the Au pore size with increasing thickness, which excludes the high MW polyelectrolytes from penetrating the microgel-based layer. This is supported by other investigations conducted in the group [63]. Figure 11b shows the shift of λ for pNIPAm-co-AAc etalons in pH 6.5 solution after addition of poly (diallyldimethylammonium chloride) (pDADMAC) solution of different molecular weights (MW). From this observation, we then developed pNIPAm microgel-based etalons and etalon arrays to determine the molecular weights of polymers in solution [64]. These devices show promise as MW selective sensors and biosensors.

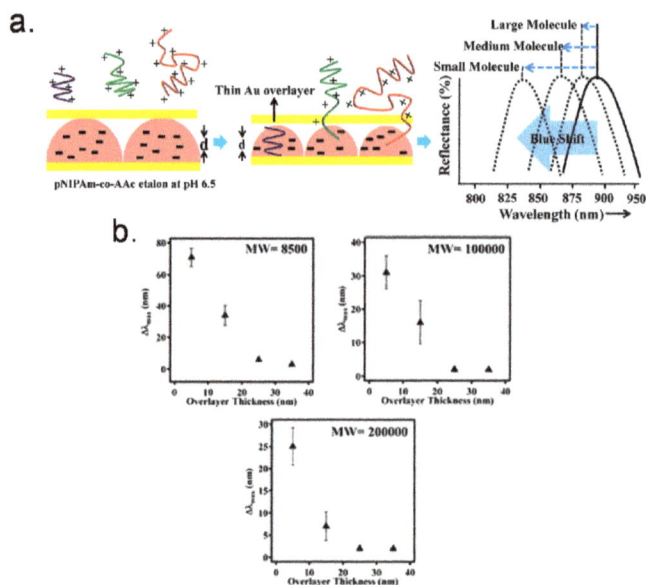

Figure 11. (a) Schematic depiction of polyelectrolyte penetration through the porous Au overlayer of an etalon and a schematic of the anticipated response; (b) Shift of λ_{max} for pNIPAm-co-AAc etalons in pH 6.5 after addition of pDADMAC solution with MW 8500, <100,000, and 100,000–200,000. Reprinted with the permission of [62] Copyright © 2013, American Chemical Society.

In an effort to exploit this observed phenomenon, we reported that biotinylated polycationic polymer can penetrate through the Au overlayer of a pNIPAm-co-AAc microgel-based etalon and cause the microgel layer to collapse [56,65]. The collapse results in a shift in the spectral peaks of the reflectance spectra. We found that the extent of peak shift depends on the amount of biotinylated polycation added to the etalon, which can subsequently be used to determine the concentration of streptavidin in solution at nM-pM concentrations. The sensing mechanism is shown schematically in Figure 12. As shown in Figure 13, the blue shift in the spectral peaks depends linearly on the amount of streptavidin, and hence PAH-biotin, added to the etalon. We were able to detect streptavidin concentrations in the nM range without any system optimization. We also note that this response is unique due to the fact that the response is highest for the lowest streptavidin concentration—this is counterintuitive based on the fact that other analytical approaches show small responses for low analyte concentrations.

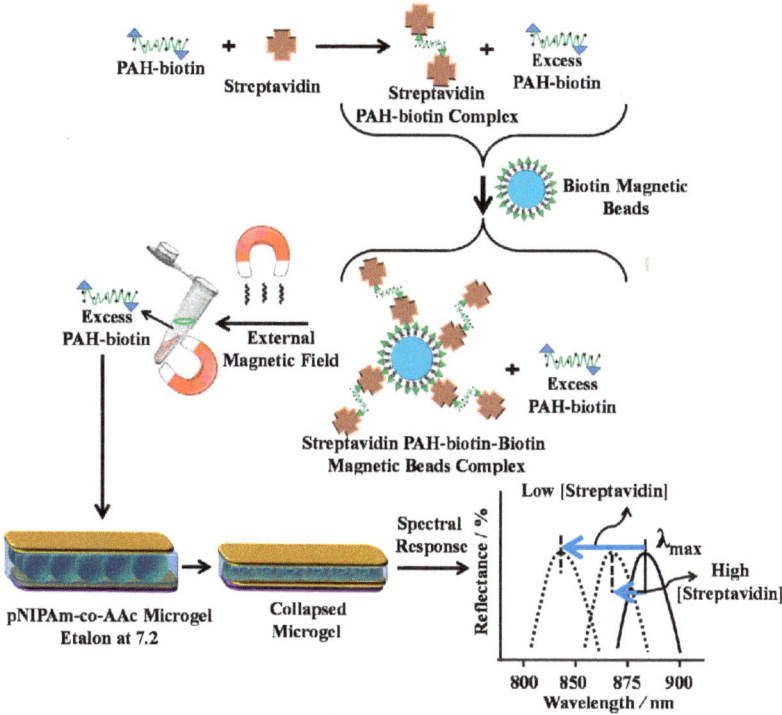

Figure 12. Streptavidin (the analyte) is added to an excess amount of biotin-modified poly (allylamine hydrochloride) (PAH). The streptavidin–biotin–PAH complex is then removed from solution using biotin modified magnetic particles, leaving behind free, unbound PAH. The unbound PAH is subsequently added to a pNIPAm-co-AAc microgel-based etalon immersed in aqueous solution at a pH that renders both the microgel layer and the PAH charged. As a result, the etalon's spectral peaks shift in proportion to the amount of PAH–biotin that was added. This, in turn can be related back to the original amount of streptavidin added to the PAH–biotin. Reprinted with the permission of [65] Copyright © 2013, Elsevier.

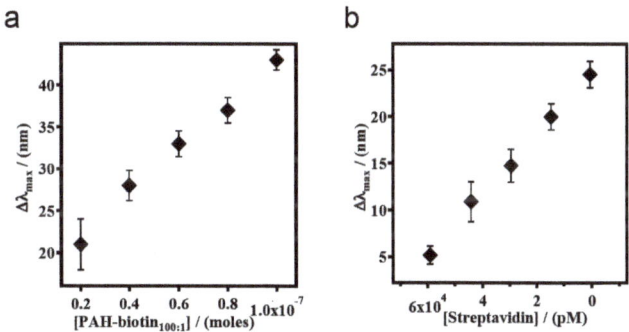

Figure 13. Cumulative shift of the etalon's $m = 3$ reflectance peak with different amounts of (**a**) PAH-biotin$_{100:1}$ (**b**) streptavidin added to PAH–biotin$_{100:1}$. The pNIPAm-co-AAc microgel-based etalon was soaked in pH 7.2 throughout the experiment, while the temperature was maintained at 25 °C. Each point represents the average of at least three independent measurements, and the error bars are standard deviation for those values. Reprinted with the permission of [65] Copyright © 2013, Elsevier.

Many groups have investigated novel stimuli-responsive polymer-based biosensing systems for detection of proteins and DNA [66–68]. As mentioned above, pNIPAm based microgels can be modified with a variety of functional groups that can be incorporated into our optical devices to sense various targets, such as glucose, proteins, and DNA [53,56]. For example, poly (*N*-isopropylacrylamide-co-*N*-(3-aminopropyl) methacrylamide hydrochloride) (pNIPAm-co-APMAH) microgels were synthesized via temperature ramp, surfactant-free, free radical precipitation polymerization [57,69]. PNIPAm-co-APMAH microgels are positively charged at pH 7.2, and when negatively charged DNA is added to the etalon composed of these positively charged microgels, the microgels are crosslinked and collapse due to electrostatics and the devices exhibit a spectral shift, as shown in Figure 14. The sensing protocol that we developed from this phenomenon is shown in Figure 15. As can be seen in Figure 15, an excess amount of PDNA is exposed to a solution containing TDNA and DNA with a completely mismatched sequence (CMMDNA), and DNA with four (4BPMMDNA) and two base mismatches (2BPMMDNA). The PDNA binds the TDNA completely, leaving behind excess, unbound PDNA in solution. Magnetic microparticles (MMPDNA) that are functionalized with the complete complement to PDNA were added to the solution to capture the excess PDNA. A magnet was then used to isolate the magnetic microparticles bound with PDNA (MMPDNA–PDNA) from the solution. After washing the MMPDNA–PDNA, the PDNA was recovered by heating the solution to melt the DNA off of the MMPDNA–PDNA, and the excess PDNA was recovered and added to the etalon. In this case, a large spectral shift from the etalon corresponds to a large excess of PDNA, which means a low concentration of TDNA was present in the initial solution. The opposite is true as well—a low concentration of PDNA left in solution yields a small spectral shift from the device, meaning there was a large amount of TDNA present in the initial solution. This illustrates the strength of the current system—low concentrations of TDNA yield large spectral shifts making the device more sensitive to low DNA concentrations.

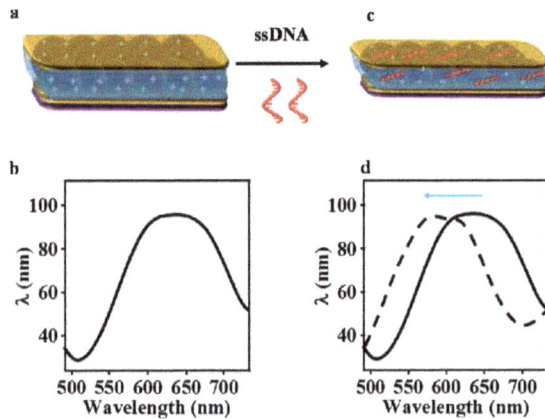

Figure 14. (**a**) The basic structure of pNIPAm-co-*N*-(3-aminopropyl) methacrylamide hydrochloride microgel-based etalon; (**b**) A schematic representation of a single reflectance peak. Here, the microgels are positively charged in water with pH < 10.0; (**c**) After addition of ssDNA, the microgels were crosslinked and collapsed, reducing the distance between the two Au layers of the device; (**d**) The peak of the reflectance spectrum shifts to a shorter wavelength. Reprinted with the permission of [57] Copyright © 2014, Springer Verlag Berlin Heidelberg.

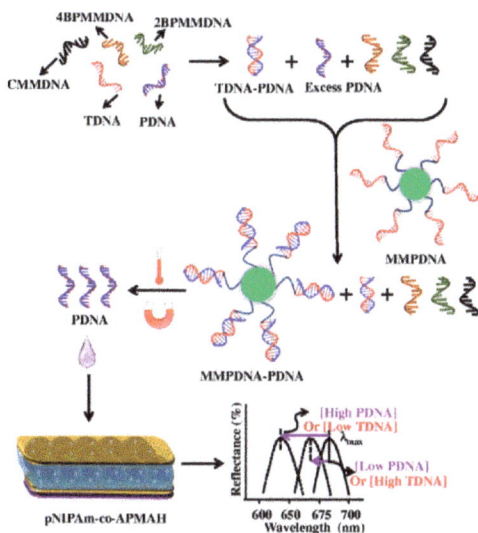

Figure 15. The protocol used for indirectly sensing target DNA (TDNA), by sensing probe DNA (PDNA). In this case excess PDNA can be related to TDNA concentration, even in the presence of DNA that is completely noncomplementary to PDNA (CMMDNA), and with 2 and 4 base pair mismatches (2BPMMDNA and 4BPMMDNA). Reprinted with the permission of [57] Copyright © 2014, Springer Verlag Berlin Heidelberg.

Recently, our group developed novel multiresponsive pNIPAm-based microgels by incorporation of the molecule triphenylmethane leucohydroxide (TPL) into their structure. Figure 16 shows the schematic depiction of TPL-modified microgels and their response to various stimuli. These microgels were subsequently used to fabricate etalons, and the optical properties investigated in response to ultraviolet and visible irradiation, solution pH changes, and the presence of a mimic of the nerve agent Tabun was characterized [58]. We also clearly showed that the optical properties of the device depended dramatically on these stimuli and show great promise for remote actuation and sensing.

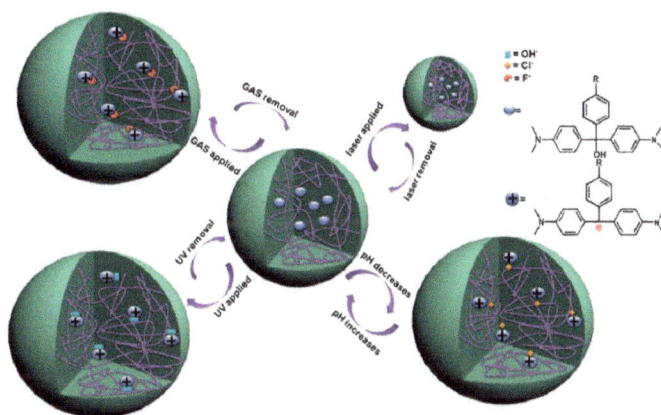

Figure 16. The various responses expected from TPL-modified microgels. The TPL structure is shown on the right. Reprinted with the permission of [58] Copyright © 2014, WILEYVCH Verlag GmbH & Co. KGaA, Weinheim.

Etalon-based systems were also prepared that are capable of changing their optical properties in response to light by employing a photoacid combined with pH responsive microgels [70]. Specifically, a photoacid is a molecule that is capable of generating protons when exposed to UV irradiation, which can decrease the pH of a solution. Figure 17 shows the relationship between UV irradiation times, pH change and wavelength shift that also yields a visible color change—these results were achieved using the photoacid o-nitrobenzaldehyde (o-NBA). The color of this device could be visibly changed in less than 3 min. Light responsivity can easily be initiated/stopped by simply switching the excitation source on/off, while the magnitude of the response can be tuned by modulating the excitation source intensity, and/or wavelength.

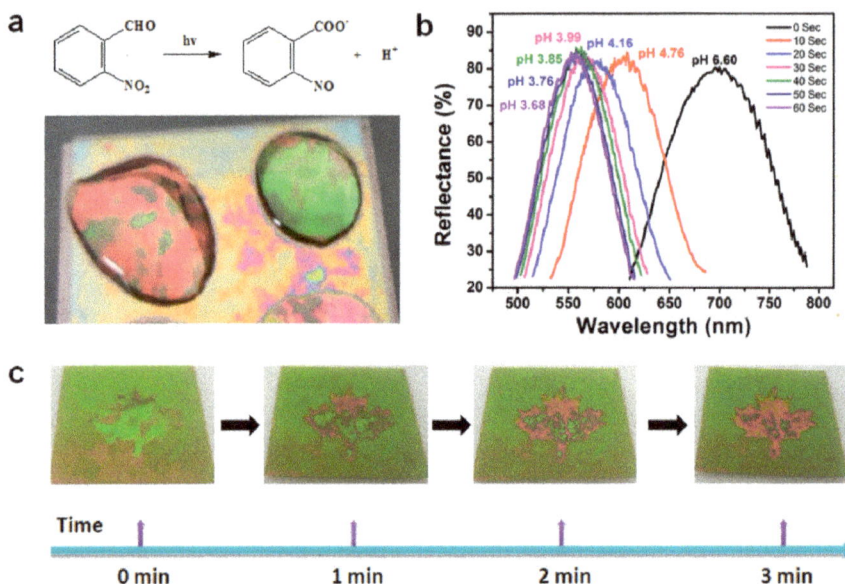

Figure 17. (**a**) Schematic depiction of the photoacid generation process upon exposure to UV light, and a pH responsive etalon that changes color upon light exposure and acid generation; (**b**) The dependence of the device reflectance spectrum at different pH values that were changed by UV exposure; (**c**) Photographs of patterned etalons (maple leaf pattern is pH responsive) after the indicated irradiation times. Reproduced with permission from the authors of [70], Copyright © 2014, Royal Society of Chemistry.

Etalons that were sensitive to the application of an electric field were also fabricated and tested, and the response is shown in Figure 18. In this study, by applying a certain voltage (~3 V), a pH responsive etalon could exhibit visible and reversible color changes [59]. This is due to the hydrolysis of water at the applied potential, which subsequently changes the pH of the environment. Hence, if pH responsive etalons were exposed to this system, the etalon optical properties will likewise depend on applied electrical potential. The peak shift is related to the potential applied as shown in Figure 18b, and this phenomenon is completely reversible. We show that the etalon's optical properties (color) are stable for many hours, until an appropriate potential is applied to bring the solution pH back to its initial value. In this example, we constructed an etalon with a maple leaf pattern that was pH responsive—the color of the maple leaf changed when an electrical potential was applied, and was reversible over many cycles, as shown in Figure 18d.

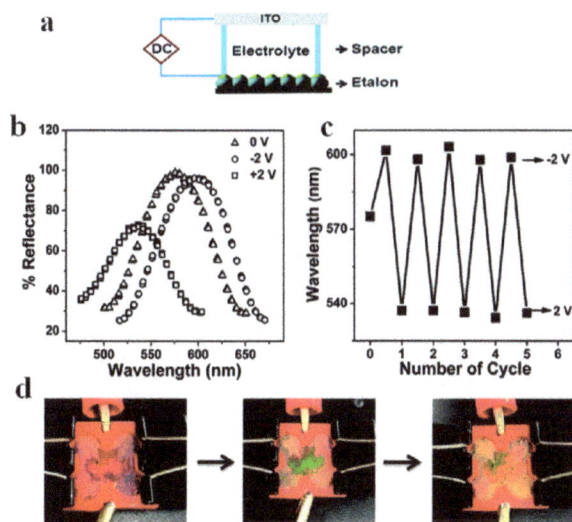

Figure 18. (**a**) Schematic of the experimental setup; (**b**) Reflectance spectra collected from an etalon after the application of the indicated voltages; (**c**) Final peak positions after application of the indicated potentials to the etalon over many cycles; (**d**) Photographs of a patterned etalon in an electrochemical cell at: 0 V, 2 V, −2 V, from left to right. Reproduced with permission from the authors of [59], Copyright © 2014, Royal Society of Chemistry.

While we have shown that microgel-based etalons can be fabricated on various planar substrates, there are some substrates that are difficult to coat with etalons, e.g., curved surfaces, rods, and tubes. Furthermore, for our sensing and monitoring efforts, it is advantageous to fabricate devices that can be adhered to skin. To achieve this, etalons could be fabricated on planar substrates containing a previously adsorbed sacrificial layer that could be easily dissolved and the etalon desorbed from the surface. The desorbed etalon can then be adhered to any other substrate as needed. To demonstrate that this is possible, we fabricated free-standing pNIPAm-co-AAc microgel-based etalons that exhibit high quality optical properties, which are capable of being transferred to multiple substrates, as shown in Figure 19. The etalon was fabricated on a solid support that was coated with a sacrificial polymer layer, which was generated by the layer-by-layer self-assembly technique. Therefore, the etalon could be easily removed from the solid substrate upon dissolution of the sacrificial layer. The desorbed etalons exhibit similar optical properties to the substrate adhered etalons, and retain their pH and temperature responsivity. This free-standing optical device will open new applications for sensing in environments that cannot tolerate the planar etalon geometry. Furthermore, the devices can be adhered to skin for real-time monitoring of human/animal health [71].

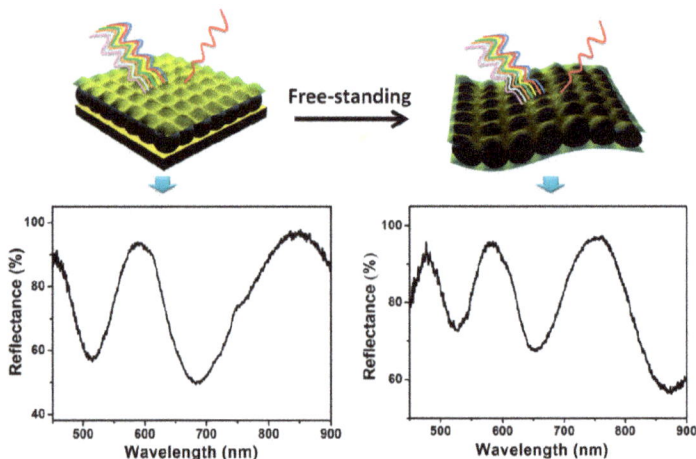

Figure 19. (**top**) Poly (*N*-isopropylacrylamide-co-acrylic acid) (pNIPAm-co-AAc) microgel-based free-standing etalons that exhibit (**bottom**) optical properties similar to substrate bound devices. Reprinted with permission from the authors of [71], Copyright © 2014, Royal Society of Chemistry.

Furthermore, we recently developed polymer-based artificial muscles, which can act as sensing devices by utilizing responsive microgel and polymer layers [72]. The structure of this device is shown in Figure 20. As can be seen, negatively charged microgels were deposited on a Au-coated plastic substrate followed by the deposition of a solution of the cationic polyelectrolyte, poly(diallyldimethyl ammonium chloride) (pDADMAC). The assembly was then allowed to dry, and the device subsequently bends. The bending was due to the strong adhesion of the pDADMAC layer to the flexible substrate (through electrostatic interaction with the microgels), and it's contraction upon drying. When the pDADMAC layer contracts upon drying the flexible substrate must bend. If the environmental humidity is increased, the pDADMAC layer rehydrates, and the device unbends, as shown schematically in Figure 20b. This bending/unbending mechanism is completely reversible over many cycles. We further showed that the device could be used to lift weights, as can be seen in Figure 20c and can move components around by lifting and dropping components in a humidity dependent fashion [72].

Furthermore, we demonstrated that the ability of the devices to lift masses can be used for sensing applications [73]. Firstly, the device and paperclips (used as weights) were added to a humidity-controlled chamber along with a top loading balance. The paperclips were allowed to rest on the balance, which recorded a mass. A photograph of the setup can be seen in Figure 21a. Initially, the chamber was held at 0% relative humidity, which caused the device to be completely curled up—this resulted in a relatively low mass on the balance. When the relative humidity was increased from 0% to 10% (usually takes ~30 min to fully stabilize) the device opened up, which allowed more of the paperclip mass to be added to the balance pan resulting in a concomitant increase in the measured mass. As the chamber humidity was incrementally increased, the device opened up, subsequently adding more of the paperclip mass to the balance pan. At 50% relative humidity, the device was completely open, bringing the mass to a maximum. The complete response of the device over the whole humidity range is shown in Figure 21b.

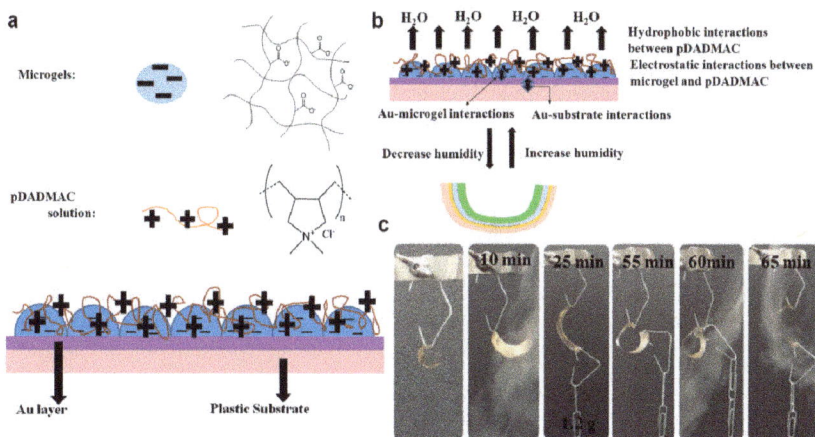

Figure 20. (**a**) Humidity sensor devices were constructed by depositing a single layer of pNIPAm-co-AAc microgels on a flexible plastic substrate (bottom layer) coated with an Au/Cr layer. After adding a solution of pDADMAC on top of the layer, a strong electrostatic interaction between the negatively charged microgels and the positively charged pDADMAC was formed; (**b**) The humidity sensing mechanism of this device. The pDADMAC layer resolvates or desolvated with the humidity change, and cause the device unbends or bend. This bending/unbending mechanism is completely reversible over many cycles; (**c**) A small curled substrate was hung from an arm and cycled between low and high humidity. Reprinted with permission from the authors of [72], Copyright © 2014, WILEY-VCH Verlag GmbH & Co. KGaA, Weinheim.

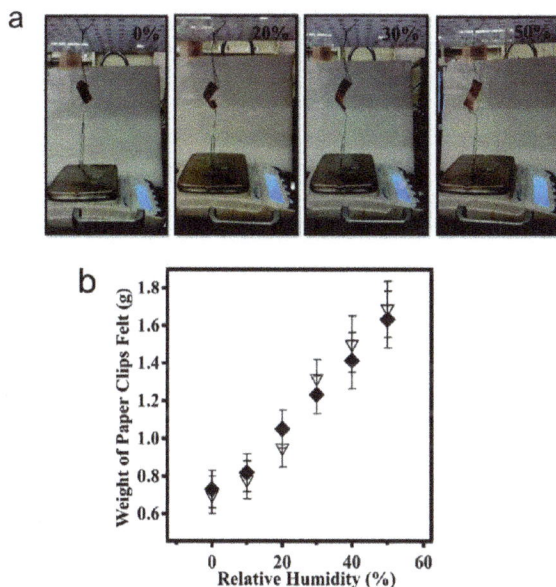

Figure 21. (**a**) Polymer-based actuator hanging in the humidity controlled chamber with paperclips attached, which were rested on the pan of a top loading balance; (**b**) The measured mass of the paperclips on the balance pan as a function of (♦) increasing and (△) decreasing humidity. Reprinted with permission from the authors of [73], Copyright © 2014, Royal Society of Chemistry.

4. Conclusions

We have briefly reviewed a few key examples of the use of responsive polymer-based photonic materials as sensors. The sensing mechanisms primarily depended on the responsive polymers changing the spacing between the PM/PC array elements, yielding the observed change in optical properties. While this is the case, changes in the lattice effective refractive index can also yield changes in their optical properties. PNIPAm microgel-based devices were subsequently introduced, and their use as sensors detailed, with a focus on pH, glucose, protein, and DNA sensing. We conclude that pNIPAm microgel-based devices have a lot of promise for sensing applications, although more work is needed to bring the technology to the market. The first challenge is the limited tunability of the PMs optical properties, which limits sensitivity; although, many examples in this review do show tunability over large wavelength ranges. Another limitation is response time; many examples in this review depend on diffusion to yield a response, which is a slow process. Yet another limitation is building reusability into the sensors; this could be addressed by exploiting the weakening of interactions with *T*, pH, and ionic strength to break signal-causing interactions to regenerate sensors. Despite the limitations, the foundations for building useful PM/PC-based sensors has been laid and further tuning of the devices can alleviate the above-mentioned drawbacks. Furthermore, new research is leading to a new understanding of how PM/PC-based sensors can be generated, which will lead to a new generation of PM/PC-based sensor technology.

Acknowledgments: Michael J. Serpe acknowledges funding from the University of Alberta (the Department of Chemistry and the Faculty of Science), the Natural Sciences and Engineering Research Council of Canada (NSERC), the Canada Foundation for Innovation (CFI), the Alberta Advanced Education & Technology Small Equipment Grants Program (AET/SEGP), Grand Challenges Canada and IC-IMPACTS. Yongfeng Gao and Xue Li acknowledge Alberta Innovates Technology Futures (AITF) for graduate student scholarships.

Author Contributions: Xue Li and Yongfeng Gao gathered the review content, organized the material, and wrote the review. Michael J. Serpe helped write the review and did complete editing.

Conflicts of Interest: The authors declare no conflict of interest.

References

1. Hamner, K.L.; Alexander, C.M.; Coopersmith, K.; Reishofer, D.; Provenza, C.; Maye, M.M. Using Temperature-Sensitive Smart Polymers to Regulate DNA-Mediated Nanoassembly and Encoded Nanocarrier Drug Release. *ACS Nano* **2013**, *7*, 7011–7020. [CrossRef] [PubMed]

2. Meng, H.; Hu, J. A Brief Review of Stimulus-Active Polymers Responsive to Thermal, Light, Magnetic, Electric, and Water/Solvent Stimuli. *J. Intell. Mater. Syst. Struct.* **2010**, *21*, 859–885. [CrossRef]

3. Ionov, L. Actively-Moving Materials Based on Stimuli-Responsive Polymers. *J. Mater. Chem.* **2010**, *20*, 3382–3390. [CrossRef]

4. Peppas, N.A.; Hilt, J.Z.; Khademhosseini, A.; Langer, R. Hydrogels in Biology and Medicine: From Molecular Principles to Bionanotechnology. *Adv. Mater.* **2006**, *18*, 1345–1360. [CrossRef]

5. Qiu, Y.; Park, K. Environment-Sensitive Hydrogels for Drug Delivery. *Adv. Drug Deliv. Rev.* **2001**, *53*, 321–339. [CrossRef]

6. Holtz, J.H.; Asher, S.A. Polymerized Colloidal Crystal Hydrogel Films as Intelligent Chemical Sensing Materials. *Nature* **1997**, *389*, 829–832. [CrossRef]

7. Reese, C.E.; Mikhonin, A.V.; Kamenjicki, M.; Tikhonov, A.; Asher, S.A. Nanogel Nanosecond Photonic Crystal Optical Switching. *J. Am. Chem. Soc.* **2004**, *126*, 1493–1496. [CrossRef] [PubMed]

8. Guan, Y.; Zhang, Y. PNIPAM Microgels for Biomedical Applications: From Dispersed Particles to 3D Assemblies. *Soft Matter* **2011**, *7*, 6375–6384. [CrossRef]

9. Bromberg, L.E.; Ron, E.S. Temperature-Responsive Gels and Thermogelling Polymer Matrices for Protein and Peptide Delivery. *Adv. Drug Deliv. Rev.* **1998**, *31*, 197–221. [CrossRef]

10. Hoare, T.; Pelton, R. Highly pH and Temperature Responsive Microgels Functionalized with Vinylacetic Acid. *Macromolecules* **2004**, *37*, 2544–2550. [CrossRef]

11. Dai, S.; Ravi, P.; Tam, K.C. Thermo- and Photo-Responsive Polymeric Systems. *Soft Matter* **2009**, *5*, 2513–2533. [CrossRef]

12. Liu, G.; Liu, W.; Dong, C.-M. UV- and NIR-Responsive Polymeric Nanomedicines for On-Demand Drug Delivery. *Polym. Chem.* **2013**, *4*, 3431–3443. [CrossRef]

13. Kwon, I.C.; Bae, Y.H.; Kim, S.W. Electrically Credible Polymer Gel for Controlled Release of Drugs. *Nature* **1991**, *354*, 291–293. [CrossRef] [PubMed]

14. Brugger, B.; Richtering, W. Magnetic, Thermosensitive Microgels as Stimuli-Responsive Emulsifiers Allowing for Remote Control of Separability and Stability of Oil in Water-Emulsions. *Adv. Mater.* **2007**, *19*, 2973–2978. [CrossRef]

15. Hendrickson, G.R.; Andrew Lyon, L. Bioresponsive Hydrogels for Sensing Applications. *Soft Matter* **2009**, *5*, 29–35. [CrossRef]

16. Su, S.; Ali, M.M.; Filipe, C.D.M.; Li, Y.; Pelton, R. Microgel-Based Inks for Paper-Supported Biosensing Applications. *Biomacromolecules* **2008**, *9*, 935–941. [CrossRef] [PubMed]

17. Oh, J.K.; Lee, D.I.; Park, J.M. Biopolymer-Based Microgels/Nanogels for Drug Delivery Applications. *Prog. Polym. Sci.* **2009**, *34*, 1261–1282. [CrossRef]

18. Gao, Y.; Zago, G.P.; Jia, Z.; Serpe, M.J. Controlled and Triggered Small Molecule Release from a Confined Polymer Film. *ACS Appl. Mater. Interfaces* **2013**, *5*, 9803–9808. [CrossRef] [PubMed]

19. Takashima, Y.; Hatanaka, S.; Otsubo, M.; Nakahata, M.; Kakuta, T.; Hashidzume, A.; Yamaguchi, H.; Harada, A. Expansion–Contraction of Photoresponsive Artificial Muscle Regulated by Host–Guest Interactions. *Nat. Commun.* **2012**, *3*, 1270. [CrossRef] [PubMed]

20. Bassil, M.; Ibrahim, M.; El Tahchi, M. Artificial Muscular Microfibers: Hydrogel with High Speed Tunable Electroactivity. *Soft Matter* **2011**, *7*, 4833–4838. [CrossRef]

21. Lee, K.Y.; Mooney, D.J. Hydrogels for Tissue Engineering. *Chem. Rev.* **2001**, *101*, 1869–1880. [CrossRef] [PubMed]

22. Hollister, S.J. Porous Scaffold Design for Tissue Engineering. *Nat. Mater.* **2005**, *4*, 518–524. [CrossRef] [PubMed]

23. Phadke, A.; Zhang, C.; Arman, B.; Hsu, C.-C.; Mashelkar, R.A.; Lele, A.K.; Tauber, M. J.; Arya, G.; Varghese, S. Rapid Self-Healing Hydrogels. *Proc. Natl. Acad. Sci. USA* **2012**, *109*, 4383–4388. [CrossRef] [PubMed]

24. Wei, Z.; Yang, J.H.; Zhou, J.; Xu, F.; Zrínyi, M.; Dussault, P.H.; Osada, Y.; Chen, Y.M. Self-Healing Gels Based on Constitutional Dynamic Chemistry and Their Potential Applications. *Chem. Soc. Rev.* **2014**, *43*, 8114–8131. [CrossRef] [PubMed]

25. Zhang, G.; Wu, C. The Water/Methanol Complexation Induced Reentrant Coil-to-Globule-to-Coil Transition of Individual Homopolymer Chains in Extremely Dilute Solution. *J. Am. Chem. Soc.* **2001**, *123*, 1376–1380. [CrossRef]

26. Wu, C.; Zhou, S. Laser Light Scattering Study of the Phase Transition of Poly(N-isopropylacrylamide) in Water. 1. Single Chain. *Macromolecules* **1995**, *28*, 8381–8387. [CrossRef]

27. Hoare, T.; Pelton, R. Engineering Glucose Swelling Responses in Poly(N-isopropylacrylamide)-Based Microgels. *Macromolecules* **2007**, *40*, 670–678. [CrossRef]

28. Gao, J.; Wu, C. The "Coil-to-Globule" Transition of Poly(N-isopropylacrylamide) on the Surface of a Surfactant-Free Polystyrene Nanoparticle. *Macromolecules* **1997**, *30*, 6873–6876. [CrossRef]

29. Zhou, M.; Xie, J.; Yan, S.; Jiang, X.; Ye, T.; Wu, W. Graphene@Poly(phenylboronic acid)s Microgels with Selectively Glucose-Responsive Volume Phase Transition Behavior at a Physiological pH. *Macromolecules* **2014**, *47*, 6055–6066. [CrossRef]

30. Dhanya, S.; Bahadur, D.; Kundu, G.C.; Srivastava, R. Maleic Acid Incorporated Poly-(N-isopropylacrylamide) Polymer Nanogels for Dual-Responsive Delivery of Doxorubicin Hydrochloride. *Eur. Polym. J.* **2013**, *49*, 22–32. [CrossRef]

31. Zhang, W.; Mao, Z.; Gao, C. Preparation of TAT Peptide-Modified Poly(N-isopropylacrylamide) Microgel Particles and Their Cellular Uptake, Intracellular Distribution, and Influence on Cytoviability in Response to Temperature Change. *J. Colloid Interface Sci.* **2014**, *434*, 122–129. [CrossRef] [PubMed]

32. Ge, J.; Yin, Y. Responsive Photonic Crystals. *Angew. Chem. Int. Ed.* **2011**, *50*, 1492–1522. [CrossRef] [PubMed]

33. Moon, J.H.; Yang, S. Chemical Aspects of Three-Dimensional Photonic Crystals. *Chem. Rev.* **2009**, *110*, 547–574. [CrossRef] [PubMed]

34. Schacher, F.H.; Rupar, P.A.; Manners, I. Functional Block Copolymers: Nanostructured Materials with Emerging Applications. *Angew. Chem. Int. Ed.* **2012**, *51*, 7898–7921. [CrossRef] [PubMed]

35. Kelly, J.A.; Shukaliak, A.M.; Cheung, C.C.; Shopsowitz, K.E.; Hamad, W.Y.; MacLachlan, M.J. Responsive Photonic Hydrogels Based on Nanocrystalline Cellulose. *Angew. Chem. Int. Ed.* **2013**, *52*, 8912–8916. [CrossRef] [PubMed]

36. Wang, X.Q.; Wang, C.F.; Zhou, Z.F.; Chen, S. Hydrogels: Robust Mechanochromic Elastic One-Dimensional Photonic Hydrogels for Touch Sensing and Flexible Displays. *Adv. Opt. Mater.* **2014**, *2*, 651–662. [CrossRef]

37. Santos, A.; Kumeria, T.; Losic, D. Nanoporous Anodic Aluminum Oxide for Chemical Sensing and Biosensors. *TrAC Trend. Anal. Chem.* **2013**, *44*, 25–38. [CrossRef]

38. Shinohara, S.-I.; Seki, T.; Sakai, T.; Yoshida, R.; Takeoka, Y. Chemical and Optical Control of Peristaltic Actuator Based on Self-Oscillating Porous Gel. *Chem. Commun.* **2008**, 4735–4737. [CrossRef] [PubMed]

39. Choi, S.Y.; Mamak, M.; Von Freymann, G.; Chopra, N.; Ozin, G.A. Mesoporous Bragg Stack Color Tunable Sensors. *Nano Lett.* **2006**, *6*, 2456–2461. [CrossRef] [PubMed]

40. Peppas, N.A.; Van Blarcom, D.S. Hydrogel-Based Biosensors and Sensing Devices for Drug Delivery. *J. Control. Release* **2015**. [CrossRef] [PubMed]

41. Burgess, I.B.; Mishchenko, L.; Hatton, B.D.; Kolle, M.; Loncar, M.; Aizenberg, J. Encoding Complex Wettability Patterns in Chemically Functionalized 3D Photonic Crystals. *J. Am. Chem. Soc.* **2011**, *133*, 12430–12432. [CrossRef] [PubMed]

42. Kuang, M.; Wang, J.; Bao, B.; Li, F.; Wang, L.; Jiang, L.; Song, Y. Inkjet Printing Patterned Photonic Crystal Domes for Wide Viewing-Angle Displays by Controlling the Sliding Three Phase Contact Line. *Adv. Opt. Mater.* **2014**, *2*, 34–38. [CrossRef]

43. Weissman, J.M.; Sunkara, H.B.; Tse, A.S.; Asher, S.A. Thermally Switchable Periodicities and Diffraction from Mesoscopically Ordered Materials. *Science* **1996**, *274*, 959–960. [CrossRef] [PubMed]

44. Lee, K.; Asher, S.A. Photonic Crystal Chemical Sensors: pH and Ionic Strength. *J. Am. Chem. Soc.* **2000**, *122*, 9534–9537. [CrossRef]

45. Alexeev, V.L.; Sharma, A.C.; Goponenko, A.V.; Das, S.; Lednev, I.K.; Wilcox, C.S.; Finegold, D.N.; Asher, S.A. High Ionic Strength Glucose-Sensing Photonic Crystal. *Anal. Chem.* **2003**, *75*, 2316–2323. [CrossRef] [PubMed]

46. Zhang, J.-T.; Smith, N.; Asher, S.A. Two-Dimensional Photonic Crystal Surfactant Detection. *Anal. Chem.* **2012**, *84*, 6416–6420. [CrossRef] [PubMed]

47. Fudouzi, H.; Sawada, T. Photonic Rubber Sheets with Tunable Color by Elastic Deformation. *Langmuir* **2005**, *22*, 1365–1368. [CrossRef] [PubMed]

48. Lee, E.; Kim, J.; Myung, J.; Kang, Y. Modification of Block Copolymer Photonic Gels for Colorimetric Biosensors. *Macromol. Res.* **2012**, *20*, 1219–1222. [CrossRef]

49. Honda, M.; Seki, T.; Takeoka, Y. Dual Tuning of the Photonic Band-Gap Structure in Soft Photonic Crystals. *Adv. Mater.* **2009**, *21*, 1801–1804. [CrossRef]

50. Karg, M.; Hellweg, T.; Mulvaney, P. Self-Assembly of Tunable Nanocrystal Superlattices Using Poly-(NIPAM) Spacers. *Adv. Funct. Mater.* **2011**, *21*, 4668–4676. [CrossRef]

51. Sorrell, C.D.; Carter, M.C.D.; Serpe, M.J. Color Tunable Poly (N-Isopropylacrylamide)-co-Acrylic Acid Microgel–Au Hybrid Assemblies. *Adv. Funct. Mater.* **2011**, *21*, 425–433. [CrossRef]

52. Sorrell, C.D.; Serpe, M.J. Reflection Order Selectivity of Color-Tunable Poly(N-isopropylacrylamide) Microgel Based Etalons. *Adv. Mater.* **2011**, *23*, 4088–4092. [CrossRef] [PubMed]

53. Sorrell, C.; Serpe, M. Glucose Sensitive Poly (N-isopropylacrylamide) Microgel Based Etalons. *Anal. Bioanal. Chem.* **2012**, *402*, 2385–2393. [CrossRef] [PubMed]

54. Hu, L.; Serpe, M.J. Color Modulation of Spatially Isolated Regions on a Single Poly(N-isopropylacrylamide) Microgel Based Etalon. *J. Mater. Chem.* **2012**, *22*, 8199–8202. [CrossRef]

55. Johnson, K.C.C.; Mendez, F.; Serpe, M.J. Detecting Solution pH Changes Using Poly (N-isopropylacrylamide)-co-Acrylic Acid Microgel-Based Etalon Modified Quartz Crystal Microbalances. *Anal. Chim. Acta* **2012**, *739*, 83–88. [CrossRef] [PubMed]

56. Islam, M.R.; Serpe, M.J. Polyelectrolyte Mediated Intra and Intermolecular Crosslinking in Microgel-Based Etalons for Sensing Protein Concentration in Solution. *Chem. Commun.* **2013**, *49*, 2646–2648. [CrossRef] [PubMed]

57. Islam, M.; Serpe, M. Polymer-Based Devices for the Label-Free Detection of DNA in Solution: Low DNA Concentrations Yield Large Signals. *Anal. Bioanal. Chem.* **2014**, *406*, 4777–4783. [CrossRef] [PubMed]

58. Zhang, Q.M.; Xu, W.; Serpe, M.J. Optical Devices Constructed from Multiresponsive Microgels. *Angew. Chem. Int. Ed.* **2014**, *53*, 4827–4831. [CrossRef] [PubMed]

59. Xu, W.; Gao, Y.; Serpe, M.J. Electrochemically Color Tunable Poly(N-isopropylacrylamide) Microgel-Based Etalons. *J. Mater. Chem. C* **2014**, *2*, 3873–3878. [CrossRef]

60. Hu, L.; Serpe, M.J. Controlling the Response of Color Tunable Poly(N-isopropylacrylamide) Microgel-Based Etalons with Hysteresis. *Chem. Commun.* **2013**, *49*, 2649–2651. [CrossRef] [PubMed]

61. Islam, M.R.; Johnson, K.C.C.; Serpe, M.J. Microgel-Based Etalon Coated Quartz Crystal Microbalances for Detecting Solution pH: The Effect of Au Overlayer Thickness. *Anal. Chim. Acta* **2013**, *792*, 110–114. [CrossRef] [PubMed]

62. Islam, M.R.; Serpe, M.J. Penetration of Polyelectrolytes into Charged Poly(N-isopropylacrylamide) Microgel Layers Confined between Two Surfaces. *Macromolecules* **2013**, *46*, 1599–1606. [CrossRef]

63. Carter, M.C.; Sorrell, C.D.; Serpe, M.J. Deswelling Kinetics of Color Tunable Poly (N-isopropylacrylamide) Microgel-Based Etalons. *J. Phys. Chem. B* **2011**, *115*, 14359–14368. [CrossRef] [PubMed]

64. Islam, M.R.; Serpe, M.J. Poly (N-isopropylacrylamide) Microgel-Based Etalons and Etalon Arrays for Determining the Molecular Weight of Polymers in Solution. *APL Mat.* **2013**, *1*, 052108. [CrossRef]

65. Islam, M.R.; Serpe, M.J. Label-Free Detection of Low Protein Concentration in Solution Using a Novel Colorimetric Assay. *Biosens. Bioelectron.* **2013**, *49*, 133–138. [CrossRef] [PubMed]

66. Taton, T.A.; Mirkin, C.A.; Letsinger, R.L. Scanometric DNA Array Detection with Nanoparticle Probes. *Science* **2000**, *289*, 1757–1760. [CrossRef] [PubMed]

67. Zhao, Y.; Zhao, X.; Tang, B.; Xu, W.; Li, J.; Hu, J.; Gu, Z. Quantum-Dot-Tagged Bioresponsive Hydrogel Suspension Array for Multiplex Label-Free DNA Detection. *Adv. Funct. Mater.* **2010**, *20*, 976–982. [CrossRef]

68. Gao, M.; Gawel, K.; Stokke, B.T. Toehold of dsDNA Exchange Affects the Hydrogel Swelling Kinetics of a Polymer-dsDNA Hybrid Hydrogel. *Soft Matter* **2011**, *7*, 1741–1746. [CrossRef]

69. Islam, M.R.; Serpe, M.J. A Novel Label-Free Colorimetric Assay for DNA Concentration in Solution. *Anal. Chim. Acta* **2014**, *843*, 83–88. [CrossRef] [PubMed]

70. Gao, Y.; Serpe, M.J. Light-Induced Color Changes of Microgel-Based Etalons. *ACS Appl. Mater. Interfaces* **2014**, *6*, 8461–8466. [CrossRef] [PubMed]

71. Gao, Y.; Xu, W.; Serpe, M.J. Free-Standing Poly (N-isopropylacrylamide) Microgel-Based Etalons. *J. Mater. Chem. C* **2014**, *2*, 5878–5884. [CrossRef]

72. Islam, M.R.; Li, X.; Smyth, K.; Serpe, M.J. Polymer-Based Muscle Expansion and Contraction. *Angew. Chem. Int. Ed.* **2013**, *52*, 10330–10333. [CrossRef] [PubMed]

73. Islam, M.R.; Serpe, M.J. Poly (N-isopropylacrylamide) Microgel-Based Thin Film Actuators for Humidity Sensing. *RSC Adv.* **2014**, *4*, 31937–31940. [CrossRef]

Technical Note

Improved PNIPAAm-Hydrogel Photopatterning by Process Optimisation with Respect to UV Light Sources and Oxygen Content

Sebastian Haefner [1], Mathias Rohn [2], Philipp Frank [1], Georgi Paschew [1], Martin Elstner [3] and Andreas Richter [1,3,*]

[1] Polymeric Microsystems, Institute of Semiconductors and Microsystems, Technische Universität Dresden, 01062 Dresden, Germany; sebastian.haefner@tu-dresden.de (S.H.); philipp.frank@tu-dresden.de (P.F.); georgi.paschew@tu-dresden.de (G.P.)

[2] Physical Chemistry of Polymers, Department of Chemistry and Food Chemistry, Technische Universität Dresden, 01062 Dresden, Germany; mathias.rohn@tu-dresden.de

[3] Center for Advancing Electronics Dresden (cfaed), Technische Universität Dresden, 01062 Dresden, Germany; martin.elstner@tu-dresden.de

* Correspondence: andreas.richter7@tu-dresden.de; Tel.: +49-351-463-36336; Fax: +49-351-463-37021

Academic Editor: Dirk Kuckling
Received: 25 January 2016; Accepted: 18 February 2016; Published: 4 March 2016

Abstract: Poly-*N*-isopropylacrylamide (PNIPAAm) hydrogels, known for their sensor and actuator capabilities, can be photolithographically structured for microsystem applications. For usage in microsystems, the preparation, and hence the characteristics, of these hydrogels (e.g., degree of swelling, size, cooperative diffusion coefficient) are key features, and have to be as reproducible as possible. A common method of hydrogel fabrication is free radical polymerisation using a thermally-initiated system or a photoinitiator system. Due to the reaction quenching by oxygen, the polymer solution has to be rinsed with protective inert gases like nitrogen or argon before the polymerisation process. In this paper, we focus on the preparation reproducibility of PNIPAAm hydrogels under different conditions, and investigate the influence of oxygen and the UV light source during the photopolymerisation process. The flushing of the polymer solution with inert gas is not sufficient for photostructuring approaches, so a glove box preparation resulting in better quality. Moreover, the usage of a wide-band UV light source yields higher reproducibility to the photostructuring process compared to a narrow-band UV source.

Keywords: photopatterning; photopolymerisation; PNIPAAm; stimuli-responsive hydrogel; reproducibility

---

1. Introduction

Hydrogels are 3D polymer networks with the capability to swell in a swelling agent (e.g., water [1]). During the swelling process, the gel can take up multiples of its dry mass of water [2]. This type of polymer can be chemically tuned regarding its chemical, physical, or mechanical properties [3]. By functionalisation, a polymeric network can act as storage containers for enzymes, antibodies, or DNA [4,5]. By changing the cross-linker concentrations, the elastic modulus [6] of the hydrogel can be altered from very soft to rubber-like so that an adjustment of mechanical requirements becomes possible. Hydrogel copolymers produced by mixing different monomers in diverse ratios also affects the hydrogel properties. So far, a variety of hydrogel compositions incorporating various functional properties have been published and commercialised [7]. Promising features have been presented for the fields of drug delivery systems [8,9], as 3D scaffolds for tissue engineering [10–12], and as an active material in microfluidic valves [13] and chemostats [14,15].

Popular   hydrogels   for   microsystem   engineering   approaches   are   PNIPAAm

(Poly-*N*-isopropylacrylamide)-based hydrogels [16,17]. Such smart hydrogels react to environmental condition changes with an altered degree of swelling [18–20]. PNIPAAm hydrogels or derived copolymers are used as microfluidic valves or active material in chemofluidic transistors, and as active components in microfluidic circuits [21]. The ability to pattern these gels by lithography makes PNIPAAm hydrogels applicable for large scale fabrication and high-integration techniques [22]. Therefore, reproducibility of its fabrication process is a key factor for system reliability.

A common method to produce polymers in general, including hydrogels is free radical polymerisation, because of its convenient and time-effective synthesis procedure. A drawback of this approach is the sensitivity to radical quenchers like oxygen. These are strongly reduced or even eliminated by degassing and rinsing the system with protective gas. However, during the process of photolithographical patterning, the influence of oxygen can hardly be avoided. Most sources of error are related to short exposures to non-oxygen-free atmosphere, to oxygen molecules attached to a material surface, or to diffusion through a tube or seal. In some cases the contamination results from oxygen encapsulated in a material slowly surfacing out. In general, a reaction chamber is filled with prepolymer solution and is selectively exposed to UV light through a photomask. We tested three methods of preparation and concentrate on two of them. The assembling (filling and closing) of the reaction chamber is most critical in the handling due to a possible contamination with oxygen: 1) A reaction chamber is filled with a polymer solution in a standard air environment and is immediately closed. 2) A tube system is used for rinsing with protective gas and filling with a prepolymer solution. 3) The chamber is filled in a glove-box under an atmosphere with less than 5 ppm oxygen. As preliminary experiments only showed significant difference between method 1) and 3), we concentrate on them. In this paper we focus on the hydrogel preparation conditions for photostructuring by a free radical polymerisation reaction. We show results for different UV light sources (continuous wide-band and pulsed narrow-band), for different exposure times and compare strictly inert and atmospheric handling procedures. The cooperative diffusion coefficients and the degree of swelling were measured as characteristics to get insights into the reproducibility of the polymerisation process and therefore the hydrogel preparation.

2. Results and Discussion

2.1. Oxygen and UV Source Influence on the Polymerisation Process

The photopolymerisation homogeneity of PNIPAAm hydrogels was investigated regarding oxygen content and UV light source. NMR tubes were filled with polymerisation solution under inert or standard conditions. After sealing, the rotation setup was assembled following the sketch in Figure 1. The NMR tube's orientation was orthogonal to the UV light radiation. A water bath (0 °C) was used for cooling the solution during the exposure time. Due to the thickness of the tube (outer diameter 5 mm, inner diameter 4.2 mm, length 178 mm), a rotation setup is necessary to avoid a gradient in the gel polymerisation.

For UV light radiation we used two different sources: one wide-band and one narrow-band source. A common wide-band source for UV light are mercury-vapour lamps (see Figure 2). A bandpass filter suppresses all wavelengths below 300 nm and all wavelengths above 400 nm, so that the effective bandwidth is between 300 nm to 400 nm. The UV/Vis absorption spectra of the photoinitiator Irgacure® is also plotted in Figure 2. The absorption maximum of the photoinitiator is below 300 nm, and therefore both UV light sources do not fit well to the absorption spectra of the photoinitiator. Nevertheless, for applications in which for instance enzymes should be incorporated in the hydrogels during the polymerisation process, the excitation wavelength should be as high as possible to avoid enzyme damage. Due to safety considerations, moderate UV wavelengths are also preferred.

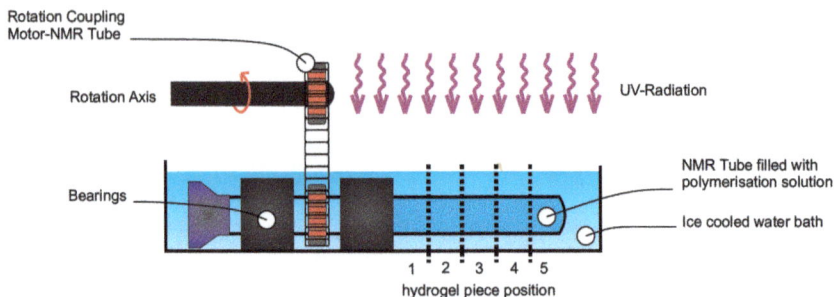

Figure 1. Diagram of the rotation setup for rotating a NMR tube filled with polymerisation solution.

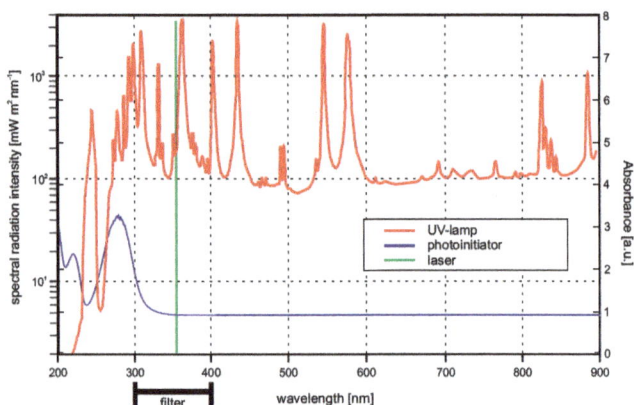

Figure 2. Plot of the spectral radiation intensity at 0.5 m proximity of the mercury-vapour lamp is demonstrated including the UV/Vis spectrum of the photoinitiator Irgacure®. The green line sketches the wavelength of the laser with no respect to one of the y-axes. The lamp spectra was reprinted from the lamp datasheet with kindly permission from LOT-Oriel Group Europe.

For the hydrogel polymerisation process we established three parameters, which will be discussed in this section. These parameters are the oxygen content, the exposure time, and the homogeneity of the intensity distribution of the UV light sources. First, the dependency on oxygen content and exposure time is discussed. For each exposure time experiment, a NMR tube was filled with polymerisation solution, coupled in the rotation setup (Figure 1), and exposed to UV light. Afterwards, the hydrogel was released from the tube, washed, cut into five pieces, and dried as described before. The degree of swelling (Q) was determined by Equation (1).

In Figure 3, the results for the exposure time variation experiments are shown. It is demonstrated that for an oxygen-free NMR tube filling inside a glove box, the required time for complete polymerisation (=gelpoint) is 10 s for the lamp (Figure 3a) and also for the laser (Figure 3b). Under standard conditions, 30 s (lamp) or 25 s (laser) exposure time is required to obtain a hydrogel. Shorter exposure times lead to non- or only partially-polymerised solutions. The degree of swelling drops with increasing exposure time. This can be explained by the fact that with longer exposure time the cross-linking density of the gel increases. With an increasing cross-linking density the degree of swelling decreases due to less hydrogen bond formation possibilities [6]. Comparing the inert condition laser and the lamp experiment, the degree of swelling for the laser is overall higher than the degree of swelling for the gels polymerised by the UV lamp. Referring to the light spectra from

Figure 2, the UV lamp provides radiation with wavelengths in regions with higher light absorption of the initiator compared to the laser. This yields higher quantum efficiency for the lamp. Therefore, at the same exposure time, more photoinitiator will be excited by the UV lamp than by the laser. This results in a higher cross-linking density for the gels fabricated with the lamp. For 30 s and 35 s exposure time, the degrees of swelling are nearly the same for both tested light sources. Here the amount of excited photoinitiator reaches saturation, which leads to comparable cross-linking densities and therefore degrees of swelling.

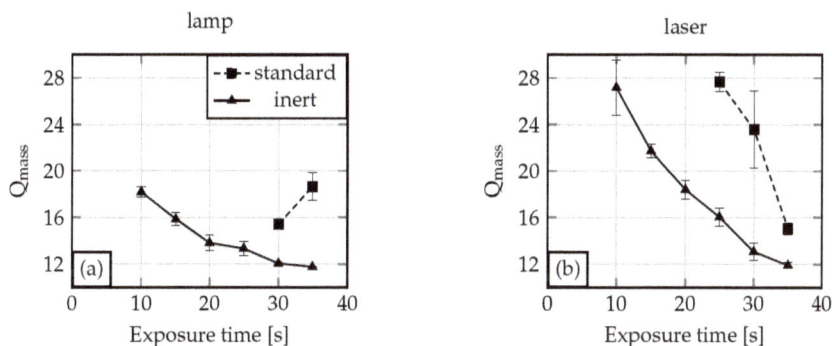

Figure 3. Degree of swelling of Poly-*N*-isopropylacrylamide (PNIPAAm) in water for different exposure times and UV light sources (lamp (**a**), laser (**b**)).

If the polymerisation solution is filled in the NMR tubes under standard conditions, the polymerisation process needs higher exposure times (see Figure 3). For the experiments conducted with the lamp, the required exposure time for the gel polymerisation was 30 s, and for the laser, 25 s. This and also the rise in the degree of swelling for the 35 s exposure time (lamp) demonstrates the influence of oxygen. Usually polymerisation processes conducted with the lamp should need less or the same exposure time as processes conducted with the laser, due to the better quantum yield. Furthermore, the degree of swelling for a hydrogel polymerised for 35 s should be smaller than the degree of swelling for a 30 s polymerised hydrogel. We explain both phenomena as the influence of undefined oxygen content in the polymerisation solution. This oxygen quenches radicals generated by the UV light and therefore leads to non-conclusive process parameters or hydrogel characteristics.

Preparing hydrogels with comparable characteristics under standard conditions becomes hardly possible due to the low reproducibility of the oxygen content in the polymerisation solution. Comparing the degrees of swelling between the systems with and without oxygen, it can be seen that systems with oxygen show overall a higher degree of swelling than the systems without oxygen. As discussed before, a smaller degree of swelling is a result of higher cross-linking density. Generated photoinitiator radicals due to UV light excitation will be quenched in oxygen-containing systems, and therefore a reduced cross-linking density is the result.

The hydrogel polymerisation process depends not only on the oxygen content and exposure time, but also on the radiation intensity of the UV light sources. Higher intensity will speed up the polymerisation process and lower intensity will slow down the process so that shorter or longer exposure times are required, respectively. Therefore, we asked how the intensity distribution of the UV light spots influences the polymerisation. If the UV light intensity distribution is too inhomogeneous, the polymerisation process will also be inhomogeneous, and therefore also the hydrogel characteristics. The field intensity distribution of the mercury-vapour lamp was investigated at 365 nm (highest intensity peak in the bandpass range) at different positions over the whole spot area with a light sensor (100S, Karl Suss, Garching, Germany). The same measurement was conducted for the laser. The results are shown by two surface plots in Figure 4. The intensity at the spot boundaries drops sharply for both sources, so that these areas were avoided during the polymerisation process. Some intense spots could also be measured, but the middle of the spot areas seems to be homogeneous. So these areas were chosen for the polymerisation process.

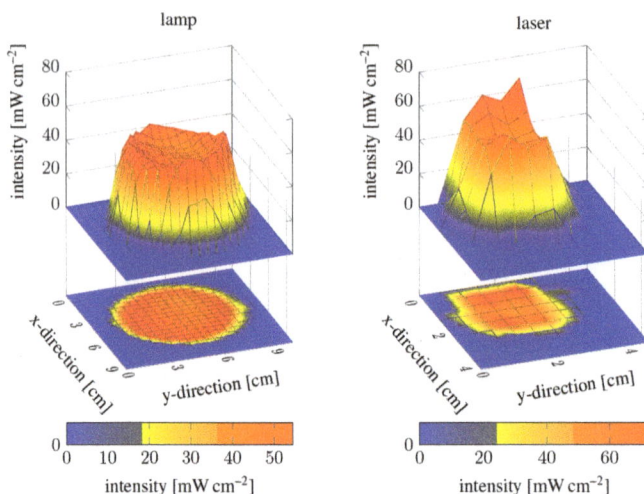

Figure 4. Intensity field distribution of the mercury-vapour lamp and the laser. Measurements were conducted at a distance of 0.5 m for the lamp and 1.5 m for the laser.

To exclude the influence of local spot intensities on the polymerisation process, the gels fabricated in the exposure time experiments were cut into five pieces per gel, washed, and dried as described before. In Figure 5, the degree of swelling for the gel pieces is plotted over the position of the gel piece in the original hydrogel body (see Figure 1). If these lines would be straight horizontal lines, this would imply that the fabricated primary hydrogel body has overall comparable characteristics independent from the tube position under the UV light in the polymerisation process. The spot homogeneities of the UV light sources are suitable in the middle, but the boundaries and the intense spots were avoided.

For inert conditions, a nearly-straight horizontal line course is observed. Especially for the lamp with an increasing polymerisation time, this progress becomes obvious. For the laser as the UV light source, the 10 s exposure time shows high alterations in the degree of swelling. Also, the results for the laser look overall more inhomogeneous than the results for the lamp. The reason for the high alteration at the 10 s exposure time with the laser may be the wavelength and therefore quantum yield or the pulsed light. With exposure time elongation, this drawback compared to the lamp will be reduced.

For the systems assembled under standard conditions, higher variations for the degree of swelling are demonstrated in Figure 5. This is explained by a non-uniform oxygen concentration distribution in the systems, which affects the cross-linking density locally in different ways.

We show that the polymerisation process depends on the exposure time and also the oxygen content. By measuring the UV light spots regarding their local intensities and comparing these results with the degree of swelling of hydrogel pieces from the original gel bodies, we demonstrate that the middle of the UV light spots are homogeneous and do not significantly influence the polymerisation process. These findings are important for investigating the reproducibility of the polymerisation process.

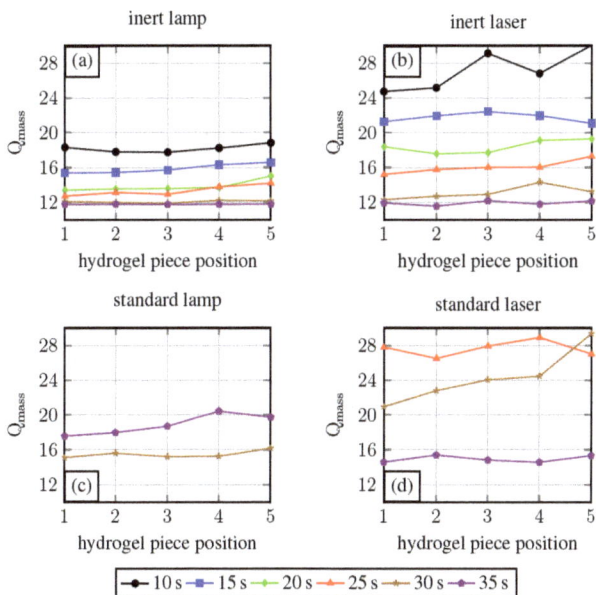

Figure 5. Localised resolution of the polymerisation process. The primary gel were cut into five pieces per body and the degrees of swelling were measured.

2.2. Polymerisation Process Reproducibility

As mentioned in the beginning, the reproducibility of gel characteristics is one of the key conditions for hydrogel applications in microsystem engineering. To investigate the reproducibility, NMR tubes were filled under different conditions (inert or standard) and exposed to UV light (lamp or laser) for 30 s. At this exposure time an overall hydrogel formation could be observed (see Figure 3). Afterwards DLS measurements were conducted with the gels in those NMR tubes to supply the cooperative diffusion coefficient D_{coop} for each hydrogel. The diffusion coefficients and the degree of swelling were compared (see Figure 6). For the degree of swelling measurements, gel bodies were cut into pieces and degrees of swelling were averaged.

For the evaluation of the reproducibility, the percentage on the mean for each experimental condition and characterisation parameter were used. The polymerisation process under these experimental conditions are for systems under inert conditions—the value is always significantly smaller than for systems under standard conditions (see Table 1). Also, the value for polymerisation under inert conditions with the mercury-vapour lamp is smaller for polymerisation with the laser. Therefore we got the highest reproducibility for an oxygen-free polymerisation with a UV lamp as radiation source.

Table 1. Hydrogel characteristics cooperative diffusion coefficient D_{coop} and degree of swelling Q_{mass} at different polymerisation conditions (standard or inert, lamp or laser).

Condition	D_{coop}	$SD_{D_{coop}}$		Q_{mass}	SD_Q	
Unit	$(cm^{-2} \cdot s^{-1})$	$(cm^{-2} \cdot s^{-1})$	%	(-)	(-)	%
lamp, inert	4.79×10^{-7}	5.64×10^{-9}	1.18	12.50	0.58	4.61
lamp, standard	2.57×10^{-7}	3.05×10^{-8}	11.84	17.81	2.87	16.11
laser, inert	3.97×10^{-7}	1.04×10^{-8}	2.61	12.76	1.37	10.70
laser, standard	2.48×10^{-7}	6.01×10^{-8}	24.23	19.02	4.06	21.33

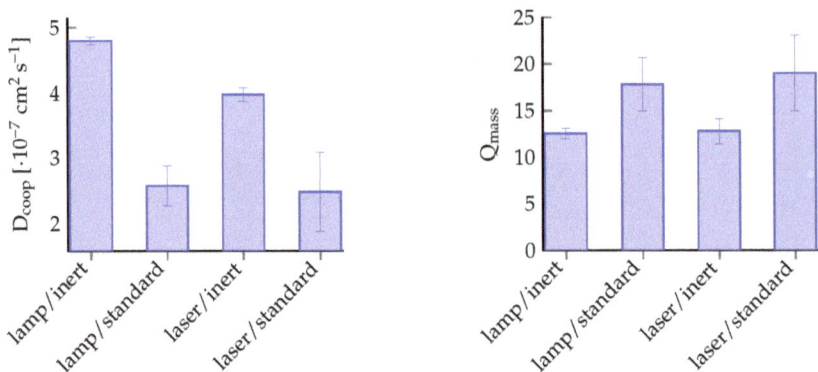

Figure 6. Process reproducibility with the cooperative diffusion coefficient D_{coop} and the degree of swelling Q as characterisation parameters.

Furthermore, inert polymerisation with the lamp yields the highest hydrogel cross-linking densities. With increasing cross-linking density, the degree of swelling decreases and the cooperative diffusion coefficient increases [23]. Following this interpretation, the polymerisation (independent of oxygen content) with the laser causes a smaller degree of cross-linking than the mercury-vapour lamp, as shown in Figure 6. Therefore a polymerisation process initiated by the laser used herein is favorable for hydrogels with high degrees of swelling.

2.3. Improved Hydrogel Microstructuring

To apply our previous findings, we performed photostructuring experiments to show the improvement in hydrogel photopatterning in the micrometer range. Following our previous findings we decided to use the mercury-vapour lamp as UV light source.

Figure 7 displays the results from hydrogel microstructuring experiments for different mask apertures ranging from 400 µm to 800 µm. The hydrogel diameter for glove box-assembled systems (inert conditions) are close to the mask aperture (see Figure 7). For the systems assembled under standard conditions, the hydrogel diameter is always smaller than the size of the mask aperture. Interestingly, in the case of the 400 µm aperture size in standard conditions, the resulting gel has a diameter of only 270 µm. The reason for this sparse result is that the polymerisation process gets quenched by oxygen. This becomes more critical with reduction of the photomask aperture and therefore hydrogel downscaling. With a reduction of photomask aperture, scattering effects at the mask boundaries become intensified, hence less light excites photoinitiator molecules. If there are excited photoinitiator molecules forming radicals for polymerisation, these radicals will be quickly quenched by oxygen, resulting in smaller hydrogel diameters than the mask aperture prompts.

For microsystems applications, these results demonstrate in the best way the necessity to conduct the polymerisation process under inert conditions. If hydrogel downscaling becomes essential, e.g., for high integration densities of hydrogel actuators, the polymerisation has to be conducted under inert conditions to guarantee pattern fidelity. Consequently, working under inert conditions not only improves the reproducibility of the polymerisation process but also the resolution of the photopatterning process.

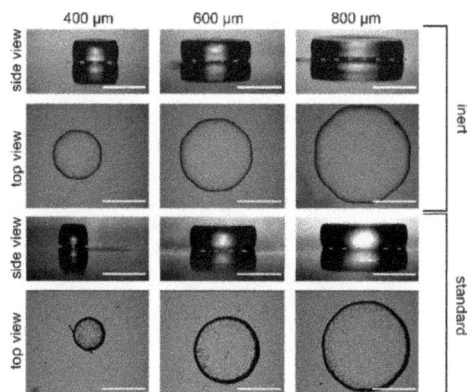

Figure 7. Results from polymerisation processes under different conditions. Hydrogels were covalently cross-linked to the glass surface during the polymerisation process. Therefore, a reflection of the gels in the side views is observable with the plane of reflection indicated by the red dashed lines (white bar = 400 μm).

3. Conclusions

In this paper we discussed the polymerisation procedure for PNIPAAm hydrogels under different conditions. We investigated the influence of different oxygen content in the polymerisation setup and also the influence of different UV radiation sources, where we introduce a pulsed laser as UV source (51 kHz, λ = 355 nm). We found that assembling the polymerisation setup under inert gas conditions improves the reproducibility of hydrogel characteristics and therefore the hydrogel itself. By comparing the percentage of the standard deviations (standard and inert conditions) for both UV sources, the height of the improvement was 10 times (lamp) or 12 times (laser) for the diffusion coefficient, and 4 times (lamp) or 2 times (laser) for the degree of swelling, respectively. From an engineer point of view, the reproducibility of hydrogel characteristics (e.g., size, degree of swelling, cooperative diffusion coefficient) is one of the key features to use hydrogels in microsystem engineering. We could also demonstrate an improvement in the photopatterning process by using these findings. These result in better downscaling behaviour, and therefore in a resolution enhancement, which solves the limitation of decreasing pattern size. With the results shown here, we recommend a polymerisation setup assembled under inert gas conditions. Results at this optimisation state indicate the advantages of continous wide-band UV lamp with wavelengths in the range of 300 nm to 400 nm.

4. Outlook

For further improvement of the polymerisation process, we consider some process parameters. We believe that with a laser optimisation time mode (pulse frequency), another wavelength, or another photoinitiator whose absorption spectra fits better to the spectra of the laser used herein a further optimisation of the polymerisation process should be possible.

5. Experimental Section

5.1. Chemicals

N-Isopropylacrylamide (NIPAAm), *N*,*N'*-Methylene-bis-acrylamide (BIS), photoinitiator Irgacure® 2959, and 3-(trimethoxysilyl) propyl methacrylate were purchased from Sigma-Aldrich (Sigma-Aldrich, St. Louis, USA). NIPAAm was recrystallised from n-hexane (VWR International GmbH, Darmstadt, Germany), and other chemicals were used as received. All experiments were conducted in deionised water with an electrical conductivity of 0.056 µS·cm^{-1} (Barnstead GENPURE PRO, Thermo Scientific, Langenselbold, Germany), generated by ion exchange.

5.2. Preparation of Polymerisation Solution A

Polymerisation solution was prepared by dissolving 2.122 g NIPAAm and 0.058 g BIS in 15 mL water. The solution was flushed by bubbling argon through the solution for 1 h to remove oxygen, afterwards sealed with a septum, and brought into a glove box. 0.042 g photoinitiator was added, and the solution was thoroughly mixed.

5.3. Polymerisation Process Characterisation

For oxygen-free polymerisation systems preparation, solution A (1 mL) was filtered with a syringe filter (pore size 0.45 µm, VWR International GmbH, Darmstadt, Germany), and filled in NMR tubes (Deutero GmbH, Kastellaun, Germany) inside of the argon filled glove box (= inert conditions). The oxygen content in the glove box was less than 0.5 ppm. Otherwise, the solution was filtered in the glove box (mBraun, LAB Star, Garching, Germany) and subsequently filled in NMR tubes outside of the glove box (= standard conditions). Exposure times for the photopolymerisation process ranged from 10 s to 35 s in 5 s steps.

To avoid inhomogeneities in the polymer due to tube thickness, the NMR tubes were rotated during exposure time. For rotation of the NMR tube during the polymerisation process, a rotating system was developed (see Figure 1), forced by a coupling between an electrically-driven motor and the NMR tube realised by a metal chain and two gears. The motor rotation is brought to the NMR tube by the chain and a further gear. Here the NMR tube acts as a second shaft with two bearings. To exclude thermal effects, the reaction mixture is cooled to 0 °C with an ice bath.

UV light sources were a mercury-vapour lamp (1000 W Hg(Xe), L.O.T.-Oriel Group Europe, Darmstadt, Germany) (spectra in Figure 2) or a laser with a wavelength of λ = 355 nm and a pulse frequency of 51 kHz (AVIA 355-5, Coherent, Dieburg, Germany).

5.4. Hydrogel Characteristics Determination—Degree of Swelling

Weight measurements were done with a precision balance (BP 210 S, Sartorius, Goettingen, Germany). For mass measurement, hydrogels were cut into pieces and degrees of swelling were determined following Equation (1). The following three-step cycle was repeated five times. First, the hydrogels were swollen in water for 24 h (step 1). After the incubation, the hydrogels were deswollen at 60 °C for 2 h in a convection oven (Venti-Line, VWR International GmbH, Darmstadt, Germany) (step 2). Following the deswelling, the water was exchanged to fresh water (step 3), and incubation was again carried out for 24 h (step 1). These cycles, including swelling and deswelling steps, are necessary to wash out non-polymerised parts. After the fifth cycle, hydrogels were again deswollen, water was decanted, and the hydrogels were dried for 24 h at 60 °C in a convection oven. The lingering water content was removed by drying the hydrogels for 60 h at 60 °C in a vacuum oven (Heraeus, Thermo Scientific, Langenselbold, Germany). Hydrogels were weighed and consecutively incubated for 72 h in water to calculate the degree of swelling afterwards.

$$Q_{mass} = \frac{mass_{swollen}}{mass_{dried}} \tag{1}$$

The characteristic of choice for evaluation of the reproducibility was the standard deviation (SD). For a better comparison, the percentage of the standard deviation on the mean was calculated.

5.5. Determination of Cooperative Diffusion Coefficient

The cooperative diffusion coefficient was measured by dynamic light scattering (DLS, also known as photon correlation spectroscopy). With these measurements, the diffusion coefficient was determined for each hydrogel prepared in a NMR tube [24,25].

DLS measurements were carried on an ALV-5000 compact goniometer system (ALV, Langen, Germany) equipped with a helium-neon laser (λ = 632.8 nm) and coupled with an ALV photon correlator. Samples were measured at a scattering angle of θ = 90°. A toluene bath was used to match the refractive index and control the temperature at 25 °C. The time-averaged scattering intensities $< I >_T$ and the time-averaged intensity correlation functions (ICF, $g_T^{(2)}(q, \tau) - 1$) were determined at 50 different sample positions selected by randomly rotating the cuvette before each run. The time for each run was 30 s. From the ICFs measured at each position, apparent diffusion coefficients D_A were estimated according to Equation (2).

$$D_A = -\frac{1}{2q^2} \lim_{\tau \to 0} \frac{\delta}{\delta t} \ln \left(g_T^{(2)}(q, \tau) - 1 \right) \tag{2}$$

here, $q = (4\Pi n/\lambda_0) \sin(\theta/2)$ is the scattering vector, with θ being the scattering angle, λ_0 the wavelength of the incident light in a vacuum, and n the refractive index of the medium. For different sample positions, different values of D_A and the local scattering intensities $< I >_T$ were obtained. The relationship between D_A and the cooperative diffusion coefficient D_{coop} is given by Equation (3).

$$\frac{< I >_T}{D_A} = \frac{2}{D_{coop}} < I >_T - \frac{< I_F >_T}{D_{coop}} \tag{3}$$

Plotting $< I >_T / D_A$ *versus* $< I >_T$, the data formed essentially a straight line, from whose slope and intercept the fluctuating component of the scattering intensity, $< I_F >_T$, as well as D_{coop}, were obtained.

5.6. Hydrogel Microstructuring

Photostructuring experiments were conducted under inert gas conditions or atmosphere. First, cover slips (VWR International GmbH, Darmstadt, Germany) were cleaned with acetone, isopropanol, as well as water and dried under nitrogen stream. The glass surface was activated by oxygen plasma for 2 min at 50 W (Hochvakuum Dresden GmbH, Dreva Clean 450, Dresden, Germany) for subsequent silanisation. The activated glass slides were immersed in a 2 % *v/v* ethanolic solution with 3-(trimethoxysilyl) propyl methacrylate for 2 h to get a methacrylic surface modification of the slides for covalent bonding of the polymerised hydrogels to the glass surface. Afterwards, the slides were rinsed with water and dried under a stream of nitrogen. Following that, a polymerisation chamber was assembled and the polymerisation solution was added. This assembling procedure was conducted inside (inert conditions) or outside (standard conditions) of the glove box, respectively. Finally, the polymerisation was directly carried out for 60 s (inert conditions) or 90 s (standard conditions) under a mercury-vapour lamp through a polymer film mask.

Acknowledgments: The authors would like to thank Gerald Hielscher (Electronic Packaging Laboratory, Technische Universität Dresden) for preparing the photomasks. We also acknowledge the support from David Simon (Leibniz Institute of Polymer Research Dresden) for UV/Vis measurements. S.H. and P.F. thank the Deutsche Forschungsgemeinschaft (Research Training Group 1865 "Hydrogel based Microsystems") for a PhD scholarship. This work is partially financed by the cluster of excellence Center for Advancing Electronics Dresden (cfaed).

Author Contributions: Sebastian Haefner and Philipp Frank conducted the experiments and the data analysis. Matthias Rohn conducted the DLS experiments. Georgi Paschew built up the mercury-vapour lamp and the laser setup. Martin Elstner and Andreas Richter supervised the work and supported the scientific discussion. All authors reviewed the manuscript.

Conflicts of Interest: The authors declare no conflict of interest.

References

1. Mandal, B.B.; Kapoor, S.; Kundu, S.C. Silk fibroin/polyacrylamide semi-interpenetrating network hydrogels for controlled drug release. *Biomaterials* **2009**, *30*, 2826–2836.
2. Hoffman, A.S. Hydrogels for biomedical applications. *Adv. Drug Deliv. Rev.* **2012**, *64*, 18–23.

3. Ahmed, E.M. Hydrogel: Preparation, characterization, and applications: A review. *JARE* **2015**, *6*, 105–121.
4. Mitchell, H.T.; Schultz, S.A.; Costanzo, P.J.; Martinez, A.W. Poly(N-isopropylacrylamide) hydrogels for storage and delivery of reagents to paper-based Analytical devices. *Chromatography* **2015**, *2*, 436–451.
5. Hynd, M.R.; Frampton, J.P.; Burnham, M.R.; Martin, D.L.; Dowell-Mesfin, N.M.; Turner, J.N.; Shain, W. Functionalized hydrogel surfaces for the patterning of multiple biomolecules. *J. Biomed. Mater. Res. A.* **2007**, *81*, 347–354.
6. Okay, O. General Properties of Hydrogels. In *Hydrogel Sensors and Actuators*; Urban, G., Gerlach, G., Arndt, K.-F., Eds.; Springer Series on Chemical Sensors and Biosensors 6 Berlin Heidelberg; Springer: Berlin, Germany, 2009; pp. 3–8.
7. Caló, E.; Khutoryanskiy, V.V. Biomedical applications of hydrogels: A review of patents and commercial products. *Eur. Polym. J.* **2015**, *65*, 252–267.
8. Vashist, A.; Vashist, A.; Gupta, Y.K.; Ahmad, S. Recent advances in hydrogel based drug delivery systems for the human body. *J. Mater. Chem. B* **2014**, *2*, 147–166.
9. Cully, M. Hydrogel drug delivery for inflammatory bowel disease. *Nat. Rev. Drug Discov.* **2015**, *14*, 678–679.
10. Nichol, J.W.; Khademhosseini, A. Modular tissue engineering: Engineering biological tissues from the bottom up. *Soft Matter* **2009**, *5*, 1312–1319.
11. Slaughter, B.V.; Khurshid, S.S.; Fisher, O.Z.; Khademhosseini, A.; Peppas, N.A. Hydrogels in Regenerative Medicine. *Adv. Mater.* **2009**, *21*, 3307–3329.
12. Khademhosseini, A.; Langer, R. Microengineered hydrogels for tissue engineering. *Biomaterials* **2007**, *8*, 5087–5092.
13. Beebe, D.J.; Moore, J.S.; Bauer, J.M.; Yu, Q.; Liu, R.H.; Devadoss, C.; Jo, B.-H. Functional hydrogel structures for autonomous flow control inside microfluidic channels. *Nature* **2000**, *404*, 588–590.
14. Richter, A.; Wenzel, J.; Kretschmer, K. Mechanically adjustable chemostats based on stimuli-responsive polymers. *Sens. Actuators B* **2007**, *125*, 569–573.
15. Richter, A.; Türke, A.; Pich, A. Controlled Double-Sensitivity of Microgels Applied to Electronically Adjustable Chemostats. *Adv. Mater.* **2007**, *19*, 1109–1112.
16. Richter, A.; Paschew, P.; Klatt, S.; Lienig, J.; Arndt, K.-F.; Adler, H.-J.P. Review on Hydrogel-based pH Sensors and Microsensors. *Sensors* **2008**, *8*, 561–581.
17. Richter, A.; Klenke, C.; Arndt, K.-F. Adjustable low dynamic pumps based on hydrogels. *Macromol. Symp.* **2004**, *210*, 377–384.
18. Richter, A.; Klatt, S.; Paschew, G.; Klenke, C. Micropumps operated by swelling and shrinking of temperature-sensitive hydrogels. *Lab Chip* **2009**, *9*, 613–618.
19. Dong, L.; Jiang, H. Autonomous microfluidics with stimuli-responsive hydrogels. *Soft Matter* **2007**, *3*, 1223–1230.
20. Boyko, V.; Lu, Y.; Richter, A.; Pich, A. Preparation and characterization of acetoacetoxyethyl methacrylate-based gels. *Macromol. Chem. Phys.* **2003**, *204*, 2031–2039.
21. Paschew, G.; Schreiter, J.; Voigt, A.; Pini, C.; Chavez, J.P.; Allerdißen, M.; Marschner, U.; Siegmund, R.; Schüffny, R.; Jülicher, F.; *et al.* Autonomous chemical oscillator circuit based on bidirectional chemical-microfluidic coupling. *Adv. Mater. Technol.* **2015**, accepted.
22. Greiner, R.; Allerdissen, M.; Voigt, A.; Richter, A. Fluidic microchemomechanical integrated circuits processing chemical information. *Lab Chip* **2012**, *12*, 5034–5044.
23. Arndt, K.-F.; Krahl, F.; Richter, S.; Steiner, G. Swelling-related processes in hydrogels. In *Hydrogel Sensors and Actuators*; Urban, G., Gerlach, G., Arndt, K.-F., Eds.; Springer Series on Chemical Sensors and Biosensors 6 Berlin Heidelberg; Springer: Berlin, Germany, 2009; pp. 90–96.
24. Shibayama, M. Spatial inhomogeneity and dynamic fluctuations of polymer gels. *Macromol. Chem. Phys.* **1998**, *199*, 1–30.
25. Joosten, J.G.H.; Geladé, E.T.F.; Pusey, P.N. Spatial inhomogeneity and dynamic fluctuations of polymer gels. *Phys. Rev. A* **1990**, *42*, 2161–2173.

gels

MDPI

Article

Temperature-Responsive Hydrogel-Coated Gold Nanoshells

Hye Hun Park [1], La-ongnuan Srisombat [1], Andrew C. Jamison [1], Tingting Liu [1],
Maria D. Marquez [1], Hansoo Park [2], Sungbae Lee [3], Tai-Chou Lee [4] and T. Randall Lee [1,*]

[1] Department of Chemistry and the Texas Center for Superconductivity, University of Houston, Houston, TX 77204-5003, USA; hhpark95@gmail.com (H.H.P.); slaongnuan@yahoo.com (L.S.); andrewcjamison@yahoo.com (A.C.J.); sdzbltt@gmail.com (T.L.); mdmarqu2@gmail.com (M.D.M.)
[2] School of Integrative Engineering, Chung-Ang University, Seoul 156-756, Korea; heyshoo@cau.ac.kr
[3] Departments of Physics and Photon Science, Gwangju Institute of Science and Technology, 123 Chemdan-gwagiro (Oryong-dong), Buk-gu, Gwangju 500-712, Korea; jaylinlee@gist.ac.kr
[4] Department of Chemical and Materials Engineering, National Central University, 300 Jhongda Road, Jhongli City 32001, Taiwan; taichoulee@ncu.edu.tw
* Correspondence: trlee@uh.edu; Tel.: +1-713-743-2724; Fax: +1-281-754-4445

Received: 15 January 2018; Accepted: 16 March 2018; Published: 26 March 2018

Abstract: Gold nanoshells (~160 nm in diameter) were encapsulated within a shell of temperature-responsive poly(N-isopropylacrylamide-co-acrylic acid) (P(NIPAM-co-AA)) using a surface-bound rationally-designed free radical initiator in water for the development of a photothermally-induced drug-delivery system. The morphologies of the resultant hydrogel-coated nanoshells were analyzed by scanning electron microscopy (SEM), while the temperature-responsive behavior of the nanoparticles was characterized by dynamic light scattering (DLS). The diameter of the P(NIPAM-co-AA) encapsulated nanoshells decreased as the solution temperature was increased, indicating a collapse of the hydrogel layer with increasing temperatures. In addition, the optical properties of the composite nanoshells were studied by UV-visible spectroscopy. The surface plasmon resonance (SPR) peak of the hydrogel-coated nanoshells appeared at ~800 nm, which lies within the tissue-transparent range that is important for biomedical applications. Furthermore, the periphery of the particles was conjugated with the model protein avidin to modify the hydrogel-coated nanoshells with a fluorescent-tagged biotin, biotin-4-fluorescein (biotin-4-FITC), for colorimetric imaging/monitoring.

Keywords: drug delivery; temperature responsive; gold nanoshell; hydrogel coating

---

1. Introduction

A major goal of nanotechnology has been the development of nanoscale materials with functional properties. To this end, gold nanoparticles (AuNPs) are particularly attractive for use in medicinal applications due to their biocompatibility [1] and optical properties, especially their surface plasmon resonance (SPR) [2]. SPR-based biosensors are a popular tool for the detection of biomolecules. For example, a Cy5.5-substrated/AuNP system was reported as a multi-quenched near-infrared-fluorescence probe, providing a visual means to monitor a target protease and its inhibitor [3]. Additionally, AuNPs have been used as sensors for the colorimetric detection of the serum protein homocystamide, a biomarker that aids in the diagnosis of cardiovascular disease [4]. Furthermore, in a recent report, an aptazyme-AuNP sensor was developed as the first example of a sensor that allows for the amplified detection of biomolecules inside living cells [5].

Significant biomedical applications can be realized in vivo when AuNP resonances are tuned to the near infrared (NIR) region by changing the particle shape into hollow structures [6–8], gold nanorods [9], or gold nanoshells [10]. Among the different types of AuNPs, gold-silica nanoparticles, which consist of a silica core surrounded by a thin gold layer, have been well investigated fundamentally, as well as their use in a wide range of technically important applications [11–13]. In contrast to solid gold nanoparticles, the plasmon resonance of nanoshells can be precisely and systemically varied to specific wavelengths ranging from the UV-Vis to the near IR (NIR) region of the spectrum, by simply adjusting the size of the dielectric core and particularly the thickness of the gold layer [14]. Monodisperse silica nanoparticle cores can be synthesized over extremely broad size ranges, with the core providing one method of tuning the optical properties of the completed nanoshells [14,15]. If the shell thickness is varied on a silica core of constant size, the resulting optical properties are shown to shift as a function of shell thickness; specifically, as the thickness of the gold layer is decreased, the SPR is shifted to longer wavelengths. Absorption in the NIR region, generated through these methods, becomes particularly relevant for biomedical applications mainly due to the transparency of both blood and human tissue in that region of the spectrum, a span of wavelengths called the "phototherapeutic window"; light emitted at wavelengths between 800 and 1200 nm can pass through human tissue and then be absorbed/scattered by the nanoshells [16]. As NIR light interacts with the nanoshells, the radiation is converted efficiently into heat due to electron-phonon and phonon-phonon processes, providing rapid and efficient photothermal heating to the medium surrounding the nanoshells [17]. Using these features, nanoshells have been utilized in photothermal cancer therapy in vitro and in vivo [18]. The nanoshells accumulated at the tumor deliver a therapeutic dose of heat to the cancer cells upon irradiation with NIR light, thereby giving rise to localized photothermal ablation of the tumor tissue.

Hydrogels are three-dimensional, cross-linked polymer networks composed of more than 90% water. Due to the high porosity of the hydrogel and its ability to swell in aqueous environments, hydrogels play an important role in biological applications. For example, novel inverse opal hydrogel particles were recently developed as enzymatic carriers, making them viable for biocatalytic applications [19]. Notably, a reversible glucose sensor has also been developed, which can sense in the millimolar range via a combination of gold nanoantennas and a boronic acid-functionalized hydrogel [20]. Similarly, phenylboronic acids embedded in a hydrogel network have been reported for glucose sensing [21]. Moreover, a unique technology that embeds nanoparticle-stabilized liposomes into a hydrogel shows promising results toward topical antimicrobial delivery [22].

In this current study, gold nanoshells were incorporated into temperature-responsive poly(*N*-isopropylacrylamide-*co*-acrylic acid) (P(NIPAM-*co*-AA)) hydrogels for the development of a multifunctional material with potential use as a photothermally-modulated drug-delivery system (Scheme 1). To afford enhanced structural and compositional control, the hydrogel layer was grown using surface-immobilized radical initiators anchored onto the nanoshells through covalent S-Au bonds. Furthermore, the terminal PEG moieties were added to provide colloidal stability and biocompatibility to the composite nanoparticle system. *N*-isopropylacrylamide (NIPAM)-based hydrogels have been thoroughly studied due to their unique thermal behavior in aqueous solution [23–28]. At low temperatures, the poly(*N*-isopropylacrylamide) (PNIPAM) hydrogel exhibits hydrophilic behavior in water, due to hydrogen bonding between the PNIPAM functional groups and water, which causes the material to swell. However, at higher temperatures, there is an entropically-favored release of water molecules embedded in the hydrogel structure that leads to the collapse of the structure. The temperature at which the phase change occurs is called the lower critical solution temperature (LCST), and PNIPAM shows a LCST at 32 °C [29]. Introducing acrylic acid (AA) into the PNIPAM backbone shifts the LCST of the copolymer from 32 °C up to as high as 60 °C, depending on the amount of incorporated AA [30]. Within that temperature range, PNIPAM-based hydrogels can undergo completely reversible swelling-collapsing volume changes in response to temperature changes in the environment. The temperature-responsive properties of PNIPAM-based hydrogels can be combined with the physical and chemical properties of the gold nanoshells by incorporating the nanoshells into the hydrogel network [31]. The embedded nanoshells absorb/scatter NIR light, which will in turn heat the hydrogel layer, causing the structure to collapse [31]. When the temperature of the polymer exceeds the LCST, the hydrogels can release soluble materials held within the hydrogel matrix by collapsing their polymer network, leading to optically-triggerable drug release [32].

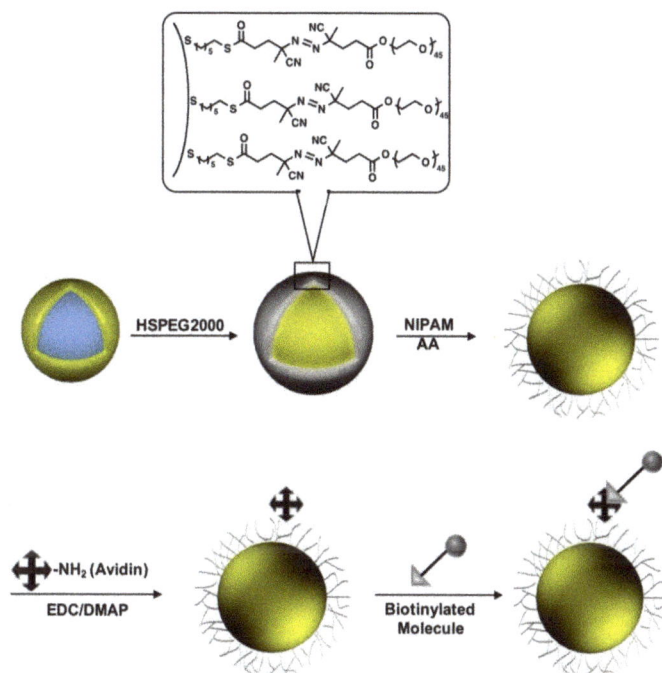

Scheme 1. Schematic diagram of hydrogel growth on gold nanoshells using radical polymerization followed by modification of the hydrogel periphery with avidin and biotin-4-FITC molecules. **HSPEG2000**, (*E*)-PEG2000-4-cyano-4-((*E*)-((*R*)-2-cyano-5-(6-mercaptohexylthio)-5-oxopentan-2-yl) diazenyl)pentanoic acid; NIPAM, *N*-isopropylacrylamide; AA, acrylic acid; EDC, *N*-(3-dimethylaminopropyl)-*N*′-ethylcarbodiimide hydrochloride; DMAP, 4-dimethylaminopyridine.

The P(NIPAM-*co*-AA) shell was generated through the free radical polymerization of the rationally-designed initiator **HSPEG2000** (see Scheme 2), synthesized as described previously [33]. The morphology of the resultant hydrogel-coated nanoshells was studied with scanning electron microscopy (SEM), and the temperature-responsive behavior of the particles was characterized by dynamic light scattering (DLS). After the polymer was formed on the surface of the nanoshells, the periphery of the hydrogel layer was further functionalized with avidin, a tetrameric protein that can bind with high affinity to four biotin molecules ($K_d = 10^{-13}\sim10^{-16}$ M) [34]. We chose to use the avidin/biotin system in this study for at least three reasons. First, biotin is a naturally-occurring vitamin found in every living cell [35]; specifically, the tissues with the highest amounts of biotin are found in the kidney, pancreas and liver [35]. Cancerous tumors are also known to have more biotin than normal tissue [35]. For drug-delivery applications, the latter feature should act as a driving force for avidin-conjugated particles to reach specific sites with cancerous tissue [35]. Second, a wide variety of biotinylated molecules that can withstand a variety of environments are commercially available. Third, the biotinylated molecules can be readily linked to avidin and, consequently, to the surface of the P(NIPAM-*co*-AA) hydrogel layer via the pendant AA moieties. In the work presented here, the presence of avidin was confirmed by complexing the particles with a fluorescent biotin (biotin-4-FITC). The successful application of this method shows that the hydrogel-coated nanoshells can be modified with essentially any biotinylated target molecule, providing targeted delivery in medical applications. Furthermore, the heat produced from the absorption of NIR light by the nanoparticles can be controlled to promote the collapse of the hydrogel layer and release specific payloads (e.g., drugs) on command.

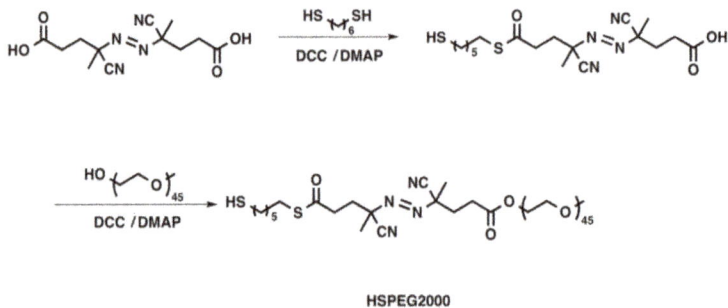

HSPEG2000

Scheme 2. Synthetic strategy used to prepare the free radical initiator **HSPEG2000** [33]. DCC, *N*,*N'*-dicyclohexylcarbodiimide; DMAP, 4-dimethylaminopyridine.

2. Results and Discussion

2.1. Characterization of the Gold-Coated Silica Nanoshell Particles

Gold nanoshells were successfully fabricated with a ~120 nm silica core and a shell thickness of ~20 nm (see the Experimental Section for the synthetic procedure); representative SEM images of the SiO$_2$ nanoparticles and the gold nanoshells are shown in Figure 1A,B, respectively. The diameters of the gold nanoshells are notably larger than those of the solid SiO$_2$ nanoparticles. In addition, the surfaces of the nanoshells exhibit topological roughness on the nanometer scale, consistent with that reported previously [13]. The resultant gold nanoshells were then used as templates to grow temperature-responsive P(NIPAM-*co*-AA) hydrogel layers on the surface as described below.

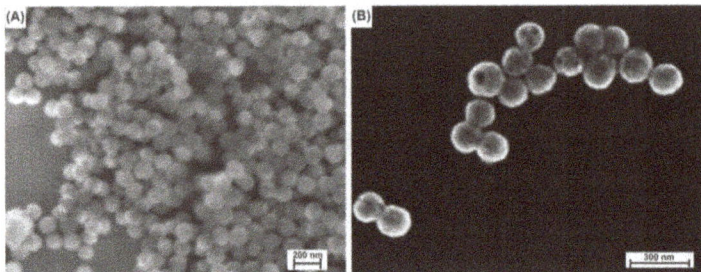

Figure 1. SEM images of (**A**) SiO$_2$ cores (~120 nm in diameter) and (**B**) gold nanoshells (~160 nm in diameter). The scale bars are 200 and 300 nm, respectively, where the magnified image on the right illustrates the increase in diameter arising from the gold coating.

2.2. Immobilization of Initiators and Copolymerization of NIPAM and AA on the Surface of the Nanoshells

To grow the hydrogel layer around the gold nanoshells, we first immobilized the free radical initiator **HSPEG2000** on the surface of the nanoshells. The strong sulfur-gold bond of ~50 kcal/mole drives the immobilization of the initiator molecule on the surface of the nanoshell [36]. The sulfur atom of the thiol acts as the headgroup by binding to the gold surface of the nanoshell while exposing the bulky PEG2000 (tailgroup) toward the surrounding solution. The tailgroup of the initiator monolayer also provides steric hindrance between the nanoshell particles; the resulting steric hindrance disrupts the inter-particle interaction, which typically leads to nanoshell aggregation, allowing the nanoshells to be stable in aqueous solution for up to six months under ambient conditions [33].

After immobilization of **HSPEG2000**, the newly-formed Au-S covalent bond was verified through analysis by X-ray photoelectron spectroscopy (XPS). In analyses by XPS, the binding energy (BE) of core electrons is greatly affected by the oxidation state of the atom of interest and its surroundings [37]. In particular, the S 2p region of the XPS spectra can be used to evaluate the nature of the S atom in self-assembled monolayers (SAMs) on gold [38]. The binding energy will be different for a bound thiolate, an unbound thiolate, and a thioester. However, spin-orbit coupling can inhibit an accurate analysis [38,39]. For example, the binding energy of the S $2p_{3/2}$ peak for a bound thiol onto an Au surface is 162 eV, whereas the same peak for an unbound thiol appears at ~164 eV in the XPS spectra [38]; at the same time, the peak position for the S $2p_{3/2}$ of a thioester also appears at ~164 eV [40]. For the S 2p photoelectrons, spin-orbit coupling gives rise to a doublet with an energy difference of 1.2 eV [38]. Figure 2 shows the high-resolution XPS spectra for the Au 4f and S 2p regions of the initiator-functionalized nanoshells. Figure 2A shows the Au 4f region, and Figure 2B shows the S 2p region for the **HSPEG2000**-functionalized nanoshells, confirming the presence of a bound thiolate at 162 eV and a thioester at 164 eV. The S 2p peak for the thioester is more intense than that for the thiol due to attenuation of the thiol photoelectrons by the intervening methylene groups.

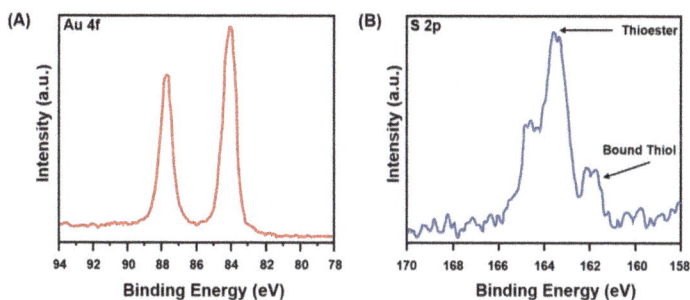

Figure 2. XPS spectra for **HSPEG2000**-functionalized nanoshells for (**A**) the Au 4f region and (**B**) the S 2p region.

Another major focus of concern regarding the functionalized nanoshells is the evaluation of the presence of unbound thiols versus bound ones on the surface of the nanoshells. For example, the S $2p_{3/2}$ doublet at ~164 eV in Figure 2B can be attributed to either unbound **HSPEG2000** species or to the sulfur of the thioester. We believe this is the latter case given the relative attenuation of the photoelectrons within the **HSPEG2000** initiator and the location of the various sulfur moieties around the nanoshells. The sulfur of the bound thiol is located deeper within the monolayer, closer to the nanoshell-**HSPEG2000** interface, while the sulfur of the thioester is closer to the outer interface. In a layered structure, attenuation tends to underestimate the elements buried deeper (Au-S) relative to those near the outer surface (thioester sulfur), giving a smaller intensity for the buried atoms in the XPS spectrum [40,41]. The attenuation of the bound sulfur can also explain the higher thioester to bound thiol ratio observed in the spectrum compared to the theoretical value of 1:1. Overall, the XPS data show that the initiator molecules were immobilized on the surface of the nanoshells. Due to the close proximity of the BE of an unbound thiol and a thioester, it is possible that the nanoshells have unbound **HSPEG2000**. However, based on the above discussion, we believe that most of the initiator molecules were immobilized on the nanoshell surface.

Following the immobilization of the initiator molecule, the nanoshells where then encapsulated within the P(NIPAM-*co*-AA) hydrogel by thermally-activating the initiator in the presence of the monomers (see the Experimental Section for details). Figure 3 displays SEM images of the hydrogel-coated nanoshells. The imaged composite consists of a nanoshell core having a diameter of ~160 nm encased by a hydrogel layer; the pale halos surrounding the brighter nanoshell centers provide evidence of the existence of the hydrogel coating on the nanoshells. The SEM images also show that most of the nanoshells were encapsulated with the P(NIPAM-*co*-AA) layers, and more importantly, the nanoshells that have the coating were completely covered, as reflected in the contrast of the SEM images in Figure 3B. We note, however, that these images cannot be used to determine the thickness of the hydrogel coating due to artifacts arising from drying and/or charging; instead, we determined the hydrogel thicknesses (i.e., hydrodynamic dimensions) using dynamic light scattering (DLS) as described below.

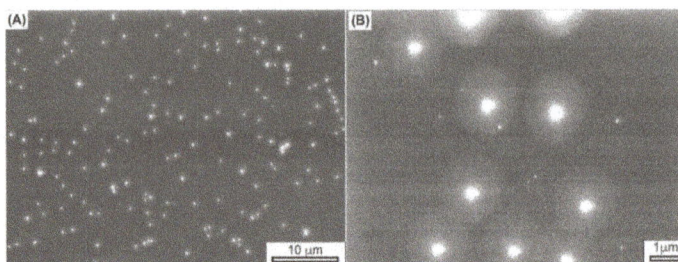

Figure 3. SEM images of (**A**) the hydrogel-coated nanoshells and (**B**) an enlarged-view of the same sample.

2.3. Temperature-Responsive Behavior of the Hydrogel-Coated Gold Nanoshells

The temperature-responsive behavior of the bare gold nanoshells and hydrogel-coated nanoshells was studied by DLS at various temperatures. Figure 4 shows the hydrodynamic diameter of the nanoparticles as a function of solution temperature. The hydrodynamic diameter of the hydrogel-coated nanoshells changed systemically with an increase or decrease in the temperature, between 25 and 40 °C; however, the diameters of the bare nanoshells remained constant under the tested conditions. More specifically, the diameters of the hydrogel-coated nanoshells decreased by ~70 nm upon heating to a temperature of 40 °C, which signified the collapse of the hydrogel structure. Nonetheless, upon cooling to 25 °C, the hydrogel structure swelled back to its original size. The changes in the hydrodynamic diameter of the hydrogel-coated nanoshells were completely reversible with repeated heating and cooling cycles. It should also be noted that the temperature of the phase transition (lower critical solution temperature (LCST)) of our system, containing 5 wt % AA, is 34 °C (see Figure 4). The LCST of our system, determined from three separate temperature profiles including that shown in Figure 4, is about 2 °C higher than the value of a pure NIPAM polymer system and is in good agreement with the results of a previous report [42]. The incorporation of ionizable groups, such as AA or amide groups, into the hydrogel network provides more hydrophilic sites along the polymer backbone, which lead to extensive hydrogen bonding between water and the hydrogel network. By taking into consideration that the collapse of the hydrogel structure above the LCST results from a disruption of the hydrogen bonding between water and the hydrophilic sites of the hydrogel network [43], the higher LCST for our system appears to be reasonable when ionizable groups are incorporated into the polymer backbone. The ability to raise the LCST upon addition of AA monomers presents the possibility of tuning the LCST of the polymer system to be close to the physiological body temperature of humans. Access to a higher LCST renders these P(NIPAM-*co*-AA) hydrogel-coated nanoshells more attractive for potential applications as dynamic materials in the human body.

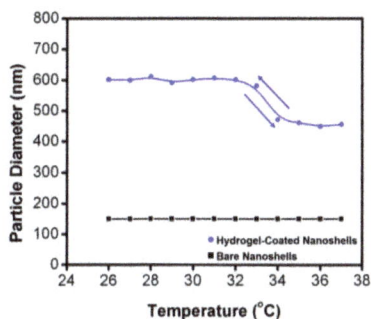

Figure 4. The hydrodynamic diameter of bare nanoshells (black square) and hydrogel-coated nanoshells (blue circle) as a function of temperature. The data shown correspond to a representative set taken from three separate experiments.

2.4. Optical Properties of the Hydrogel-Coated Gold Nanoshells

The optical properties of 40-nm AuNPs (included for comparison), the bare nanoshells (**NSs**), initiator-functionalized (**HSPEG2000-NSs**), and hydrogel-coated nanoshells (**P(NIPAM-co-AA)-NSs**) were studied by UV-Vis spectroscopy (see Figure 5). For the solid gold nanoparticles, the extinction maximum appears at ~530 nm, which is characteristic of small gold nanoparticles [44]. In contrast, the plasmon resonance of the bare gold nanoshells shifts to a much longer wavelength, showing an extinction maximum at ~800 nm. Previous studies have shown that the plasmon resonance of gold nanoshells can be tuned by varying the size of the silica core or the thickness of the gold shell [14]. For our system, gold nanoshells were fabricated with a ~20 nm-thick gold layer and a ~120 nm-diameter silica core, which led to a strong, broad absorption maximum at ~800 nm. A large red-shift in the plasmon resonance and the SEM images shown in Figure 1 can indirectly support the successful formation of the nanoshell structure.

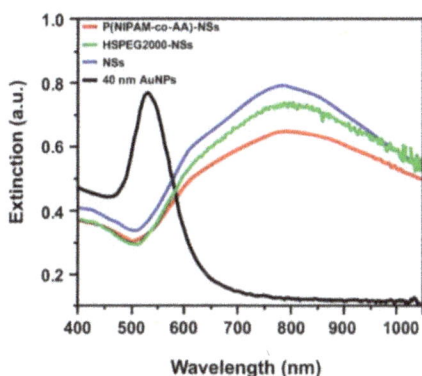

Figure 5. UV-Vis spectra of the 40-nm AuNPs (black), bare nanoshells (blue), **HSPEG2000**-modified nanoshells (green) and the hydrogel-coated nanoshells (red).

The influence of the organic layers (**HSPEG2000** and the hydrogel) on the plasmon resonance of the gold nanoshells was also examined. Despite the fact that the plasmon resonance of a noble metal is sensitive to the medium in contact with the surface of the metal [45,46], the band positions observed for the initiator-functionalized and hydrogel-coated nanoshells were similar to those observed for the bare nanoshells. The optical properties of the temperature-responsive hydrogel-coated nanoshells make them a potential candidate for photothermally-modulated drug delivery; as the light is absorbed by the nanoshells embedded in the hydrogel network, the absorbed light can be converted to heat, resulting in a higher temperature around the nanoshells and the release of encapsulated drugs [32]. Combining the optical properties of the nanoshells and the temperature-responsive behavior of the P(NIPAM-*co*-AA) hydrogel layer could make this new type of system ideal for applications involving nanoscale drug-delivery vehicles.

2.5. Modification of the Hydrogel-Coated Gold Nanoshells with Avidin

To demonstrate the capacity for surface functionalization and the ultimate use in targeted therapeutics, the temperature-responsive hydrogel-coated nanoshells were further conjugated with the protein avidin to create a surface that can bind with high affinity to biotinylated species. The covalent coupling of avidin to the nanoparticle surface was facilitated by activating the carboxyl groups of the P(NIPAM-*co*-AA) hydrogel layer (details are provided in the Experimental Section). The activated carboxyl groups were conjugated to avidin via the ε-amino groups present in the protein's lysine, which leads to the formation of amide bonds with avidin [47]. After bioconjugation, the modified diameter of the nanoshells increased by ~7 nm, as measured by DLS, which is consistent with the dimensions of an avidin layer, 6.0 nm × 5.5 nm × 4.0 nm, around the hydrogel-coated nanoshells [48].

In addition to obtaining the diameter of the conjugated nanoshells, we also verified the attachment of avidin by using a fluorescently-tagged biotin (biotin-4-FITC). The hydrogel-coated nanoshells were mixed with biotin-4-FITC dissolved in a buffer solution followed by three washing steps after the reaction to ensure the removal of unbound biotin-4-FITC from the solution. As a control experiment, hydrogel-coated nanoshells, which were not modified with avidin molecules, were mixed with biotin-4-FITC using the same procedure. The fluorescence spectra for the avidin-nanoshells complexed with biotin-4-FITC and the control experiment are shown in Figure 6. For the avidin-nanoshells, the emission maximum occurs at 524 nm, providing a similar spectrum to that of biotin-4-FITC dissolved in PBS buffer. Although gold nanoparticles are known to serve as ultra-efficient quenchers of the molecular excitation energy in a chromophore via their surface-energy-transfer properties [49], the fluorescence properties of the biotin-4-FITC were still in a detectable range, probably due to the presence of the thick hydrogel layer around the gold nanoshells. In contrast, for the control experiment, almost no fluorescence properties were detected, indicating that all the biotin-4-FITC molecules were washed out during the washing process.

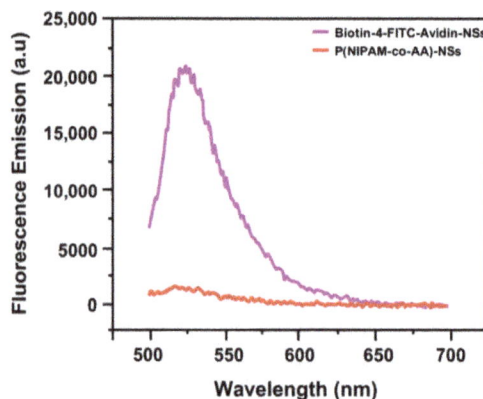

Figure 6. Fluorescence spectra of avidin-nanoparticles complexed with biotin-4-FITC (magenta) and the control experiment (hydrogel nanoshells, red).

To support this observation, the conjugated nanoshells suspended in PBS were visualized by confocal microscopy in normal brightfield and fluorescence modes. Figure 7 shows the brightfield and fluorescence confocal images for avidin-nanoshells complexed with biotin-4-FITC along with images for the control experiment. Fluorescence was only observed for the avidin-nanoshells complexed with biotin-4-FITC, in agreement with the fluorescence spectroscopy. From these experiments, we can conclude that avidin was successfully conjugated to the surface of the hydrogel-coated nanoshells, providing potential utilization of the immobilized avidin to attach various types of biotinylated target molecules around the hydrogel-coated gold nanoshells. Additionally, the results also provide evidence that the avidin molecules on the nanoshell surface remain active, as demonstrated by the ability to bind biotin even after conjugation.

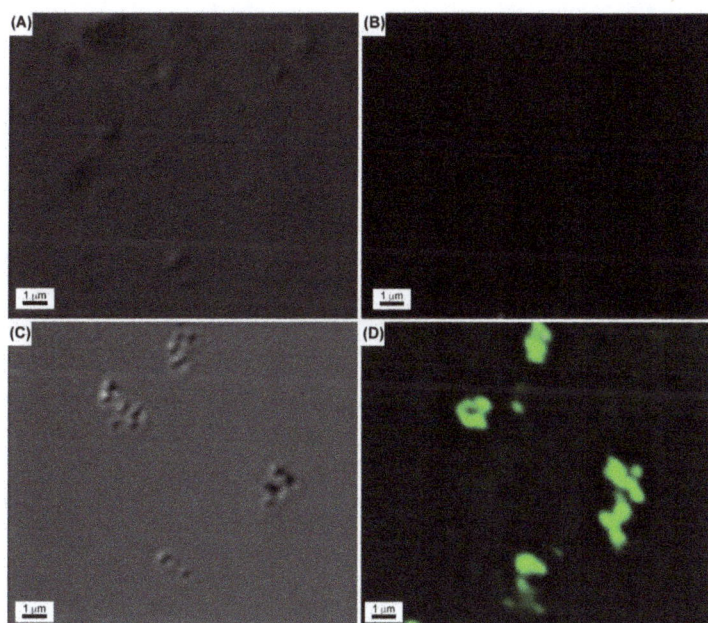

Figure 7. Confocal microscope images of: (**A**) control experiment in normal brightfield mode; (**B**) control experiment in fluorescence mode; (**C**) avidin-nanoparticles complexed with biotin-4-FITC in normal brightfield mode; and (**D**) avidin-nanoparticles complexed with biotin-4-FITC in fluorescence mode.

In previous studies [31,32,50–52], we demonstrated the potential of various hydrogel-coated gold nanoparticles as vehicles for loading and thermally-triggered payload release by illustrating their dynamic behavior in response to systematic changes in temperature. Chemically cross-linked PNIPAM-*co*-AA hydrogel particles have a porous network structure, which is particularly suitable to trap small molecules [53]. The advantages of the system reported here lie not only in the capacity for activation by NIR modulation as described above, but also in the covalent attachment of the hydrogels to the nanoshell cores and the ability to grow relatively thick hydrogel overlayers for optimal drug loading and delivery.

3. Conclusions

Temperature-responsive hydrogel-coated gold nanoshells were prepared using a rationally-designed surface-bound initiator. The gold nanoshells used for this study were ~160 nm in diameter, and the thickness of the P(NIPAM-*co*-AA) hydrogel was ~200 nm, as characterized by SEM and DLS. The temperature-responsive behavior of the hydrogel-coated nanoshells was demonstrated with DLS, showing a decrease in the particle diameter with increasing solution temperature. The surface plasmon resonance of the hydrogel-coated nanoshells appeared at ~800 nm, which is a particularly important spectral range for further biomedical applications. Furthermore, the periphery of the hydrogel-coated nanoshells was conjugated with the model protein avidin. The successful bioconjugation of the nanoparticles indicates that coupling of biotinylated targeting moieties and other biomolecules (e.g., proteins, DNA, and antibodies) is possible with this system. Combining all of the highly functionalized properties, our temperature-responsive hydrogel-coated gold nanoshells offer considerable promise for use in various biotechnological applications.

4. Materials and Methods

4.1. Materials

All reagents were purchased from the indicated suppliers and used without further purification unless indicated otherwise: ammonium hydroxide (30% NH_3), tetraethyl orthosilicate (TEOS, 99%), 3-aminopropyltrimethoxysilane (3-APTMS, 97%), tetrakis(hydroxymethyl)phosphonium chloride (THPC, 80% in H_2O), potassium carbonate (K_2CO_3, 99%), 4,4'-azobis(4-cyanovaleric acid) (75%), 1,6-hexanedithiol (96%), *N*-(3-dimethylaminopropyl)-*N'*-ethylcarbodiimide hydrochloride (EDC, 98%), poly(ethylene glycol) methyl ether (Mn 2000) and avidin from egg white (avidin) (Aldrich, St. Louis, MO, USA); sodium hydroxide (NaOH, 98%) and formaldehyde (HCOH, 37%) (EM Sciences, Hatfield, PA, USA); *N*-isopropylacrylamide (NIPAM, 99%), acrylic acid (AA, 99.5%), *N,N'*-methylenebisacrylamide (BIS, 96.0%); 4-dimethylaminopyridine (DMAP, 99.0%) (Acros); *N,N'*-dicyclohexylcarbodiimide (DCC, Fluka, 99.0%); absolute ethanol (McCormick Distilling Co., Weston, MO, USA); hydrogen tetrachloroaurate (HAuCl₄, Strem, Newburyport, MA, USA); biotin-4-fluorescein (biotin-4-FITC, >99%, Biotum, Fremont, CA, USA); tetrahydrofuran, hexane, ethyl acetate, chloroform, dichloromethane and methanol (Avantor, Center Valley, PA, USA). NIPAM was recrystallized in hexane and dried under vacuum before use. Tetrahydrofuran was freshly distilled over calcium hydride (Sigma Aldrich, St. Louis, MO, USA) and collected immediately prior to use. All water used in the reactions was purified to a resistance of 10 MΩ (Milli-Q Reagent Water System, Millipore Corporation, Burlington, MA, USA) and filtered through a 0.2 μm filter to remove any particulate matter.

4.2. Preparation of Gold Nanoshells

4.2.1. Preparation of Gold-Seeded SiO₂ Nanoparticles

Silica nanoparticles with a diameter of ~120 nm were prepared by the Stöber method [15]. In order to functionalize the surface of the SiO₂ particles with -NH₂ groups, the method by Waddell et al. [54] was followed by adding an excess amount of APTMS (~50 μL) to 100 mL of the SiO₂ nanoparticle solution. The mixture was allowed to stir for 24 h at room temperature and then refluxed at 80 °C for 1 h. The resulting amine-functionalized SiO₂ nanoparticles were isolated by centrifugation at 2500 rpm for 1 h and redispersed in absolute ethanol.

To attach colloidal gold to the amino-functionalized SiO₂ particles, a method described by Duff et al. [55] was followed. Briefly, 0.5 mL of a 1 M NaOH solution and 1 mL of a THPC solution (prepared by adding 12 μL of 80% THPC in H_2O to 1 mL of Milli-Q water) were added to 45 mL of Milli-Q water. The mixture was stirred for 5 min after which 2 mL of 1 wt % HAuCl₄ in water was added to the solution. The mixture was then stirred for 30 min and stored in the refrigerator for at least 3 days before use. Afterwards, 2 mL of the APTMS-coated SiO₂ nanoparticles solution and 25 mL of concentrated THPC-gold solution were mixed overnight. The gold-seeded SiO₂ nanoparticles were isolated by centrifugation and redispersed in Milli-Q water (100 mL).

4.2.2. Nanoshell Growth

To grow the gold layer on the THPC-gold seeded SiO$_2$ nanoparticles, we prepared a solution containing a reducible gold salt (K-gold solution). To prepare the K-gold solution, 0.025 g of K$_2$CO$_3$ were dissolved in 100 mL of Milli-Q water. Afterwards, 2 mL of 1 wt % HAuCl$_4$ solution were added. The K-gold solution was stirred at room temperature for 1 h and stored in the refrigerator overnight before use. Gold shells were grown by adding 2 mL of the THPC-gold-seeded SiO$_2$ nanoparticles to 40 mL of the K-gold solution. After the reaction mixture was stirred for 10 min, 0.2 mL of formaldehyde were added to the solution to reduce the K-gold. The gold nanoshells (~160 nm in diameter) were centrifuged at 2300 rpm and redispersed in Milli-Q water before use.

4.3. Functionalization of the Gold-Nanoshells with P(NIPAM-co-AA)

To functionalize the gold nanoshells with the initiator molecule, 20 mL of the gold nanoshell solution was mixed with 2 mL of a 1 mM solution of **HSPEG2000** (in ethanol) for 30 min and allowed to stand at room temperature for 24 h. The initiator-functionalized nanoshells were washed by centrifugation at 2300 rpm with ethanol and water (each twice) before use. The hydrogel-coated nanoshells were prepared by free radical polymerization in aqueous solution by the following procedure: 10 mL of the initiator-functionalized nanoshells were dispersed in 20 mL of water followed by the addition of 2 mL of NIPAM (0.01 M; 0.20 mmol), 0.2 mL of AA (0.01 M; 0.02 mmol) and 0.2 mL of BIS (0.01 M; 0.02 mmol) [44,56,57]. The solution was stirred and bubbled with argon for 1 h to remove oxygen, which can intercept radicals. To initiate the polymerization, the solution was heated to 65 °C in an oil bath and stirred for 6 h under argon. At the end of the reaction time, the solution was cooled to 20 °C. The hydrogel-coated particles were purified by dialysis (Spectra/Por Dialysis Membrane, MWCO 12–14000, VWR, Radnor, PA, USA) over the course of one week at room temperature; the water used for dialysis was changed daily.

4.4. Modification of the Hydrogel-Coated Nanoshells with Avidin and Biotin-4-FITC

The protein avidin was immobilized on the hydrogel layer by linking it to the carboxyl group present on the hydrogel layer. The covalent coupling of avidin to the nanoparticle surface was facilitated by the crosslinker EDC by activating the carboxyl groups of the P(NIPAM-*co*-AA) hydrogel layer. The activated carboxyl groups were conjugated to avidin via the ε-amino groups of the protein's lysine, which leads to the formation of an amide bond with the avidin. The reaction procedure required 1 mg of DMAP and 2 mg of EDC to be dissolved in 15 mL of the hydrogel-coated nanoshell solution. The avidin, dissolved in 0.25 mL of PBS buffer (0.01 M, pH 7.2), was then added to the nanoshell solution, and the solution was stirred overnight to ensure the amide bond formation between the hydrogel layer and the avidin.

To verify the attachment of avidin to the hydrogel layer, we also performed a study utilizing biotin linked to a fluorescent organic dye (biotin-4-FITC). Specifically, 1 mg of biotin-4-FITC was dissolved in a mixture of 0.5 mL of PBS buffer (0.01M, pH 7.2) and 0.5 mL of a 0.1 M NaOH solution. Thirty microliters of the biotin solution were added to 7 mL of the avidin-conjugated nanoshell solution while stirring. As a control experiment, the same amount of biotin solution was also added to the nanoshell solution, which was not conjugated with the avidin molecule. The solutions were stirred for 30 min and washed by centrifugation three times at 5000 rpm to remove the unbound biotin-4-FITC and stored in the refrigerator until use.

4.5. Characterization Methods

To characterize the SiO$_2$ nanoparticles, bare gold nanoshells, initiator-functionalized nanoshells, and hydrogel-coated nanoshells X-ray photoelectron spectroscopy (XPS), scanning electron microscopy (SEM) dynamic light scattering (DLS) and ultraviolet-visible (UV-Vis) spectroscopy were used.

Avidin-biotin conjugated nanoparticles were characterized by fluorescence spectroscopy and confocal microscopy.

4.5.1. X-ray Photoelectron Spectroscopy

XPS spectra of the initiator-functionalized nanoshells were collected using a PHI 5700 X-ray photoelectron spectrometer (Physical Electronics, Chanhassen, MN, USA) equipped with a monochromatic Al Kα X-ray source (hυ = 1486.7 eV) incident at 90° relative to the axis of a hemispherical energy analyzer. The initiator-functionalized nanoshells were deposited onto a silicon wafer, and the solvent was allowed to evaporate before analysis. The spectrometer was operated at high resolution with a pass energy of 23.5 eV, a photoelectron takeoff angle of 45° from the surface and an analyzer spot diameter of 2 nm. The base pressure in the chamber during the measurements was 3×10^{-9} Torr, and the spectra were collected at rt. After collecting the data, the binding energies of the S and C peaks were referenced by setting the Au $4f_{7/2}$ binding energy to 84 eV.

4.5.2. Scanning Electron Microscopy

Analysis by SEM was performed using a LEO-1525 Scanning Electron Microscope (Carl Zeiss AG, Thorwood, NY, USA) with 20 kV of accelerating voltage and a JEOL JSM 6400 Scanning Electron Microscope (JEOL, Peabody, MA, USA) with 10 kV of accelerating voltage during the measurements. Bare SiO_2 nanoparticles, gold-silica nanoshells and hydrogel-coated nanoshells were deposited on silicon wafers and dried at room temperature to collect the images. The overall morphology of the particles was examined with SEM.

4.5.3. Dynamic Light Scattering

For the DLS measurements, an ALV-5000 Multiple Tau Digital Correlation instrument (ALV-Laser Vertriebsgesellschaft mbH, Langen, Hesse, Germany) was used, operating at a light source wavelength of 514.5 nm and a fixed scattering angle of 90°. The hydrodynamic diameters of the bare nanoshells and the hydrogel-coated nanoshells were measured as a function of temperature in water. The samples were analyzed at dilute concentrations, and all of the collected data showed good Gaussian distribution curves.

4.5.4. UV-Vis Spectroscopy

The optical properties of the 40 nm AuNPs, initiator-functionalized nanoshells and the hydrogel-coated nanoshells were monitored at room temperature using a Cary 50 Scan UV-Vis optical spectrometer (Varian, Palo Alto, CA, USA) in conjunction with Cary Win UV software (Varian, Palo Alto, CA, USA). UV-Vis spectra of the prepared nanoparticles were collected in solution over a wavelength range of 400–1100 nm in a quartz cuvette having a 1 cm optical path length.

4.5.5. Fluorescence Measurements

For the fluorescence study, the fluorescence spectra of the avidin-nanoparticles complexed with biotin-4-FITC and the control experiment were measured by a PTI (Photon Technology International) Xe lamp and steady-state fluorescence (Photon Technology International, Birmingham, NJ, USA). The collected data were analyzed by the software FeliX32 Analysis (Version 1.2, Photon Technology International, Birmingham, NJ, USA). The particles were also visualized by a Leica TCS SP2 confocal microscope (Leica, Mannheim, Germany) equipped with 488 nm argon and 543 nm HeNe lasers in normal brightfield mode and fluorescence mode.

Acknowledgments: We thank the Air Force Office of Scientific Research and the Asian Office of Aerospace Research and Development (AFOSR/AOARD FA2386-16-1-4067 and FA2386-17-1-4028), the Robert A. Welch Foundation (E-1320), the National Science Foundation (CHE-1411265 and CHE-1710561) and the Texas Center for Superconductivity at the University of Houston for supporting this research.

Author Contributions: T. Randall Lee designed the project. Hye Hun Park and La-ongnuan Srisombat performed the experiments and the characterizations. Hye Hun Park, La-ongnuan Srisombat, Andrew C. Jamison, Tingting Liu, Maria D. Marquez, Hansoo Park, Sungbae Lee, Tai-Chou Lee and T. Randall Lee interpreted the data and wrote the paper.

Conflicts of Interest: The authors declare no conflict of interest.

References

1. Sun, T.; Zhang, Y.S.; Pang, B.; Hyun, D.C.; Yang, M.; Xia, Y. Engineered nanoparticles for drug delivery in cancer therapy. *Angew. Chem. Int. Ed.* **2014**, *53*, 12320–12364. [CrossRef]

2. Liz-Marzan, L.M. Tailoring surface plasmons through the morphology and assembly of metal nanoparticles. *Langmuir* **2006**, *22*, 32–41. [CrossRef] [PubMed]

3. Lee, S.; Cha, E.-J.; Park, K.; Lee, S.-Y.; Hong, J.-K.; Sun, I.-C.; Kim, S.Y.; Choi, K.; Kwon, I.K.; Kim, K.; et al. A near-infrared-fluorescence-quenched gold-nanoparticle imaging probe for in vivo drug screening and protease activity determination. *Angew. Chem. Int. Ed.* **2008**, *47*, 2804–2807. [CrossRef] [PubMed]

4. Gates, A.T.; Fakayode, S.O.; Lowry, M.; Ganea, G.M.; Murugeshu, A.; Robinson, J.W.; Strongin, R.M.; Warner, I.M. Gold nanoparticle sensor for homocysteine thiolactone-induced protein modification. *Langmuir* **2008**, *24*, 4107–4113. [CrossRef] [PubMed]

5. Yang, Y.; Huang, J.; Yang, X.; Quan, K.; Wang, H.; Ying, L.; Xie, N.; Ou, M.; Wang, K. Aptazyme–gold nanoparticle sensor for amplified molecular probing in living cells. *Anal. Chem.* **2016**, *88*, 5981–5987. [CrossRef] [PubMed]

6. Vongsavat, V.; Vittur, B.M.; Bryan, W.W.; Kim, J.-H.; Lee, T.R. Ultra-small hollow gold-silver nanoshells with extinctions strongly red-shifted to the near infrared. *ACS Appl. Mater. Interfaces* **2011**, *3*, 3616–3624. [CrossRef] [PubMed]

7. Yin, Z.; Zhang, X.; Zhou, D.; Wang, H.; Xu, W.; Chen, X.; Zhang, T.; Song, H. Enhanced upconversion luminescence on the plasmonic architecture of Au–Ag nanocages. *RSC Adv.* **2016**, *6*, 86297–86300. [CrossRef]

8. Shakiba, A.; Shah, S.; Jamison, A.C.; Rusakova, I.; Lee, T.C.; Lee, T.R. Silver-free gold nanocages with near-infrared extinctions. *ACS Omega* **2016**, *1*, 456–463. [CrossRef]

9. Zheng, Y.; Xiao, M.; Jiang, S.; Ding, F.; Wang, J. Coating fabrics with gold nanorods for colouring, UV-protection, and antibacterial functions. *Nanoscale* **2013**, *5*, 788–795. [CrossRef] [PubMed]

10. Zhang, X.; Ye, S.; Zhang, X.; Wu, L. Optical properties of SiO$_2$@M (M = Au, Pd, Pt) core–shell nanoparticles: Material dependence and damping mechanisms. *J. Mater. Chem. C* **2015**, *3*, 2282–2290. [CrossRef]

11. Pham, T.; Jackson, J.B.; Halas, N.J.; Lee, T.R. Preparation and characterization of gold nanoshells coated with self-assembled monolayers. *Langmuir* **2002**, *18*, 4915–4920. [CrossRef]

12. Bishnoi, S.W.; Lin, Y.J.; Tibudan, M.; Huang, Y.; Nakaema, M.; Swarup, V.; Keiderling, T.A. SERS biodetection using gold-silica nanoshells and nitrocellulose membranes. *Anal. Chem.* **2011**, *83*, 4053–4060. [CrossRef] [PubMed]

13. Khantamat, O.; Li, C.-H.; Yu, F.; Jamison, A.C.; Shih, W.-C.; Cai, C.; Lee, T.R. Gold nanoshell-decorated silicone surfaces for the near infrared (NIR) photothermal destruction of the pathogenic bacterium *E. faecalis*. *ACS Appl. Mater. Interfaces* **2015**, *7*, 3981–3993. [CrossRef] [PubMed]

14. Huang, X.; El-Sayed, M.A. Gold nanoparticles: Optical properties and implementations in cancer diagnosis and photothermal therapy. *J. Adv. Res.* **2010**, *1*, 13–28. [CrossRef]

15. Stöber, W.; Fink, A.; Bohn, E. Controlled growth of monodisperse silica spheres in the micron size range. *J. Colloid Interface Sci.* **1968**, *26*, 62–69. [CrossRef]

16. Simpson, C.R.; Kohl, M.; Essenpreis, M.; Cope, M. Near-infrared optical properties of ex vivo human skin and subcutaneous tissues measured using the Monte Carlo inversion technique. *Phys. Med. Biol.* **1998**, *43*, 2465–2478. [CrossRef] [PubMed]

17. Von Maltzahn, G.; Centrone, A.; Park, J.H.; Ramanathan, R.; Sailor, M.J.; Alan Hatton, T.; Bhatia, S.N. SERS-coded gold nanorods as a multifunctional platform for densely multiplexed near-infrared imaging and photothermal heating. *Adv. Mater.* **2009**, *21*, 3175–3180. [CrossRef] [PubMed]

18. Xuan, M.; Shao, J.; Dai, L.; Li, J.; He, Q. Macrophage cell membrane camouflaged Au nanoshells for in vivo prolonged circulation life and enhanced cancer photothermal therapy. *ACS Appl. Mater. Interfaces* **2016**, *8*, 9610–9618. [CrossRef] [PubMed]

19. Wang, H.; Gu, H.; Chen, Z.; Shang, L.; Zhao, Z.; Gu, Z.; Zhao, Y. Enzymatic inverse opal hydrogel particles for biocatalyst. *ACS Appl. Mater. Interfaces* **2017**, *9*, 12914–12918. [CrossRef] [PubMed]
20. Mesch, M.; Zhang, C.; Braun, P.V.; Giessen, H. Functionalized hydrogel on plasmonic nanoantennas for noninvasive glucose sensing. *ACS Photonics* **2015**, *2*, 475–480. [CrossRef]
21. Zhang, C.; Losego, M.D.; Braun, P.V. Hydrogel-based glucose sensors: Effects of phenylboronic acid chemical structure on response. *Chem. Mater.* **2013**, *25*, 3239–3250. [CrossRef]
22. Gao, W.; Vecchio, D.; Li, J.; Zhu, J.; Zhang, Q.; Fu, V.; Li, J.; Thamphiwatana, S.; Lu, D.; Zhang, L. Hydrogel containing nanoparticle-stabilized liposomes for topical antimicrobial delivery. *ACS Nano* **2014**, *8*, 2900–2907. [CrossRef] [PubMed]
23. Wu, C.; Zhou, S. Laser light scattering study of the phase transition of poly(N-isopropylacrylamide) in water. 1. Single chain. *Macromolecules* **1995**, *28*, 8381–8387. [CrossRef]
24. Wu, C.; Zhou, S. Internal motions of both poly(N-isopropylacrylamide) linear chains and spherical microgel particles in water. *Macromolecules* **1996**, *29*, 1574–1578. [CrossRef]
25. Nayak, S.; Lyon, L.A. Photoinduced phase transitions in poly(N-isopropylacrylamide) microgels. *Chem. Mater.* **2004**, *16*, 2623–2627. [CrossRef]
26. Debord, J.D.; Lyon, L.A. Synthesis and characterization of pH-responsive copolymer microgels with tunable volume phase transition temperatures. *Langmuir* **2003**, *19*, 7662–7664. [CrossRef]
27. Wang, J.; Gan, D.; Lyon, L.A.; El-Sayed, M.A. Temperature-jump investigations of the kinetics of hydrogel nanoparticle volume phase transitions. *J. Am. Chem. Soc.* **2001**, *123*, 11284–11289. [CrossRef] [PubMed]
28. Kim, J.; Nayak, S.; Lyon, L.A. Bioresponsive hydrogel microlenses. *J. Am. Chem. Soc.* **2005**, *127*, 9588–9592. [CrossRef] [PubMed]
29. Prevot, M.; Déjugnat, C.; Möhwald, H.; Sukhorukov, G.B. Behavior of temperature-sensitive PNIPAM confined in polyelectrolyte capsules. *ChemPhysChem* **2006**, *7*, 2497–2502. [CrossRef] [PubMed]
30. Snowden, M.J.; Chowdhry, B.Z.; Vincent, B.; Morris, G.E. Colloidal copolymer microgels of N-isopropylacrylamide and acrylic acid: pH, ionic strength and temperature effects. *J. Chem. Soc. Faraday Trans.* **1996**, *92*, 5013–5016. [CrossRef]
31. Kim, J.-H.; Lee, T.R. Thermo-responsive hydrogel-coated gold nanoshells for in vivo drug delivery. *J. Biomed. Pharm. Eng.* **2008**, *2*, 29–35.
32. Kim, J.-H.; Park, H.H.; Chung, S.Y.; Lee, T.R. Hydrogel-coated shell/core nanoparticles for in vivo drug delivery. *PMSE Preprints* **2008**, *99*, 730–731.
33. Park, H.H.; Park, H.; Jamison, A.C.; Lee, T.R. Colloidal stability evolution and completely reversible aggregation of gold nanoparticles functionalized with rationally designed free radical initiators. *Colloid Polym. Sci.* **2014**, *292*, 411–421. [CrossRef]
34. Green, N.M. Avidin. *Adv. Protein Chem.* **1975**, *29*, 85–133. [CrossRef] [PubMed]
35. Savage, M.D. *Avidin-Biotin Chemistry: A Handbook*, 2nd ed.; Pierce Chemical Co.: Rockford, IL, USA, 1992; ISBN 978-0935940114.
36. Love, J.C.; Estroff, L.A.; Kriebel, J.K.; Nuzzo, R.G.; Whitesides, G.M. Self-assembled monolayers of thiolates on metals as a form of nanotechnology. *Chem. Rev.* **2005**, *105*, 1103–1170. [CrossRef] [PubMed]
37. Vickerman, J.C.; Gilmore, I.S. *Surface Analysis: The Principal Techniques*, 2nd ed.; John Wiley & Sons, Ltd.: Hoboken, NJ, USA, 2009; ISBN 978-0470017647.
38. Castner, D.G.; Hinds, K.; Grainger, D.W. X-ray photoelectron spectroscopy sulfur 2p study of organic thiol and disulfide binding interactions with gold surfaces. *Langmuir* **1996**, *12*, 5083–5086. [CrossRef]
39. Laibinis, P.E.; Whitesides, G.M.; Allara, D.L.; Tao, Y.T.; Parikh, A.N.; Nuzzo, R.G. Comparison of the structures and wetting properties of self-assembled monolayers of n-alkanethiols on the coinage metal surfaces, copper, silver, and gold. *J. Am. Chem. Soc.* **1991**, *113*, 7152–7167. [CrossRef]
40. Wenzler, L.A.; Moyes, G.L.; Raikar, G.N.; Hansen, R.L.; Harris, J.M.; Beebe, T.P.; Wood, L.L.; Saavedra, S.S. Measurements of single-molecule bond-rupture forces between self-assembled monolayers of organosilanes with the atomic force microscope. *Langmuir* **1997**, *13*, 3761–3768. [CrossRef]
41. Lu, H.B.; Campbell, C.T.; Castner, D.G. Attachment of functionalized poly(ethylene glycol) films to gold surfaces. *Langmuir* **2000**, *16*, 1711–1718. [CrossRef]
42. Morris, G.E.; Vincent, B.; Snowden, M.J. Adsorption of lead ions onto N-isopropylacrylamide and acrylic acid copolymer microgels. *J. Colloid Interface Sci.* **1997**, *190*, 198–205. [CrossRef] [PubMed]

43. Shibayama, M.; Mizutani, S.; Nomura, S. Thermal properties of copolymer gels containing *N*-isopropylacrylamide. *Macromolecules* **1996**, *29*, 2019–2024. [CrossRef]
44. Park, H.H.; Lee, T.R. Thermo-and pH-responsive hydrogel-coated gold nanoparticles prepared from rationally designed surface-confined initiators. *J. Nanopart. Res.* **2011**, *13*, 2909–2918. [CrossRef]
45. Li, C.H.; Jamison, A.C.; Rittikulsittichai, S.; Lee, T.C.; Lee, T.R. In situ growth of hollow gold-silver nanoshells within porous silica offers tunable plasmonic extinctions and enhanced colloidal stability. *ACS Appl. Mater. Interfaces* **2014**, *6*, 19943–19950. [CrossRef] [PubMed]
46. Zhou, X.; Liu, G.; Yu, J.; Fan, W. Surface plasmon resonance-mediated photocatalysis by noble metal-based composites under visible light. *J. Mater. Chem.* **2012**, *22*, 21337. [CrossRef]
47. Costioli, M.D.; Fisch, I.; Garret-Flaudy, F.; Hilbrig, F.; Freitag, R. DNA purification by triple-helix affinity precipitation. *Biotechnol. Bioeng.* **2003**, *81*, 535–545. [CrossRef] [PubMed]
48. Meiser, F.; Cortez, C.; Caruso, F. Biofunctionalization of fluorescent rare-earth-doped lanthanum phosphate colloidal nanoparticles. *Angew. Chem. Int. Ed.* **2004**, *43*, 5954–5957. [CrossRef] [PubMed]
49. Dubertret, B.; Calame, M.; Libchaber, A.J. Single-mismatch detection using gold-quenched fluorescent oligonucleotides. *Nat. Biotechnol.* **2001**, *19*, 365–370. [CrossRef] [PubMed]
50. Kim, J.H.; Lee, T.R. Thermo- and pH-responsive hydrogel-coated gold nanoparticles. *Chem. Mater.* **2004**, *16*, 3647–3651. [CrossRef]
51. Kim, J.H.; Lee, T.R. Discrete thermally responsive hydrogel-coated gold nanoparticles for use as drug-delivery vehicles. *Drug Dev. Res.* **2006**, *67*, 61–69. [CrossRef]
52. Kim, J.H.; Lee, T.R. Thermo-responsive hydrogel-coated gold nanoshells. In Proceedings of the International Conference on Biomedical and Pharmaceutical Engineering (ICBPE), Singapore, 11–14 December 2006; pp. 271–275. [CrossRef]
53. Wu, W.; Mitra, N.; Yan, E.C.Y.; Zhou, S. Multifunctional hybrid nanogel for integration of optical glucose sensing and self-regulated insulin release at physiological pH. *ACS Nano* **2010**, *4*, 4831–4839. [CrossRef] [PubMed]
54. Waddell, T.G.; Leyden, D.E.; DeBello, M.T. The nature of organosilane to silica-surface bonding. *J. Am. Chem. Soc.* **1981**, *103*, 5303–5307. [CrossRef]
55. Duff, D.G.; Baiker, A.; Edwards, P.P. A new hydrosol of gold clusters. 1. Formation and particle size variation. *Langmuir* **1993**, *9*, 2301–2309. [CrossRef]
56. Wu, T.; Ge, Z.; Liu, S. Fabrication of Thermoresponsive Cross-Linked Poly(*N*-isopropylacrylamide) Nanocapsules and Silver Nanoparticle-Embedded Hybrid Capsules with Controlled Shell Thickness. *Chem. Mater.* **2011**, *23*, 2370–2380. [CrossRef]
57. Cao, Z.; Ziener, U.; Landfester, K. Synthesis of Narrowly Size-Distributed Thermosensitive Poly(*N*-isopropylacrylamide) Nanocapsules in Inverse Miniemulsion. *Macromolecules* **2010**, *43*, 6353–6360. [CrossRef]

![gels logo] *gels*

MDPI

Article

Tuning the Size of Thermoresponsive Poly(*N*-Isopropyl Acrylamide) Grafted Silica Microgels

Nils Nun [1], Stephan Hinrichs [1], Martin A. Schroer [2,3,†], Dina Sheyfer [2,3], Gerhard Grübel [2,3] and Birgit Fischer [1,*]

[1]   Institute of Physical Chemistry, University of Hamburg, 20146 Hamburg, Germany;
      nils.nun@studium.uni-hamburg.de (N.N.); Stephan.Hinrichs@chemie.uni-hamburg.de (S.H.)
[2]   Deutsches Elektronen-Synchrotron DESY, Notkestr. 85, 22607 Hamburg, Germany;
      mschroer@embl-hamburg.de (M.A.S.); dina.sheyfer@desy.de (D.S.); gerhard.gruebel@desy.de (G.G.)
[3]   The Hamburg Centre for Ultrafast Imaging (CUI), Luruper Chaussee 149, 22761 Hamburg, Germany
*   Correspondence: birgit.fischer@chemie.uni-hamburg.de; Tel.: +49-40-42838-8347
†   Present Address: European Molecular Biology Laboratory EMBL c/o DESY, Notkestr. 85,
      22607 Hamburg, Germany.

Received: 13 July 2017; Accepted: 13 September 2017; Published: 17 September 2017

Abstract: Core-shell microgels were synthesized via a free radical emulsion polymerization of thermoresponsive poly-(*N*-isopropyl acrylamide), pNipam, on the surface of silica nanoparticles. Pure pNipam microgels have a lower critical solution temperature (LCST) of about 32 °C. The LCST varies slightly with the crosslinker density used to stabilize the gel network. Including a silica core enhances the mechanical robustness. Here we show that by varying the concentration gradient of the crosslinker, the thermoresponsive behaviour of the core-shell microgels can be tuned. Three different temperature scenarios have been detected. First, the usual behaviour with a decrease in microgel size with increasing temperature exhibiting an LCST; second, an increase in microgel size with increasing temperature that resembles an upper critical solution temperature (UCST), and; third, a decrease with a subsequent increase of size reminiscent of the presence of both an LCST, and a UCST. However, since the chemical structure has not been changed, the LCST should only change slightly. Therefore we demonstrate how to tune the particle size independently of the LCST.

Keywords: hydrogel; pNipam; core-shell particle; lower critical solution temperature (LCST); thermoresponsive

---

1. Introduction

There is a large number of hydrogels that responds to external stimuli like temperature and pH [1–3]. These so called stimuli-responsive hydrogels offer a broad range of applications in the biomedical (e.g., controlled drug delivery systems [4]) as well as technical fields (e.g., in catalysis [2,5]). One of the most investigated stimuli-responsive hydrogels is poly(*N*-isopropyl acrylamide) (pNipam) due to its lower critical solution temperature (LCST) in aqueous solution of about 32 °C [2,6]. Being close to human body temperature, it is an ideal candidate for biomedical applications [7]. Above the LCST pNipam expels water and undergoes a coil-to-globule transition. This is similar to the cold denaturation of proteins [8], and can therefore be used as a model system for proteins to study the influence of osmolytes [9,10]. An upper critical solution temperature (UCST) in aqueous solution is less common for polymers. With a UCST polymers are solubilized above the UCST in aqueous solution [11]. The change of the solution behaviour depends on the hydrophobic nature of the solvent. pNipam exhibits a UCST behaviour using a solvent mixture like water/ethanol or water/dimethyl sulfoxide with a low water content [12,13].

One synthesis method for the preparation of pNipam is the free radical emulsion polymerization [6]. Sodium dodecyl sulphate (SDS) can be used as surfactant and methylene-*bis*-acrylamide (BIS) as crosslinker connecting two pNipam chains. By switching the temperature the pNipam chains elongate below the LCST and make a coil-to-globule transition above the LCST. Without the crosslinker, the pNipam chains randomly form globular structures above the LCST to minimize their contact with water. In presence of the crosslinker the coil-to-globule-transition of the pNipam chains gets reversibly. The addition of crosslinker increases the LCST slightly [14]. The crosslinker itself has a higher reaction rate than *N*-isopropyl acrylamide (Nipam) [5], therefore most of the prepared microgels have a heterogeneous distribution of the crosslinker within the pNipam network decreasing from the inside to the outside of the microgel [5,15–17]. This distribution or internal structure has a high impact on the swelling behaviour of the microgel [18], that is, the degree of swelling of a microgel with homogeneously distributed crosslinker is higher than the degree of swelling of a microgel with heterogeneously distributed crosslinker [18].

A disadvantage of pNipam microgels is its relatively high softness in the swollen state which limits the mechanical stability. Interpenetrating networks of different polymers or incorporated nanoparticles have been created to make pNipam hydrogels more interesting in terms of mechanical investigations [1]. In particular, silica nanoparticles are of special interest as these can be easily and highly reproducibly synthesized. In addition, the synthesis can be directly used for nanoparticles covered with a silica shell, which is an frequently used method [19]. By incorporating nanoparticles inside the silica core, multi-responsive hydrogels can be prepared with additional optical or magnetic properties [20–25].

Several methods exist to synthesize core-shell particles containing a silica core and a pNipam shell [26,27]. Here, we report on a synthesis route for pNipam core-shell microgel by controlling the growth process of the pNipam shell together with the crosslinker BIS on the silica surface. We show that depending on the growth process we can alter the thermo-responsive properties of the pNipam shell. The pNipam microgel seems to behave like it has either an LCST, a UCST, or even both. However, since the chemical structure has not been changed compared to the literature [5,28], pNipam should still have an LCST. This means by altering the synthesis route slightly we control the internal structure of the pNipam shell, and with this we can tune the hydrodynamic size independently from the coil-to-globule transition of pNipam.

2. Results and Discussion

To demonstrate the three different scenarios of thermal responsivity described above we will show one example for each case. Each system consists of a silica core and a pNipam shell. Due to the difference in the synthesis procedures the distribution of the crosslinker densities differs in each system. For each synthesis a new core was synthesized. First we discuss the silica core for each system. Then we compare the different microgel systems.

2.1. Characterization of the Silica Core

For each microgel a silica core has been synthesized via the Stöber process. The Stöber process results in spherical particles [29]. The hydrodynamic sizes of the silica particles were measured by dynamic light scattering. For this, a dilute sample of each silica core was measured at 25 °C. The samples Si-1–Si-3 have a radius of 67 nm, 31 nm, and 36 nm, respectively. Exemplarily for sample Si-2, the scanning electron microscopy (SEM) image is shown (Figure 1a).

To link the pNipam onto the particle surface, the silica surface was modified with 3-(trimethoxysilyl)-propyl-methacrylate (TPM). The addition of TPM does not affect the size of the particles much. However, it does change the surface properties and make them more hydrophobic. After the successful addition of TPM the particles are no longer dispersible in water, but they can be well dispersed with ethanol. In addition it allows for chemical coupling between the silica surface and the pNipam chains [26]. Thermogravimetric analysis shows (Figure S1 and Table S1), that a small organic content is present in the sample compared to pure silica particles. Additionally, Fourier transformed infrared spectroscopy (FTIR) confirms the presence of TPM bound to the silica surface (Figure 1b). From the silica a broad and almost featureless band is visible between 2500 and 3800 cm^{-1}, corresponding to the stretching vibrations of the surface silanols Si-OH perturbed by hydrogen, bonding either intramolecularly or with adsorbed water [30]. For the TPM-modified silica particles, additionally two small bands are visible between 2800 and 3000 cm^{-1} from the stretching vibration of C-H bonds. The bands at 1710, 1450 and 1410 cm^{-1} correspond to the stretching vibration of C-O, methylene C-H and vinyl C-H bending vibration of TPM [31].

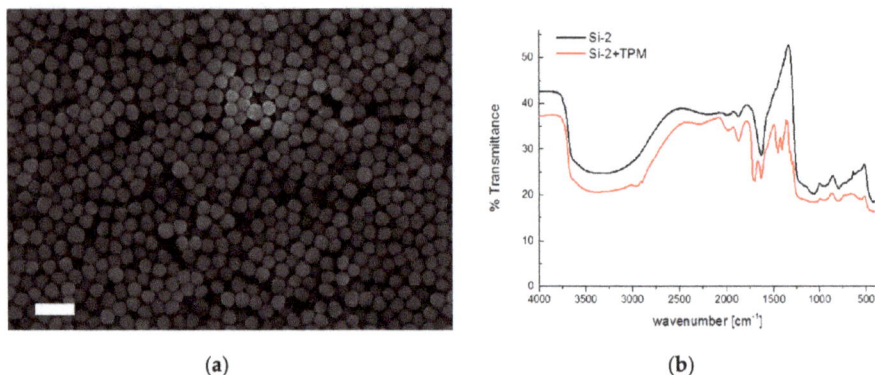

(a)   (b)

Figure 1. (**a**) Scanning electron microscopy (SEM) image of silica particles (Si-2). The scale bar is 200 nm; (**b**) Fourier transformed infrared spectroscopy (FTIR) spectra of Si-2 before and after (Si-2+TPM) surface modification with TPM.

2.2. Temperature Behaviour of the Hybrid Hydrogel

2.2.1. Sample SiPN-1

For all three samples the pNipam shell was made out of Nipam/BIS with about 3–4 mol % BIS, for SiPN-1 about 3.2 mol % was used. In case of SiPN-1 both monomers are added at the beginning of the polymerization process. Pure pNipam has an LCST of about 32 °C. By using a copolymerization with BIS the LCST can be slightly shifted to higher temperatures. The SiPN-1 particles collapse above a temperature of about 33.2 °C during heating and swell below 32.4 °C during cooling (Figure 2a). The LCST T^* is determined by fitting the data with a Boltzmann function [32]:

R_{min} is the minimum hydrodynamic radius and was about 68 nm for heating and cooling, which is close to the diameter of the silica core, which is about 67 nm (Si-1). The LCST T^* is at about 33.4 ± 0.1 °C (32.3 ± 0.1 °C) for heating (cooling). The swelling and deswelling behaviour are reversible as is found by cycling the sample (Figure 2b).

$$R_{hyd} = \frac{R_{min} - R_{max}}{1 + \exp\left(-\dfrac{T - T^*}{dT}\right)} \tag{1}$$

In all dynamic light scattering data only a single decay has been observed, this is a clear indication of just one particle species. For the Transmission electron microscopy (TEM) measurements, the particles have been dried on a carbonated copper grid. By this preparation method for TEM the particles resemble the state above the LCST. In this phase the particles are collapsed and all water is squeezed out of the microgel particles. A darker core—which results from the silica core—and a bright corona—which corresponds to a thin rough surface—are visible in TEM (Figure 3). No pure pNipam particles are visible in the TEmicrograph.

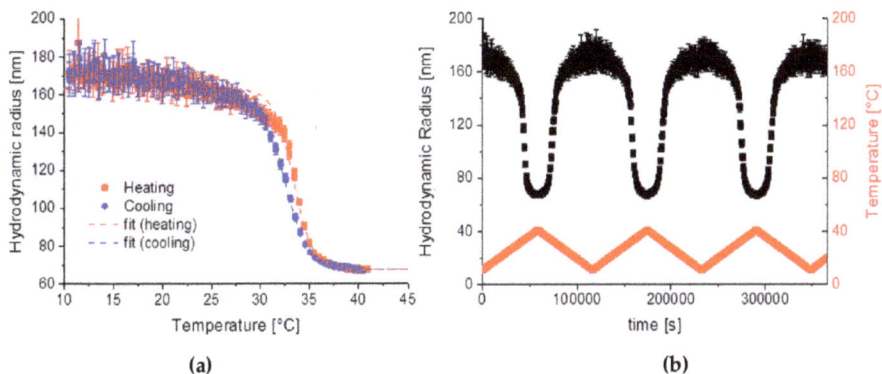

Figure 2. (a) Hydrodynamic radius of SiPN-1 for a heating (black cubes) and cooling cycle (red dots) with an empirical fitting function Equation (1); (b) Hydrodynamic radius for several heating and cooling cycles for SiPN-1. In red the temperature gradient is shown.

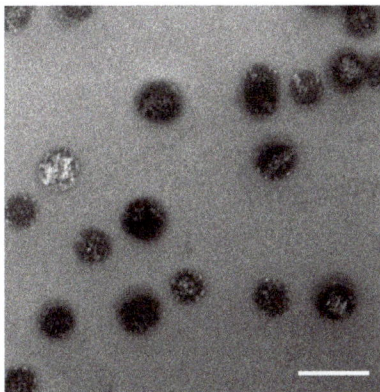

Figure 3. TEmicrograph of SiPN-1. Scale bar is 200 nm. Each particle has a darker core and a rough surface resulting from the dried microgel shell.

2.2.2. Sample SiPN-2

While for SiPN-1 the crosslinker and Nipam were added simultaneously, for SiPN-2 the whole amount of crosslinker BIS was added at the beginning of the encapsulation, followed by Nipam. The molar ratio of the crosslinker was about 4 mol %, a little higher amount than for SiPN-1. The pNipam chains at the surface of the silica particles are more densely crosslinked than at the periphery. Compared to the reaction of SiPN-1, the initiator concentration was reduced as well as the redoxsystem concentration. The SDS concentration as well as the initiator concentration was the same like for SiPn-1.

Like for SiPN-1 neither pure pNipam particles are visible in TEM nor a second decay could be observed by dynamic light scattering. The hydrodynamic radius of the microgel particles increases from 58 nm to about 108 nm by increasing the temperature (Figure 4a). The guide line for cooling and heating was done by Equation (1). Here the core size of the silica particles (Si-2) is 31 nm, which results in a pNipam shell of at least 30 nm. The transition temperature was about 38.0 °C for heating and 37.0 °C for cooling. The TEmicrograph shows a similar picture like for SiPN-1, a darker image for the silica core and a brighter surface corona in the collapsed state (Figure 4b).

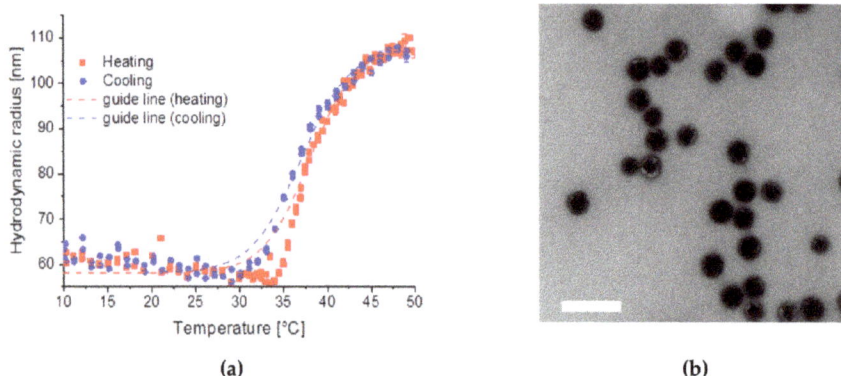

Figure 4. (a) Hydrodynamic radius of SiPN-2 as a function of temperature for heating and cooling; (b) TEmicrograph of SiPN-2. Scale bar is 200 nm.

2.2.3. Sample SiPN-3

For the third sample the monomers Nipam and BIS were grown in three steps, which results in three shells with a heterogeneous distribution relatively to sample SiPN-1; therefore, the density of the crosslinker always decreases from the inside to the outside within a shell. The molar ratio of the crosslinker was ~3.2% for the first two shells, similar to SiPN-1 and 3.7% for the third shell, which was similar to SiPN-2. For the synthesis the same concentration of SDS, Na_2SO_3, $K_2S_2O_8$ and $(NH_4)_2Fe(SO_4)_2$ as for SiPN-1 were used. The reaction temperature was lowered to 50 °C, which slows down the reaction compared to 60 °C.

Measuring the temperature-dependent hydrodynamic radius of the microgel particles shows a scenario such as for the presence of both an LCST as well as a UCST at higher temperatures as the hydrodynamic radius exhibits a minimum size (Figure 5a). Additionally, here only a single decay has been observed by dynamic light scattering, which indicates only one particle size. The minimal hydrodynamic radius is about 56 nm for heating at 39.6 °C and, for cooling, 58 nm at 38.6 °C. The silica core (Si-3) has a diameter of about 36 nm which results in a pNipam shell size of at least 20 nm. In the TEM images darker spheres show the electron denser silica particles which have some brighter spheres on their surface, which result from the collapsed pNipam chains (Figure 5b).

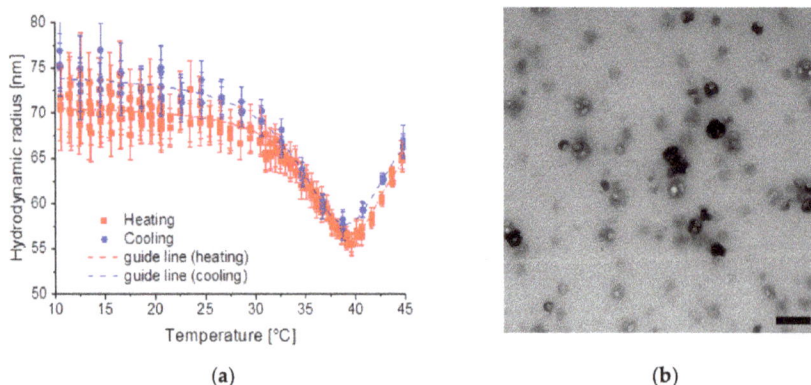

Figure 5. (a) Hydrodynamic radius of SiPN-3 as a function of temperature for heating and cooling; (b) TEmicrograph of SiPN-3. Scale bar is 200 nm.

However, it is also visible that due to the drying process agglomerates are formed which have not been observed in the dynamic light scattering experiments, and which are very sensitive to large particles.

2.3. Discussion

Below, the LCST the pNipam hydrogel swells with water and the pNipam chains tend to elongate. At temperatures above the LCST pNipam gets more hydrophobic and expels water. Due to the surface charge resulting from the initiator, the pNipam is still dispersible in water and does not precipitate. This is the usual behaviour of pNipam microgel dispersions. Sample SiPN-1 shows this typical behaviour of pNipam microgels (Figure 2a). At low temperatures the hydrodynamic radius is larger than at high temperatures because the particle is swollen with water. Above the LCST the particle radius shrinks due to the squeezing out the water. This temperature behaviour is repeatable over several cycles (Figure 2b). TEM shows a thin pNipam skin of less than 5 nm covering the silica particles (Figure 3). The microgel particles in the collapsed state have a similar hydrodynamic radius as the pure silica particles in accordance with this observation by TEM.

Since Si-1 has twice the radius of Si-2, a smaller amount of Nipam was used for the synthesis to reach a similar shell thickness. In addition, a lower amount of initiator was used. This leads, on the one hand, to lower surface charges and, on the other hand, to fewer but longer pNipam chains. Thus, SiPN-2 shows a completely different temperature scenario. By increasing the temperature the hydrodynamic radius increases, resembling the presence of a UCST, although pNipam still becomes more hydrophobic, which is visible on the changing turbidity. At high temperature the sample SiPN-2, as well as SiPN-3, turns turbid as a result of the change in the refractive index of the microgel after squeezing out the water. Here, the minimum radius of the microgel particles reaches about 60 nm and is larger than the one of the pure silica particle with a hydrodynamic radius of about 31 nm (Si-2). In contrast, in TEM the sample does not look different to SiPN-1, only the pNipam shell seemed to be thicker (~30 nm).

SiPN-3 shows a scenario like for the presence of both an LCST as well as a UCST. A possible explanation is visible in the TEmicrograph (Figure 5b). Here, instead of a rough surface, microgel balls are visible. Since the hydrophobicity of the pNipam chains increases at high temperature a spherical form is advantageous because of its minimized surface area and with it the contact to water. Several studies of pNipam grafted silica particles or gold nanoparticles demonstrate that pNipam does not change its chemical behaviour due to grafting and still has an LCST around 32 °C [22,27,28,33–37]. Variation of the crosslinker density only has an effect on the swelling degree and the LCST temperature [14], however, not on the solvation behaviour. So far only through the copolymerization of Nipam with a small amount of methacrylic acid and nitrocatechol monomers in phosphate-buffer in saline medium was UCST behaviour found [38]. In addition, by adding salts of the Hofmeister series [39] and by using mixtures out of water and co-non-solvents like ethanol, the thermo-responsive behaviour changes [12,13,40,41].

All SiPN samples studied here were dialysed against water, so no salt or any other solvents were present during the measurement. Thus, the observed response is not due to any additional solute. The organic content lies for all three systems at about 70–80 wt % (see TGA measurements Figure S1 and Table S1). Therefore, we suggest that the difference in the temperature behaviour occurs due to the internal structure of the microgel.

To understand why the samples SiPN-2 and SiPN-3 show a behaviour which can be compared to a UCST behaviour, we tried to sketch the different scenarios in Figure 6. The pNipam chains of sample SiPN-1 are well crosslinked (Figure 6a). Due to this internal crosslinking the pNipam shell homogeneously shrinks and the pNipam shell gets thinner due to the expulsion of water and the hydrodynamic radius decreasing at higher temperature. The pNipam hydrogel in SiPN-2 is only densely crosslinked near the surface of the silica particles (Figure 6b). Due to the addition of BIS only at the beginning of the synthesis, the crosslinker concentration decreases, which means that single pNipam chains grow from a thin crosslinked pNipam shell on the silica surface. This makes it possible that the pNipam chains stay at the silica particle. The lower amount of the initiator used for SiPN-2 also favours the growth of long chains. If the pNipam chains collapse, these chains form a globular structure on the surface of the silica particles (Figure 6b). Since spheres have the smallest surface area they form microgel balls (highlighted by a black circle in Figure 6b) onto the surface to minimize the contact with water. This leads to an increase of the hydrodynamic radius in total. It might also explain the thicker pNipam shell of SiPN-2 at high temperatures compared to SiPN-1, because here, instead of a thin homogeneous pNipam shell, several thick microgel balls are present. However, since the microgel balls seem to cover the complete surface these are not clearly visible in TEM (Figure 4b).

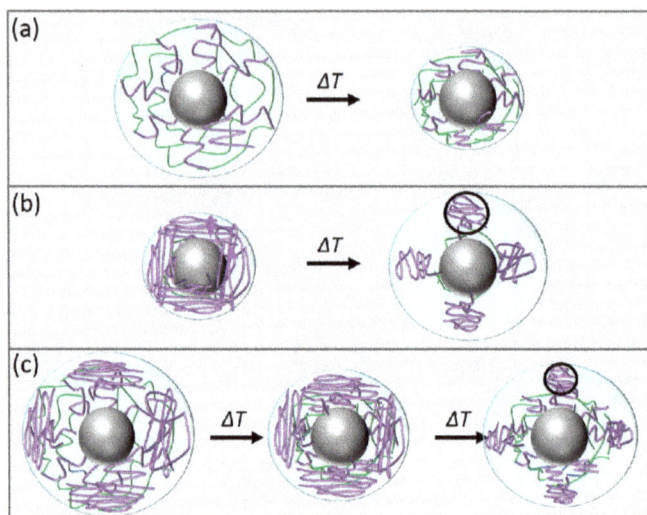

Figure 6. Different scenarios for the core-shell particles during the coil-to-globule transition. The silica core is shown as a grey sphere. The poly-(*N*-isopropyl acrylamide) (pNipam) chains are represented in purple and the crosslinking chains between two pNipam chains are visualized in green. The light blue circles indicate the hydrodynamic volume of the particles. Scenario (**a**): Below the lower critical solution temperature (LCST): The microgel particle is swollen with water and the pNipam chains are elongated. The pNipam chains are crosslinked from the inner to the outer shell. Above the LCST the pNipam shell collapses. Due to the internal crosslinking the pNipam shell homogeneously shrinks and the pNipam shell gets thinner due to the expulsion of water. Scenario (**b**): Below the LCST the pNipam chains are elongated and linked to the silica surface. Only near the surface of the silica particles are the pNipam chains crosslinked. Therefore, above the LCST the pNipam chains collapse into small spheres onto the surface of the silica particles. These small spheres we call microgel balls—we highlighted one with a black circle. Scenario (**c**): Below the LCST the microgel is swollen with water and the pNipam chains are elongated. By increasing the temperature the microgel structure expels water and shrinks. Due to external crosslinking the shell gets thinner like in scenario (**a**). However, since long not-crosslinked pNipam chains are also presented, these pNipam chains show up as microgel balls on the surface.

For the last sample the pNipam network is not as crosslinked as in the first sample SiPN-1, but also not as loose as for sample SiPN-2, where the pNipam chains are only crosslinked at the silica surface. Here it seems both scenarios play a role (Figure 6c).

In this case first the crosslinked pNipam chains collapse and expel water, the size of the shell decreases. However, since not all chains are crosslinked, the pNipam chains build microgel balls linked to the silica surface. This is clearly visible in the TEmicrogaph (Figure 5b).

The temperature behaviours for SiPN-2 and SiPN-3 are repeatable; the hydrodynamic radii were measured for both samples for several heating and cooling cycles (Figure 7).

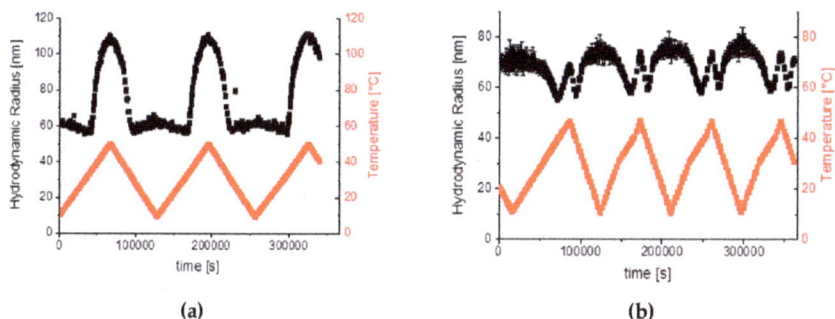

(a)   (b)

Figure 7. Hydrodynamic radius for several heating and cooling cycles for sample SiPN-2 (**a**) and SiPN-3c (**b**).

3. Conclusions

Three different pNipam grafted silica microgels have been prepared. In general the synthesis can be divided into three steps [26]. First the silica core is synthesized by a modified Stöber process; second, the surface was modified by TPM; third, silica cores were grafted with pNipam.

By changing the way of adding the crosslinker the temperature-responsive behaviour can be tuned. Three different temperature scenarios have been detected. The first scenario has the typical behaviour of pNipam with an LCST temperature of about 33.4 ± 0.1 °C (32.3 ± 0.1 °C) for heating (cooling). The second resembles a USCT-like behaviour.

In the last scenario both LCST and UCST-like behaviour seem to be combined. However, the chemical composition has not been changed compared to other studies [22,27,33–37], therefore pNipam should still have an LCST temperature. However, the swelling and collapsing has the opposite effect on the hydrodynamic radius due to the formation of pNipam microballs on the surface of the silica particles.

The incorporating of a silica core helps to improve the mechanical properties of pNipam. Growing pNipam onto a silica surface makes it possible to tune the hydrodynamic particle radius differently. This enables us to tune the particle size independently of the swelling and deswelling behaviour.

4. Materials and Methods

4.1. Materials

Tetraethylorthosilicate (TEOS) for synthesis ($\geq$99%), potassium peroxodisulphate ($K_2S_2O_8$) for analysis ($\geq$99%), sodium sulphite (Na_2SO_3) for analysis ($\geq$97%), and N,N'-methylenebisacrylamide (BIS) for synthesis ($\geq$98%) were purchased from Merck (Darmstadt, Germany).

3-(trimethoxysilyl)-propyl-methacrylate (TPM) ($\geq$98%) was purchased from Aldrich (Munic, Germany).

Ethanol (99.5%, dried over molecular sieves) and N-isopropyl acrylamide (Nipam) (stabilized, 99%) were purchased from ACROS Organics (Nidderau, Germany).

Ammonium hydroxide (NH_3; 25%) for analysis was purchased from VWR Chemicals (Darmstadt, Germany).

Sodium dodecyl sulphate (SDS) for biochemistry ($\geq$99%) was purchased from Carl Roth GmbH (Karlsruhe, Germany).

Ammonium iron (II) sulphate hexahydrate ((NH_4)$_2$Fe(SO_4)$_2$) American Chemical Society (ACS) reagent (99%) and technical ethanol (92.6–93.8%) were purchased from Sigma Aldrich (Munic,

Germany). The technical ethanol was distilled twice prior to use. The other chemicals were used as obtained.

4.2. Synthesis

The synthesis method of silica particles grafted with pNipam can be divided into three steps. First, a silica core is prepared, and then a methacrylate group is grafted onto the surface of the silica particles, which finally could be used as a starting point for growing of the pNipam/BIS chains. The different synthesis steps are given in detail below.

4.2.1. TPM-Grafted Silica Particles

The synthesis of silica particles is derived from the Stöber-process [29]. Ethanol was mixed with ammonium hydroxide (A, Table 1) and stirred for a few minutes. TEOS was added, followed by substantial stirring. After another 12 h of stirring additional ammonium hydroxide (B, Table 1) was added together with TPM similar to the synthesis reported by Karg et al. [26]. Afterwards, the ethanol/ammonium hydroxide was removed until only a few millilitres of ethanol were left. The sample was kept dispersed with ethanol and dialyzed against ethanol to remove the residual TPM. The specific amounts are displayed in Table 1.

Table 1. Specific amounts of chemicals used for the Synthesis of the TPM-grafted silica particles.

Name	Ethanol	NH$_3$ (A)	TEOS	NH$_3$ (B)	TPM
	mL	mL	mL	mL	mL
Si-1	200	15	5	10	1
Si-2	450 *	20	7.5	15	2
Si-3	750	35	5	10	1

* Here 99% ethanol was use. Tetraethylorthosilicate (TEOS); 3-(trimethoxysilyl)-propyl-methacrylate (TPM).

4.2.2. pNipam-Grafted Silica Microgels

The pNipam shell was synthesized similar to Karg et al. [26]. In a typical synthesis SDS and Na$_2$SO$_3$ were dissolved in deionized water in a three neck round flask. TPM-grafted silica particles were added and have been stirred for one hour under nitrogen atmosphere. Potassium peroxodisulphate was added as initiator and one crystal of ammonium iron (II)-sulphate, which acts together with Na$_2$SO$_3$ as a redox-system to decrease the reaction temperature.

To change the internal structure of the microgel shell the method to add Nipam and BIS was varied, that is, the order of the monomers and the addition rate as well as the reaction temperature.

The microgel particles were directly dialyzed against deionized water for a week to remove the excess of SDS. Then, the particle dispersions were concentrated with a rotary evaporator under reduced pressure until only a few millilitres were left. The samples were always kept in water. For purification the concentrated dispersions were dialyzed again against water for several days and filtered afterwards.

To ensure that pNipam/BIS grow onto the surface of the silica particles diluted dispersions, especially low-monomer concentrations are used. In this case the probability for the monomers to find each other to build additional pNipam particles is reduced. This effect can be further reduced by pumping the monomer to the dispersions. In case of unwanted side products, pure pNipam particles are visible in TEM and can be observed as a second correlation peak by dynamic light scattering (DLS).

SiPN-1

Si-1 was used as the silica core. The sample was diluted with water to 1.5 L. Afterwards, the addition of Nipam and BIS was done in two steps. First 44.19 mmol Nipam and 1.43 mmol BIS (3.2 mol %) were added. After one hour of stirring another 44.19 mmol Nipam and 1.43 mmol BIS (3.2 mol %)

were added. The reaction has been left stirring for six hours at 60 °C. The used amounts of the chemicals are 0.87 mmol/L SDS, 2.20 mmol/L Na_2SO_3, 1 mmol/L $K_2S_2O_8$, and 1 crystal $(NH_4)_2Fe(SO_4)_2$.

SiPN-2

Si-2 was used as the silica core and diluted to a volume of about 1.5 L with water. The addition of 20 mMol Nipam and 0.80 mmol BIS (4 mol %) was done by pumping with a pumping speed of 1 mL/min. For this purpose, BIS was dissolved in 10 mL ethanol and Nipam was dissolved in 40 mL ethanol. After 5 min of pumping the monomers into the solution, the influx was stopped for 15 min, and afterwards this process was repeated until all monomer was inserted. At the beginning only BIS was added. After 40 min the Nipam was pumped to the reaction.

The specific amounts of each chemical are 0.88 mmol/L SDS, 1 mmol/L Na_2SO_3, 0.25 mmol/L $K_2S_2O_8$ and 1 crystal $(NH_4)_2Fe(SO_4)_2$. The reaction has been stirred at 60 °C for about 15 h.

SiPN-3

Si-3 was used as the silica core. The addition of Nipam and BIS was done in three steps; the specific amounts of each chemical are displayed in Table 2. Each time the synthesis procedure was repeated and the mixture has been stirred at 50 °C for about 6 h. During the reaction the SDS concentration was kept constant at 0.87 mMol/L, the Na_2SO_3 at 2.2 mmol/L and the $K_2S_2O_8$ at 1 mmol/L.

Table 2. Amounts of the chemicals used for the synthesis of SiPn-3 (a–c). Methylene-bis-acrylamide (BIS).

Chemical	Nipam mmol	BIS mmol
First addition	17.67	0.57
Second addition	17.67	0.57
Third addition	17.67	0.65
sum	53.01	1.79

4.3. Characterization Methods

Dynamic light scattering (DLS) measurements were conducted on an ALV/CSG-3 Compact Goniometer-System using an ALV/LSE-5003 Multiple Tau Digital Correlator working with pseudo cross-correlation and the ALV Digital Correlator Software 3.0. The measuring angle was set to 90° for all measurements. As a light source, an Nd:YAG laser emitting at 532 nm was used. A toluene bath which was used for matching of the index of refraction was tempered by a Julabo F25 thermostat working with a mixture of water and ethylene glycol and delivering a temperature accuracy of 0.01 °C. For the measurements diluted samples were used and the temperature was varied between 10 and 45 °C (50 °C for SiPN-2). Around the LCST transition the temperature was increased in 0.5° steps above 40 °C and below 30 °C in 1 °C steps. Each temperature was measured at least three times. After each temperature was reached, the sample has been given 3 min time to equilibrate. The acquisition time was 120 s for sample SiPN-1, 60 s for SiPN-2 and SiPN-3 per temperature step. During the cooling cycle less temperature steps were done. Prior to the measurement all samples were diluted to have a volume fraction of about 0.1%. Below the LCST the samples were clear dispersions. Above the LCST the dispersions were a slightly turbid.

Transmission electron microscopy (TEM) was performed on a Philips CM-300 microscope operating at 300 kV and on a FEI Tecnai 12 transmission electron microscope operating at 120 kV. TEM samples were prepared by dropping diluted solutions transferred in ethanol onto a 400-mesh carbon coated copper grid with the excessive solvent being immediately evaporated.

Scanning electron microscopy (SEM) was obtained on a high resolution SEM (Gemini Leo 1550) at an acceleration voltage of 5 keV. SEM samples were prepared by dropping diluted solutions on a silicon wafer. The excessive solvent was evaporated afterwards.

Thermogravimetric analysis (TGA) was conducted on a Netzsch TGA 209 F1 Iris. The experiments were run with a heating rate of 10 K/min in a range of 25–500 °C and conducted in an inert nitrogen atmosphere. Therefore, about 6–8 mg of the specimen was weighed into an 85 µL aluminium oxide crucible (purchased from Netzsch, Selb, Germany). The data processing was performed by Netzsch Proteus Software afterwards (4.7.0, Selb, Germany).

Fourier Transform infrared spectra were performed on a VERTEX-70/FT-IR spectrometer (Bruker Optics, Ettlingen, Germany). Therefore the samples were dried and placed in between two NaBr windows. As background the empty windows were measured. The measurements were done under nitrogen atmosphere.

Supplementary Materials: The following are available online at www.mdpi.com/2310-2861/3/3/34/s1, Figure S1: TGA measurements (a) and its derivative (b) of Si-2, SiTPM-2, SiPN-1, SiPN-2 and SiPn3, Table S1: Weight loss of the sample:Si2, SiTPM-2, SiPN-1, SiPN-2 and SiPn3 within the temperature range 25–150 °C (region 1) and within the temperature range 150–500 °C (region 2) and residue at 500 °C. In the first region the weight loss corresponds to the solvent evaporation. In the second region the weight loss corresponds to the thermal decomposition of the organic content.

Acknowledgments: The authors thank Andreas Kornowski, Ulf König and Daniela Weinert for TEM measurements, Elisabeth Wittenberg for TGA measurements and Robert Schön for SEM measurements. Birgit Fischer, Nils Nun and Stephan Hinrichs thank the Deutsche Forschungsgemeinschaft for financial support within the Priority Program 1681 ("Field controlled particle matrix interaction: synthesis, multi-scale modelling and application of magnetic hybrid-materials", grant: FI 1235/2-1). This work has been supported by the excellence cluster "The Hamburg Centre for Ultrafast Imaging, Structure, Dynamics, and Control of Matter at the Atomic Scale" of the DFG.

Author Contributions: Birgit Fischer, Martin A. Schroer and Gerhard Grübel conceived and designed the experiments. Birgit Fischer wrote the paper; Nils Nun, Stephan Hinrichs and Dina Sheyfer performed the experiments. Birgit Fischer analyzed the data.

Conflicts of Interest: The authors declare no conflict of interest.

References

1. Haq, M.A.; Su, Y.; Wang, D. Mechanical properties of pnipam based hydrogels: A review. *Mater. Sci. Eng. C* **2017**, *70*, 842–855. [CrossRef] [PubMed]

2. Doring, A.; Birnbaum, W.; Kuckling, D. Responsive hydrogels—Structurally and dimensionally optimized smart frameworks for applications in catalysis, micro-system technology and material science. *Chem. Soc. Rev.* **2013**, *42*, 7391–7420. [CrossRef] [PubMed]

3. Lyon, L.A.; Meng, Z.; Singh, N.; Sorrell, C.D.; St John, A. Thermoresponsive microgel-based materials. *Chem. Soc. Rev.* **2009**, *38*, 865–874. [CrossRef] [PubMed]

4. De Las Heras Alarcon, C.; Pennadam, S.; Alexander, C. Stimuli responsive polymers for biomedical applications. *Chem. Soc. Rev.* **2005**, *34*, 276–285. [CrossRef] [PubMed]

5. Varga, I.; Gilányi, T.; Mészáros, R.; Filipcsei, G.; Zrínyi, M. Effect of cross-link density on the internal structure of poly(*N*-isopropylacrylamide) microgels. *J. Phys. Chem. B* **2001**, *105*, 9071–9076. [CrossRef]

6. Pelton, R. Temperature-sensitive aqueous microgels. *Adv. Colloid Int. Sci.* **2000**, *85*, 1–33. [CrossRef]

7. Guan, Y.; Zhang, Y. Pnipam microgels for biomedical applications: From dispersed particles to 3D assemblies. *Soft Matter* **2011**, *7*, 6375. [CrossRef]

8. Tiktopulo, E.I.; Uversky, V.N.; Lushchik, V.B.; Klenin, S.I.; Bychkova, V.E.; Ptitsyn, O.B. "Domain" coil-globule transition in homopolymers. *Macromolecules* **1995**, *28*, 7519–7524. [CrossRef]

9. Schroer, M.A.; Michalowsky, J.; Fischer, B.; Smiatek, J.; Grubel, G. Stabilizing effect of tmao on globular pnipam states: Preferential attraction induces preferential hydration. *Phys. Chem. Chem. Phys.* **2016**, *18*, 31459–31470. [CrossRef] [PubMed]

10. Micciulla, S.; Michalowsky, J.; Schroer, M.A.; Holm, C.; von Klitzing, R.; Smiatek, J. Concentration dependent effects of urea binding to poly(*N*-isopropylacrylamide) brushes: A combined experimental and numerical study. *Phys. Chem. Chem. Phys.* **2016**, *18*, 5324–5335. [CrossRef] [PubMed]

11. Zhang, Q.; Hoogenboom, R. Polymers with upper critical solution temperature behavior in alcohol/water solvent mixtures. *Prog. Polym. Sci.* **2015**, *48*, 122–142. [CrossRef]

12. Liu, L.; Wang, T.; Liu, C.; Lin, K.; Liu, G.; Zhang, G. Specific anion effect in water-nonaqueous solvent mixtures: Interplay of the interactions between anion, solvent, and polymer. *J. Phys. Chem. B* **2013**, *117*, 10936–10943. [CrossRef] [PubMed]

13. Costa, R.O.R.; Freitas, R.F.S. Phase behavior of poly(N-isopropylacrylamide) in binary aqueous solutions. *Polymer* **2002**, *43*, 5879–5885. [CrossRef]

14. Kratz, K.; Hellweg, T.; Eimer, W. Structural changes in pnipam microgel particles as seen by sans, dls, and em techniques. *Polymer* **2001**, *42*, 6631–6639. [CrossRef]

15. Fernandez-Barbero, A.; Fernandez-Nieves, A.; Grillo, I.; Lopez-Cabarcos, E. Structural modifications in the swelling of inhomogeneous microgels by light and neutron scattering. *Phys. Rev. E* **2002**, *66*, 051803. [CrossRef] [PubMed]

16. Schmid, A.J.; Dubbert, J.; Rudov, A.A.; Pedersen, J.S.; Lindner, P.; Karg, M.; Potemkin, I.I.; Richtering, W. Multi-shell hollow nanogels with responsive shell permeability. *Sci. Rep.* **2016**, *6*, 22736. [CrossRef] [PubMed]

17. Hertle, Y.; Zeiser, M.; Hasenöhrl, C.; Busch, P.; Hellweg, T. Responsive P(NIPAM-co-NTBAM) microgels: Flory–Rehner description of the swelling behaviour. *Colloid Polym. Sci.* **2010**, *288*, 1047–1059. [CrossRef]

18. Acciaro, R.; Gilanyi, T.; Varga, I. Preparation of monodisperse poly(N-isopropylacrylamide) microgel particles with homogenous cross-link density distribution. *Langmuir* **2011**, *27*, 7917–7925. [CrossRef] [PubMed]

19. Graf, C.; Vossen, D.L.J.; Imhof, A.; van Blaaderen, A. A general method to coat colloidal particles with silica. *Langmuir* **2003**, *19*, 8. [CrossRef]

20. Rittikulsittichai, S.; Kolhatkar, A.G.; Sarangi, S.; Vorontsova, M.A.; Vekilov, P.G.; Brazdeikis, A.; Randall Lee, T. Multi-responsive hybrid particles: Thermo-, ph-, photo-, and magneto-responsive magnetic hydrogel cores with gold nanorod optical triggers. *Nanoscale* **2016**, *8*, 11851–11861. [CrossRef] [PubMed]

21. Messing, R.; Schmidt, A.M. Perspectives for the mechanical manipulation of hybrid hydrogels. *Polym. Chem.* **2011**, *2*, 18. [CrossRef]

22. Dagallier, C.; Dietsch, H.; Schurtenberger, P.; Scheffold, F. Thermoresponsive hybrid microgel particles with intrinsic optical and magnetic anisotropy. *Soft Matter* **2010**, *6*, 2174. [CrossRef]

23. Hinrichs, S.; Nun, N.; Fischer, B. Synthesis and characterization of anisotropic magnetic hydrogels. *J. Magn. Magn. Mater.* **2017**, *431*, 237–240. [CrossRef]

24. Dulle, M.; Jaber, S.; Rosenfeldt, S.; Radulescu, A.; Forster, S.; Mulvaney, P.; Karg, M. Plasmonic gold-poly(N-isopropylacrylamide) core-shell colloids with homogeneous density profiles: A small angle scattering study. *Phys. Chem. Chem. Phys.* **2015**, *17*, 1354–1367. [CrossRef] [PubMed]

25. Karg, M.; Pastoriza-Santos, I.; Perez-Juste, J.; Hellweg, T.; Liz-Marzan, L.M. Nanorod-coated pnipam microgels: Thermoresponsive optical properties. *Small* **2007**, *3*, 1222–1229. [CrossRef] [PubMed]

26. Karg, M.; Pastoriza-Santos, I.; Liz-Marzan, L.M.; Hellweg, T. A versatile approach for the preparation of thermosensitive pnipam core-shell microgels with nanoparticle cores. *Chem. Phys. Chem.* **2006**, *7*, 2298–2301. [CrossRef] [PubMed]

27. Karg, M.; Wellert, S.; Prevost, S.; Schweins, R.; Dewhurst, C.; Liz-Marzán, L.M.; Hellweg, T. Well defined hybrid pnipam core-shell microgels: Size variation of the silica nanoparticle core. *Colloid Polym. Sci.* **2010**, *289*, 699–709. [CrossRef]

28. Contreras-Cáceres, R.; Pacifico, J.; Pastoriza-Santos, I.; Pérez-Juste, J.; Fernández-Barbero, A.; Liz-Marzán, L.M. Au@pnipam thermosensitive nanostructures: Control over shell cross-linking, overall dimensions, and core growth. *Adv. Funct. Mater.* **2009**, *19*, 3070–3076. [CrossRef]

29. Stöber, W.; Fink, A.; Bohn, E. Controlled growth of monodisperse silica spheres in the micron size range. *J. Colloid Interface Sci.* **1968**, *26*, 62–69. [CrossRef]

30. Roldan, P.S.; Alcântara, I.L.; Rocha, J.C.; Padilha, C.C.F.; Padilha, P.M. Determination of copper, iron, nickel and zinc in fuel kerosene by faas after adsorption and pre-concentration on 2-aminothiazole-modified silica gel. *Eclet. Quím. São Paulo* **2004**, *29*. [CrossRef]

31. Buga, M.R.; Zaharia, C.; Balan, M.; Bressy, C.; Ziarelli, F.; Margaillan, A. Surface modification of silk fibroin fibers with poly(methyl methacrylate) and poly(tributylsilyl methacrylate) via raft polymerization for marine antifouling applications. *Mater. Sci. Eng. C Mater. Biol. Appl.* **2015**, *51*, 233–241. [CrossRef] [PubMed]

32. Popescu, M.-C.; Filip, D.; Vasile, C.; Cruz, C.; Rueff, J.M.; Marcos, M.; Serrano, J.L.; Singurel, G. Characterization by fourier transform infrared spectroscopy (FT-IR) and 2D IR correlation spectroscopy of PAMAM dendrimer. *J. Phys. Chem. B* **2006**, *110*, 14198–14211. [CrossRef] [PubMed]

33. Zhang, F.; Hou, G.; Dai, S.; Lu, R.; Wang, C. Preparation of thermosensitive pnipam microcontainers and a versatile method to fabricate pnipam shell on particles with silica surface. *Colloid Polym. Sci.* **2012**, *290*, 1341–1346. [CrossRef]

34. Cejkova, J.; Hanus, J.; Stepanek, F. Investigation of internal microstructure and thermo-responsive properties of composite pnipam/silica microcapsules. *J. Colloid Interface Sci.* **2010**, *346*, 352–360. [CrossRef] [PubMed]

35. Zhang, K.; Ma, J.; Zhang, B.; Zhao, S.; Li, Y.; Xu, Y.; Yu, W.; Wang, J. Synthesis of thermoresponsive silica nanoparticle/pnipam hybrids by aqueous surface-initiated atom transfer radical polymerization. *Mater. Lett.* **2007**, *61*, 949–952. [CrossRef]

36. Park, J.-H.; Lee, Y.-H.; Oh, S.-G. Preparation of thermosensitive pnipam-grafted mesoporous silica particles. *Macromol. Chem. Phys.* **2007**, *208*, 2419–2427. [CrossRef]

37. Guo, J.; Yang, W.; Deng, Y.; Wang, C.; Fu, S. Organic-dye-coupled magnetic nanoparticles encaged inside thermoresponsive pnipam microcapsules. *Small* **2005**, *1*, 737–743. [CrossRef] [PubMed]

38. Marcelo, G.; Areias, L.R.P.; Viciosa, M.T.; Martinho, J.M.G.; Farinha, J.P.S. Pnipam based microgels with a ucst response. *Polymer* **2017**, *116*, 261–267. [CrossRef]

39. Wang, Q.; Biswas, C.S.; Galluzzi, M.; Wu, Y.; Du, B.; Stadler, F.J. Random copolymer gels of N-isopropylacrylamide and N-ethylacrylamide: Effect of synthesis solvent compositions on their properties. *RSC Adv.* **2017**, *7*, 9381–9392. [CrossRef]

40. Bischofberger, I.; Calzolari, D.C.E.; Trappe, V. Co-nonsolvency of pnipam at the transition between solvation mechanisms. *Soft Matter* **2014**, *10*, 8288–8295. [CrossRef] [PubMed]

41. Bischofberger, I.; Calzolari, D.C.; De Los Rios, P.; Jelezarov, I.; Trappe, V. Hydrophobic hydration of poly-N-isopropyl acrylamide: A matter of the mean energetic state of water. *Sci. Rep.* **2014**, *4*, 4377. [CrossRef] [PubMed]

gels

MDPI

Communication

Self-Assembly of Colloidal Nanocomposite Hydrogels Using 1D Cellulose Nanocrystals and 2D Exfoliated Organoclay Layers

Takumi Okamoto [1,2], Avinash J. Patil [1,*], Tomi Nissinen [1] and Stephen Mann [1]

[1] Centre for Organized Matter for Chemistry and Centre for Protolife Research, School of Chemistry, University of Bristol, Bristol BS8 1TS, UK; okamoto.bristol@gmail.com (T.O.); tomi.nissinen84@gmail.com (T.N.); s.mann@bristol.ac.uk (S.M.)
[2] Denso Corporation, 1-1, Showa-cho, Kariya, Aichi 448-8661, Japan
* Correspondence: avinash.patil@bristol.ac.uk; Tel.: +44-117-3317215

Academic Editor: Dirk Kuckling
Received: 12 December 2016; Accepted: 13 March 2017; Published: 17 March 2017

Abstract: Stimuli-responsive colloidal nanocomposite hydrogels are prepared by exploiting non-covalent interactions between anionic cellulose nanocrystals and polycationic delaminated sheets of aminopropyl-functionalized magnesium phyllosilicate clays.

Keywords: cellulose nanocrystals; organoclay; nanocomposite hydrogels

1. Introduction

Hydrogels have emerged as an important class of soft materials that have found numerous applications ranging from cosmetics and personal care products to biotechnological and biomedical applications. A typical hydrogel comprises a physically or chemically cross-linked 3D network of natural or synthetic building blocks, which has the ability to encapsulate an extremely high percentage of water compared with their dry weight. Significantly, the capacity to retain water and degree of swelling can be regulated by tailoring the cross-linked network of building blocks in the gel matrix. As a result, the physico-chemical (porosity and hydrophilicity) and mechanical (viscoelastic) properties of hydrogels can be modulated for desired applications [1].

Traditionally, hydrogel assembly involves cross-linking of components such as synthetic or natural polymers [2] and biological molecules [3], or self-assembly of low molecular weight gelators [4–7], or both. Alternatively, recent studies have indicated that self-supporting nanocomposite hydrogels can be prepared by integrating nanoparticles or nanostructures into 3D hydrated cross-linked polymer or biopolymer networks [8,9]. This approach has opened up new opportunities to introduce novel physical, chemical, mechanical, electrical, magnetic, and optical properties into soft materials. For example, materials such as carbon nanotubes and graphene [10,11], polymeric nanoparticles [12,13], inorganic nanoparticles (hydroxyapatite, calcium phosphate, synthetic clays) [14–16], and metallic/metal-oxide nanoparticles (gold, silver, iron oxide, titania) [17–20] have been physically or chemically incorporated within polymeric networks to produce nanocomposite hydrogels with reinforced properties. In this study, we present a new type of stimuli-responsive nanocomposite hydrogel based on the cooperative assembly of two aqueous colloidal sols comprising 1D cellulose nanocrystals (CNCs) and 2D exfoliated sheets of an organically modified magnesium phyllosilicate clay (organoclay). Materials based on CNCs have received considerable attention in recent years due to their applicability in diverse areas such as composites [21], wound dressing and medical implants [22], and as chiral templates for synthesis of inorganic materials [23–25]. To the best of our knowledge, we demonstrate the first example of an organic–inorganic hybrid hydrogel in which non-covalent interactions between 1D and

2D nanoparticles spontaneously form a 3D cross-linked matrix, which expedites the entrapment of water to produce a self-supported colloidal nanocomposite hydrogel matrix. Notably, the CNC–organoclay nanocomposite hydrogels can be disassembled and reconstructed by exposing the soft materials to ammonia and carbon dioxide gases, respectively, suggesting that they could be developed as environmentally sensitive soft materials. We also show that the nanocomposite hydrogels are capable of storing and releasing small drug molecules such as ibuprofen. Significantly, introduction of ibuprofen facilitates co-assembly of the organoclay sheets to produce mesolamellar domains within the hydrogel matrix.

2. Results and Discussion

CNCs were prepared by sulfuric acid hydrolysis of microcrystalline cellulose powder (see Experimental section) [26]. Transmission electron microscopy images of uranyl acetate-stained CNCs revealed the presence of rod-like morphologies, with lengths of 100–300 nm and a thickness of 10–30 nm (Supporting Information, Figure S1a). Aminopropyl-functionalized magnesium phyllosilicate was synthesized using our previously reported studies (see Experimental section) [27]. For delamination, organoclay powders were dispersed in distilled water using ultrasonication. As a consequence, protonation of the aminopropyl-functional groups associated with the inorganic framework facilitated exfoliation of stacks of organoclay sheets and produced a clear suspension containing 50–300 nm sized delaminated organoclay particles (Supporting Information, Figure S1b). CNC–organoclay nanocomposite hydrogels were prepared by adding aqueous dispersions of exfoliated aminopropyl-functionalized magnesium phyllosilicate clay (1–10 wt %) to a 3 wt % colloidal sol of CNCs. The resulting mixtures showed gradual increase in viscosity with increased loadings of the exfoliated organoclay. TEM images of unstained CNC–organoclay hydrogel samples showed the presence of large aggregates comprising cross-linked networks of CNC nanoparticles (Figure 1a). Energy dispersive X-ray (EDX) analysis of the hybrid gel samples showed the presence of Si, Mg, and Cl (counter ion) associated with the organoclay particles and S from sulfonic functional groups of the CNC nanoparticles (Supporting Information, Figure S2). Typically, weight ratios of CNC:organoclay in the range of 1:0.03, 1:0.06. 1:0.13, 1:0.2, and 1:0.26 produced opaque self-supported hydrogels (Figure 1b), and showed no gravity-mediated flow upon inversion of the hydrogels. In contrast, mixing CNC and organoclay dispersions at weight ratios of 1:0.015 and 1:0.033 produced free-flowing liquid and viscous suspensions, respectively.

(a)   (b)

Figure 1. (a) Unstained TEM image of cellulose nanocrystal (CNC)-organoclay hydrogel; (b) photograph showing self-supported CNC:organoclay (1:0.13) colloidal nanocomposite hydrogel, inset showing schematic illustration of cross-linked network formed by non-covalent interactions between CNC (grey) and exfoliated organoclay sheets.

To probe the interactions between the exfoliated organoclay particles and CNCs, we characterized the colloidal nanocomposite hydrogels prepared at a CNC:organoclay weight ratio of 1:0.13 using zeta potential, powder X-ray diffraction (PXRD), and rheometry techniques. Zeta potential studies on the CNCs, exfoliated organoclay sheets, and CNC–organoclay nanocomposite hydrogel gave values of -43 mV, $+20$ mV, and -16 mV, respectively (Supporting Information, Figure S3). The significant decrease in the overall surface charge of the CNCs in the hydrogels was consistent with strong electrostatic interactions between the cationic aminopropyl-functionalities associated with the magnesium phyllosilicate framework and sulfonic functional groups of CNCs. Low-angle PXRD studies of the as-synthesized aminopropyl-functionalized magnesium phyllosilicate showed a broad reflection at $2\theta = 5.9°$, which was attributed to an interlayer spacing (d_{001}) of 1.5 nm, consistent with covalently anchored aminopropyl-functionalities present in the interlamellar regions [22]. In contrast, CNC–organoclay nanocomposite hydrogel samples displayed no reflections below $2\theta = 10°$, confirming that the organoclay sheets remained exfoliated within the hydrogel matrix (Supporting Information, Figure S4a). High-angle PXRD pattern showed that characteristic peaks associated with CNC nanoparticles were retained at $2\theta = 12.6°$ and $22.6°$ (Supporting Information, Figure S4b). Frequency sweep experiments showed a linear viscoelastic region where parallel storage ($G' = 240$ Pa) remained higher than loss ($G'' = 25$ Pa) moduli, which is a typical characteristic for solid-like viscoelastic hydrogels (Figure 2a). Oscillatory amplitude sweeps at a constant frequency of 1 Hz of hydrogels showed a linear viscoelastic region with parallel G' (290 Pa) and G'' (30 Pa) moduli up to a shear strain of 10%. Two crossover points observed above 10% shear strain were assigned to deformation of the hydrogel into a quasi-liquid state, indicating that the colloidal nanocomposite hydrogel exhibited responsive properties (Figure 2b).

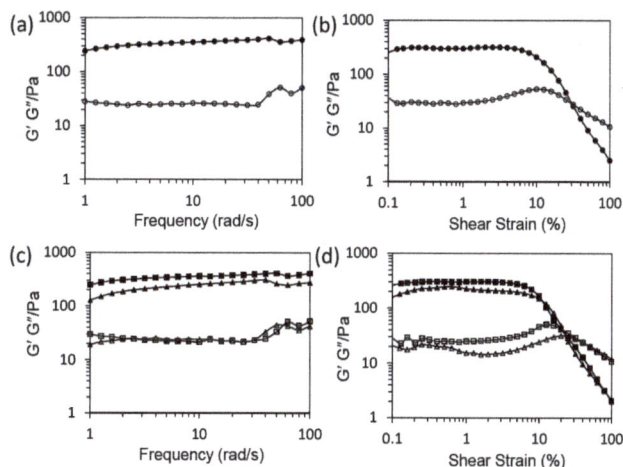

Figure 2. Rheometry studies showing (**a**) frequency sweep and (**b**) amplitude sweep curves for an as-prepared CNC–organoclay colloidal nanocomposite hydrogel (ratio 1:0.13); values for storage G' (filled circles) and loss G'' moduli (open circles); (**c**) and (**d**) show frequency and amplitude sweep profiles, respectively, for a CNC–organoclay nanocomposite hydrogel (ratio 1:0.13) after exposure to gaseous ammonia (G', filled triangles; G'', open triangles) and carbon dioxide (G', filled squares; G'', open squares).

We also investigated the gas-sensing properties of the colloidal nanocomposite hydrogels. For this, a hybrid hydrogel (CNC:organoclay, 1:0.13) was carefully exposed to ammonia gas to increase the pH of the bulk hydrogel from 9.4 to 10.4. Interestingly, the inversion tests carried out after change in the pH showed gravity-induced flow of the viscous hydrogel. Significantly, exposing the above viscous fluids to carbon dioxide gas lowered the pH value of the bulk hydrogel back to 9.4, which re-instigated cross-linking of the organic and inorganic building blocks to yield a self-supported hydrogel. Corresponding frequency sweep measurements on the ammonia-treated colloidal nanocomposite hydrogels at pH 10.4 showed a decrease in G' values to 127 Pa, whilst lowering the pH back to 9.4 using gaseous carbon dioxide increased the G' value to 240 Pa (Figure 2c), consistent with a decrease and restoration of the solid-like viscoelastic behavior of the hydrogel, respectively. In contrast, loss moduli (G'') values of the as-prepared hydrogel and hydrogels prepared at pH 10.4 or 9.4 remained unchanged. The oscillatory amplitude sweep experiments were also consistent with above observations (Figure 2d). Ammonia- and carbon dioxide-treated colloidal nanocomposite hydrogels maintained linear viscoelastic regions up to a shear strain of 10%, above which the samples became deformed.

We also employed cross-polarized (CP) optical microscopy to investigate the effect of gas-induced reassembly and disassembly of CNC–organoclay colloidal nanocomposite hydrogel networks (Figure 3). CP microscopy images of the as-prepared CNC:organoclay nanocomposite hydrogels showed birefringence, indicating the presence of anisotropic CNC–organoclay domains dispersed within an isotropic phase. Exposure to ammonia gas caused significant reduction in the birefringence of the hybrid gels, consistent with an increase in the isotropic phase due to loss of the interactions between the CNC and organoclay particles. Notably, birefringence was regained when the above samples were carefully exposed to the carbon dioxide gas, confirming that the non-covalent interactions between the organic and inorganic particles were restored.

Figure 3. Cross-polarized microscopy images of (**a**) as-prepared (**b**) ammonia- and (**c**) carbon dioxide-treated CNC–organoclay colloidal nanocomposite hydrogels, scale bar = 100 μm.

Taken together, the above observations suggest that pH-dependent electrostatic interactions between the CNC and exfoliated organoclay sheets were primarily responsible for the formation of the self-supported hydrogels. Gas-induced disassembly and reconstruction of the CNC–organoclay nanocomposite hydrogels was attributed to deprotonation and re-protonation of aminopropyl-functionalities (pKa of primary amine ~10.5) associated with the organoclay sheets. As a consequence, environmentally induced changes in pH are able to alter the columbic interactions between the CNC and organoclay layers, and, therefore, strongly influence the mechanical and optical properties of the colloidal nanocomposite hydrogels.

The potential use of CNC–organoclay colloidal nanocomposite hydrogels for the encapsulation and controlled release of functional small molecules such as the anti-inflammatory drug, ibuprofen, was investigated. Encapsulation of the drug was carried out by mixing an aqueous solution of ibuprofen (10 wt %) with a CNC sol (6 wt %), followed by addition of an aqueous dispersion of the exfoliated organoclay sheets. The resulting mixture produced a self-supported hydrogel with a CNC:organoclay:ibuprofen weight ratio of 1:0.13:1.6. Frequency sweeps revealed a linear viscoelastic region in which the parallel storage moduli were higher than the loss moduli. However, the storage moduli (G' = 121 Pa) and loss moduli (G'' = 13.8) values were significantly lower than that of parent CNC–organoclay hybrid hydrogel (Supporting Information, Figure S5). Oscillatory amplitude sweeps at a constant frequency of 1 Hz revealed that G' and G'' remained parallel and deformed above 10% shear strain (Supporting Information, Figure S6). The marked decrease in solid-like viscoelastic properties of the drug-loaded hybrid hydrogel was attributed to charge screening of the columbic interactions between the CNC and exfoliated clay layers due to the presence of the anionic drug molecules.

The concentration of ibuprofen released over time from the colloidal nanocomposite CNC–organoclay hydrogel was determined by using UV–vis spectroscopy. As a control sample, an ibuprofen–organoclay composite was prepared by adding an aqueous solution of ibuprofen to a freshly exfoliated suspension of organoclay. This allowed spontaneous restacking of the exfoliated sheets in the presence of the drug molecules. The resulting precipitates were isolated, dried, and pressed into pellet for the release studies. Both the ibuprofen-loaded colloidal nanocomposite hydrogel and organoclay–ibuprofen control nanocomposite showed a steady release of the drug molecules over a period of 6 h. Comparison of the initial release profiles obtained within an hour indicated that the extraction of ibuprofen molecules from the drug-loaded CNC–organoclay hydrogel was approximately 10% slower than that from the ibuprofen–organoclay nanocomposite pellet (Figure 4a). To further elucidate the origin of these differences, structural investigations on the nature of the ibuprofen–organoclay nanocomposite and CNC–organoclay–ibuprofen hydrogel were undertaken by using PXRD studies (Figure 4b). PXRD profiles recorded from the nanocomposite indicated that the exfoliated organoclay sheets were spontaneously restacked into a mesolamellar bulk phase in the presence of the drug molecules. As a result, the interlamellar spacing for the parent organoclay was increased from 1.5 nm ($2\theta = 5.9°$) to 2.3 nm ($2\theta = 3.8°$), which suggested that ibuprofen molecules were intercalated within the interlayer regions of the reconstituted organoclay. Significantly, PXRD patterns of the ibuprofen-containing CNC–organoclay hydrogels also exhibited a low-angle reflection at $2\theta = 3.8°$ corresponding to an interlayer distance of 2.3 nm, indicating that the colloidal nanocomposite hydrogels comprised mesolamellar domains containing intercalated drug molecules [22]. The slower displacement of the drug molecules observed from the nanocomposite hydrogels could be therefore attributed to a combination of charge-mediated and diffusion-limited processes associated with interactions between the aminopropyl side chains of the organoclay sheets and carboxylic acid moieties of ibuprofen, and physical immobilization of the drug molecules within the interconnected CNC network, respectively.

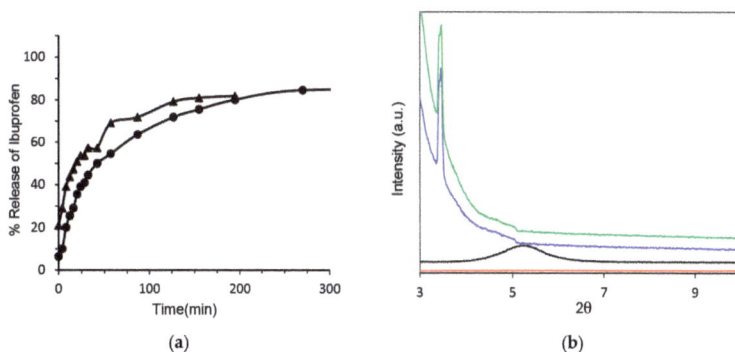

Figure 4. (**a**) Ibuprofen release profiles from organoclay–ibuprofen nanocomposite pellet (triangles) and CNC–organoclay–ibuprofen colloidal nanocomposite hydrogels (circles); (**b**) powder X-Ray diffraction (PXRD) pattern of CNC–organoclay nanocomposite hydrogel (red), as-synthesized organoclay (black), organoclay–ibuprofen nanocomposite (blue), and CNC–organoclay–ibuprofen colloidal nanocomposite hydrogel (green).

3. Conclusions

In summary, we have demonstrated a simple methodology for construction of colloidal nanocomposite hydrogels by using aqueous sols of 1D cellulose nanocrystals and 2D delaminated sheets of aminopropyl-functionalized magnesium phyllosilicate clay. The results indicate that electrostatic interactions between the negatively charged cellulose nanocrystals and polycationic exfoliated sheets of the organoclay spontaneously generate a 3D interconnected matrix that encapsulates water molecules to yield self-supported hydrogels. We also show the colloidal CNC–organoclay nanocomposite hydrogels are responsive to external stimuli such that gas-induced changes in pH-triggered reversible changes in viscoelastic properties. Moreover, inclusion of ibuprofen in the colloidal nanocomposite hydrogels facilitates reassembly of the cationic organoclay sheets to produce mesolamellar domains comprising intercalated drug molecules. As a consequence, the hydrogel matrices show slower drug release rates. Thus, the CNC–organoclay colloidal nanocomposite hydrogels should offer an excellent opportunity to fabricate new types of stimuli-responsive hydrogels that are bioactive and biocompatible, and of potential use in applications such as wound dressing and consumer care products.

4. Experimental Section

4.1. Materials and Methods

All chemicals were obtained from Sigma-Aldrich (St. Louis, MO, USA) and used without purification. Rheology experiments were performed by using Malvern Kinexus Pro Rheometer (Malvern Instruments Ltd, Malvern, UK) equipped with parallel plate (diameter of 20 mm). Hydrogel samples aged for 1 day were added to the base plate to minimize shear. The top plate was then lowered to a set gap width of 150 μm, and the normal force was then measured; once the normal force reached an equilibrium, the measurements were performed at room temperature. CNC–organoclay hydrogel samples obtained from different batches showed similar rheological behavior and the data were within ±10% error. We ascribe the differences to changes in hydration levels and shear-induced effects during sample loading. Cross-polarized optical microscopy images were obtained by using Leica DMI3000 B manual inverted fluorescence microscope (Leica Microsystems Ltd., UK). Transmission electron microscopy analysis was undertaken on JEOL1400 operating (JEOL Ltd., Tokyo, Japan) at 120 keV in bright field mode. Energy dispersive X-ray analysis (EDX) analysis was performed by using Oxford Instrument Aztec microanalysis SDD detector (Oxford Instruments, Abingdon, UK) attached to JEOL 2100F-STEM microscope (JEOL Ltd.). CNC samples were prepared by mounting 5 μL of the 1 wt % CNC dispersions onto carbon-coated grids and left to dry at room temperature. Negative staining was carried out using an aqueous solution of 1% uranyl acetate. Delaminated sheets of organoclay were imaged by mounting freshly exfoliated dispersion of organoclay (0.01 mg/mL). TEM micrographs of hydrogel samples were obtained by dispersing 5–10 μL of gel samples in 500 μL distilled water prior to mounting onto carbon-coated grids. Zeta-potential measurements for aqueous dispersions of CNCs (1 wt %), freshly exfoliated organoclay (1 mg/mL), and hydrogels (20 μL of gel dispersed in 5 mL distilled water) were performed using Malvern Zetasizer Nano ZS (Malvern Instruments Ltd., Malvern, UK). Powder diffraction patterns were obtained by using Bruker Advanced D8 powder diffractometer (Bruker Corporation, Billerica, MA, USA). Controlled release experiments were performed by using Lambda-35 Perkin-Elmer UV–vis spectrophotometer (PerkinElmer Inc., Waltham, MA, USA).

4.2. Synthesis and Exfoliation of Aminopropyl-Functionalized Magnesium Phyllosilicate (Organoclay)

Magnesium chloride hexahydrate (0.84 g) was dissolved in 20 g ethanol. To this solution, 1.3 mL of 3-aminopropyltriethoxysilane was added dropwise with continuous stirring. The white slurry obtained after 5 min was stirred overnight and the precipitate was isolated by centrifugation, washed with ethanol (50 mL), and dried at 40 °C. A stable clear aqueous dispersions containing exfoliated organoclay layers (1–10 wt %) were prepared by dispersing desired amounts of organoclay powders in distilled water followed by ultrasonication for 5 min.

4.3. Synthesis of Crystalline Nanocellulose (CNC)

Fibrous cellulose powder (10.0 g, Whatman–CF11, Sigma-Aldrich Co.) was hydrolyzed by sulfuric acid (87.5 mL, 64%) for 40 min at 45 °C with continuous stirring. The hydrolysis was quenched by adding a large amount of water (500 mL) to the reaction mixture. The resulting mixture was cooled to room temperature 25 °C and centrifuged (4000 rpm) for 10 min at room temperature. The supernatant was decanted. Distilled water (500 mL) was added to the precipitate and the mixture was then stirred vigorously to form a new suspension. This centrifugation process was repeated three times. The newly generated suspension was dialyzed with an ultrafiltration membrane (30,000 molecular weight cutoff) until the pH of the suspension reached a constant value (pH ~6). Finally, the suspension was collected into glass vessel, then evaporated at 40 °C to obtain the desired concentration of CNC.

4.4. Fabrication of Colloidal CNC–Organoclay Nanocomposite Hydrogels

Typically, nanocrystalline cellulose-containing nanocomposite hydrogels were prepared by adding an aqueous solution of exfoliated organoclay at low concentrations (1–10 wt %) to a colloidal sol of CNC with low concentration (3 wt %). The mixture was allowed to stand for several minutes to obtain a self-supported CNC–organoclay hybrid hydrogel.

4.5. Synthesis of CNC–Organoclay–Ibuprofen Nanocomposite Hydrogels

Typically, 0.1 mL of a colloidal sol of CNC at 6 wt % concentration was added to 0.1 mL of an aqueous ibuprofen (10 wt %) solution with stirring. Freshly exfoliated organoclay (5 wt %, 0.016 mL) dispersions were added and the reaction mixture was allowed to stand to obtain the CNC–organoclay–ibuprofen hydrogel.

4.6. Synthesis of Organoclay–Ibuprofen Composites

Ibuprofen solution (0.1 mg/mL) was added to a freshly prepared exfoliated dispersion of organoclay (0.05 mg/mL). The resulting precipitates were centrifuged, dried, and pressed into pellets for drug-release experiments.

4.7. Controlled Release Studies

Organoclay–ibuprofen pellets or CNC–organoclay–ibuprofen hydrogels were immersed in a known volume of distilled water. Aliquots were removed at regular time intervals and time-dependent release of ibuprofen was measured by monitoring changes in the intensity of the absorption peak at 264 nm using UV–vis spectroscopy. Release profiles were plotted by averaging absorption values obtained over three runs.

Supplementary Materials: The following are available online at www.mdpi.com/2310-2861/3/1/11/s1. Supporting figures of TEM, zeta potential measurements, powder X-ray diffraction patterns, and rheology data.

Acknowledgments: Takumi Okamoto and Stephen Mann thank Denso Corporation for research funds and Avinash J. Patil is grateful to University of Bristol for financial support. Authors would like to thank J. Jones for assistance with TEM.

Author Contributions: Avinash J. Patil conceived and designed the experiments; Takumi Okamoto and Tomi Nissinen performed the experiments; Takumi Okamoto, Tomi Nissinen and Avinash J. Patil analyzed the data; Avinash J. Patil and Stephen Mann wrote the paper.

Conflicts of Interest: The authors declare no conflict of interest.

References

1. Loh, X.J.; Scherman, O.A. *Polymeric and Self-Assembled Hydrogels: From Fundamental Understanding to Applications*; RSC Publishing: Cambridge, UK, 2013.
2. Haraguchi, K. Nanocomposite, Hydrogels. *Curr. Opin. Solid State Mater. Sci.* **2007**, *11*, 47–54. [CrossRef]
3. Jonker, A.M.; Lowik, D.W.P.M.; van Hest, J.C.M. Peptide and Protein-Based Hydrogels. *Chem. Mater.* **2012**, *24*, 759–773. [CrossRef]
4. Kopecek, J.; Yang, J. Smart Self-Assembled Hybrid Hydrogel Biomaterials. *Angew. Chem. Int. Ed.* **2012**, *51*, 7396–7417. [CrossRef] [PubMed]
5. Raeburn, J.; Cardoso, A.Z.; Adams, D.J. The Importance of the Self-Assembly Process to Control Mechanical Properties of Low Molecular Weight Hydrogels. *Chem. Soc. Rev.* **2013**, *42*, 5143–5156. [CrossRef] [PubMed]
6. Du, X.; Zhou, J.; Shi, J.; Xu, B. Supramolecular Hydrogelators and Hydrogels: From Soft Matter to Molecular Biomaterials. *Chem. Rev.* **2015**, *115*, 13165–13307. [CrossRef] [PubMed]

7. Gaharwar, A.K.; Peppas, N.A.; Khademhosseini, A. Nanocomposite Hydrogels for Biomedical Applications. *Biotechnol. Bioeng.* **2014**, *111*, 441–453. [CrossRef] [PubMed]
8. Merino, S.; Martin, C.; Kostarelos, K.; Prato, M.; Vazquez, E. Nanocomposite Hydrogels: 3D Polymer-Nanoparticle Synergies for On-Demand Drug Delivery. *ACS Nano* **2015**, *9*, 4686–4697. [CrossRef] [PubMed]
9. Shin, S.R.; Jung, S.M.; Zalabany, M.; Kim, K.; Zorlutuna, P.; Kim, S.B.; Nikkhah, M.; Khabiry, M.; Azize, M.; Kong, J.; et al. Carbon Nanotube-Embedded Hydrogel Sheets for Engineering Cardiac Constructs and Bioactuators. *ACS Nano* **2013**, *7*, 2369–2380. [CrossRef] [PubMed]
10. Giri, A.; Bhowmick, M.; Pal, S.; Bandopadhyay, A. Polymer Hydrogel from Caboxymethyl Guar Gum and Carbon Nanotube for Sustained Trans-dermal Release of Diclofenac Sodium. *Int. J. Biol. Macromol.* **2011**, *49*, 885–893. [CrossRef] [PubMed]
11. Goenka, S.; Sant, V.; Sant, S. Graphene-based Nanomaterials for Drug Delivery and Tissue Engineering. *J. Control Release* **2014**, *173*, 75–88. [CrossRef] [PubMed]
12. Zhong, S.P.; Yung, L.Y.L. Enhanced Biological Stability of Collagen with Incorporation of PAMAM Dendrimer. *J. Biomed. Mater. Res. A* **2009**, *91A*, 114–122. [CrossRef] [PubMed]
13. Appel, E.A.; Tibbitt, M.W.; Webber, M.J.; Mattix, B.A.; Veiseh, O.; Langer, R. Self-Assembled Hydrogels Utilizing Polymer-Nanoparticle Interactions. *Nat. Commun.* **2015**. [CrossRef] [PubMed]
14. Schexnailder, P.; Schmidt, G. Nanocomposite Polymer Hydrogels Colloid. *Polym. Sci.* **2009**, *287*, 1–11.
15. Wang, Q.; Mynar, J.L.; Yoshida, M.; Lee, E.; Lee, M.; Okura, K.; Kinbara, K.; Aida, T. High-Water content Mouldable Hydrogels by Mixing Clay and a Dendritic Molecular Binder. *Nature* **2010**, *463*, 339–343. [CrossRef] [PubMed]
16. Martin, J.E.; Patil, A.J.; Butler, M.F.; Mann, S. Guest-Molecule Directed Assembly of Mesostructured Nanocomposite Polymer/Organoclay Hydrogels. *Adv. Funct. Mater.* **2011**, *21*, 674–681. [CrossRef]
17. Liu, T.Y.; Hu, S.H.; Liu, D.M.; Chen, S.Y. Magnetic-sensitive Behaviour of Intelligent Ferrogels for Controlled Release of Drug. *Langmuir* **2006**, *22*, 5974–5978. [CrossRef] [PubMed]
18. Dvir, T.; Timko, B.P.; Brigham, M.D.; Naik, S.R.; Karajangai, S.S.; Levy, O.; Jin, H.; Parker, K.K.; Langer, R.; Kohane, D.S. Nanowired Three-Dimensional Cardiac Patches. *Nat. Nanotechnol.* **2011**, *6*, 720–725. [CrossRef] [PubMed]
19. Liu, Y.; Ma, W.; Liu, W.; Li, C.; Liu, Y.; Jiang, X.; Tang, Z. Silver (I)-Glutathione Biocoordination Polymer Hydrogel: Effective Antibacterial Activity and Improved Cytocompatibility. *J. Mater. Chem.* **2011**, *21*, 19214–19218. [CrossRef]
20. Thoniyot, P.; Tan, M.J.; Karim, A.A.; Young, D.J.; Loh, X.J. Nanoparticle-Hydrogel Composites: Concept, Design, and Applications of These Promising, Multi-Functional Materials. *Adv. Sci.* **2015**. [CrossRef] [PubMed]
21. Dufresne, A. Comparing the Mechanical Properties of High Performance Polymer Nancomposites from Biological Resources. *J. Nanosci. Nanotechnol.* **2006**, *6*, 322–330. [CrossRef] [PubMed]
22. Thomas, S. A Review of the Physical Biological and Clinical Properties of a Bacterial Cellulose Wound. *J. Wound Care* **2008**, *17*, 349–352. [CrossRef] [PubMed]
23. Shopsowitz, K.E.; Qi, H.; Hamad, W.Y.; MacLachlan, M.J. Free-Standing Mesoporous Silica Films with Tuneable Chiral Nematic Structures. *Nature* **2010**, *468*, 422–425. [CrossRef] [PubMed]
24. Shopsowitz, K.E.; Hamad, W.Y.; MacLachlan, M.J. Chiral Nematic Mesoporous Carbon Derived from Nanocrystalline Cellulose. *Angew. Chem. Int. Ed.* **2011**, *50*, 10991–10995. [CrossRef] [PubMed]
25. Shopsowitz, K.E.; Stahl, A.; Hamad, W.Y.; MacLachlan, M.J. Hard Templating of Nanocrystalline Titanium Dioxide with Chiral Nematic Ordering. *Angew. Chem. Int. Ed.* **2012**, *51*, 6886–6890. [CrossRef] [PubMed]
26. Dujardin, E.; Blaseby, M.; Mann, S. Synthesis of Mesoporous Silica by Sol-Gel Mineralisation of Cellulose Nanorod Nematic Suspensions. *J. Mater. Chem.* **2003**, *13*, 696–699. [CrossRef]
27. Patil, A.J.; Muthusamy, E.; Mann, S. Synthesis and Self-Assembly of Organoclay-Wrapped Biomolecul. *Angew. Chem. Int. Ed.* **2004**, *43*, 4928–4933. [CrossRef] [PubMed]

Article

Synthesis and Properties of New "Stimuli" Responsive Nanocomposite Hydrogels Containing Silver Nanoparticles

G. Roshan Deen [†,*] and Vivien Chua [†]

Soft Materials Laboratory, Natural Sciences and Science Education AG, National Institute of Education, Nanyang Technological University, 1-Nanyang Walk, Singapore 637616, Singapore; vi.chua938@gmail.com
* Author to whom correspondence should be addressed; roshan.gulam@nie.edu.sg;
  Tel.: +65-6790-3816; Fax: +65-6896-9414.
† These authors contributed equally to this work.

Academic Editor: Dirk Kuckling
Received: 10 June 2015; Accepted: 21 August 2015; Published: 28 August 2015

Abstract: Hydrogel nanocomposites containing silver nanoparticles of size 15–21 nm were prepared by diffusion and *in-situ* chemical reduction in chemically crosslinked polymers based on N-acryloyl-N′-ethyl piperazine (AcrNEP) and N-isopropylacrylamide (NIPAM). The polymer chains of the hydrogel network offered control and stabilization of silver nanoparticles without the need for additional stabilizers. The presence of silver nanoparticles and their size was quantified by UV-Vis absorption spectroscopy and scanning electron microscopy. The nanocomposite hydrogels were responsive to pH and temperature changes of the external environment. The equilibrium weight swelling ratio of the hydrogel nanocomposite was lower in comparison with the precursor hydrogel. Silver nanoparticles present in the nanocomposite offered additional physical crosslinking which influenced media diffusion and penetration velocity. The release of silver nanoparticles from the hydrogel matrix in response to external pH changes was studied. The rate of release of silver nanoparticles was higher in a solution of pH 2.5 due to maximum swelling caused by ionization of the gel network. No significant release of nanoparticles was observed in a solution of pH 7.

Keywords: hydrogels; silver nanoparticles; absorption; nanocomposites; swelling

1. Introduction

Hydrogels are three-dimensional crosslinked (chemical or physical) polymer networks that can absorb large amounts of water and yet remain insoluble. "Stimuli" responsive hydrogels respond to changes in external stimuli such as pH, temperature, ionic strength, electric field, pressure, magnetic field *etc.* [1–5]. The potential responses to these stimuli are changes in shape, volume, phase, and optical properties. The stimuli-responsive volume change of the gels is a result of many factors such as type of monomers, hydrophilic-hydrophobic balance, crosslink density, osmotic pressure, conformation of chemical groups *etc.* These materials are widely applied in targeted drug delivery systems, nanocomposites, protein purification, sensor technology, and enzyme immobilizations [6–10].

The synthesis of nanoparticles is a major area of research, and many methods such as chemical reduction, micro-wave assisted methods, radiation–chemical reductions *etc.* have been widely reported. The chemical reduction process by the methods of co-precipitation, polymer stabilization, and microemulsion has been reported to yield nanoparticles with narrow size distribution [11–13]. In recent years the synthesis of nanoparticles within a crosslinked polymer network (hydrogels) by *in-situ* and *ex-situ* chemical reduction has gained considerable research focus in the development of polymer-metal hybrid materials [14–17]. Such hybrid materials (hydrogel nanocomposites) containing colloidal nanoparticles have wide applications in catalysis, drug-delivery systems, anti-bacterial systems, chemical sensors *etc.* The hydrogel networks provide easy nucleation and growth of nanoparticles without much aggregation or agglomeration which is undesirable for any application [15–17]. The presence of nanoparticles in the hydrogel matrix acts at nano-fillers which improves the mechanical properties of the nanocomposites.

Among metal nanoparticles, silver nanoparticles have attracted much research attention in recent years owing to their excellent properties such as, electrical conductivity, bacterial action, optical properties, and catalysis [18–20]. They are considered as a non-toxic and environmentally friendly anti-bacterial material. Controlled formation of silver nanoparticles in microgels based on N-isopropylacrylamide, acrylic acid, and hydroxyethyl acrylate for the development of photonic crystals has been reported by Kumacheva [20–22]. The synthesis of silver and gold nanoparticles in microgels and hydrogels using various chemical methods has been documented in the literature [22–25]. Thus the development of non-toxic nanocomposite hydrogels containing silver nanoparticles is an active area of research.

In this article, we report the synthesis and characterization of pH- and temperature-responsive silver nanocomposite hydrogels by a facile *in-situ* method. The chemically crosslinked hydrogel is composed of N-acryloyl-N'-ethyl piperazine (AcrNEP) a pH responsive component [26] and N-isopropylacrylamide (NIPAM) a temperature-responsive component [27–30]. The nanocomposite hydrogel was synthesized by incorporating silver nanoparticles in the gel matrix by diffusion and *in-situ* chemical reduction process. This *in-situ* method of synthesis facilitates the formation of nanoparticles in hydrogels with a low size polydispersity index. The low mesh size of the hydrogel matrix facilitated the growth of silver nanoparticles, and the polymer chain offered stability against aggregation. This study provides the possibility to control the size and quantity of silver nanoparticles in the hydrogel matrix by varying the amount of cationic monomer (AcrNEP) and crosslinker. This type of nanocomposite material is non-toxic and envisaged to possess excellent optical and anti-microbial properties. The influence of external stimuli such as changes in pH on the release of silver nanoparticles from the nanocomposite hydrogel matrix is also described in this report.

2. Results and Discussion

2.1. Synthesis of Hydrogel and General Characteristics

The hydrogels were prepared by free-radical solution polymerization at 23 °C using an accelerator viz. Tetramethylethylene diamine (TEMED). The solution became viscous and gelled within 10 min after the addition of free-radical initiator potassium persulfate (KPS) and the accelerator. The gels were transparent in appearance and soft in texture. The gel A7N3 was softer than the gel A1N9. The conversion of monomer (C^{mon}) was estimated based on the following equation and the results are presented in Table 1.

$$C^{mon} = \frac{W_d}{W_i} \times 100 \qquad (1)$$

where W_i and W_d are the weight of monomers before polymerization and dry weight of the gel respectively.

Table 1. Monomer feed compositions, conversion, and physical appearance of gels.

Gel	Monomer Feed (mol %)			C^{mon} (%)	Appearance	
	AcrNEP [a]	NIPAM [b]	MBA		Before polymerization	After polymerization
A1N9	6.90	91.60	1.50	93.45	Clear solution	Clear gel
A7N3	68.95	29.55	1.50	94.26	Clear solution	Clear gel

[a] molar mass of N-acryloyl-N'-ethyl piperazine (AcrNEP) = 168 g·mol^{-1}; [b] molar mass of N-isopropylacrylamide (NIPAM) = 113 g·mol^{-1}. $C^{mon.}$ = Conversion of monomer. MBA = N,N'-methylenebisacrylamide

The conversion of monomers is more than 90% which indicates an efficient crosslinking of the monomers with crosslinker *N,N'*-methylenebisacrylamide (MBA). Due to similar reactivity of the acrylic groups of the monomer and crosslinker (CH$_2$=CH–CO–) an efficient crosslinking reaction is achieved. Similar observations have been reported for the crosslinking reaction of vinyl caprolactam and hydroxyl ethylmethacrylate with ethylene glycol methacrylate as the crosslinker [31,32]. The chemical structure of the hydrogel is shown in Figure 1.

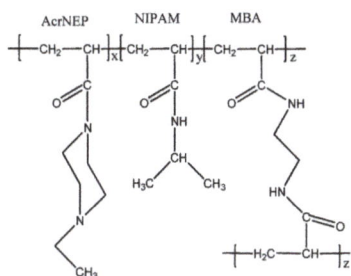

Figure 1. Chemical structure of crosslinked hydrogel.

2.2. Fabrication of Silver Nanoparticles by In-Situ Chemical Reduction Method

The use of hydrogels as templates is an interesting and easy approach in the preparation of nanoparticles/nanocomposites [14–16]. Hydrogel silver nanocomposites were prepared by a two-step method at 23 °C. In the first step, the gel was soaked in aqueous AgNO$_3$ solution until swelling equilibrium was attained. During this process the silver ions diffused into the gel and formed a 1:1 complex [2,15,16] with the tertiary amine nitrogen of AcrNEP. In the second step, the silver salt loaded gel was chemically reduced by a strong reducing agent, NaBH$_4$. During the process of chemical reduction with NaBH$_4$ the visual appearance of the gel changed from colorless to yellowish-brown indicating the formation of silver nanoparticles in the gel matrix. The general scheme of the two-step method and the digital image of the gel A7N3 before and after chemical reduction are shown in Figure 2.

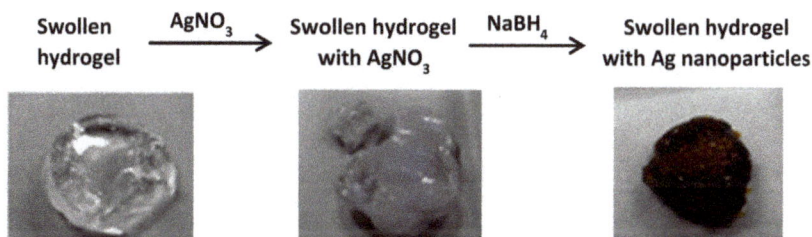

Figure 2. General scheme for the synthesis of silver nanocomposite hydrogels.

The hydrogels reported in this work contain random distribution of AcrNEP and NIPAM repeating units along the copolymer chain with a number of MBA crosslink points. The presence of AcrNEP (ionic monomer) with tertiary amine nitrogen and NIPAM with secondary amine functionality and their random distribution along the copolymer chain acts as nano-templates or nano-reactors for the formation of silver nanoparticles [15,16]. The polymer chains of the hydrogel network offer control and stabilization of the nanoparticles without the need of any further stabilizer.

Further it has been demonstrated that uniform dispersion of metal nanoparticles can easily be synthesized in copolymer hydrogels. The role of AcrNEP in the chemical reduction process cannot be ruled out; however such a reduction process is effective only at higher temperatures [16]. In general, large polymer chains can either act as stabilizing or encapsulating agents for nanoparticle stabilization [33]. The use of natural and synthetic polymers as stabilizers for metal nanoparticles has been reported [34,35].

2.3. UV-Vis Absorption Spectra of Hydrogel Nanocomposites

The presence of silver nanoparticles in the gels was confirmed by UV-Vis absorption spectroscopy, and the absorption spectra are shown in Figure 3. A distinct absorption maximum is observed at wavelengths 408 nm and 396 nm for the gels A1N9 and A7N3 respectively. This absorption maximum is a characteristic surface Plasmon resonance feature that arises from the quantum size silver nanoparticles present in the gel [36]. Similar absorption peaks have been reported for silver nanoparticles of quantum size [15,16,36–38]. The difference in absorption peak position between the gels is attributed to the size polydispersity of silver nanoparticles. In general, very small nanoparticles of uniform size shift the absorption maximum to lower wavelength [36–38]. The absorption maximum and the area under the peak exhibited by the gel A7N3 is intense and narrow compared to that of A1N9 which is less intense and significantly broad.

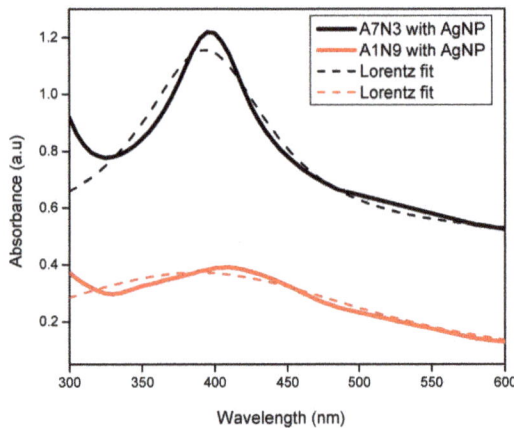

Figure 3. UV-Vis spectra of silver nanoparticles in hydrogel nanocomposites.

This difference in absorption intensity can be explained as follows: The gel A7N3 contains 68.95 mol % of the ionizable monomer (AcrNEP) units, while A1N9 contains only 6.90 mol %. The presence of higher amount of AcrNEP makes the gel hydrophilic and allows easy diffusion of silver ions and their subsequent reduction by the borohydride ions. The width of absorption peak for gel A1N9 containing silver nanoparticles is comparatively broad which is attributed to the large size polydispersity.

The presence of lower amount of AcrNEP (6.90 mol %) hinders easy diffusion of silver ions and does not offer stabilization to silver nanoparticles against aggregation. The increase in absorption intensity and large peak area therefore represents the formation of a large number of silver nanoparticles. Thus the presence of ionic monomer plays an important role in controlling the size and size distribution of silver nanoparticles in the hydrogel nanocomposites.

Assuming the free particle behavior of electrons the size of silver nanoparticles was calculated using the following formula which is based on modified Mie scattering theory [39],

$$d = \frac{h\,v_f}{\pi\,\Delta E_{1/2}} \tag{2}$$

where h is the Plank's constant, v_f is the Fermi velocity of electrons in bulk silver (1.39×10^6 m s^{-1}), and $\Delta E_{1/2}$ is the full width at half maxima (FWHM) of the absorption peak. The measured absorbance was fitted with Lorentz distribution fit to obtain the FWHM as shown in Figure 3. The above equation is valid only if the silver nanoparticles are smaller than the mean free path of electrons in the bulk metal. The mean free path of electrons is about 27 nm at room temperature for bulk silver [40]. Using the above formula the average size of silver nanoparticles at 23 °C in A7N3-AgNp nanocomposite was calculated to be ~20 nm.

2.4. Surface Morphology of Hydrogel Silver Nanocomposites

The size and distribution of silver nanoparticles in the gel matrix was studied by scanning electron microscopy, and the micrograph of gel A7N3-AgNP is shown in Figure 4 as a representative example. The presence of finely dispersed silver nanoparticles throughout the gel matrix is clearly seen. The average size of the silver nanoparticles is in the range 15–21 nm as calculated from the micrograph. The porous morphology of the gel network and the presence of charged monomer units along the copolymer chain are believed to be the factors in stabilizing the silver nanoparticles against aggregation. Such domains of nanometer length scale allow the growth of silver nanoparticles. The porosity of gel network and hydrophilicity of polymer chain allow easy diffusion of small molecules (silver nitrate, sodium borohydride, and water) into the network for rapid reduction. The hydrogels facilitate easy diffusion of small ions for complexation and subsequent chemical reduction and these important properties are widely accepted in sensor technology and catalysis [15,16,20].

Figure 4. Scanning electron micrograph of gel A7N3-AgNp.

The immobilization of silver nanoparticles throughout the hydrogel matrix is due to strong location of silver ions in the network. This is possible due to complexation of silver ions with nitrogen and oxygen atoms of the polymer chain. The tertiary amine nitrogen of the piperazine ring of AcrNEP is known to form co-ordinate complexes with both univalent and divalent metal ions [41]. The crosslinker content of the gels was fixed at 1.5 mol % to allow easy diffusion of small ions such as borohydride into the gel for chemical reduction. Gels with high crosslinker content are rigid and restrict diffusion of ions leading to non-uniform distribution of nanoparticles in the gel matrix.

2.5. Effect of pH on Swelling of Hydrogel Nanocomposite

The gels prepared in this work were responsive to pH changes of the external solution. The gels swelled considerably more in acidic solutions than in basic solutions. The equilibrium swelling ratio as a function of the pH of the external solution is summarized in Table 2. Maximum equilibrium swelling is observed for the gels in a solution of pH 2.5, which is attributed to complete protonation of the tertiary amine ($^+$N–CH$_2$–CH$_3$) of AcrNEP. The protonation leads to the formation of fixed electrical charges on the polymer network that repel by Coulombic interactions. This leads to an osmotic pressure gradient between the gel matrix and external solution which causes the gel to swell. The pK$_a$ of AcrNEP is 4 and maximum swelling is expected in solutions of pH lower than 4.

Table 2. Equilibrium weight swelling ratio (WSR^{Eq}) of gels as function of solution pH at 23 °C.

Gel	WSR^{Eq}					WSR$^{Eq\,pH\,=\,2.5}$/WSR^{Eq} pH = 12
	pH = 2.5	4.0	6.5	10.0	12.0	
A1N9	7.14	5.81	5.20	4.52	3.54	2.02
A7N3	34.25	13.21	11.63	7.02	6.26	5.47

Cationic gels prepared from vinylimidazole, dimethyl aminoethylmethacrylate, vinyl pyridine *etc.* have been reported to show maximum equilibrium swelling below their respective pK$_a$ values [42–44]. The visual appearance of the swollen gels in acidic solution was transparent and homogeneous. This indicates a favorable thermodynamic interaction between the polymer chains and water molecules. In solutions of basic pH the tertiary amines are not ionized and the gels do not show significant swelling. Therefore by directly varying the pH of the external solution the swelling capacity of these cationic gels can be easily modulated.

The influence of ionizable monomer viz. AcrNEP on the equilibrium swelling of gels is clearly observed in Table 2. The equilibrium swelling ratio of the gel A7N3 that contains 68.95 mol % of ionizable group is 35 while for the gel A1N9 that contains only 6.90 mol % of ionizable groups it is only 7.1. An increase in the ionizable group increases the protonation sites on the gel network and hence the charge density leading to a large increase in swelling in solution of low pH. The swelling transition defined by WSR$^{Eq\,pH\,=\,2.5}$/WSR$^{Eq\,pH\,=\,12}$ also increases with increase in ionic group content of the gels which is attributed to the pH dependent protonation and de-protonation in acidic and basic solutions respectively.

2.6. Effect of Silver Ions and Silver Nanoparticles on the Swelling of Nanocomposite

The swelling behavior of the gels in water (pH = 6.5) before and after chemical reduction process was also studied by measuring the weight swelling ratio at equilibrium. The swelling experiment was conducted in deionized water to eliminate any possible interference by ions present in buffer solutions on the swelling of gels.

The inorganic ions can reduce the hydrogen bonding sites in the gel by acting as physical crosslinkers. This additional physical crosslinking effect can decrease the swelling capacity of the gel.

The change in equilibrium weight swelling ratio of the gels before and after chemical reduction is shown in Figure 5. The influence of silver salt and the silver nanoparticles on the bulk properties of the gels is observed, and the order of swelling ratio is as follows: hydrogel > gel with silver nanoparticles > gel with silver salt.

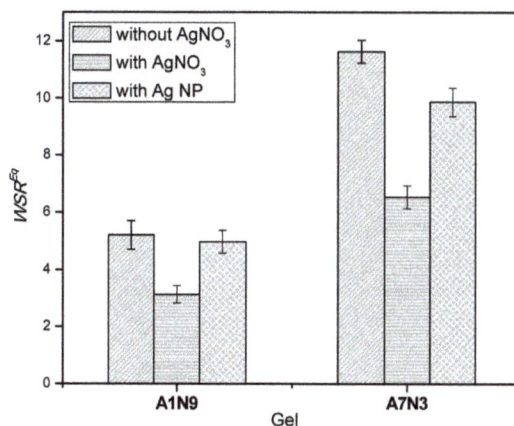

Figure 5. Equilibrium weight swelling ratio of gels and silver nanocomposites in water (pH = 6.5) at 23 °C.

It is evident that upon incorporation of $AgNO_3$ within the gel by the diffusion method, the swelling ratio of the gels decreases considerably. The swelling ratio for the gel A7N3 decreases from 11.63 to 6.91, and for the gel A1N9 it decrease from 5.20 to 3.12 when loaded with silver salt. This reduction in swelling ratio for both gels is ~40%. This marked decrease in swelling ratio is attributed to the additional physical crosslinking (ionic crosslinking) of the gel network caused by the silver ions. This behavior is similar to the influence of chemical crosslinker on the swelling behavior of gels *i.e.*, with an increase in the amount of chemical crosslinker the swelling of gels decreases.

Interestingly, the swelling ratio of gels increases after the chemical reduction processes indicating the loss of ionic crosslinking caused by the silver salt. The swelling ratio of the gels containing the silver nanoparticles is lower than that of the gels without the nanoparticles. This again is a consequence of the silver nanoparticles that hinders the diffusion of water into the gel matrix. Similar reduction in swelling of silver nanocomposites hydrogels has been reported earlier [15,16]. Therefore, hydrogel nanocomposites with the desired degree of swelling can be developed by varying the amount of metal nanoparticles in the gel matrix.

2.7. Effect of External Temperature on Swelling of Hydrogel Nanocomposite

The effect of external temperature on the swelling of gels and nanocomposite gels in water (pH = 6.5) was studied and the results are shown in Figure 6.

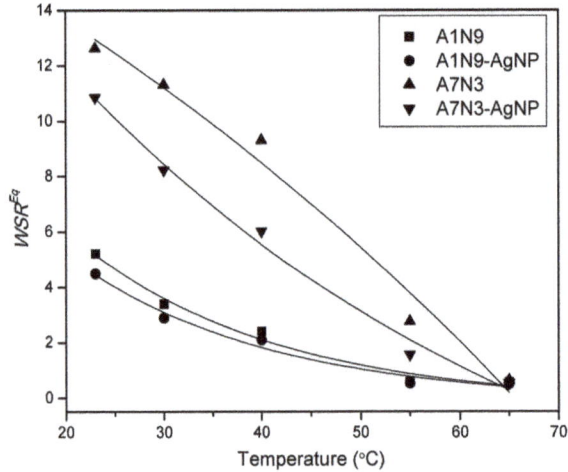

Figure 6. Effect of temperature on the swelling of gels and nanocomposite gels in water. Lines are an eye guide.

With increasing temperature from 23 °C to 65 °C, the swelling ratio of the gels and nanocomposite gels drops continuously. This decrease in swelling at high temperature is a signature of the thermo-responsive property of hydrogels. The gels contain two types of polymer components *viz.* PNIPAM which is a temperature sensitive component with lower critical solution temperature (LCST) around 32 °C in water [28], and PAcrNEP which is a pH sensitive component with no LCST. At 23 °C the gels and the nanocomposite gels are in the fully swollen state with a thermodynamically good interaction with the surrounding water molecules. As the temperature is increased from 23 °C to 65 °C, the interaction with water molecules weakens which leads to enhanced polymer-polymer interaction (hydrophobic interactions) in line with Gibbs' free energy equation as [28].

$$\Delta G = \Delta H^{mix} - T\Delta S \tag{3}$$

where H^{mix}, S, and T are the enthalpy of mixing, entropy, and temperature in K.

The volume phase transition temperature (VPT) of the gels A1N9 and A7N3 were measured to be 36 °C and 64.8 °C respectively which were in agreement with the LCST of the corresponding linear polymers [45]. This increase in VPT of A7N3 is due to the presence of a high amount of AcrNEP which is a hydrophilic comonomer. In general, hydrophilic comonomers increase the VPT of gels while hydrophobic comonomers decrease it [27]. Therefore above the VPT, complete shrinking of the gels is expected owing to hydrophobic interactions [27,28]. Interestingly, the nanocomposite gels also exhibit a decrease in swelling ratio with increase in temperature and the trend is more pronounced for A7N3-AgNp than A1N9-AgNp. This is because nanocomposite A7N3-AgNp contains a high amount of silver nanoparticles which act like additional physical crosslinkers that facilitate the collapse of the gel [15,24].

2.8. Effect of Silver Nanoparticles on the Water Diffusion Mechanism of Nanocomposite

The influence of silver nanoparticles on the water sorption kinetics of the gel A7N3 was studied to understand the water transport mechanism. This gel and the corresponding nanocomposite gel were chosen for this study due to their good swelling properties. The normalized water uptake (W_t/W_∞) isotherms measured in water at 23 °C are shown in Figure 7. In general, the gel and the nanocomposite (A7N3-AgNp) show a steep water uptake during the first 20 min followed by a gradual slow increase.

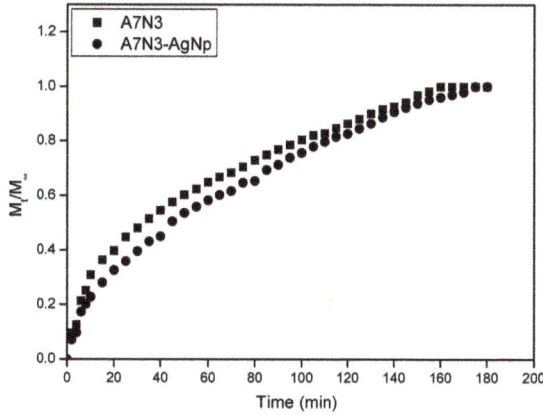

Figure 7. Normalized water sorption isotherm of hydrogel and hydrogel nanocomposite.

Two different processes describe the diffusion of solvent into the gel *viz.* (i) diffusion of solvent and (ii) advancement of the swollen-deswollen boundary. In order to determine the type of water sorption mechanism of the gel and the nanocomposite gel, the initial swelling data was fitted to the following empirical equation [46] in the range $M_t/M_\infty \leq 0.6$:

$$\frac{M_t}{M_\infty} = k\,t^n \tag{4}$$

where M_t and M_∞ are the mass of water measured at time t and time infinity (equilibrium swelling) respectively, k is a characteristic constant of the gel, and n is the diffusion exponent that describes the transport mechanism of penetrate. This generalized expression combines both Fickian and non-Fickian mechanisms for thin polymer slabs or discs.

The values of n and k were directly calculated from the gradient and intercepts of a log (M_t/M_∞) *versus* log t plots and the results are presented in Table 3. For the gel without the silver nanoparticles the value of n is equal to 0.63. This corresponds to non-Fickian type diffusion of solvent molecules into the gel in which the rate of solvent diffusion and polymer relaxations is comparable. For the nanocomposite hydrogel the value of n is lower and is equal to 0.53. This value almost corresponds to Fickian type of diffusion in which the rate of diffusion of solvent is lower than polymer relaxation [47]. This change in diffusion process for the nanocomposite hydrogel is expected due to the presence of silver nanoparticles. The silver nanoparticles acts as additional physical crosslinks and retard the solvent diffusion. Similar behavior of solvent diffusion has been reported for silver nanocomposite hydrogels based on *N*-isopropylacrylamide and sodium acrylate [15].

Table 3. Water transport characteristic parameters measured at 23 °C.

Gel	n	k	D $(10^{-8}/cm^2 \cdot s^{-1})$	v $(10^{-3}/cm \cdot g^{-1})$
A7N3	0.62	0.17	15.72	2.59
A7N3-Ag Np	0.53	0.10	13.23	1.50

n: mode of transport of penetrant, *k*: characteristic constant of the gel, *D*: diffusion coefficient, *v*: *penetration velocity.*

The role of silver nanoparticles as additional crosslinks in the nanocomposite hydrogel was further confirmed by calculating the diffusion coefficient and penetration velocity and the results are also presented in Table 3. The diffusion coefficient D for the short-time scale was evaluated from the slope of $M_t/M_\infty \le 0.6$ *versus* square-root of t plot using the following expression [46,48,49]:

$$\frac{M_t}{M_\infty} = \frac{4}{l}\sqrt{\frac{Dt}{\pi}} \tag{5}$$

where l is the initial length of the dry gel and t is the time.

The penetration velocity [50] of water into the hydrogel and the nanocomposites hydrogel was calculated using the following expression:

$$v = \frac{1}{2\,\rho_w A}\frac{M_t}{M_\infty} \tag{6}$$

where ρ_w is the density of water at 23 °C (0.9975 g·cm^{-3}), and A is the area of one face of the thin hydrogel. The factor 2 accounts for water diffusion through both faces of the disc.

The diffusion coefficient of water for the nanocomposite hydrogel is lower by 16% compared to the hydrogel without silver nanoparticles. The penetration velocity of water in the nanocomposite hydrogel also shows a marked decrease by 35% compared to hydrogel without silver nanoparticles. The marked decrease in diffusion and penetration velocity of water in the nanocomposite hydrogel is attributed to the presence of silver nanoparticles which hinder water penetration into the gel. It is believed that the silver nanoparticles increase the entropy of the system and hinder the hydrogen bonding sites causing lower expansion of the gel network.

2.9. State of Water in Swollen Hydrogel Nanocomposite

The effect of silver nanoparticles on the states of water in fully swollen hydrogel nanocomposite was also studied. In general, the state of water in the hydrogels can be divided into free water, freezing bound water, and non-freezing bound water [51]. Free water exists in hydrogel networks without any hydrogen bonding with polymer chains and has a similar transition temperature, enthalpy, and DSC thermogram similar to that of pure water. Freezing bound or intermediate water exhibits weak interactions with the polymeric network of the gels. The non-freezing water or bound water forms hydrogen bonds with the polymeric network and does not show any endothermic peak in the temperature range −70 °C to 0 °C.

The precursor hydrogel A7N3 and nanocomposite A7N3-AgNp showed a melting peak in the range 0 °C to 7 °C (DSC thermogram not shown). This indicates the existence of free and freezing bound water in the swollen hydrogel nanocomposite network in addition to bound water. The presence of unbound water which is comprised of free and freezing water was estimated by following the standard equation [52],

$$W_{bound}\,(\%) = EWC - (W_f + W_{fb}) \tag{7}$$

where EWC, W_f, and W_{fb} are the equilibrium water content, free water, and freezing bound water respectively.

The amount of free water content in the total water present in the gels was calculated as the ratio of the DSC endothermic peak area of the swollen gel to that of the melting endothermic heat of fusion of pure water (333.3 J·g^{-1}) using the following equation. This equation assumes that the heat of fusion of free water in the gel is similar to that of ice.

$$W_{bound}\,(\%) = EWC - \left(\frac{\Delta H_{endo}}{\Delta H_{ice}}\right) \tag{8}$$

where ΔH_{endo} and ΔH_{ice} are the heat of fusion of free water in the gels and ice respectively. The results are summarized in Table 4.

Table 4. Water content (free and bound) in gel and nanocomposite gel.

Sample	Unbound Water (%)	Bound Water (%)	Unbound Water/Bound Water
A7N3	82.15	17.85	4.60
A7N3-AgNp	75.60	24.40	3.10

The amount of unbound water in hydrogel nanocomposite is lower than the precursor hydrogel A7N3 that does not contain silver nanoparticles. The presence of silver nanoparticles in the nanocomposite acts as physical crosslinkers which increase the rigidity of the gel network. This increase in rigidity resembles that of hydrogels with a high crosslink density. As a result of this, a decrease in unbound water content of the hydrogel nanocomposite is observed. This is also further highlighted by a decrease in the ratio of unbound water/bound water content of the hydrogel nanocomposite.

2.10. Effect of External pH on the Leaching of Silver Nanoparticles from Nanocomposite

The influence of external pH on the leaching of silver nanoparticles from the gel A7N3 was studied by UV-Vis absorption spectroscopy at 23 °C. The gel containing the silver nanoparticles was soaked in solutions of pH 2.5 and pH 7 under very slow stirring (200 rpm) and the absorbance of solution at 406 nm was recorded at intervals of 15 min. The results are shown in Figure 8. The insert in Figure 8 shows the change in absorbance of solution for the first 300 min.

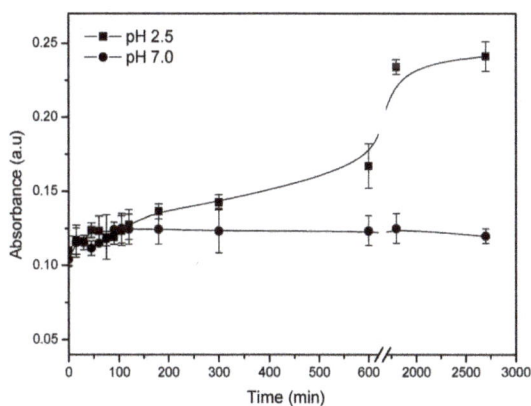

Figure 8. Effect of pH on the release of silver nanoparticles from hydrogel nanocomposite.

No significant change in absorbance is observed for the gel soaked in a solution of pH 7 even after 24 h. This is attributed to the weak swelling of the gel which does not favor diffusion of silver nanoparticles from the gel to the external solution. However, for the gel soaked in a solution of pH 2.5, a significant and gradual increase in absorbance of the external solution is clearly observed. This increase in absorbance is due to the presence of silver nanoparticles in the external solution as a result of leaching from the gel. In a solution of pH 2.5, the gel is in the fully swollen state due to the fixed charge repulsions of the polymer network.

In this swollen state, the silver nanoparticles in the gel easily diffuse out to the external medium through the many pores in the swollen gel. This process of diffusion is illustrated in Figure 6. The ionic monomer content in the gel and the nature of the swelling medium thus play a vital role in the leaching of nanoparticles from the gel. This type of leaching could be prevented by decreasing the mesh size of the gel by increasing the amount of chemical crosslinker.

3. Conclusions

A dual-responsive hydrogel nanocomposite containing finely dispersed silver nanoparticles was successfully prepared by a facile two-step approach. The silver nanoparticles were stabilized by the polymer network. The nanocomposite hydrogel was responsive to changes in external pH and temperature. The presence of silver nanoparticles acts as additional physical crosslinkers and influenced the media diffusion and penetration velocity. By this study, it has been demonstrated that stimuli-responsive nanocomposite hydrogel systems of the desired swelling capacity can be fabricated by a facile *in-situ* chemical reduction process, and stability of the nanoparticles can be achieved by fine-tuning the hydrogel network. The developed hydrogel nanocomposites are non-toxic and could find applications as anti-bacterial surfaces.

4. Experimental Section

4.1. Materials

N-isopropylacrylamide (NIPAM) (Aldrich, Saint Louis, USA) was purified by recrystallization in toluene/hexane (80:20 v/v) mixture. *N,N*-methylenebisacrylamide (MBA) (Aldrich), *N*-ethylpiperazine (Aldrich), triethylamine (Aldrich), potassium persulfate (KPS) (Aldrich), tetramethylethylenediamine (TEMED) (Aldrich), sodium borohydride (NaBH$_4$) (Aldrich), and silver nitrate (AgNO$_3$) (Aldrich) were used as received. Tetrahydrofuran (THF) (Merck, Darmstadt, Germany) was purified by distillation and stored over molecule sieves of pore size 3 Å (Aldrich). Deionized water was obtained from Barnstead purification system and was used for all aqueous sample preparations. The pH of the deionized water was 6.5. The buffer solutions of pH 2.5, 4.0, and 6.5 were prepared by adjusting 0.1 M sodium acetate solution with acetic acid to the desired pH. The buffer solutions of pH 10.0 and 12.0 were similarly prepared using sodium hydrogen phosphate and disodium hydrogen phosphate. The ionic strength of all the buffer solutions was fixed at 0.1 M.

4.2. Synthesis of Acryloyl Chloride and N-Acryloyl-N'-ethylpiperazine (AcrNEP)

Acryloyl chloride and AcrNEP were synthesized according to the procedure described in the literature [2,26].

4.3. Synthesis of Crosslinked Copolymer Hydrogels

Copolymer hydrogels of AcrNEP and NIPAM with two different monomer feed were synthesized by free radical solution polymerization at 23 °C. The crosslinker content in the gels was fixed at 1.5 mol %, and the total monomer concentration was kept at 11 wt %. The synthesis of gel A1N9 is described as follows: AcrNEP (6.90 mol %), NIPAM (91.60 mol %), and MBA (1.5 mol %) were dissolved in deionized water (8 g) in a clean test tube. The content was degassed by bubbling dry nitrogen gas for 10 min to expel any dissolved oxygen. KPS (0.01 g/L in 1 mL water), and TEMED (0.01 g/L in 1 mL water) were added to the monomer solution and the test tube was sealed using Para film. Polymerization was conducted at 23 °C. The solution turned viscous and finally to a clear gel in about 10 min and the reaction was allowed to continue for 24 h. The transparent soft solid gel was cut into small discs (thickness = 0.3 cm and diameter 0.7 cm) and washed in water for 1 day to remove any unreacted monomers. The gel discs were dried in a vacuum oven at 40 °C for 2 days until constant weight was attained. The gel A7N3 was similarly prepared. The monomer feed compositions and the physical appearances of gels are summarized in Table 1.

4.4. Fabrication of Silver Nanoparticles in the Hydrogels

Silver nanoparticles were fabricated in the hydrogel by the following *in-situ* method: The dry hydrogel disc (for example A7N3, thickness = 0.3 cm, diameter = 0.7 cm) was soaked in 30 mL of aqueous AgNO$_3$ solution (5×10^{-3} M) until equilibrium swelling was achieved (usually 1 day). The silver nitrate loaded gel was then soaked in 30 mL of aqueous solution of $NaBH_4$ (10×10^{-3} M) for 30 min. During this *in-situ* chemical reduction process, the silver salt in the hydrogel was reduced to colloidal silver particles. The transparent gel turned yellowish brown in color during this reduction process. The gel was repeatedly washed in water and dried in a vacuum oven at 40 °C until constant weight was recorded.

4.5. Characterization of Hydrogels

Equilibrium swelling of the gels and nanocomposites was studied gravimetrically to determine their response to pH and temperature changes. The samples were soaked and equilibrated in buffer solutions of various pH (2.5, 4.0, 6.5, 10.0, and 12.0) at 23 °C until swelling equilibrium was reached (~1 day). The swollen weight was recorded after carefully removing the excess surface water by using a damp Kim-wipe towel. The equilibrium weight swelling ratio WSR^{Eq} was calculated using the following equation,

$$WSR^{Eq} = \frac{W_s - W_d}{W_d} \tag{9}$$

where W_s and W_d are the equilibrium swollen weight and dry weight of the sample respectively.

For water uptake and diffusion study, the samples were cut into thin rectangular slabs (area ~1.5 cm^2, thickness ~0.3 cm) and placed in glass vials containing 5 mL of water (pH = 6.5). The vials were placed in a thermostatic water bath at 23 °C, and the weight of samples was recorded at different time intervals (W_t). All measurements were repeated three times and the average value was recorded.

The UV-Vis absorption spectra of the gels containing silver nanoparticles were recorded on a Cary-1E UV-Vis spectrophotometer. Samples for this measurement were prepared as follows: About 5 mg of the dry hydrogel disc was dispersed in 5 mL of water at room temperature. The dispersion (~3 mL) was transferred into a clean quartz cell (Helma, path-length = 1 cm), and the absorbance was recorded in the wavelength range 300–700 nm.

The morphology of freeze-fractured gels was characterized using a Jeol JSM 6700 Field Emission Scanning Electron Microscope (FESEM). The freeze-fractured and dried hydrogel samples were coated with a thin layer of gold in vacuum for 10–20 s for the FESEM measurements. Samples containing the silver nanoparticles were not coated with gold.

The amount of free and bound water content in the fully swollen gels was determined using a Perkin-Elmer Diamond Differential Scanning Calorimeter (DSC). The swollen gel samples were heated from −20 °C to 20 °C with a ramp rate of 2 °C under a nitrogen atmosphere.

Acknowledgments: The authors thank National Institute of Education, Nanyang Technological University, and Nanyang Research Programme (NRP 2010) for the financial support under NIEAcRF RI8/09GRD and NIEAcRF RI6/12GRD.

Author Contributions: All authors contributed equally for this research work.

Conflicts of Interest: The authors declare no conflict of interest.

References

1. Peppas, N.A.; Bures, P.; Leobandung, W.; Ichikawa, H. Hydrogels in pharmaceutical formulaations. *Eur. J. Pharm. Biopharm.* **2000**, *50*, 27–46. [CrossRef]
2. Roshan Deen, G.; Gan, L.H. Determination of reactivity ratios and swelling characteristics of stimuli responsive copolymers of *N*-acryloyl-*N'*-ethyl piperazine and MMA. *Polymer* **2006**, *47*, 5025–5034. [CrossRef]
3. Dill, K.A. Strengthening biomedicine roots. *Nature* **1999**, *400*, 309–310. [CrossRef] [PubMed]
4. Peppas, N.A.; Langer, R. New challenges in biomaterials. *Science* **1994**, *263*, 1715–1720. [CrossRef] [PubMed]
5. Hoare, T.R.; Kohane, D.S. Hydrogels in drug delivery: Progress and challenges. *Polymer* **2008**, *49*, 1993–2007. [CrossRef]
6. Ferruti, P.; Marchisio, M.A.; Duncan, R. Poly(amido-amines)s: Biomedical applications. *Macromol. Rapid Commun.* **2002**, *23*, 332–355. [CrossRef]

7. Ratner, B.D.; Hoffman, A.S. Synthetic hydrogels for biomedical applications. In *Hydrogels for Medical and Related Applications*; Andrade, J.D., Ed.; ACS Symposium Series No. 31, American Chemical Society: Washington, DC, USA, 1976; pp. 1–36.

8. Hu, Z.; Lu, X.; Gao, J. Hydrogel opals. *Adv. Mater.* **2001**, *13*, 1708–1712. [CrossRef]

9. Peppas, N.A. *Hydrogels in Medicine*; CRC Press: Boca Raton, FL, USA, 1986.

10. Gil, E.S.; Hudson, S.M. Stimuli-responsive polymers and their bioconjugates. *Prog. Polym. Sci.* **2004**, *29*, 1173–1222. [CrossRef]

11. Matejka, L.; Dukh, O.; Kolarik, J. Reinforcement of crosslinked rubbery epoxies by *in-situ* formed silica. *Polymer* **2000**, *41*, 1449–1459. [CrossRef]

12. Okomoto, M.; Morita, S.; Taguchi, H.; Kim, Y.H.; Kotaka, T.; Tateyama, H. Synthesis and structure of smectic clay/poly(methyl methacrylate) and clay/polystyrene nanocomposites via *in situ* intercalative polymerization. *Polymer* **2000**, *41*, 3887–3890. [CrossRef]

13. Chou, K.S.; Lai, Y.S. Effect of polyvinyl pyrrolidone molecular weights on the formation of nanosized silver colloids. *Mater. Chem. Phys.* **2004**, *84*, 82–88. [CrossRef]

14. Lee, W-F.; Tsao, K-T. Preparation and properties of nanocomposite hydrogels containing silver nanoparticles by *ex-situ* polymerization. *J. Appl. Polymer Sci.* **2006**, *100*, 3653–3661. [CrossRef]

15. Roshan Deen, G.; Santha, S. Influence of a new stiff crosslinker on the swelling of poly(N-isopropylacrylamide-co-sodium acrylate) hydrogels and silver nanocomposite. *Int. J. Polymer Mater. Polymer Biomater.* **2013**, *62*, 1–7.

16. Mohan, M.Y.; Premkumar, T.; Lee, K.; Geckeler, K.E. Fabrication of silver nanoparticles in hydrogel networks. *Macromol. Rapid Commun.* **2006**, *27*, 1346–1354. [CrossRef]

17. Ho, C.H.; Tobis, J.; Sprich, C.; Thomann, R.; Tiller, J.C. Nanoseparated polymeric networks with multiple antimicrobial properties. *Adv. Mater.* **2004**, *16*, 957–961. [CrossRef]

18. Zhang, Z.; Zhang, L.; Wang, S.; Chen, W.; Lei, Y. A convenient route to polyacrylonitrile/silver nanoparticle composite by simultaneous polymerization-reduction approach. *Polymer* **2001**, *42*, 8315–8318. [CrossRef]

19. Son, W.K.; Youk, J.H.; Lee, T.S.; Park, W.H. Preparation of antimicrobial ultrafine cellulose acetate fibers with silver nanoparticles. *Macromol. Rapid Commun.* **2004**, *25*, 1632–1637. [CrossRef]

20. Zhang, J.; Xu, S.; Kumacheva, E. Polymer microgels: Reactors for semiconductor, metal and magnetic nanoparticles. *J. Am. Chem. Soc.* **2004**, *126*, 7908–7914. [CrossRef] [PubMed]

21. Zhang, J.; Xu, S.; Kumacheva, E. Photogeneration of fluorescent silver nanoclusters in polymer microgels. *Adv. Mater.* **2005**, *19*, 2336–2340. [CrossRef]

22. Biffis, A.; Orlandi, N.; Corain, B. Microgel-stabilized metal nanoclusters: Size control by microgel nanomorphology. *Adv. Mater.* **2003**, *15*, 1551–1555. [CrossRef]

23. Li, Y.Y.; Cunin, F.; Link, J.R.; Gao, T.; Betts, R.E.; Reiver, S.H.; Chin, V.; Bhatia, S.N.; Sailor, M.J. Polymer replicas of photonic porous silicon for sensing and drug delivery applications. *Science* **2003**, *299*, 2045–2047. [CrossRef] [PubMed]

24. Pich, A.; Karak, A.; Lu, Y.; Ghosh, A.K.; Adler, P.H.-J. Preparation of hybrid microgels functionalized by silver nanoparticles. *Macromol. Rapid Commun.* **2006**, *27*, 344–350. [CrossRef]

25. Wang, C.; Flynn, N.T.; Langer, R. Controlled structure and properties of thermoresponsive nanoparticle-hydrogel composites. *Adv. Mater.* **2004**, *16*, 1074–1079. [CrossRef]

26. Roshan Deen, G.; Lim, E.K.; Mah, C.H.; Heng, K.M. New cationic linear copolymers and hydrogels of N-vinyl caprolactam and N-acryloyl-N'-ethyl piperazine: Synthesis, reactivity, influence of external stimuli on the LCST and swelling properties. *Ind. Eng. Chem. Res.* **2012**, *51*, 13354–13365. [CrossRef]

27. Kim, J.H.; Ballauff, M. The volume transitions in thermosensitive core-shell latex particles containing charge groups. *Colloid Polymer Sci.* **1999**, *277*, 1210–1214. [CrossRef]

28. Schild, H.G. Poly(N-isopropylacrylamide): Experiment, theory and application. *Prog. Polymer Sci.* **2002**, *16*, 69–74. [CrossRef]

29. Dhara, D.; Rathan, G.V.N.; Chatterji, P.R. Volume phase transition in interpenetrating networks of poly(N-isopropylacrylamide) with gelatin. *Langmuir* **2000**, *16*, 2424–2429. [CrossRef]

30. Muniz, E.C.; Geuskens, G. Influence of temperature on the permeability of polyacrylamide hydrogels and semi-IPNs with poly(N-isopropylacrylamide). *J. Membr. Sci.* **2000**, *172*, 287–293. [CrossRef]

31. Chen, K.S.; Ku, Y.A.; Lin, H.R.; Tan, T.R.; Sheu, D.C.; Chen, T.M.; Lin, F.H. Preparation and characterization of pH sensitive poly(*N*-vinyl-2-pyrrolidone/itaconic acid) copolymer hydrogels. *Mater. Chem. Phys.* **2005**, *91*, 484–489. [CrossRef]

32. Cakal, E.; Cavvus, S. Novel poly(*N*-vinylcaprolactam-co-2-(diethylamino) ethyl methacrylate) gels: Characterization and detailed investigation on their stimuli-sensitive behaviors. *Ind. Eng. Chem. Res.* **2010**, *49*, 11741–11751. [CrossRef]

33. Behrens, S.; Habicht, W.; Wagner, K.; Unger, E. Assembly of nanoparticle ring structures based on protein templates. *Adv. Mater.* **2006**, *18*, 284–289. [CrossRef]

34. Henglein, A.; Ershov, B.G.; Malow, M. Absorption spectrum and some chemical reactions of colloidal platinum in aqueous solution. *J. Phys. Chem. B* **1995**, *99*, 14129–14136. [CrossRef]

35. Shiraishi, Y.; Toshima, N. Oxidation of ethylene catalyzed by colloidal dispersions of poly (sodium acrylate)-protected silver nanoclusters. *Colloids Surf. A* **2000**, *169*, 59–66. [CrossRef]

36. Hussain, I.; Brust, M.; Papworth, A.J.; Cooper, A.I. Preparation of acrylate-stabilized gold and silver hydrosols and gold-polymer composite films. *Langmuir* **2003**, *19*, 4831–4835. [CrossRef]

37. Gao, J.; Fu, J.; Lin, C.; Lin, J.; Han, Y.; Yu, X.; Pan, C. Formation and photoluminescence of silver nanoparticles stabilized by a two-armed polymer with a crown ether core. *Langmuir* **2004**, *20*, 9775–9781. [CrossRef] [PubMed]

38. Patel, K.; Kapoor, S.; Dave, D.P.; Mukherjee, T. Synthesis of nanosized silver colloids by microwave dielectric heating. *J. Chem. Sci.* **2005**, *117*, 53–60. [CrossRef]

39. Xu, X.; Asher, S.A. Synthesis and utilization of monodisperse hollow polymeric particles in photonic crystals. *J. Am. Chem. Soc.* **2004**, *126*, 7940–7945. [CrossRef] [PubMed]

40. Rozra, J.; Saini, I.; Aggarwal, S.; Sharma, A. Spectroscopic analysis of Ag nanoparticles embedded in glass. *Adv. Mater. Lett.* **2013**, *4*, 598–604.

41. Kittel, C. *Introduction to Solid State Physics*, 8th ed.; Wiley: Eastern, India, 2007.

42. Roshan Deen, G.; Gan, L.H. Influence of amino group pKa on the properties of stimuli-responsive piperazine-based polymers and hydrogels. *J. Appl. Polymer Sci.* **2008**, *107*, 1449–1458. [CrossRef]

43. Zhao, Y.; Yang, Y.; Yang, X.; Xu, H. Preparation and pH-sensitive swelling behavior of physically crosslinked polyampholyte gels. *J. Appl. Polymer Sci.* **2006**, *102*, 3857–3861. [CrossRef]

44. Kakinoki, S.; Kaetsu, I.; Nakayama, M.; Sutani, K.; Uchida, K.; Yukutake, K. Temperature and pH responsiveness of poly(DMAA-co-unsaturated carboxylic acid) hydrogels synthesized by UV-irradiation. *Radiat. Phys. Chem.* **2003**, *67*, 685–693. [CrossRef]

45. Roshan Deen, G.; Chua, V.; Ilyas, U. Synthesis, swelling properties and network structure of new stimuli responsive poly(*N*-acryloyl-*N'*-ethyl piperazine-co-*N*-isopropylacrylamide) hydrogels. *J. Polymer Sci. Polymer Chem.* **2012**, *50*, 3363–3372. [CrossRef]

46. Franson, N.M.; Peppas, N.A. Influence of copolymer composition on non-Fickian water transport through glassy copolymers. *J. Appl. Polym. Sci.* **1983**, *28*, 1299–1310. [CrossRef]

47. Afif, A.E.; Germe, M. Non-Fickian mass transport in polymers. *J. Rheol.* **2002**, *26*, 591–628. [CrossRef]

48. Crank, J. *The Mathematics of Diffusion*; Oxford University: London, UK, 1975.

49. Berens, A.R.; Hopfenberg, H.B. Diffusion and relaxation in glassy polymer powders. *Polymer* **1978**, *19*, 489–496. [CrossRef]

50. Mohan, Y.M.; Premkumar, T.; Joseph, D.K.; Geckeler, K.E. Stimuli-responsive poly(*N*-isopropylacrylamide-co-sodium acrylate) hydrogels: A swelling study in surfactant and polymer solutions. *React. Funct. Polymer* **2007**, *67*, 844–858. [CrossRef]

51. Kim, S.J.; Park, S.J.; Kim, S.I. Synthesis and characteristics of interpenetrating polymer network hydrogels composed of poly(vinyl alcohol) and poly(*N*-isopropylacrylamide). *React. Funct. Polymer* **2003**, *55*, 61–67. [CrossRef]

52. Ahmad, M.B.; Huglin, M.B. DSC studies on states of water in crosslinked poly(methyl methacrylate-co-*N*-vinyl-2-pyrrolidone) hydrogels. *Polymer Int.* **1994**, *33*, 273–277. [CrossRef]

Article

Surface Morphology at the Microscopic Scale, Swelling/Deswelling, and the Magnetic Properties of PNIPAM/CMC and PNIPAM/CMC/Fe₃O₄ Hydrogels

Marianna Uva [1,2] and Andrea Atrei [1,2,*]

1   Dipartimento di Biotecnologie, Chimica e Farmacia, Università di Siena, Via A. Moro 2, 53100 Siena, Italy; marianna.uva@unisi.it
2   CRISMA, Via. Matteotti 15 Colle di Val d'Elsa, 53034 Siena, Italy
*   Correspondence: andrea.atrei@unisi.it; Tel.: +39-577-234371

Academic Editors: Dirk Kuckling and David Díaz Díaz
Received: 19 September 2016; Accepted: 5 December 2016; Published: 13 December 2016

Abstract: Poly(*N*-isopropylacrylamide) (PNIPAM) hydrogels containing carboxymethylcellulose (CMC) and CMC/Fe₃O₄ nanoparticles were prepared. Free-radical polymerization with BIS as cross-linker was used to synthesize the hydrogels. The morphology at the microscopic scale of these materials was investigated using field emission scanning electron microscopy (FESEM). The images show that CMC in the PNIPAM hydrogels induces the formation of a honeycomb structure. This surface morphology was not observed for pure PNIPAM hydrogels prepared under similar conditions. The equilibrium swelling degree of the PNIPAM/CMC hydrogels (5200%) is much larger than that of the pure PNIPAM hydrogels (2500%). The water retention of PNIPAM/CMC hydrogels above the volume phase transition temperature is strongly reduced compared to that of pure PNIPAM hydrogel. Both PNIPAM/Fe₃O₄ and PNIPAM/CMC/Fe₃O₄ hydrogels exhibit a superparamagnetic behavior, but the blocking temperature (104 K) of the former is higher than that of the latter (83 K).

Keywords: hydrogels; thermoresponsive materials; magnetic nanoparticles; poly(*N*-isopropylacrylamide) (PNIPAM); field emission scanning electron microscopy (FESEM)

---

1. Introduction

Poly(*N*-isopropylacrylamide) (PNIPAM) hydrogels are the most investigated thermoresponsive hydrogels [1]. Because of their capability to change volume reversibly at a transition temperature, hydrogels of PNIPAM or co-polymer of PNIPAM have applications in several fields ranging from drug delivery to environmental remediation [2,3]. The introduction of magnetic nanoparticles (NPs) into PNIPAM hydrogels allows to heat remotely the hydrogels by applying alternating magnetic fields of suitable frequencies [2]. In most of these applications the rate of the deswelling/swelling processes across the volume phase transition temperature (VPTT) is important [4]. Unfortunately, the response of macroscopic PNIPAM hydrogels to temperature changes is rather slow [5–7]. The collapse of the swollen hydrogel (with expulsion of water molecules out of the polymer matrix) has a complex dynamics involving diffusion of polymer segments and conformation changes of the polymer chains [8]. A contribution to slow the thermal response of PNIPAM hydrogels comes from the so-called "skin effect", which is the formation of a dense, vitrous layer which prevents the exchange of water molecules between the hydrogel and the surrounding [9]. Several methods have been introduced to obtain PNIPAM hydrogels featuring a faster response to temperature variations. Some of these methods use the copolymerization of NIPAM with hydrophilic molecules to eliminate the skin effect and form hydrophilic channels for the diffusion of water molecules [9,10]. Other methods are based on promoting the formation of interconnected pores or use PNIPAM microgels embedded in a macrogel [11,12]. It has been shown that the interpenetrating polymer network (IPN) or semi-IPN hydrogels of NIPAM with a hydrophilic polysaccharide such as sodium alginate have a faster response to temperature variation [13]. In the present work, we prepared PNIPAM hydrogels containing carboxymethylcellulose (CMC) and magnetite (Fe$_3$O$_4$) NPs coated with CMC. CMC, a hydrophilic, biocompatible derivative of cellulose, appears to be a suitable candidate to prepare PNIPAM hydrogels exhibiting a faster thermal response. The choice of CMC was also motivated by its ability to stabilize aqueous dispersions of magnetite NPs [14]. One of the aims of this study was to investigate the effect of CMC and CMC/Fe$_3$O$_4$ NPs on the morphology at micrometric scale and on the swelling/deswelling properties of PNIPAM hydrogels. The other aim was to study the effect of coating magnetite NPs with CMC on the magnetic properties of PNIPAM hydrogels containing Fe$_3$O$_4$ NPs. The morphology of the PNIPAM hydrogels was investigated by means of field emission scanning electron microscopy (FESEM). The magnetic properties of PNIPAM/Fe$_3$O$_4$ and PNIPAM/CMC/Fe$_3$O$_4$ hydrogels were studied by measuring magnetization versus magnetic field curves at various temperatures, and zero field cooling (ZFC) and field cooling (FC) curves.

2. Results and Discussion

The attenuated total reflectance–Fourier transform infrared spectroscopy (ATR-FTIR) spectrum measured for the PNIPAM hydrogel containing 20 wt % of CMC (PNIPAM/CMC20) can be fitted by summing the spectra of PNIPAM dry hydrogel and CMC. The best fit is obtained by optimizing the weight factors of the intensities of the spectra of PNIPAM hydrogel and CMC (Figure 1). This result suggests that there are no large interactions between the PNIPAM network and the CMC chains which, otherwise, would produce variations of the peak positions and widths in the spectrum of PNIPAM/CMC hydrogel with respect to those of the pure components. Because of partial overlapping of the main peaks of PNIPAM and CMC and the relatively small contribution of CMC to the intensity of the PNIPAM/CMC20 spectrum, Fourier transform infrared spectroscopy (FTIR) is not able to reveal subtle changes in the spectrum, resulting from possible interactions between PNIPAM and CMC or cross-linking of CMC chains. Hence, attenuated total reflectance (ATR)-FTIR results are consistent with the formation of a semi-IPN, but the cross-linking of CMC chains, leading to a full IPN [15], cannot be ruled out.

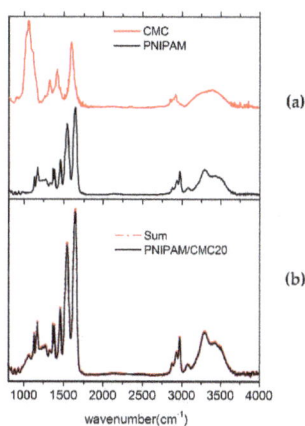

Figure 1. (a) Attenuated total reflectance–Fourier transform infrared spectroscopy (ATR-FTIR) spectra of carboxymethylcellulose (CMC) sodium salt and poly(*N*-isopropylacrylamide) (PNIPAM) hydrogel. The spectra were measured under the same instrumental conditions. (b) Fitting of the spectrum of the PNIPAM/CMC20 sample as a sum of the spectra of CMC and PNIPAM hydrogels weighted by optimized scaling factors. The spectra were measured for the lyophilized hydrogels.

FESEM images show that the microstructures of pure PNIPAM and PNIPAM/CMC lyophilized hydrogels are significantly different. Pure PNIPAM hydrogels have a morphology consisting of relatively flat foils over which large craters (size around 100 nm) are visible (Figure 2a). Even at higher magnifications, PNIPAM hydrogels exhibit a non-porous structure (Figure 2b). The images shown here are representative of the morphology of the whole samples.

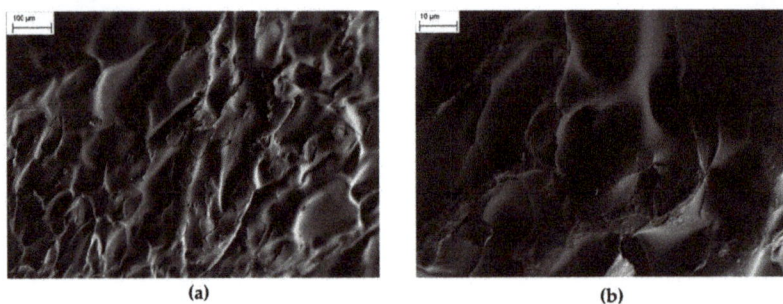

(a)   (b)

Figure 2. (a) Field emission scanning electron microscopy (FESEM) image collected for a pure PNIPAM hydrogels. (b) FESEM image collected for a pure PNIPAM hydrogel at a larger magnification.

On the other hand, FESEM images of PNIPAM/CMC are characterized by a honeycomb-like structure (Figure 3a,b), with features smaller than those observed in the images of pure PNIPAM. Images of internal cross sections of the PNIPAM/CMC hydrogels show that a network of interconnected pores exists inside the hydrogels. The observed honeycomb-like structure is probably due to the emergence of micropores at the surface of the PNIPAM/CMC hydrogels as suggested by the images measured for the inner cross-sections of the hydrogel (Figure 3c). The surface morphology of the PNIPAM/CMC hydrogels does not change significantly in the explored range of CMC concentration (that is from 5 wt % to 20 wt %) as far as the presence of pores is concerned. However, the size of the micropores tends to be smaller for the lower CMC concentration (Figure 3d).

The surface morphology of the hydrogels is not significantly affected by the presence of CMC/Fe_3O_4 NPs as shown by the FESEM images of the PNIPAM/CMC hydrogel containing 4.3 wt % magnetite (PNIPAM/CMC5/Fe_3O_4) samples (Figure 4a,b).

The attainment of the equilibrium swelling degree (SD_{eq}) was checked by measuring swelling kinetics starting from lyophilized samples of the hydrogels (Figure 5). These data indicate that the swelling process is slightly faster for the PNIPAM/CMC20 hydrogel than for the pure PNIPAM hydrogel. The SD_{eq} measured at 295 K of the PNIPAM/CMC20 sample (5200% ± 200%) is much larger than that of the PNIPAM hydrogel (2500% ± 200%). On the contrary, the addition of 5 wt % of CMC to the PNIPAM hydrogels slightly reduces its SD_{eq} (1900% ± 100%). Similar SD_{eq} values (ca. 2000%) were measured for the PNIPAM/CMC5/Fe_3O_4 hydrogel. These results can be explained considering that the interactions of the hydrophilic groups of PNIPAM with those of CMC are more favorable than with water at low CMC concentration; thus, a decrease of the swelling degree is observed [15].

The VPTT estimated from the temperature at which the hydrogel samples start to be opalescent is not affected (within an uncertainty of ±2 K) by CMC. This result was confirmed by the swelling degree measurements as a function of temperature of PNIPAM/CMC20 hydrogels which show a drastic change from 303 to 307 K. Moreover, the presence of CMC/Fe_3O_4, at the concentrations used in this work, does not influence, within the accuracy of the measurements, the VPTT of the PNIPAM hydrogels.

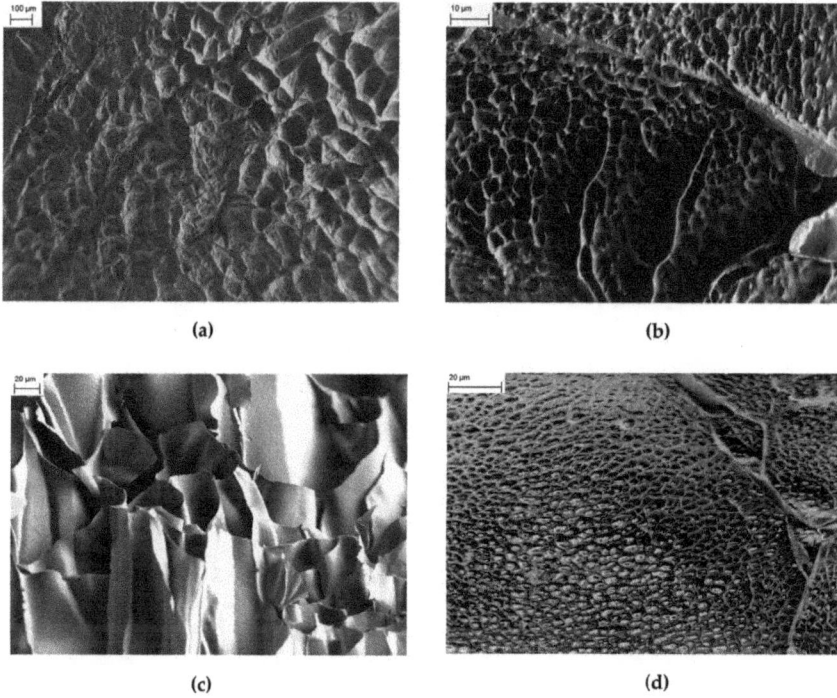

Figure 3. (a) FESEM image collected for a PNIPAM/CMC20 sample. (b) FESEM image collected for a PNIPAM/CMC20 sample at a larger magnification. (c) FESEM image of an inner cross section of the PNIPAM/CMC20 sample. (d) FESEM image of a PNIPAM hydrogel containing 5 wt % of CMC.

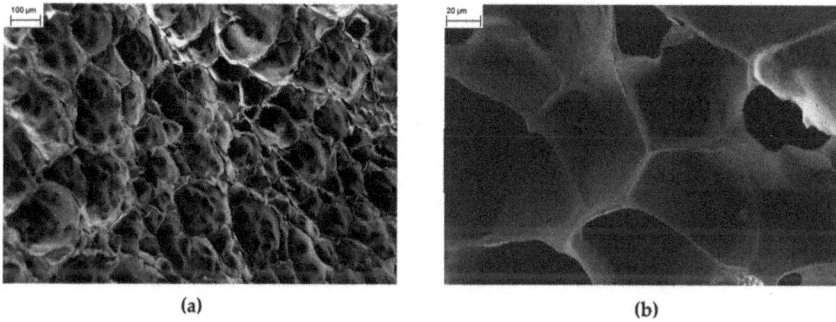

Figure 4. (a) FESEM image of the PNIPAM/CMC5/Fe$_3$O$_4$ hydrogel. (b) FESEM image collected for the NIPAM/CMC5/Fe$_3$O$_4$ hydrogel at a larger magnification. FESEM image of a PNIPAM/CMC sample containing 5 wt % of CMC.

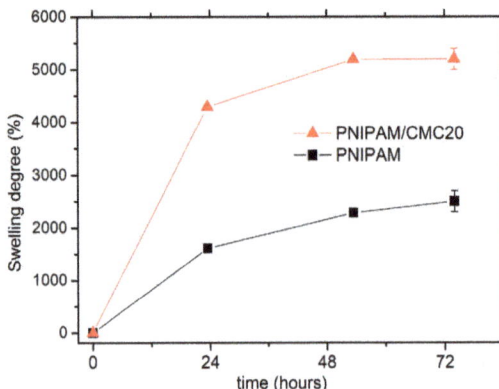

Figure 5. Swelling kinetics of PNIPAM and PNIPAM/CMC20 hydrogels in water at 295 K.

The deswelling kinetics of PNIPAM/CMC20 and of pure PNIPAM hydrogel samples, previously swollen in water at 295 K, were measured by monitoring the water retention % as a function of the immersion time in water at 313 K. After 30 min, the water retention of the PNIPAM/CMC20 hydrogel is reduced to about 35%, whereas that of pure the PNIPAM hydrogel reaches a steady-state value of 70% (Figure 6). Similar results were reported for PNIPAM/sodium alginate semi-IPNs [13]. The higher water retention of the pure PNIPAM hydrogel is due to the formation of a compact, vitreous layer at the surface of the samples above the VPTT, which hampers the release of water. The addition of CMC prevents the formation of this impermeable layer. This interpretation is supported by the different macroscopic morphology of the two kinds of hydrogels in water at 40 °C. The bubbles (full of water) that form at the surface of pure PNIPAM hydrogels in water above the VPTT [7] are not observed for the PNIPAM/CMC samples.

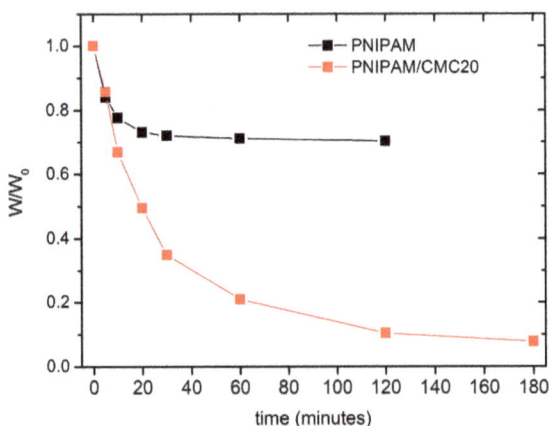

Figure 6. Water retention (defined as the weight W of the hydrogel divided by weight of the hydrogel swollen at 295 K W_0) as a function of immersion time in water at 315 K for the pure PNIPAM and the PNIPAM/CMC20 hydrogels.

Magnetization vs. magnetic field intensity curves were measured for PNIPAM/CMC5/Fe$_3$O$_4$ and PNIPAM/Fe$_3$O$_4$ hydrogels to investigate the effect of CMC on the magnetic properties of the hybrid organic–inorganic hydrogels. Magnetization vs. magnetic field intensity curves of PNIPAM/CMC5/Fe$_3$O$_4$ and PNIPAM/Fe$_3$O$_4$ hydrogels show that the magnetic NPs are superparamagnetic. The magnetization vs. magnetic field intensity curves measured at 2.5 and 300 K for the PNIPAM/CMC5/Fe$_3$O$_4$ hydrogel are shown in Figure 7a. The curves measured at room temperature do not show any remanence. On the contrary, a hysteresis loop is observed in the curves measured at 2.5 K, with a remanent magnetization of 18 emu/g Fe$_3$O$_4$ and a coercive field of ca. 350 Oe (Figure 7a). The saturation magnetizations (63 and 52 emu/g Fe$_3$O$_4$ at 2.5 and 300 K, respectively) measured for the PNIPAM/CMC5/Fe$_3$O$_4$ hydrogel are higher than those of PNIPAM/Fe$_3$O$_4$ (46 and 33 emu/gFe$_3$O$_4$ at 2.5 and 300 K, respectively). These values are lower than that of bulk magnetite (92 emu/g at room temperature) but comparable with those of magnetite NPs of similar size [16]. The ZF and ZFC curves confirm that magnetite NPs in the PNIPAM/CMC5/Fe$_3$O$_4$ hydrogel are superparamagnetic with a blocking temperature (T_b) of 83 K (Figure 7b). T_b measured for the PNIPAM/Fe$_3$O$_4$ hydrogel is 104 K, higher than that of the PNIPAM/CMC5/Fe$_3$O$_4$ hydrogel. Although many parameters can affect T_b (particle size, degree of oxidation, etc.), an increase in the number of particles in the clusters enhances the dipole–dipole interactions leading to an increase of T_b [16]. Hence, the observed difference of T_b suggests that the clusters of magnetite NPs are smaller in PNIPAM/CMC/Fe$_3$O$_4$ hydrogels than in PNIPAM/Fe$_3$O$_4$ hydrogels.

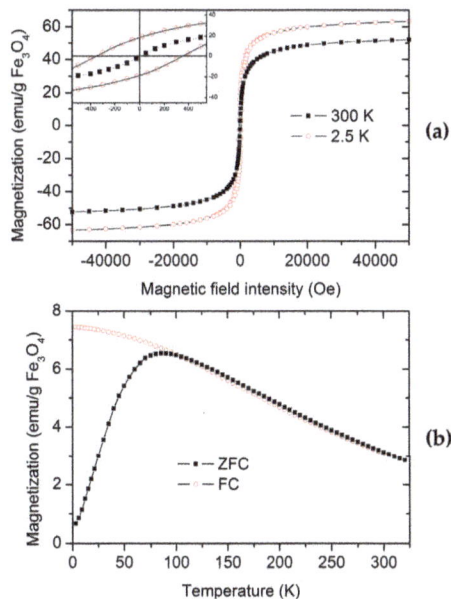

Figure 7. (a) Magnetization versus field intensity measured for the PNIPAM/CMC5/Fe$_3$O$_4$ lyophilized hydrogel at 2.5 K and 300 K. In the inset the curves at low intensity magnetic field are shown; (b) panel: zero field cooling (ZFC) and field cooling (FC) curves measured for the PNIPAM/CMC5/Fe$_3$O$_4$ sample. These curves were measured with an applied field of 50 Oe.

3. Conclusions

The surface morphology at micrometric scale of PNIPAM/CMC hydrogels is characterized by a porous network structure which is not observed for pure PNIPAM hydrogels. These features are probably the emergence of micropores at the surface of the material. With an equal cross-linking degree, PNIPAM hydrogels containing 20 wt % of CMC have a much larger equilibrium swelling degree than pure PNIPAM hydrogels. The observed reduction of water retention for the PNIPAM/CMC hydrogels above the VPTT compared to pure PNIPAM hydrogels can reasonably be attributed to their surface morphology. The microporosity of the PNIPAM/CMC hydrogels prevents the formation of the dense layer impermeable to water as occurs for pure PNIPAM hydrogels. Moreover, the surface morphology of the PNIPAM/CMC5/Fe_3O_4 hydrogel is characterized by a honeycomb-like structure similar to PNIPAM/CMC hydrogels. Fe_3O_4 NPs prepared by co-precipitation in the presence of CMC embedded in PNIPAM hydrogels exhibit a superparamagnetic behavior. The lower blocking temperature measured for the PNIPAM/CMC5/Fe_3O_4 hydrogel compared to the PNIPAM/Fe_3O_4 hydrogel suggests that smaller aggregates of NPs are present in the PNIPAM/CMC5/Fe_3O_4 hydrogel. The capability of the PNIPAM/CMC hydrogels to absorb and release larger volumes of water at higher rates compared to pure PNIPAM hydrogels is important for many applications.

4. Materials and Methods

NIPAM, N,N'-methylenebisacrylamide (BIS), $N,N,N'N'$-tetramethylenediamine (TEMED), $K_2S_2O_8$, CMC sodium salt (molecular weight 700 kDa, 0.8 substitution degree), $FeCl_3 \cdot 6H_2O$, and $FeSO_4 \cdot (NH_4)_2SO_4 \cdot 6H_2O$ were purchased from Sigma-Aldrich (St. Louis, MO, USA) and used as received. Deionized water was used for the experiments. PNIPAM hydrogels were prepared according to a free-radical polymerization procedure. The molar ratios of BIS, TEMED, and $K_2S_2O_8$ with respect to NIPAM were 2%, 10%, and 0.25%, respectively. For the preparation of PNIPAM/CMC hydrogels, the reaction was carried out in water solutions of CMC. Hydrogels containing 5 and 20 wt % CMC with respect to NIPAM are indicated as PNIPAM/CMC5 and PNIPAM/CMC20, respectively. Fe_3O_4 NPs were prepared by co-precipitation from Fe(II)/Fe(III) (1:2 stoichiometric ratio) solutions by adding NaOH in the presence of CMC at 333 K [14]. The size of the Fe_3O_4/CMC NPs was 9 ± 1 nm, as determined by X-ray diffraction in a previous work [14]. The composition of the CMC/magnetite dispersions was 50 wt %. The preparation of the PNIPAM hydrogels containing 5 wt % CMC and 5 wt % Fe_3O_4 (with respect to PNIPAM) was carried out in aqueous dispersions of CMC/Fe_3O_4 NPs. The PNIPAM sample with embedded Fe_3O_4 NPs was prepared by carrying out the PNIPAM hydrogel synthesis in a water dispersion of Fe_3O_4 NPs, previously prepared using the coprecipitation method. The concentrations of magnetite in the samples (as determined by UV-visible spectrophotometry [14]) was 4.3 ± 0.2 wt % and 1.5 ± 0.2 wt % (with respect to PNIPAM) for the PNIPAM/CMC/Fe_3O_4 and PNIPAM/Fe_3O_4 samples, respectively. These samples will be indicated as PNIPAM/CMC5/Fe_3O_4 and PNIPAM/Fe_3O_4. After prolonged washing in deionized water to remove unreacted species, the hydrogels were frozen and lyophilized.

The equilibrium swelling degree (SD_{eq}), defined as $(w - w_d)/w_d$ %, where w and w_d are the weights of the swollen and of the dried hydrogel, was measured in deionized water at 295 K. Attenuated total reflectance–Fourier transform infrared (ATR-FTIR) spectra were collected using a FTS6000 spectrometer (Bio-Rad, Cambridge, MA, USA) at a resolution of 4 cm^{-1}. A sigma VP FESEM microscope (Zeiss, Germany) was used for the FESEM measurements. The energy of the electron beam was in the range 1–10 keV, and the "in lens" detector was used to collect the secondary electrons. Measurements were performed under high vacuum conditions on the lyophilized samples without metallization or graphitization.

Magnetization versus field intensity and zero field cooling (ZFC) and field cooling (FC) curves were measured by means of a Superconducting Quantum Interference Device magnetometer (Quantum Design Ltd. San Diego, CA, USA).

Acknowledgments: This work was financially supported by MIUR (Ministero Istruzione Università e Ricerca) under the FIRB project RBAP11ZJFA, 2010. The authors are indebted to Claudia Innocenti (INSTM and Dipartimento di Chimica U. Shiff, Università di Firenze) for the magnetic measurements.

Author Contributions: Both authors conceived, designed, and performed the experiments. Both authors revised, read, and approved the final manuscript.

Conflicts of Interest: The authors declare no conflict of interest.

References

1. Nguyen, H.H.; Payre, B.; Fitremann, J.; Lauth-de Viguerie, N.; Marty, J. Thermoresponsive Properties of PNIPAM-Based Hydrogels: Effect of Molecular Architecture and Embedded Gold Nanoparticles. *Langmuir* **2015**, *31*, 4761–4768. [CrossRef] [PubMed]
2. Satarkar, N.S.; Hilt, J.Z. Magnetic hydrogel nanocomposite for remote controlled pulsative drug release. *J. Control. Release* **2008**, *130*, 246–251. [CrossRef] [PubMed]
3. Jing, G.; Wang, L.; Yu, H.; Amer, W.A.; Zhang, L. Recent progress on study of hybrid hydrogels for water treatment. *Colloids Surf. A* **2015**, *416*, 86–94. [CrossRef]
4. Imaran, A.B.; Seki, T.; Takeoka, Y. Recent advances in hydrogels in terms of fast stimuli responsiveness and superior mechanical performance. *Polym. J.* **2010**, *42*, 839–851. [CrossRef]
5. Depa, K.; Strachota, A.; Slouf, M.; Hromadkova, J. Fast temperature-responsive nanocomposite PNIPAM hydrogels with controlled pore wall thickness: Force and rate of T-response. *Eur. Polym. J.* **2012**, *48*, 1997–2007. [CrossRef]
6. Hirose, H.; Shibayama, M. Kinetics of Volume Phase Transition in Poly(*N*-isopropylacrylamide-*co*-acrylic acid) Gels. *Macromolecules* **1998**, *31*, 5336–5342. [CrossRef]
7. Shibayama, M.; Nagai, K. Shrinking Kinetics of Poly(*N*-isopropylacrylamide) Gels T-jumped across Their Volume Phase Transition Temperature. *Macromolecules* **1999**, *32*, 7461–7468. [CrossRef]
8. Sun, S.; Hu, J.; Tang, H.; Wu, P. Chain Collapse and Revival Thermodynamics of Poly(*N*-isopropylacrylamide) Hydrogel. *J. Phys. Chem. B* **2010**, *114*, 9761–9770. [CrossRef] [PubMed]
9. Kaneko, Y.; Nakamura, S.; Kiyotaka, S.; Aoyagi, T.; Kikuchi, A.; Sakurai, Y.; Okano, T. Rapid Deswelling Response of Poly(*N*-isopropylacrylamide) Hydrogels by the Formation of Water Release Channels Using Poly(ethylene oxide) Graft Chains. *Macromolecules* **1998**, *31*, 6099–6105. [CrossRef]
10. Gil, E.S.; Hudson, S.M. Effect of Silk Fibroin Interpenetrating Networks on Swelling/Deswelling Kinetics and Rheological Properties of Poly(*N*-isopropylacrylamide) Hydrogels. *Biomacromolecules* **2007**, *8*, 258–264. [CrossRef] [PubMed]
11. Zhao, Z.X.; Li, Z.; Xia, Q.B.; Bajalis, E.; Xi, H.X.; Lin, Y.S. Swelling/deswelling kinetics of PNIPAAm hydrogels synthesized by microwave irradiation. *Chem. Eng. J.* **2008**, *142*, 263–270. [CrossRef]
12. Xia, L.; Xie, R.; Ju, X.; Wang, Q.; Chen, Q.; Chu, L. Nano-structured smart hydrogels with rapid response and high elasticity. *Nat. Commun.* **2013**, *4*, 2226. [CrossRef] [PubMed]
13. Zhang, G.; Zha, L.; Zhou, M.; Ma, M.; Liang, B. Rapid deswelling of sodium alginate/poly(*N*-isopropylacrylamide) semi-interpenetrating polymer network hydrogels in response to temperature and pH changes. *Colloid Polym. Sci.* **2005**, *283*, 431–438. [CrossRef]
14. Maccarini, M.; Atrei, A.; Innocenti, C.; Barbucci, R. Interactions at the CMC/magnetite interface: Implications for the stability of aqueous dispersions and magnetic properties of magnetite nanoparticles. *Colloids Surf. A* **2014**, *462*, 107–114. [CrossRef]
15. Ekici, S. Intelligent poly(*N*-isopropylacrylamide)-carboxymethyl cellulose full interpenetrating polymer networks for protein adsorption studies. *J. Mater. Sci.* **2011**, *46*, 2843–2850. [CrossRef]
16. Mandel, K.; Hutter, F.; Gellermann, C.; Sextl, G. Stabilisation effects of superparamagnetic nanoparticles on clustering in nanocomposite microparticles and on magnetic behavior. *J. Magn. Magn. Mater.* **2013**, *331*, 269–275. [CrossRef]

MDPI
St. Alban-Anlage 66
4052 Basel
Switzerland
Tel. +41 61 683 77 34
Fax +41 61 302 89 18
www.mdpi.com

Gels Editorial Office
E-mail: gels@mdpi.com
www.mdpi.com/journal/gels

www.ingramcontent.com/pod-product-compliance
Lightning Source LLC
Chambersburg PA
CBHW051720210326
41597CB00032B/5548